Numerical Computation
of
INTERNAL AND EXTERNAL
FLOWS

Volume 1: Fundamentals of Numerical Discretization

WILEY SERIES IN
NUMERICAL METHODS IN ENGINEERING
Consulting Editors
R. H. Gallagher, *Worcester Polytechnic Institute,*
Worcester, Massachusetts, USA
and
O. C. Zienkiewicz, *Department of Civil Engineering,*
University College of Swansea

Numerical Computation
of
INTERNAL AND EXTERNAL FLOWS

Volume 1: Fundamentals of Numerical Discretization

Charles Hirsch

Department of Fluid Mechanics,
Vrije Universiteit Brussel,
Brussels, Belgium

A Wiley–Interscience Publication

JOHN WILEY & SONS
Chichester · New York · Brisbane · Toronto · Singapore

Library of Congress Cataloging in Publication Data:

Hirsch, Ch.
    Numerical computation of internal & external flows.
    (Wiley series in numerical methods in engineering)
    'A Wiley–Interscience publication.'
    Contents: v. 1. Fundamentals of numerical
discretization.
    1. Fluid dynamics—Mathematical models.   I. Title.
II. Title: Numerical computation of internal and
external flows.   III. Series.
TA357.H574   1988   620.1'064   87-23116

ISBN 0 471 91762 1 (cased)
ISBN 0 471 92385 0 (pbk)

British Library Cataloguing in Publication Data:

Hirsch, Charles
    Numerical computation of internal &
    external flows.—(Wiley interscience
    series in numerical methods in
    engineering; 1).
    Vol. 1: Fundamentals of numerical
    discretization
    1. Fluid dynamics—Mathematical models
    I. Title
    532'.051'0724   QA911

ISBN 0 471 91762 1 (cased)
ISBN 0 471 92385 0 (pbk)

Typeset by MCS Ltd, Salisbury, Wilts
Printed and bound in Great Britain by
Biddles Ltd, Guildford and King's Lynn

To the memory of

> *Leon Hirsch*
> *and*
> *Czipa Zugman,*

my parents,
struck by destiny

CONTENTS

ix

Preface

This book, which is published in two volumes, aims at introducing the reader to the essential steps involved in the numerical simulation of fluid flows by providing a guide from the initial step of the selection of a mathematical model to practical methods for their numerical discretization and resolution.

The first volume, divided into four parts, is devoted to the fundamentals of numerical discretization techniques and attempts a systematic presentation of the successive steps involved in the definition and development of a numerical simulation. The second, on the other hand, presents the applications of numerical methods and algorithms to selected flow models, from the full potential flow model to the systems of Euler and Navier–Stokes equations.

Part I, covering Chapters 1 to 3, introduces the mathematical models corresponding to various levels of approximation of a flow system. We hope hereby to draw, if necessary, the reader's attention to the range of validity and limitations of the different available flow models so that the user will be in a position to make a choice in full awareness of its implications. Part II is devoted to a presentation of the essentials of the most frequently applied discretization methods for differential equations, the finite difference (Chapter 4), finite element (Chapter 5) and finite volume methods (Chapter 6). Part III introduces the next step in the development of an algorithm, namely the methods for the analysis of the stability, convergence and accuracy properties of a selected discretization. This is covered in Chapters 7 and 10, dealing, respectively, with basic definitions, the Von Neumann method, the method of the equivalent differential equation and the matrix method. Finally, Part IV covers the resolution methods for discretized equations. More particularly, integration methods which can be applied to systems of ordinary differential equations (in time) are discussed in Chapter 11 and iterative methods for the resolution of algebraic systems are discussed in Chapter 12.

No attempt has been made towards an exhaustive presentation of the material covered and several important topics are not treated in the text for objective as well as subjective reasons. To explain a few of them, spectral discretization methods applied to flow problems are an important technique, which is treated in existing textbooks, but also we have no practical experience with the method. Stability analysis methods, such as the energy method, require a mathematical background which is not often found in the engineering community, and it was not felt appropriate to introduce this subject in a text which is addressed mainly to engineers and physicists with an interest in flow problems. The computational techniques for boundary layers are largely covered in recent textbooks, and we thought that there was not much to add to the existing, well-documented material.

This text is directed at students at the graduate level as well as at scientists and engineers already engaged, or starting to be engaged, in computational fluid dynamics. With regard to the material for a graduate course, we have aimed at allowing a double selection. For an introductory course, one can consider an 'horizontal' reading, by selecting subsections of different chapters in order to cover a wider range of topics. An alternative 'vertical' reading would select fewer chapters, with a more complete treatment of the selected topics.

Parts of this book have been written while holding the NAVAIR Research Chair at the Naval Postgraduate School in Monterey, during the academic year 1983–4, for which I am particularly indebted to Ray Shreeve, Professor at the Areonautical Department and Director of the Turbopropulsion Laboratory. The pleasant and encouraging atmosphere during this period and during subsequent summer stays at NPS, where some additional writing could partly be done, is, for a large part, the basis of having brought this task to an end.

Some sections on Euler equations were written during a summer stay at ICASE, NASA Langley and I would like to acknowledge particularly Dr Milton Rose, former Director of ICASE, for his hospitality and the stimulating atmosphere. I have also had the privilege of benefiting from results of computations performed, at my request, on different test cases by several groups, and I would like to thank D. Caughey at Cornell University, T. Holst at NASA Ames, A. Jameson at Princeton University, M. Salas at NASA Langley, and J. South and C. Gumbert also at NASA Langley, for their willingness and effort.

Finally, I would like to thank my colleagues S. Wajc and G. Warzee as well as present and former coworkers H. Deconinck, C. Lacor and J. Peuteman for various suggestions, comments and contributions. I have also the pleasure to thank my secretaries L. Vandenbossche and J. D'haes for the patience and the effort of typing a lengthy manuscript.

Ch. HIRSCH
BRUSSELS, JANUARY 1987

Nomenclature

a	convection velocity of wave speed
A	Jacobian of flux function
c	speed of sound
c_p	specific heat at constant pressure
c_v	specific heat at constant volume
C	discretization operator
D	first derivative operator
e	internal energy per unit mass
e	vector (column matrix) of solution errors
E	total energy per unit volume
E	finite difference displacement (shift) operator
f	flux function
$\vec{f_\mathrm{e}}$	external force vector
$\vec{F}(f, g, h)$	flux vector with components f, g, h
g	gravity acceleration
G	amplification factor/matrix
h	enthalpy per unit mass
H	total enthalpy
I	rothalpy
J	Jacobian
k	coefficient of thermal conductivity
k	wavenumber
L	differential operator
M	Mach number
M_x, M_y, M_z	Mach number of cartesian velocity components
n	normal distance
$\vec{n}$	normal vector
N	finite element interpolation function
p	pressure
P	convergence or conditioning operator
Pr	Prandtl number
q	non-homogeneous term
q_H	heat source
Q	source term; matrix of non-homogeneous terms
r	gas constant per unit mass
R	residual of iterative scheme
R	mesh Reynolds (Peclet) number
Re	Reynolds number

s	entropy per unit mass
S	characteristic surface
S	space-discretization operator
$\vec{S}$	surface vector
t	time
T	temperature
u	dependent variable
$\vec{u}$	entrainment velocity
U	vector (column matrix) of dependent variables
U	vector of conservative variables
$\vec{v}(u, v, w)$	velocity vector with cartesian components u, v, w
V	eigenvectors of space-discretization matrix
$\vec{w}$	relative velocity
W	weight function
x, y, z	Cartesian co-ordinates
z	amplification factor of time-integration scheme
α	diffusivity coefficient
β	dimensionless diffusion coefficient $\beta = \alpha \Delta t / \Delta x$
γ	ratio of specific heats
Γ	circulation; boundary of domain Ω
δ	central-difference operator
$\bar{\delta}$	central-difference operator (equation (4.2.1e))
δ^+	forward difference operator
δ^-	backward difference operator
Δ	Laplace operator
Δt	time step
ΔU	variation of solution U between levels $n + 1$ and n
$\Delta x, \Delta y$	spatial mesh size in x and y directions
η	non-dimensional difference variable in local co-ordinates
ε	error of numerical solution
ε_v	turbulence dissipation rate
ε_D	dissipation or diffusion error
ε_ϕ	dispersion error
$\vec{\zeta}$	vorticity vector
ζ	magnitude of vorticity vector
θ	parameter controlling type of difference scheme
$\vec{\varkappa}$	wavenumber vector; wave-propagation direction
λ	eigenvalue of amplification matrix
μ	coefficient of dynamic viscosity
μ	averaging difference operator
ξ	non-dimensional distance variable in local co-ordinates
ρ	density
ρ	spectral radius
σ	Courant number
$\bar{\bar{\sigma}}$	shear stress tensor
τ	relaxation parameter
$\bar{\bar{\tau}}$	stress tensor
ν	kinematic viscosity
ϕ	velocity potential

ϕ	phase angle in Von Neumann analysis
Φ	phase angle of amplification factor
ψ	rotational function
ω	time frequency of plane wave
ω	overrelaxation parameters
Ω	eigenvalue of space discretization matrix
Ω	volume
$\vec{1}_x, \vec{1}_y, \vec{1}_z$	unit vectors along the x, y, z directions
$\vec{1}_n$	unit vector along the normal direction

Subscripts

e	external variable
i, j	mesh point locations in x, y directions
I, J	nodal point index
J	eigenvalue number
min	minimum
max	maximum
n	normal or normal component
o	stagnation values
v	viscous term
x, y, z	components in x, y, z directions
x, y, z	partial differentiation with respect to x, y, z
∞	freestream value

Superscripts

n	iteration level
n	time-level
$\tilde{}$	exact solution of discretized equation
$\bar{}$	exact solution of differential equation

Symbols

$\times$	vector product of two vectors
$\otimes$	tensor product of two vectors

PART I: THE MATHEMATICAL MODELS FOR FLUID FLOW SIMULATIONS AT VARIOUS LEVELS OF APPROXIMATION

Introduction

The invention of the digital computer and its introduction into the world of science and technology has led to the development, and increased awareness, of the concept of approximation. This concerns the theory of the numerical approximation of a set of equations, taken as a mathematical model of a physical system. However, it also concerns the notion of the approximation involved in the definition of this mathematical model with respect to the complexity of the physical world. We are concerned here with physical systems for which is is assumed that the basic equations describing their behaviour is known theoretically but for which no analytical solutions exist, and consequently an approximate numerical solution will be sought instead. For various reasons, the first of these being the great complexity of certain systems, it is often not practically possible to describe completely the evolution of the system in its full complexity. Of course, the definition of these limits is relative to a given time and environment, and these are being extended with the evolution of computer technology. However, at a given period, it is necessary to define mathematical models which will reduce the complexity of the original basic equations and make them tractable within fixed limits.

We can state that a mathematical model for the behaviour of a physical system, and in particular the system of fluid flows, can only be defined after consideration of the level of the approximation required in order to achieve an acceptable accuracy on a defined set of dependent and independent variables. This set contains the basic variables needed to completely describe the system

1

considered, as well as all other variables which characterize its behaviour. For instance, the evolution of a fluid can be considered to depend on all three space variables or, at the other end, the dominant influence of one space variable is sufficiently strong with regard to the two others to justify a one-dimensional approximation. This defines what we could call the 'spatial level' of the description: one-, two- or three-dimensional.

Actually, the first level to be defined is the 'scale of reality' level. Physicists propose various levels of description of our physical world, ranging from subatomic, atomic or molecular, microscopic or macroscopic (defined roughly as the scale of classical mechanics) up to the astronomical scale.

As is well known, in the statistical description of a gas the motion of the individual atoms or molecules are taken into consideration and their behaviour is basically ruled by the Boltzmann equation. This description leads, for instance, to the definitions of temperature as a measure of the mean kinetic energy of the gas molecules; of pressure as a result of the impulse of molecules on the walls of the body containing the gas; of viscosity connected to the momentum exchange due to the thermal molecular motion, and so on. At this microscopic level of description the fundamental variables are molecule velocities, number of particles per volume and other variables defining the motion of the individual molecules, while pressure, temperature and viscosity, for example, are mean properties which are deduced from other variables, more basic at this level of reality. Hence we may consider that each level of reality can be associated with a set of fundamental variables, from which other variables can be defined as measures of certain mean properties. Continuing the line of this example, we have, beyond the molecular level of statistical mechanics, the atomic, nuclear and subnuclear levels, which we do not plan to discuss here, since they are outside the domain of the definition of a fluid. The mere existence of a fluid implies that a sufficiently high number of associated atoms form molecules in order to obtain a minimal amount of interaction between these molecules.

Actually, fluid dynamics starts to exist as soon as the interaction between a sufficiently high number of particles affects and dominates, at least partly, the motion of each individual particle. Hence fluid dynamics is essentially the study of the interactive motion and behaviour of a large number of individual elements.

The limit between individual motions of isolated particles or elements and their interactive motion is of significance in the study of rarefied gases. It is known that the interaction between the particles becomes negligible if the mean-free path length attains a magnitude of the order of the length scale of the system considered. From this stage on, and for higher values of the mean-free path length, the particles behave essentially as individual elements. These limit situations will not be considered here since they are also outside the field of fluid dynamics, and we will focus on the level of reality in which the density of elements is high enough, so that we can make the approximation of considering the system of interacting elements as a *continuum*. This expresses that a continuity or a closeness exists between the elements such that their mutual interaction dominates over the individual motions, although these are not supressed. What actually happens is that a collective motion is superimposed on the motion of the isolated elements as a consequence of the large number of these elements co-existing within the same domain.

From this point of view we understand easily why the concepts of fluid mechanics can be applied to a variety of systems consisting of a large number of interacting individual elements. This is the case for the current fluids and gases, where the individual 'element', or fluid particle, is actually not a single molecule but consists of a large number of molecules occupying a small region with respect to the scale of the domain considered, but still sufficiently large in order to be able to define a meaningful and non-ambiguous average of the velocities and other properties of the individual molecules and atoms occupying this volume. It implies that this elementary volume contains a sufficiently high number of molecules, with, for instance, a well-defined mean velocity and mean kinetic energy, allowing us to define velocity, temperature, pressure, entropy and so on at each point. Hence associated fields, which will become basic variables for the description of the system, can be defined, although the temperature, pressure or entropy of an individual atom or molecule is not defined and is generally meaningless.

In the classical interpretation of turbulence each fluid particle as defined above enters into a stochastic motion, and in defining mean turbulent variables, such as a mean turbulent velocity field, an average is performed (in this case an average in time) over the motion of the fluid particles themselves. A still higher level of averaging occurs in the description of flows through porous media such as soils. In the description of groundwater flows an 'element' is the set of fluid particles, as defined above, contained in a volume large enough to contain a great number of soil and fluid particles such that a meaningful average can be performed but still small with respect to the dimensions of the region to be analysed. Such a volume is considered as a 'point' at this level of description, and the fields are attached to these points, implying that groundwater flow theories do study the behaviour of collection of fluid particles.

Following this line, the movement or overall displacement of crowds at the exits of railway or subway stations during rush hours can be analysed by fluid mechanical concepts. In this case an 'element' is the set of persons contained in a region small with regard to the dimensions of the station, for instance, but still containing a sufficiently high number of individuals in order to define non-ambiguous average values, such as velocity and other variables. In this description the displacement of an individual is not considered, only the motion of groups of individuals.

A similar analysis can be defined for heavy traffic studies, where an 'element' is defined as a set of cars (in the one-dimensional space formed by the road). Obviously in light traffic, the isolated car behaves as a single particle, but collective motion occurs when a certain intensity of traffic has been reached such that the speed of an individual car is influenced by the presence of other cars. This is actually to be considered as the onset of a 'fluid mechanical' description.

Finally, on a still larger scale, astrophysical fluid dynamics can be defined for the study of aspects of the interstellar medium or of the shape and evolution of galaxies. In this later case, for instance, an 'element' consists of a set of stellar objects, including one or several solar systems, and the dimensions of a 'point' can be of the order of light-years.

In conclusion, we can say that fluid mechanics is essentially the study of the behaviour of averaged quantities and properties of a large number of interacting elements. The same is true for another domain of scientific knowledge, namely thermodynamics. It is therefore not surprising that thermodynamics is, with the exception of incompressible isothermal media, tightly interconnected with fluid mechanics and plays an important role in the description of the evolution of 'fluid mechanical' systems such as those mentioned above.

An essential step in fluid dynamics is therefore the averaging process. We have to decide, in front of a given system, which level of averaging will be performed as a function of the quantities to be predicted, the significant variables which can be defined in a meaningful way and the precision and degree of accuracy to be achieved in the description of the system's behaviour. This is a basic task for the scientists in charge of the analysis, which requires a great understanding of the physics of the system, judgement and a sense of compromise between the required level of accuracy and the degree of sophistication of the chosen mathematical model.

The next step in the definition of the levels of approximation is to define a time or *steadiness level*. This implies an estimation of the various time constants of the flow situation being considered and the choice of the lowest time constant to be taken into consideration in the modelling of this flow system. Then a time averaging will be performed with regard to the time constants lower than the chosen minimal value. The best-known example of this procedure is the system of time-averaged Navier–Stokes equations for mean turbulent flow variables. An averaging is performed over turbulent fluctuations, since we are concerned in that case with variations of the flow slower than turbulent fluctuations and, hence, with time constants much larger than the time constant of these fluctuations. Through this procedure, extra terms appear in the equation, for example Reynolds stresses, which are averaged products of fluctuation and for which external information will have to be provided.

The *spatial level* of approximation defines the number of space variables used in the model. We have to decide, in function of certain assumptions concerning the physical behaviour of the system to be described, if a one- or two-dimensional description will provide sufficiently accurate information about the flow behaviour. It is of importance here to note that, the basic flow equations being three-dimensional, any description with less than three space variables will be obtained by disregarding the flow variations with respect to the corresponding space co-ordinate, and this can be formulated mathematically by averaging out the equations over that space variable. Therefore the averaging process, here over space, is again essential. In this space averaging we will obtain equations in a two- or a one-dimensional region, which contain terms describing the averaged influence of the full three-dimensional motion. These terms, analogous to the Reynolds stresses, will generally be neglected, due to lack of information required to estimate them, although they can, or could, be estimated in certain cases.

Since the averaging procedure implies a loss of information in the averaged space variables, this information will, in many cases, have to be provided from 'outside' the model; for instance, through empirical data. It is also clear, therefore, that simple models (for example, one-dimensional flow descriptions) may require more empirical or external input than a viscous three-dimensional description if contributions from the three-dimensionality are to be taken into account.

The next level of approximation, the *dynamical level*, is linked to an estimation of the relative influence of the various forces and their components on the system's behaviour. The dynamical evolution of a flow system is determined by the equilibrium of the different forces acting on it, but it seldom occurs that all the force components are equally important. Therefore one very basic step in the setting up of a mathematical model for the description of a system is an estimation of the dominant force components in order to simplify the model as strongly as possible.

For instance, although gravity forces are always present, in many cases these forces have only a negligible influence on flow behaviour. The detailed study of the influence of viscosity by Prandtl, which led to the boundary layer concept, is perhaps the most fascinating example of the consequences of a profound analysis of the relative influence of forces. As is well known, the considerable simplification of the Navier–Stokes equations introduced through this analysis allowed the practical calculation of many flow situations which were largely untractable by the full Navier–Stokes equations. This boundary layer concept led to the definition of the regions of validity of inviscid flows, in which the viscosity forces could be neglected. Therefore inviscid flow approximations play an important role in fluid mechanics, although their range of validity in internal flows is more restricted than in external aerodynamics.

The different levels of approximation considered here can strongly interact with each other. For instance, on a rotating blade of a turbomachine the centrifugal forces will create a radial migration of the boundary layer fluid along the blade, leading to an increased spanwise mixing of the flow, and hence will limit the validity of a purely two-dimensional description of the blade-to-blade flow. In all cases, however, the final word with regard to the validity of a given model is the comparison with experimental data or with computations at a higher level of approximation.

These comments are presented here in order to introduce the methodology

6

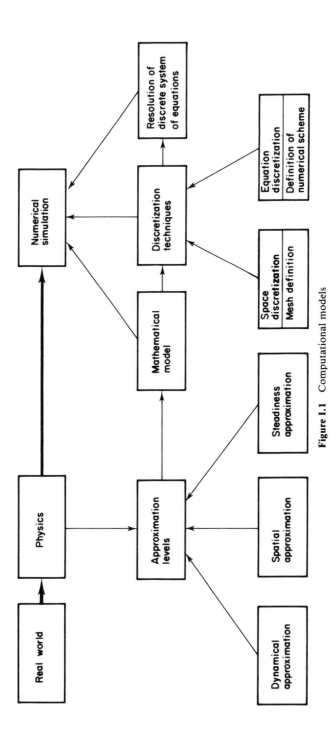

Figure I.1 Computational models

to be followed in the chapters that follow. After having summarized the basic flow equations in Chapter 1, a systematic presentation of various mathematical models describing the most current approximations will be given in Chapter 2. Chapter 3 discusses the structure of the equations defining the mathematical models and presents definitions and criteria for distinguishing between elliptic, parabolic and hyperbolic sets of partial differential equations.

Figure I.1 shows a block diagram of the interaction between the different levels of approximations, defining the mathematical model of the flow system and the numerical simulation. The latter consists of the combination of various discretizations: the space discretization, which defines the mesh, and the equation discretization, which sets up the numerical scheme. The numerical flow simulation is obtained after resolution of the discretized set of equations.

Chapter 1

The Basic Equations of Fluid Dynamics

The laws of fluid dynamics are well established and can be formulated in many equivalent ways. For instance, they can be deduced from the observation that the behaviour of a physical system is completely determined by conservation laws. This corresponds to the statement that during the evolution of a fluid a certain number of properties, such as mass, generalized momentum and energy, are 'conserved'. The significance of this expression needs, of course, to be explained.

The awareness of this fact has been one of the greatest achievements of modern science, due to the high level of generality and degree of abstraction involved. Indeed, no matter how complicated the detailed evolution of a system might be, not only are the basic properties of mass, momentum and energy conserved during the whole process at all times (in the sense to be defined later) but these three conditions completely determine the behaviour of the system without any additional dynamical law. This is a very remarkable property indeed. The only additional information concerns the specification of the nature of the fluid (for example, incompressible fluid, perfect gas, viscous fluid, viscoelastic material, etc.). Of course, an important level of knowledge implied in these statements has to be defined before the mathematical expression of these laws can be written and used to predict and describe the system's behaviour.

A fluid flow is considered as known if, at any instant in time, the velocity field and a minimum number of static properties are known at any point. The number of static properties to be known is dependent on the nature of the fluid. This number will be equal to one for an incompressible fluid (the pressure) and two (for example, pressure and density) for a perfect gas or any real compressible fluid in thermodynamic equilibrium.

We will assume that a separate analysis has provided the necessary knowledge enabling the nature of the fluid to be defined. This is obtained from the study of the behaviour of the various types of continua and the corresponding information is summarized in the constitutive laws and in other parameters such as viscosity and heat conduction coefficients. This study also provides information on the nature and properties of the internal forces acting on the fluid, since, by definition, a deformable continuum, such as a fluid, requires the existence of internal forces connected to the nature of the constitutive law. Separate studies are also needed to distinguish the various external forces which are able to influence the motion of the system in addition to internal ones. These external forces could be gravity, buoyancy, Coriolis and centrifugal forces in rotating systems or electromagnetic forces in electrical conducting fluids.

The concept of conservation, as mentioned above, means that the variation of a conserved (intensive) flow quantity within a given volume is due to the net effect of some internal sources and of the amount of the quantity which is crossing the boundary surface. This amount is called the *flux*, and its expression results from the mechanical and thermodynamical properties of the fluid. Similarly, the sources attached to a given flow quantity are also assumed to be known from basic studies. The fluxes and sources are, in general, dependent on the space–time co-ordinates as well as on fluid motion.

The associated fluxes are vectors for a scalar quantity and tensors for a vector quantity, such as momentum. The fluxes are generated from two contributions: one due to the convective transport of the fluid and another due to the molecular motion, which is always present even when the fluid is at rest. The effect of the molecular motion expresses the tendency of a fluid towards equilibrium and uniformity, since differences in the intensity of the quantity being considered create a transfer in space such as to reduce the non-homogeneity. This contribution to the total flux is proportional to the gradient of the corresponding quantity, since it has to vanish for a homogeneous distribution and therefore acts as a diffusive effect.

Diffusive fluxes do not always exist; for instance, from an analysis of the physical properties of fluid it is known that in a single-phase fluid at rest no diffusion of specific mass is possible since any variation of specific mass implies a displacements of fluid particles. Therefore, there will be no diffusive flux contribution to the mass conservation equation.

The general laws of physics tell us that certain quantities do not obey conservation laws. For instance, pressure (or entropy) does not satisfy any conservative equation while, as mentioned above, the motion of a fluid is completely described by the conservation laws for the three basic properties: mass, momentum and energy.

1.1 GENERAL FORM OF A CONSERVATION LAW

1.1.1 Scalar conservation law

Let us consider a scalar quantity per unit volume U, acting in an arbitrary volume Ω, fixed in space, bounded by a closed surface S (see Figure 1.1.1). The local intensity of U varies through the effect of *fluxes*, which express the contributions from the surrounding points to the local value and through sources Q. The flux vector $\vec{F}$ contains two components, a diffusive contribution $\vec{F}_{\rm D}$ and a convective part $\vec{F}_{\rm C}$. The general form of a conservation law is

10

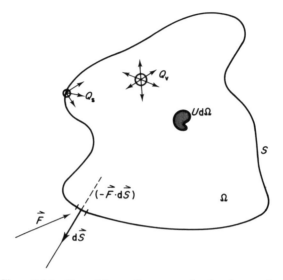

Figure 1.1.1 General form of a conservation law for a scalar
quantity

expressed in stating that the variation per unit time of the quantity U *within
the volume* Ω;

$$\frac{\partial}{\partial t} \int_\Omega U \, d\Omega$$

should be equal to the net contribution from the incoming fluxes *through the
surface S*, with the surface element vector $d\vec{S}$ pointing outward:

$$- \oint_S \vec{F} \cdot d\vec{S}$$

plus contributions from the sources of the quantity U.

These sources can be divided into volume and surface sources, Q_v and $\vec{Q}_s$,
and the total contribution is

$$\int_\Omega Q_v \, d\Omega + \oint_S \vec{Q}_s \cdot d\vec{S}$$

Hence the general form of the conservation equation for the quantity U is

$$\frac{\partial}{\partial t} \int_\Omega U \, d\Omega + \oint_S \vec{F} \cdot d\vec{S} = \int_\Omega Q_v \, d\Omega + \oint_S \vec{Q}_s \cdot d\vec{S} \qquad (1.1.1)$$

or, with Gauss's theorem, for continuous fluxes and surface sources:

$$\int_\Omega \frac{\partial U}{\partial t} \, d\Omega + \int_\Omega \vec{\nabla} \cdot \vec{F} \, d\Omega = \int_\Omega Q_v \, d\Omega + \int_\Omega \vec{\nabla} \cdot \vec{Q}_s \, d\Omega \qquad (1.1.2)$$

This last form leads to the differential form of the conservation law, since equation (1.1.2) is written for an arbitrary volume Ω:

$$\frac{\partial U}{\partial t} + \vec{\nabla} \cdot \vec{F} = Q_v + \vec{\nabla} \cdot \vec{Q}_s \qquad (1.1.3)$$

or

$$\frac{\partial U}{\partial t} + \vec{\nabla} \cdot (\vec{F} - \vec{Q}_s) = Q_v \qquad (1.1.4)$$

It is seen from equations (1.1.3) and (1.1.4) that the surface sources have the same effect on the system as a flux term, and therefore we might as well consider them from the start as an additional flux. However, we would favour the present classification in fluxes and sources, since it allows a clear physical interpretation of all the contributions to the evolution of the quantity U.

As essential aspect of the conservation law (1.1.1) lies in the observation that the internal variations of U, in the absence of volume sources, depend only on the flux contributions *through* the surface S and *not on the flux values inside the volume* Ω. Separating the flux vector into its two components $\vec{F}_C$ and $\vec{F}_D$ we obtain a more precise form of the equation. Indeed, the convective part of the flux vector $\vec{F}_C$, attached to the quantity U in a flow of velocity $\vec{v}$, is the amount of U transported with the motion, and is given by

$$\vec{F}_C = \vec{v} U \qquad (1.1.5)$$

The diffusive flux is defined as the contribution present in fluids at rest, due to the molecular, thermal agitation. It can be expressed by the generalized, gradient law of Fick:

$$\vec{F}_D = -\varkappa\rho \ \vec{\nabla} u \qquad (1.1.6)$$

where u is the quantity U per unit mass, i.e. $U = \rho u$, ρ is the specific mass of the fluid and $\varkappa$ a diffusivity constant. Equation (1.1.3) then becomes

$$\frac{\partial \rho u}{\partial t} + \vec{\nabla} \cdot (\rho \vec{v} u) = \vec{\nabla} \cdot (\varkappa \rho \vec{\nabla} u) + Q_v + \vec{\nabla} \cdot \vec{Q}_s \qquad (1.1.7)$$

This equation is the general form of a *transport* equation for the quantity $U = \rho u$. Observe that the diffusivity constant $\varkappa$ has units of m^2/s for any quantity U.

1.1.2 Vector conservation law

If the conserved property is described by a vector quantity $\vec{U}$ then the flux becomes a tensor $\bar{\bar{F}}$, the volume source term a vector $\vec{Q}_v$ and the conservation equation (1.1.2) becomes

$$\frac{\partial}{\partial t} \int_\Omega \vec{U} \, d\Omega + \oint_S \bar{\bar{F}} \cdot d\vec{S} = \int_\Omega \vec{Q}_v \, d\Omega + \oint_S \bar{\bar{Q}}_s \cdot d\vec{S} \qquad (1.1.8)$$

where the surface source term $\bar{\bar{Q}}_s$ can also be written as a tensor. Applying

Gauss's theorem, if the fluxes and the surface sources are continuous, we obtain

$$\frac{\partial}{\partial t} \int_\Omega \vec{U}\, d\Omega + \int_\Omega \vec{\nabla} \cdot \bar{\bar{F}}\, d\Omega = \int_\Omega \vec{Q}_v\, d\Omega + \int_\Omega \vec{\nabla} \cdot \bar{\bar{Q}}_s\, d\Omega \qquad (1.1.9)$$

and the equivalent differential form

$$\frac{\partial \vec{U}}{\partial t} + \vec{\nabla} \cdot (\bar{\bar{F}} - \bar{\bar{Q}}_s) = \vec{Q}_v \qquad (1.1.10)$$

Here again, the surface sources have the same effect as the flux term.

The convective component of the flux tensor is given by

$$\bar{\bar{F}}_C = \vec{v} \otimes \vec{U} \qquad (1.1.11)$$

where $\otimes$ denotes the tensor product of the vectors $\vec{v}$ and $\vec{U}$. In tensor notation, equation (1.1.11) becomes

$$F_{C_{ij}} = v_i U_j \qquad (1.1.12)$$

and the diffusive component of the flux takes the following form for an homogeneous system

$$F_{D_{ij}} = -\rho\varkappa\, \frac{\partial u_j}{\partial x_i} = -\rho\varkappa\, \partial_i u_j \qquad (1.1.13)$$

with

$$U_j = \rho u_j \qquad (1.1.14)$$

The general forms (1.1.1) or (1.1.8) are to be considered as the *basic formulation of a conservation law* and are indeed the most generally valid expressions, since they remain valid in the presence of discontinuous variations of the flow properties such as inviscid shock waves or contact discontinuities. Therefore it is important to note that the physical reality is described in the most straightforward and general way by the integral formulation of the conservation laws. Only if continuity of the flow properties can be assumed will equations (1.1.2) or (1.1.9) and their fully equivalent differential forms (1.1.3) or (1.1.10) be valid.

1.2 THE EQUATION OF MASS CONSERVATION

The law of mass conservation is a general statement of kinematic nature, that is, independent of the nature of the fluid or of the forces acting on it. It expresses the empirical fact that, in a fluid system, mass cannot disappear from the system nor be created. The quantity U is, in this case, the specific mass. As noted above, no diffusive flux exists for the mass transport, which means that mass can only be transported through convection. On the other hand, we will not consider multiphase fluids and hence no sources due to chemical reactions will have to be introduced.

The general conservation equation then becomes

$$\frac{\partial}{\partial t} \int_{\Omega} \rho \, d\Omega + \oint_{S} \rho \vec{v} \cdot d\vec{S} = 0 \qquad (1.2.1)$$

and in differential form:

$$\frac{\partial \rho}{\partial t} + \vec{\nabla} \cdot (\rho \vec{v}) = 0 \qquad (1.2.2)$$

An equivalent form to equation (1.2.2) is obtained by working out the divergence operator and introducing the material or convective derivative:

$$\frac{d}{dt} := \frac{\partial}{\partial t} + \vec{v} \cdot \vec{\nabla} \qquad (1.2.3)$$

This leads to the following form for the conservation law of mass:

$$\frac{d\rho}{dt} + \rho \vec{\nabla} \cdot \vec{v} = 0 \qquad (1.2.4)$$

Although both equations (1.2.2) and (1.2.4) are fully equivalent from a mathematical point of view they will not necessarily remain so when a numerical discretization is performed. Equation (1.2.2) corresponds to the general form of a conservation law and is said to be formally written in *conservation* or in *divergence* form. The importance of the conservative form in a numerical scheme lies in the fact that, if not properly taken into account, a discretization of equation (1.2.4) will lead to a numerical scheme in which all the mass fluxes through the mesh–cell boundaries will not cancel, and hence the numerical scheme will not keep the total mass constant. The importance of a conservative discretization of the flow equations has also been stressed by Lax (1954), who demonstrated that this condition is necessary in order to obtain correct jump relations through a discontinuity in the numerical scheme. We will return to this very important point in Chapter 6, where the finite volume method is presented, and in later chapters when discussing the discretization of Euler and Navier–Stokes equations in Volume 2.

Alternative form of a general conservation equation

The differential form of the general conservation equation (1.1.7) can be written in another way. If equation (1.2.2) multiplied by u is subtracted from the left-hand side of equation (1.1.7) we obtain

$$\rho \frac{\partial u}{\partial t} + \rho \vec{v} \cdot \vec{\nabla} u = \vec{\nabla} \cdot (\varkappa \rho \vec{\nabla} u) + Q_v + \vec{\nabla} \cdot \vec{Q}_s \qquad (1.2.5)$$

or

$$\rho \frac{du}{dt} = -\vec{\nabla} \cdot \vec{F}_{D} + Q_v + \vec{\nabla} \cdot \vec{Q}_s \qquad (1.2.6)$$

where $\vec{F}_{D}$ is the diffusive component of the flux vector.

14

Again, the difference between equations (1.2.5) or (1.2.6) and (1.1.7) lies in the conservative form of the equations. Clearly, equation (1.2.5) is not in conservation form and a straightforward discretization of equation (1.2.5) will generally not conserve the property u in the numerical simulation. It is also important to note that this conservation property is linked to the convective term and that, in a fluid at rest, there is little difference between conservative form (1.1.7) and the non-conservative form (1.2.5).

1.3 THE CONSERVATION LAW OF MOMENTUM OR EQUATION OF MOTION

Momentum is a vector quantity and therefore the conservation law will have the general form given by equations (1.1.8) and (1.1.10). In order to determine all the terms of these equations it is necessary to define the sources influencing the variation of momentum. It is known, from Newton's laws, that the sources for the variation of momentum in a physical system are the forces acting on it. These forces consist of the external volume forces $\vec{f}_e$ and the internal forces $\vec{f}_i$. The latter are dependent on the nature of the fluid considered, and result from the assumptions made about the properties of the internal deformations within the fluid and their relation to the internal stresses. We will assume that the fluid is Newtonian, and therefore the total internal stresses $\bar{\bar{\sigma}}$ are taken to be

$$\bar{\bar{\sigma}} = -p\bar{\bar{I}} + \bar{\bar{\tau}} \tag{1.3.1}$$

where $\bar{\bar{I}}$ is the unit tensor. Here the existence of the isotropic pressure component $p\bar{\bar{I}}$ is introduced and $\bar{\bar{\tau}}$ is the viscous shear stress tensor, equal to

$$\tau_{ij} = \mu[(\partial_i v_j + \partial_j v_i) - \tfrac{2}{3}(\vec{\nabla} \cdot \vec{v})\delta_{ij}] \tag{1.3.2}$$

where μ is the dynamic viscosity of the fluid, see for instance Batchelor (1970). A kinematic viscosity coefficient ν is also defined by $\nu = \mu/\rho$. This relation is valid for a Newtonian fluid in local thermodynamic equilibrium. Otherwise the most general form for the viscous stress tensor is

$$\tau_{ij} = \mu(\partial_i v_j + \partial_j v_i) + \lambda(\vec{\nabla} \cdot \vec{v})\delta_{ij} \tag{1.3.3}$$

Up to now, with the expection of very high temperature or pressure ranges, there is no experimental evidence that the Stokes relation

$$2\mu + 3\lambda = 0 \tag{1.3.4}$$

leading to equation (1.3.2) is not satisfied. Therefore we will not consider in the following the second viscosity coefficient λ as independent from μ. As with the mass conservation equations, it is assumed that no diffusion of momentum is possible in a fluid at rest, and hence there is no diffusive contribution to the flux tensor $\bar{\bar{F}}$. The source term $\vec{Q}_v$ consists of the sum of the external volume forces per unit volume $\rho\vec{f}_e$ and the sum of all the internal forces.

By definition, integral forces cancel two per two in every point inside the volume. Therefore the remaining internal forces within the volume V are those acting on the points of the boundary surface, since they have no opposite counterpart within the considered volume. Hence the internal forces act as surface sources, with the intensity

$$\int_S \bar{\bar{\sigma}} \cdot d\vec{S}$$

and the momentum conservation equation becomes

$$\frac{\partial}{\partial t} \int_\Omega \rho\vec{v} \, d\Omega + \oint_S \rho\vec{v}(\vec{v} \cdot d\vec{S}) = \int_\Omega \rho\vec{f_e} \, d\Omega + \oint_S \bar{\bar{\sigma}} \cdot d\vec{S}$$

$$= \int_\Omega \rho\vec{f_e} \, d\Omega - \oint_S p \, d\vec{S} + \oint_S \bar{\bar{\tau}} \cdot d\vec{S}$$

(1.3.5)

Applying Gauss's theorem, we obtain

$$\int_\Omega \frac{\partial}{\partial t}\rho\vec{v} \, d\Omega + \int_\Omega \vec{\nabla} \cdot (\rho\vec{v} \otimes \vec{v}) \, d\Omega = \int_\Omega \rho\vec{f_e} \, d\Omega + \int_\Omega \vec{\nabla} \cdot \bar{\bar{\sigma}} \, d\Omega$$

(1.3.6)

which leads to the differential form of the equation of motion:

$$\frac{\partial}{\partial t}(\rho\vec{v}) + \vec{\nabla} \cdot (\rho\vec{v} \otimes \vec{v} + p\bar{\bar{I}} - \bar{\bar{\tau}}) = \rho\vec{f_e}$$

(1.3.7)

An equivalent, non-conservative form is obtained after subtracting from the left-hand side the continuity equation (1.2.2) multiplied by $\vec{v}$:

$$\rho\frac{d\vec{v}}{dt} = -\vec{\nabla}p + \vec{\nabla} \cdot \bar{\bar{\tau}} + \rho\vec{f_e}$$

(1.3.8)

where the material derivative d/dt has been introduced.

When the form (1.3.2) of the shear stress tensor for a Newtonian viscous fluid is introduced into equations (1.3.7) or (1.3.8) we obtain the *Navier–Stokes equations of motion*. For constant viscosity coefficients, it reduces to

$$\rho\frac{d\vec{v}}{dt} = -\vec{\nabla}p + \mu\left[\Delta\vec{v} + \frac{1}{3}\vec{\nabla}(\vec{\nabla} \cdot \vec{v})\right] + \rho\vec{f_e}$$

(1.3.9)

For an ideal fluid without internal shear stresses (that is, for an inviscid fluid) the momentum equation reduces to the *Euler equation of motion*:

$$\rho\frac{d\vec{v}}{dt} = \rho\frac{\partial\vec{v}}{\partial t} + \rho(\vec{v} \cdot \vec{\nabla})\vec{v} = -\vec{\nabla}p + \rho\vec{f_e}$$

(1.3.10)

The vorticity equation

The equations of motion can be written in many equivalent forms, one of them being obtained through the introduction of the vorticity vector $\vec{\zeta}$:

$$\vec{\zeta} = \vec{\nabla} \times \vec{v}$$

(1.3.11)

and the vector identity

$$(\vec{v} \cdot \vec{\nabla})\vec{v} = \vec{\nabla}\left(\frac{\vec{v}^2}{2}\right) - \vec{v} \times (\vec{\nabla} \times \vec{v}) \tag{1.3.12}$$

in the inertia term $d\vec{v}/dt$. Equation (1.3.8) becomes

$$\frac{\partial \vec{v}}{\partial t} - (\vec{v} \times \vec{\zeta}) = -\frac{1}{\rho}\vec{\nabla}p - \vec{\nabla}\left(\frac{\vec{v}^2}{2}\right) + \frac{1}{\rho}\vec{\nabla} \cdot \bar{\bar{\tau}} + \vec{f}_e \tag{1.3.13}$$

This equation will be transformed further by the introduction of thermodynamical relations after having discussed the conservation law for energy.

An important equation for the vorticity $\vec{\zeta}$ can be obtained by taking the curl of the momentum equation. This leads to the *Helmholtz* equation:

$$\frac{\partial \vec{\zeta}}{\partial t} + (\vec{v} \cdot \vec{\nabla})\vec{\zeta} = (\vec{\zeta} \cdot \vec{\nabla})\vec{v} - \vec{\zeta}(\vec{\nabla} \cdot \vec{v}) + \vec{\nabla}p \times \vec{\nabla}\frac{1}{\rho} + \vec{\nabla} \times \left(\frac{1}{\rho}\vec{\nabla} \cdot \bar{\bar{\tau}}\right) + \vec{\nabla} \times \vec{f}_e \tag{1.3.14}$$

For a Newtonian incompressible fluid with constant kinematic viscosity coefficient ν, the shear stress term reduces to the Laplacian of the vorticity:

$$\left(\vec{\nabla} \times \frac{1}{\rho}\vec{\nabla} \cdot \bar{\bar{\tau}}\right) = \nu\,\Delta\vec{\zeta} \tag{1.3.15}$$

1.4 ROTATING FRAME OF REFERENCE

In many applications such as geophysical flows, turbomachinery problems or flows around helicopter blades, propellers and windmills we have to deal with rotating systems, and it is necessary to be able to describe the flow behaviour relatively to a rotating frame of reference.

We will assume that the moving system is *rotating steadily* with angular velocity $\vec{\omega}$ around an axis along which a co-ordinate z is aligned (Figure 1.4.1). Defining $\vec{w}$ as the velocity field relative to the rotating system and $\vec{u} = \vec{\omega} \times \vec{r}$ as the entrainment velocity, the composition law holds:

$$\vec{v} = \vec{w} + \vec{u} = \vec{w} + \vec{\omega} \times \vec{r} \tag{1.4.1}$$

Since the entrainment velocity does not contribute to the mass balance, the continuity equation remains invariant and can be written in the relative system:

$$\frac{\partial \rho}{\partial t} + \vec{\nabla} \cdot (\rho \vec{w}) = 0 \tag{1.4.2}$$

With regard to the momentum conservation law, observers in the two systems of reference will not see the same field of forces since the inertia term $d\vec{v}/dt$ is not invariant when passing from one system to the other. It is known that we have to add in the rotating frame of reference two forces, the *Coriolis force* per unit mass $\vec{f}_C$:

$$\vec{f}_C = -2(\vec{\omega} \times \vec{w}) \tag{1.4.3}$$

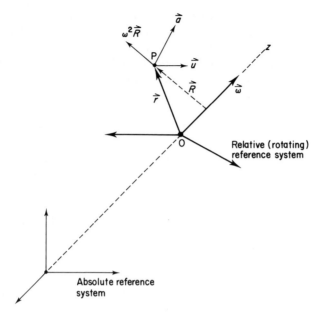

Figure 1.4.1 Relative rotating frame of reference

and the *centrifugal force* per unit mass $\vec{f}_c$:

$$\vec{f}_c = -\vec{\omega} \times (\vec{\omega} \times \vec{r}) = \omega^2 \vec{R} \tag{1.4.4}$$

if $\vec{R}$ is the component of the position vector perpendicular to the axis of rotation. Hence, additional force terms appear in the right-hand side of the conservation law (1.3.5) if this equation is written directly in the rotating frame of reference. These two forces, acting on a fluid particle in the rotating system, play a very important role in rotating flows, especially when the velocity vector $\vec{w}$ has large components in the direction perpendicular to $\vec{\omega}$.

The conservation law for momentum in the relative system then becomes

$$\frac{\partial}{\partial t} \int_\Omega \rho \vec{w} \, d\Omega + \oint_S \rho \vec{w}(\vec{w} \cdot d\vec{S})$$

$$= \int_\Omega \rho \vec{f}_e \, d\Omega + \int_\Omega \rho \vec{f}_C \, d\Omega + \int_\Omega \rho \vec{f}_c \, d\Omega - \oint_S p \, d\vec{S} + \oint_S \bar{\bar{\tau}} \cdot d\vec{S} \tag{1.4.5}$$

and the transformation of the surface integrals into volume integrals leads to the differential form:

$$\frac{\partial}{\partial t} (\rho \vec{w}) + \vec{\nabla} \cdot (\rho \vec{w} \otimes \vec{w})$$

$$= \rho \vec{f}_e - \rho \vec{\omega} \times (\vec{\omega} \times \vec{r}) - 2\rho(\vec{\omega} \times \vec{w}) - \vec{\nabla} p + \vec{\nabla} \cdot \bar{\bar{\tau}} \tag{1.4.6}$$

The shear stress tensor $\bar{\bar{\tau}}$ is to be expressed as a function of the relative velocities. It is considered that the rotation of the relative system has no effect on the internal forces within the fluid, since these internal forces cannot, by definition be influenced by solid body motions of one system of reference with respect to the other. A non-conservative form of the relative momentum equation similar to equation (1.3.13) can be obtained as

$$\frac{\partial \vec{w}}{\partial t} - (\vec{w} \times \vec{\zeta}) = -\frac{1}{\rho} \vec{\nabla} p - \vec{\nabla}\left(\frac{\vec{w}^2}{2} - \frac{\vec{u}^2}{2}\right) + \frac{1}{\rho} \vec{\nabla} \cdot \bar{\bar{\tau}} + \vec{f}_e \qquad (1.4.7)$$

where the presence of the absolute vorticity vector is to be noted.

1.5   THE CONSERVATION EQUATION FOR ENERGY

It is known, from the thermodynamical analysis of continua, that the energy content of a system is measured by its internal energy per unit mass e. This internal energy is a state variable of a system and hence its variation during a thermodynamical transformation depends only on the final and initial states.

In a fluid the total energy to be considered in the conservation equation is the sum of its internal energy and its kinetic energy per unit mass $\vec{v}^2/2$. We will indicate by E this total energy per unit mass:

$$E = e + \frac{\vec{v}^2}{2} \qquad (1.5.1)$$

The first law of thermodynamics states that the sources for the variation of the total energy are the work of the forces acting on the system plus the heat transmitted to this system.

Considering the general form of the conservation law for the quantity E we have a convective flux of energy $\vec{F}_C$:

$$\vec{F}_C = \rho \vec{v}\left(e + \frac{\vec{v}^2}{2}\right) = \rho \vec{v} E \qquad (1.5.2)$$

and a diffusive flux $\vec{F}_D$, written as

$$\vec{F}_D = -\gamma \rho \varkappa \vec{\nabla} e \qquad (1.5.3)$$

since, by definition, there is no diffusive flux associated with the motion. The coefficient $\varkappa$ is the thermal diffusivity coefficient and has to be defined empirically, together with the dynamic viscosity μ. The coefficient γ is the ratio of specific heat coefficients under constant pressure and constant volume, $\gamma = c_p/c_v$.

Actually this diffusive term (1.5.3) describes the diffusion of heat in a medium at rest due to molecular thermal conduction. It is generally written in a slightly different form, i.e. under that of Fourier's law of heat conduction:

$$\vec{F}_D = -k \vec{\nabla} T \qquad (1.5.4)$$

where T is the absolute temperature and k is the thermal conductivity coefficient. We have the relation

$$k = \rho c_p \varkappa = \frac{\mu c_p}{Pr} \qquad (1.5.5)$$

where Pr is the Prandtl number:

$$Pr = \nu/\varkappa = \mu c_p/k \qquad (1.5.6)$$

With regard to the sources of energy variations in a fluid system, a distinction has to made between the surface and the volume sources. The volume sources are the sum of the work of the volume forces $\vec{f_e}$ and the heat sources other than conduction (i.e. radiation, chemical reactions) q_H. Hence we have, per unit volume, $Q_v = \rho \vec{f_e} \cdot \vec{v} + q_H$. The surface sources Q_s are the result of the work done on the fluid by the internal shear stresses acting on the surface of the volume considering that there are no surface heat sources;

$$Q_s = \bar{\bar{\sigma}} \cdot \vec{v} = -p\vec{v} + \bar{\bar{\tau}} \cdot \vec{v} \qquad (1.5.7)$$

1.5.1 Conservative formulation of the energy equation

Grouping all the contributions, the energy conservation equation in integral form becomes

$$\frac{\partial}{\partial t} \int_\Omega \rho E \, d\Omega + \oint_S \rho E \vec{v} \cdot d\vec{S}$$

$$= \oint_S k \vec{\nabla} T \cdot d\vec{S} + \int_\Omega (\rho \vec{f_e} \cdot \vec{v} + q_H) \, d\Omega + \oint_S (\bar{\bar{\sigma}} \cdot \vec{v}) \, d\vec{S} \qquad (1.5.8)$$

After transformation to volume integrals the differential form of the conservation equation for energy becomes

$$\frac{\partial}{\partial t}(\rho E) + \vec{\nabla} \cdot (\rho \vec{v} E) = \vec{\nabla} \cdot (k \vec{\nabla} T) + \vec{\nabla} \cdot (\bar{\bar{\sigma}} \cdot \vec{v}) + W_f + q_H \qquad (1.5.9)$$

where W_f is the work of the external volume forces:

$$W_f = \rho \vec{f_e} \cdot \vec{v} \qquad (1.5.10)$$

Clarifying the term $\vec{\nabla} \cdot (\bar{\bar{\sigma}} \cdot \vec{v})$ and introducing the enthalpy h of the fluid leads to the following alternative expression in differential form:

$$\frac{\partial}{\partial t}(\rho E) + \vec{\nabla} \cdot [\rho \vec{v} H - k \vec{\nabla} T - \bar{\bar{\tau}} \cdot \vec{v}] = W_f + q_H \qquad (1.5.11)$$

where the stagnation, or total, enthalpy H is introduced:

$$H = e + \frac{p}{\rho} + \frac{\vec{v}^2}{2} = h + \frac{\vec{v}^2}{2} = E + \frac{p}{\rho} \qquad (1.5.12)$$

1.5.2 The equations for internal energy and entropy

An equation for the variation of the internal energy e can be obtained after some manipulations (See Problem 1.2) and the introduction of the *dissipation term* ε_v:

$$\varepsilon_v = (\bar{\bar{\tau}} \cdot \vec{\nabla}) \cdot \vec{v} = \frac{1}{2\mu} (\bar{\bar{\tau}} \otimes \bar{\bar{\tau}}^T)$$

$$= \tau_{ij} \frac{\partial v_i}{\partial x_j} \tag{1.5.13}$$

This leads to

$$\frac{\partial}{\partial t} (\rho e) + \vec{\nabla} \cdot (\rho \vec{v} h) = (\vec{v} \cdot \vec{\nabla}) p + \varepsilon_v + q_H + \vec{\nabla} \cdot (k \vec{\nabla} T) \tag{1.5.14}$$

An alternative form is obtained after introduction of the continuity equation:

$$\rho \frac{de}{dt} = -p(\vec{\nabla} \cdot \vec{v}) + \varepsilon_v + \vec{\nabla} \cdot (k \vec{\nabla} T) + q_H \tag{1.5.15}$$

The first term is the reversible work of the pressure forces (and vanishes in an incomprehensible flow), while the other terms are being considered as heat additions, with the dissipation term ε_v acting as an irreversible heat source. This appears clearly by introducing the entropy per unit mass s of the fluid, through the thermodynamic relation

$$T \, ds = de + p \, d\left(\frac{1}{\rho}\right) = dh - \frac{dp}{\rho} \tag{1.5.16}$$

The separation between reversible and irreversible heat additions is defined by

$$T \, ds = dq + dq' \tag{1.5.17}$$

where dq is a reversible heat transfer to the fluid while dq' is an irreversible heat addition. As is known from the second principle of thermodynamics, dq' is always non-negative and hence in an abiabatic flow ($dq = 0$) with irreversible transformations, the entropy will always increase.

Introducing definition (1.5.16) into equation (1.5.15), we obtain

$$\rho T \frac{ds}{dt} = \varepsilon_v + \vec{\nabla} \cdot (k \vec{\nabla} T) + q_H \tag{1.5.18}$$

where the last two terms can be considered as reversible head additions by conduction and by other sources. Therefore, in an adiabatic flow, $q_H = 0$, without heat conduction ($k = 0$) the non-negative dissipation term ε_v behaves as a non-reversible heat source.

Equation (1.5.18) is the entropy equation of the flow. Although this equation plays an important role it is not independent from the energy equation. Only one of these has to be added to the conservation laws for mass and momentum. Note also that the entropy is not a 'conserved' quantity in the sense of the previously derived conservation equations.

1.5.3 Energy equation in a relative system

The energy conservation equation in a relative system with steady rotation is obtained by adding the work of the centrifugal forces, since the Coriolis forces do not contribute to the energy balance of the flow.

In differential form we obtain the following full conservative form of the equation corresponding to equation (1.5.11):

$$\frac{\partial}{\partial t} \rho \left(e + \frac{\vec{w}^2}{2} - \frac{\vec{u}^2}{2} \right) + \vec{\nabla} \cdot \left[\rho \vec{w} \left(h + \frac{\vec{w}^2}{2} - \frac{\vec{u}^2}{2} \right) - k \vec{\nabla} T - \bar{\bar{\tau}} \cdot \vec{w} \right] = W_f + q_H$$

$$(1.5.19)$$

with

$$W_f = \rho \vec{f}_e \cdot \vec{w} \qquad (1.5.20)$$

In non-conservative form equation (1.5.19) becomes, where d/dt and $\partial/\partial t$ are considered in the relative system,

$$\rho \frac{d}{dt} \left(h + \frac{\vec{w}^2}{2} - \frac{\vec{u}^2}{2} \right) = \frac{\partial p}{\partial t} + \vec{\nabla} \cdot (k \vec{\nabla} T) + \vec{\nabla} \cdot (\bar{\bar{\tau}} \cdot \vec{w}) + W_f + q_H$$

$$(1.5.21)$$

The quantity

$$I = h + \frac{\vec{w}^2}{2} - \frac{\vec{u}^2}{2} = H - \vec{u} \cdot \vec{v} \qquad (1.5.22)$$

appearing in the left-hand side of the above equations plays an important role, since it appears as a stagnation enthalpy term for the rotating system. This term has been called the *rothalpy*, and it measures the total energy content in a steadily rotating frame of reference.

1.5.4 Crocco's form of the equations of motion

The pressure gradient term in the equation of motion can be eliminated by making use of the entropy equation (1.5.16) written for arbitrary variations of the state of the fluid. In particular, if the flow is followed in its displacement along its (absolute) velocity line,

$$T \vec{\nabla} s = \vec{\nabla} h - \frac{\vec{\nabla} p}{\rho} \qquad (1.5.23)$$

and introducing this relation into equation (1.3.13), we obtain

$$\frac{\partial \vec{v}}{\partial t} - (\vec{v} \times \vec{\zeta}) = T \vec{\nabla} s - \vec{\nabla} H + \frac{1}{\rho} \vec{\nabla} \cdot \bar{\bar{\tau}} + \vec{f}_e \qquad (1.5.24)$$

22

where the stagnation enthalpy H has been introduced. Similarly, in the relative system we obtain from equation (1.4.7)

$$\frac{\partial \vec{w}}{\partial t} - (\vec{w} \times \vec{\zeta}) = T\vec{\nabla}s - \vec{\nabla}I + \frac{1}{\rho}\vec{\nabla}\cdot\bar{\bar{\tau}} + \vec{f}_e \qquad (1.5.25)$$

where the rothalpy I appears as well as the absolute vorticity $\vec{\zeta}$.

The introduction of entropy and stagnation enthalpy gradients into the equation of motions is due to Crocco, and equations (1.5.24) and (1.5.25) reveal important properties. A first observation is that, even in steady flow conditions, the flow will be rotational, except in very special circumstances, namely frictionless, isentropic and isoenergetic flow conditions, without external forces or with forces which can be derived from a potential function where the corresponding potential energy is added to the total energy H. An analogous statement can be made for the equation in the relative system where the total energy is measured by I. However, since the absolute vorticity appears in the relative equation of motion, even under steady relative conditions, with constant energy I and inviscid flow conditions without body forces, the relative vorticity will not be zero but equal to $(-2\vec{\omega})$. The relative motion is therefore never irrotational but will have at least a vorticity component equal to minus twice the solid body angular velocity. This shows that, under the above-mentioned conditions of absolute vorticity equal to zero, the relative flow undergoes a solid body rotation equal to 2ω in the opposite direction to the rotation of the relative system.

Summary of the basic flow equations

The equations derived in the previous sections are valid in all generality for any Newtonian compressible fluid in an absolute or a relative frame of reference with constant rotation. The various forms of these equations can be summarized in the following tables. Table 1.1 corresponds to the equations in the absolute system, while Table 1.2 contains the equations written in the steadily rotating relative frame of reference.

References

Batchelor, G. K. (1970). *An Introduction to Fluid Dynamics*, Cambridge: Cambridge University Press.
Lax, P. D. (1954). 'Weak solutions of nonlinear hyperbolic equations and their numerical computation.' *Comm. Pure and Applied Mathematics*, 7, 159–93.

PROBLEMS

Problem 1.1

Derive equation (1.4.7).

Problem 1.2

Obtain the energy equation (1.5.14) for the internal energy e.

Hint: Introduce the momentum equation multiplied by the velocity vector into equation (1.5.9).

Problem 1.3

Show that in an incompressible fluid at rest the energy equation (1.5.14) reduces to the temperature conduction equation:

$$\rho c_v \frac{\partial T}{\partial t} = \vec{\nabla}(k\vec{\nabla}T) + q_H$$

Problem 1.4

Show that the energy equation (1.5.11) reduces to a convection–diffusion balance of the stagnation enthalpy H when the Prandtl number is equal to one and when only the contribution from the work of the shear stresses related to the viscous diffusion of the kinetic energy is taken into account.

Hint: Assume constant flow properties, setting $k = \mu c_p$ in the absense of external sources, and separate the contributions to the term $\vec{\nabla}\cdot(\bar{\bar{\tau}}\cdot\vec{v})$ according to the following relations, valid for incompressible flows:

$$\vec{\nabla}(\bar{\bar{\tau}}\cdot\vec{v}) = \partial_i[\mu(\partial_i v_j + \partial_j v_i)v_j] = \vec{\nabla}\cdot[\mu\vec{\nabla}(\vec{v}^2/2)] + \vec{\nabla}\cdot[\mu(\vec{v}\cdot\vec{\nabla})\vec{v}]$$

Neglecting the second term, setting $H = c_p T + \vec{v}^2/2$ leads to

$$\frac{\partial}{\partial t}(\rho E) + \vec{\nabla}\cdot(\rho\vec{v}H) = \vec{\nabla}\cdot(\mu\vec{\nabla}H)$$

Problem 1.5

Obtain the entropy equation (1.5.18).

Problem 1.6

Prove equations (1.3.15).

Problem 1.7

Obtain equations (1.5.24) and (1.5.25).

Table 1.1. The system of flow equations in an absolute frame of reference

Equation	Integral	Conservation form	Non-conservation form
		Differential form	
Conservation of mass	$\dfrac{\partial}{\partial t}\displaystyle\int_{\Omega}\rho\,\mathrm{d}\Omega+\oint_{S}\rho\vec{v}\cdot\mathrm{d}\vec{S}=0$	$\dfrac{\partial\rho}{\partial t}+\vec{\nabla}\cdot(\rho\vec{v})=0$	$\dfrac{\mathrm{d}\rho}{\mathrm{d}t}+\rho\vec{\nabla}\cdot\vec{v}=0$
Conservation of momentum	$\dfrac{\partial}{\partial t}\displaystyle\int_{\Omega}\rho\vec{v}\,\mathrm{d}\Omega+\oint_{S}(\rho\vec{v}\otimes\vec{v}+p\bar{I}-\bar{\bar{\tau}})\cdot\mathrm{d}\vec{S}$ $=\displaystyle\int_{V}\rho\vec{f}_{e}\,\mathrm{d}V$ $\tau_{ij}=\mu[(\partial_{j}v_{i}+\partial_{i}v_{j})-\tfrac{2}{3}(\vec{\nabla}\cdot\vec{v})\,\delta_{ij}]$	$\dfrac{\partial}{\partial t}\rho\vec{v}+\vec{\nabla}\cdot(\rho\vec{v}\otimes\vec{v}+p\bar{I}-\bar{\bar{\tau}})=\rho\vec{f}_{e}$	$\rho\dfrac{\mathrm{d}\vec{v}}{\mathrm{d}t}=-\vec{\nabla}p+\vec{\nabla}\cdot\bar{\bar{\tau}}+\rho\vec{f}_{e}$ $\vec{\nabla}\cdot\bar{\bar{\tau}}=\mu[\Delta\vec{v}+\tfrac{1}{3}\vec{\nabla}\cdot(\vec{\nabla}\cdot\vec{v})]$ for constant μ *Crocco's form* $\dfrac{\partial\vec{v}}{\partial t}-\vec{v}\times\vec{\zeta}=T\vec{\nabla}s-\vec{\nabla}H+\dfrac{1}{\rho}\vec{\nabla}\cdot\bar{\bar{\tau}}+\vec{f}_{e}$
Conservation of energy	$\dfrac{\partial}{\partial t}\displaystyle\int_{\Omega}\rho E\,\mathrm{d}\Omega+\oint_{S}(\rho\vec{v}H-k\vec{\nabla}T-\bar{\bar{\tau}}\cdot\vec{v})\cdot\mathrm{d}\vec{S}$ $=\displaystyle\int_{\Omega}(\rho\vec{f}_{e}\cdot\vec{v}+q_{H})\,\mathrm{d}\Omega$	$\dfrac{\partial\rho E}{\partial t}+\vec{\nabla}\cdot(\rho\vec{v}H-k\vec{\nabla}T-\bar{\bar{\tau}}\cdot\vec{v})=W_{f}+q_{H}$	$\rho\dfrac{\mathrm{d}e}{\mathrm{d}t}=-p\vec{\nabla}\cdot\vec{v}+(\bar{\bar{\tau}}\cdot\vec{\nabla})\vec{v}+\vec{\nabla}\cdot(k\vec{\nabla}T)+q_{H}$ $\rho\dfrac{\mathrm{d}H}{\mathrm{d}t}=\dfrac{\partial p}{\partial t}+\vec{\nabla}\cdot(k\cdot\vec{\nabla}T+\bar{\bar{\tau}}\cdot\vec{v})+W_{f}+q_{H}$ $\rho\dfrac{\mathrm{d}H}{\mathrm{d}t}=\dfrac{\partial p}{\partial t}+\rho T\dfrac{\mathrm{d}s}{\mathrm{d}t}+\vec{v}\cdot(\vec{\nabla}\cdot\bar{\bar{\tau}})+W_{f}$ *Entropy equation* $\rho T\dfrac{\mathrm{d}s}{\mathrm{d}t}=\varepsilon_{v}+\vec{\nabla}\cdot(k\vec{\nabla}T)+q_{H}$
Definitions	$H=h+\dfrac{\vec{v}^{2}}{2}$ stagnation enthalpy $E=e+\dfrac{\vec{v}^{2}}{2}$ total energy	$W_{f}=\rho\vec{f}_{e}\cdot\vec{v}$ work of external forces $\vec{f}_{e}$ q_{H} heat source	$\vec{\zeta}=\vec{\nabla}\times\vec{v}$ vorticity $\varepsilon_{v}=\dfrac{1}{2\mu}(\bar{\bar{\tau}}\otimes\bar{\bar{\tau}}^{T})=(\bar{\bar{\tau}}\cdot\vec{\nabla})\cdot\vec{v}$ viscous dissipation

Table 1.2. The system of flow equations in a steadily rotating relative frame of reference

Equation	Integral form	Conservation form	Non-Conservation form
			Differential form
Conservation of mass	$\dfrac{\partial}{\partial t}\displaystyle\int_{\Omega}\rho\,\mathrm{d}\Omega+\oint_{s}\rho\vec{w}\cdot\mathrm{d}\vec{S}=0$	$\dfrac{\partial\rho}{\partial t}+\vec{\nabla}\cdot(\rho\vec{w})=0$	$\dfrac{\mathrm{d}\rho}{\mathrm{d}t}+\rho\vec{\nabla}\cdot\vec{w}=0$
Conservation of momentum	$\dfrac{\partial}{\partial t}\displaystyle\int_{\Omega}\rho\vec{w}\,\mathrm{d}\Omega+\oint_{s}(\rho\vec{w}\otimes\vec{w}+pI-\bar{\bar{\tau}})\cdot\mathrm{d}\vec{S}$	$\dfrac{\partial}{\partial t}\rho\vec{w}+\vec{\nabla}\cdot(\rho\vec{w}\otimes\vec{w}+pI-\bar{\bar{\tau}})$	$\rho\dfrac{\mathrm{d}\vec{w}}{\mathrm{d}t}=-\vec{\nabla}p+\vec{\nabla}\cdot\bar{\bar{\tau}}-2\rho(\vec{\omega}\times\vec{w})$
	$=\displaystyle\int_{\Omega}\rho[\vec{f}_e-2\vec{\omega}\times\vec{w}-\vec{\omega}\times(\vec{\omega}\times\vec{r})]\,\mathrm{d}\Omega$	$=\rho[\vec{f}_e-2\vec{\omega}\times\vec{w}-\vec{\omega}\times(\vec{\omega}\times\vec{r})]$	$-\rho\vec{\omega}\times(\vec{\omega}\times\vec{r})+\rho\vec{f}_e$
	$\tau_{ij}=\mu(\partial_i w_j+\partial_j w_i)-\tfrac{2}{3}(\vec{\nabla}\cdot\vec{w})\,\delta_{ij}$		*Crocco's form*
			$\dfrac{\partial\vec{w}}{\partial t}-\vec{w}\times\vec{\zeta}=T\vec{\nabla}s-\vec{\nabla}I+\dfrac{1}{\rho}\vec{\nabla}\cdot\bar{\bar{\tau}}+\vec{f}_e$
Conservation of energy	$\dfrac{\partial}{\partial t}\displaystyle\int_{\Omega}\rho E^{*}\,\mathrm{d}\Omega+\oint_{s}(\rho\vec{w}I-k\vec{\nabla}T-\bar{\bar{\tau}}\cdot\vec{w})\cdot\mathrm{d}\vec{S}$	$\dfrac{\partial}{\partial t}\rho E^{*}+\vec{\nabla}\cdot(\rho\vec{w}I-k\vec{\nabla}T-\bar{\bar{\tau}}\cdot\vec{w})$	$\rho\dfrac{\mathrm{d}I}{\mathrm{d}t}=\dfrac{\partial p}{\partial t}+\vec{\nabla}\cdot(k\vec{\nabla}T+\bar{\bar{\tau}}\cdot\vec{w})+W_{\mathrm{f}}+q_{\mathrm{H}}$
		$=W_{\mathrm{f}}+q_{\mathrm{H}}$	
	$=\displaystyle\int_{\Omega}(\rho\vec{f}_e\cdot\vec{w})\,\mathrm{d}\Omega$		*Entropy equation*
			$\rho T\dfrac{\mathrm{d}s}{\mathrm{d}t}=\vec{\nabla}\cdot(k\vec{\nabla}T)+\varepsilon_v+q_{\mathrm{H}}$
Definitions	$E^{*}=e+\dfrac{\vec{w}^{2}}{2}-\dfrac{\vec{u}^{2}}{2}=E-\vec{u}\cdot\vec{v}$	$W_{\mathrm{f}}=\rho\vec{f}_e\cdot\vec{w}$ work of external forces $\vec{f}_e$	$\vec{\zeta}=\vec{\nabla}\times\vec{v}$ absolute vorticity
	$I=h+\dfrac{\vec{w}^{2}}{2}-\dfrac{\vec{u}^{2}}{2}=H-\vec{u}\cdot\vec{v}$ rothalpy		$\varepsilon_v=\dfrac{1}{2\mu}(\bar{\bar{\tau}}\otimes\bar{\bar{\tau}}^{\mathrm{T}})=(\bar{\bar{\tau}}\cdot\vec{\nabla})\cdot\vec{w}$ viscous dissipation
	$\vec{u}=\vec{\omega}\times\vec{r}$		
	$\vec{v}=\vec{u}+\vec{w}$		q_{H} heat sources

Chapter 2

The Dynamic Levels of Approximation

INTRODUCTION

The system of Navier–Stokes equations, supplemented by empirical laws for the dependence of viscosity and thermal conductivity with other flow variables and by a constitutive law defining the nature of the fluid, completely describes all flow phenomena. For laminar flows no additional information is required and we can consider that any experiment in laminar flow regime can be accurately duplicated by computations. However, and we could say unfortunately from the point of view of computational fluid dynamics, most of the flow situations occurring in nature and in technology enter into a particular form of instability, called *turbulence*. This occurs in all flow situations when the velocity, or more precisely, the Reynolds number, defined as the product of representative scales of velocity and length divided by the kinematic viscosity, exceeds a certain critical value. The particular form of instability generated in the turbulent flow regime is characterized by the presence of statistical fluctuations of all the flow quantities. These fluctuations can be considered as superimposed on mean or averaged values and can attain, in many situations, the order of 10% of the mean values, although certain flow regions, such as separated zones, can attain much higher levels of turbulent fluctuations.

Clearly, the numerical description of the turbulent fluctuations is a formidable task which puts very high demands on computer resources. In the future, with increasing computer power, both in speed and memory, we could be able to simulate the large-scale turbulent fluctuations, or even the small-scale turbulent motion, from the time-dependent Navier–Stokes equations. Estimates of the computer requirements for this level of approximation can be found in Chapman (1979) and Kutler (1983).

Since this level is currently outside the reach of our computational capabilities we will consider, as the highest level of approximation, the Reynolds-averaged Navier–Stokes equations supplemented by some models for the Reynolds stresses (Section 2.2). These models can range from simple eddy viscosity or mixing length models to transport equations for the turbulent kinetic energy and dissipation rates, the so-called k-ε model, or to still more complicated models directly computing the Reynolds stresses.

Considering the various stages within the dynamical level of approximation, a first reduction in complexity can be introduced for flows with a small amount of separation or backflow and with a predominant mainstream direction at high Reynolds numbers. This allows us to neglect viscous and turbulent diffusion in the mainstream direction and hence to reduce the number of shear stress terms to be computed, considering that they have a negligible action on the flow behaviour. This is the *thin shear layer approximation* (discussed in Section 2.3).

Within the same level we can situate the *parabolic* approximations for the *steady-state* Navier–Stokes equations. In these approximations the elliptic character of the flow is put forward through the pressure field, while all other variables are considered as transported or as having a parabolic behaviour. These methods solve an elliptic equation for the pressure correction defined such as to satisfy continuity, assuming thereby that the pressure forces are dominant together with the inertia forces, while the viscous and turbulent forces are simplified and reduced to the transverse diffusion of momentum, thereby excluding any separation (Section 2.4).

The next level to be considered is the boundary layer approximation, referred to in Section 2.5. As is well known, this analysis of the effects of viscosity by Prandtl is a most spectacular example of the impact on the description of a flow system and of a careful investigation of the magnitude of force components.

For flows with no separation and thin viscous layers, that is, at high Reynolds numbers, a separation of the viscous and inviscid parts of the flow can be introduced, whereby the pressure field is decoupled from the viscous effects, showing that the influence of the viscous and turbulent shear stresses is confined to small regions close to the walls and that outside these layers the flow behaves as inviscid. This analysis, which was perhaps the greatest breakthrough in fluid mechanics since the discovery of the Navier–Stokes equations, showed that many of the flow properties can be described by the inviscid approximation (for example, determination of the pressure distributions), and that a simplified boundary layer approximation allows for the determination of the viscous effects. The calculation of the inviscid and the boundary layer parts of the flow can be performed interactively, taking into account the influence of the boundary layers on the inviscid flow.

Recently a series of approaches in this direction have been developed, i.e. the *viscid–inviscid interaction* methods, whereby attempts are made to calculate or to model separated regions in an approximate way while keeping the advantages with regard to the reduced computational effort of the boundary layer approximations (see Le Balleur, 1983, for a recent review of the subject). When this influence or interaction is neglected we enter the field of the *inviscid approximations*, which allows generally a good approximation of the pressure field and hence of lift coefficient for non-separated flows.

An intermediate level between the partially or fully viscous flow descriptions and the inviscid approximation is the *distributed loss model*, used in internal flow problems, particularly in the simulation of multi-stage turbomachinery flows. Due to the large number of flow passages and their rotation with respect

28

Table 2.1. Approximation levels

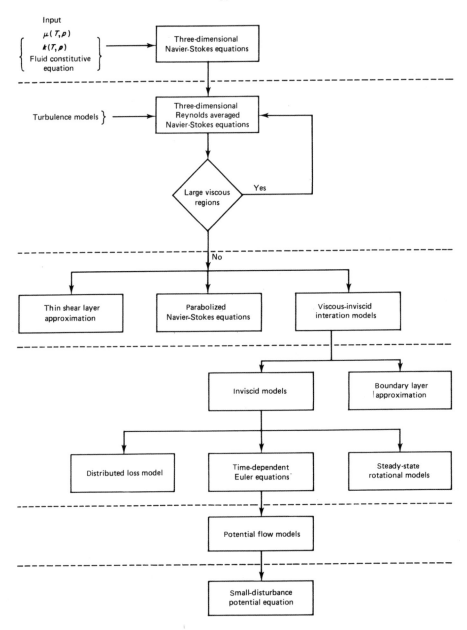

to the next blade row, we can consider that a downstream blade row sees, in a first approximation, an averaged flow in the sense that the influence of the boundary layers and the wakes of the previous blades mix out. Their overall effect on the next blade can therefore be assimilated to a distributed friction force and the implications of this approximation are presented in Section 2.6.

Within the inviscid approximations we can distinguish, next to the model of the time-dependent Euler equations (summarized in Section 2.7), a family of stationary models for rotational flows, allowing a reduction of the number of dependent variables (Section 2.8).

The potential flow model, limited to irrotational flows, is at a lower level of approximation, due to the associated assumption of isentropicity. As will be shown in Section 2.9, this leads to a description of discontinuities which deviate from the Rankine–Hugoniot relations and occasionally to problems of non-uniqueness. However, the potential flow model is equivalent to the Euler equations for continuous, irrotational flows.

Table 2.1 summarizes the various levels of approximations defined in the present approach and the corresponding mathematical models will be presented and discussed in detail in the following sections. Also, we will present practical examples of computations performed at each level of approximation as an illustration of the type of computations achieved with the model being considered.

2.1 THE NAVIER–STOKES EQUATIONS

The most general description of a fluid flow is obtained from the full system of Navier–Stokes equations. Referring to Table 1.1 of the previous chapter, the conservation laws for the three basic flow quantities $\rho, \rho\vec{v}, \rho E$ can be written in a compact form, expressing the coupled nature of the equations. We therefore obtain the following system of equations:

$$\frac{\partial}{\partial t}\begin{vmatrix}\rho\\\rho\vec{v}\\\rho E\end{vmatrix} + \vec{\nabla}\cdot\begin{vmatrix}\rho\vec{v}\\\rho\vec{v}\otimes\vec{v}+p\bar{\bar{I}}-\bar{\bar{\tau}}\\\rho\vec{v}H-\bar{\bar{\tau}}\cdot\vec{v}-k\vec{\nabla}T\end{vmatrix} = \begin{vmatrix}0\\\rho\vec{f_e}\\W_f+q_H\end{vmatrix} \tag{2.1.1}$$

The above equation defines a (5×1) column vector U of the *conservative* variables:

$$U=\begin{vmatrix}\rho\\\rho\vec{v}\\\rho E\end{vmatrix}=\begin{vmatrix}\rho\\\rho u\\\rho v\\\rho w\\\rho E\end{vmatrix} \tag{2.1.2}$$

and a generalized (5×3) flux vector $\vec{F}$:

$$\vec{F}=\begin{vmatrix}\rho\vec{v}\\\rho\vec{v}\otimes\vec{v}+p\bar{\bar{I}}-\bar{\bar{\tau}}\\\rho\vec{v}H-\bar{\bar{\tau}}\cdot\vec{v}-k\vec{\nabla}T\end{vmatrix} \tag{2.1.3}$$

with Cartesian co-ordinates f, g, h, each of these components being a (5×1) column vector. The right-hand side contains the source terms and these can be grouped into a (5×1) column vector Q, defined by

$$Q = \begin{vmatrix} 0 \\ \rho \vec{f_e} \\ W_f + q_H \end{vmatrix} \qquad (2.1.4)$$

The source terms express the effects of the external forces $\vec{f_e}$, of the heat sources q_H and of the work performed by the external forces $W_f = \rho \vec{f_e} \cdot \vec{v}$.

The group of equations (2.1.1) then takes the following condensed form:

$$\frac{\partial U}{\partial t} + \vec{\nabla} \cdot \vec{F} = Q \qquad (2.1.5)$$

Expressed in Cartesian co-ordinates, we obtain the more explicit algebraic form:

$$\frac{\partial U}{\partial t} + \frac{\partial f}{\partial x} + \frac{\partial g}{\partial y} + \frac{\partial h}{\partial z} = Q \qquad (2.1.6a)$$

or, in an alternative condensed notation,

$$\partial_t U + \partial_x f + \partial_y g + \partial_z h = Q \qquad (2.1.6b)$$

where u, v, w are the x, y, z components of the velocity vector $\vec{v}$ and the flux vector $\vec{F}$ is defined by its components f, g, h (subscripts indicate the corresponding Cartesian components):

$$f = \begin{vmatrix} \rho u \\ \rho u^2 + p - \tau_{xx} \\ \rho uv - \tau_{xy} \\ \rho uw - \tau_{xz} \\ \rho H u - (\bar{\bar{\tau}} \cdot \vec{v})_x - k \dfrac{\partial T}{\partial x} \end{vmatrix} \qquad (2.1.7a)$$

$$g = \begin{vmatrix} \rho v \\ \rho uv - \tau_{yx} \\ \rho v^2 + p - \tau_{yy} \\ \rho wv - \tau_{yz} \\ \rho H v - (\bar{\bar{\tau}} \cdot \vec{v})_y - k \dfrac{\partial T}{\partial y} \end{vmatrix} \qquad (2.1.7b)$$

$$h = \begin{vmatrix} \rho w \\ \rho uw - \tau_{zx} \\ \rho vw - \tau_{zy} \\ \rho w^2 + p - \tau_{zz} \\ \rho H w - (\bar{\bar{\tau}} \cdot \vec{v})_z - k \dfrac{\partial T}{\partial z} \end{vmatrix} \qquad (2.1.7c)$$

The system of equations (2.1.6) is written in conservation form in Cartesian co-ordinates. In practical configurations, however, the geometrical complexity of the boundaries of the flow regions calls for meshes adapted to the curved boundaries of the flow domain. This leads to curvilinear meshes, generally body fitted in the sense defined by Thompson (1982), which, even when numerically generated, can be considered as forming a family of curvilinear co-ordinate lines ξ, η, ζ.

Applying the general rules of tensor calculus, the conservative form of the equations of motions can be maintained when written in an arbitrary curvilinear co-ordinate system, as shown by Vinokur (1974) and Viviand (1974).

The system of Navier–Stokes equations has still to be supplemented by the constitutive laws and by the definition of the shear stress tensor as a function of the other flow variables. We will consider here only Newtonian fluids for which the shear stress tensor is defined by equation (1.3.2). The thermodynamic laws define the internal energy e or the enthalpy h as a function of only two other thermodynamic variables chosen between pressure p, specific mass ρ, termperature T, entropy s or any other intensive variable. For instance,

$$e = e(p, T) \tag{2.1.8}$$

\or

$$h = h(p, T) \tag{2.1.9}$$

In addition, the laws of dependence of the two fluid properties, the dynamic viscosity coefficient μ and the coefficient of thermal conductivity k, are to be given as a function of the fluid state, for instance of temperature and eventually of pressure. In particular, the viscosity coefficient μ is strongly influenced by temperature. For gases, a widely used relation is given by Sutherland's formula (for instance, for air) in the standard international, metric system:

$$\mu = \frac{1.45 \ T^{3/2}}{T + 110} \cdot 10^{-6} \tag{2.1.10}$$

where T is in degrees Kelvin. Note that for liquids, the dynamic viscosity decreases stongly with temperature and that the pressure dependence of μ, for both gases and liquids, is small. The temperature dependence of k is similar to that of μ for gases while for liquids, k is nearly constant. In any case, the temperature and pressure dependence of μ and k can only be obtained, within the framework of continuum mechanics, by experimental observation.

2.1.1 Perfect gas model

In many instances a compressible fluid can be considered as a perfect gas, even if viscous effects are taken into account, and the equation of state is written as

$$\frac{p}{\rho} = rT \tag{2.1.11}$$

32

where r is the gas constant per unit of mass and is equal to the universal gas constant divided by the molecular mass of the fluid. The internal energy e and the enthalpy h are only functions of temperature and we have the following relations, taking into account that

$$c_p = \frac{\gamma}{\gamma - 1} r \qquad (2.1.12)$$

where

$$\gamma = \frac{c_p}{c_v} \qquad (2.1.13)$$

is the ratio of specific heat coefficients under constant pressure, c_p and constant volume c_v:

$$e = c_v T = \frac{1}{\gamma - 1} \frac{p}{\rho}$$

$$\qquad (2.1.14)$$

$$h = c_p T = \frac{\gamma}{\gamma - 1} \frac{p}{\rho}$$

The entropy variation from a reference state, indicated by the subscript A, is obtained from equation (1.5.16) as

$$s - s_A = c_p \ln \frac{T}{T_A} - r \ln \frac{p}{p_A} \qquad (2.1.15)$$

or

$$s - s_A = - r \ln \frac{(p/p_A)}{(T/T_A)^{\gamma/(\gamma - 1)}} \qquad (2.1.16)$$

Introducing the equation of state, we also obtain

$$s - s_A = c_v \ln \frac{p/p_A}{(\rho/\rho_A)^\gamma} \qquad (2.1.17)$$

The stagnation variables can be derived from the total enthalpy H:

$$H = E + \frac{p}{\rho} = h + \frac{\vec{v}^2}{2} = c_p T_0 \qquad (2.1.18)$$

where the total or stagnation temperature T_0 is defined by

$$T_0 = T + \frac{\vec{v}^2}{2c_p} = T\left(1 + \frac{\gamma - 1}{2} M^2\right) \qquad (2.1.19)$$

The Mach number M has been introduced by

$$M = \frac{|\vec{v}|}{c} \qquad (2.1.20)$$

with

$$c^2 = \left(\frac{\partial p}{\partial \rho}\right)_s = \gamma r T = \frac{\gamma p}{\rho} \qquad (2.1.21)$$

being the square of the speed of sound. Similarly, we have

$$E = c_v T_0 \qquad (2.1.22)$$

Considering that the transition of the fluid from a static to a stagnation state is adiabatic and without losses of energy (that is, isentropic), we have for the stagnation pressure p_0

$$\frac{p_0}{p} = \left(\frac{T_0}{T}\right)^{\gamma/(\gamma-1)} = \left(1 + \frac{\gamma-1}{2} M^2\right)^{\gamma/(\gamma-1)} \qquad (2.1.23)$$

and hence relations (2.1.16) and (2.1.17) remain unchanged if the static variables are replaced by the stagnation variables. For instance, we have

$$s - s_A = -r \ln \frac{p_0/p_{0_A}}{(T_0/T_{0_A})^{\gamma/(\gamma-1)}} \qquad (2.1.24)$$

or

$$s - s_A = -r \ln \frac{p_0/p_{0_A}}{(H/H_A)^{\gamma/(\gamma-1)}} \qquad (2.1.25)$$

Various other forms of the relations between thermodynamic variables p, ρ, T, s, e, h can be obtained according to the choice of the independent variables.

As a function of h and s we have

$$\frac{p}{p_A} = \left(\frac{h}{h_A}\right)^{\gamma/(\gamma-1)} \cdot e^{-(s-s_A)/r} \qquad (2.1.26)$$

or, from equation (2.1.17),

$$\frac{\rho}{\rho_A} = \left(\frac{h}{h_A}\right)^{1/(\gamma-1)} \cdot e^{-(s-s_A)/r} \qquad (2.1.27)$$

Many other relations can be derived by selecting other combinations of variables.

Practical examples

Compressible flow around a circular cylinder   The compressible flow around a circular cylinder is an extremely complex flow case, since it contains the unsteadiness generated by the Von Karman vortex streets (although the incident flow conditions are stationary), separation and compressibility effects in interaction with viscous wakes. Figures 2.1.1–2.1.3 show the results from computations of this flow, with the full Navier–Stokes equations, for a perfect gas at various Mach numbers M_∞ and at Reynolds numbers ranging from 10^3 to $8\ 10^6$ (Ishii and Kuwahara, 1984; Ishii et al., 1985).

The spontaneously occurring unsteadiness of the wakes due to the periodic vortex shedding has to be captured by the numerical computation through an accurate simulation of the time behaviour. Figure 2.1.1 shows the computed

34

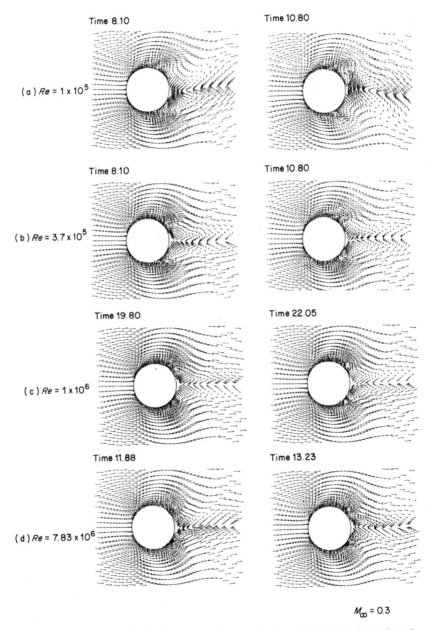

Figure 2.1.1 Computed flow field around a circular cylinder at incident Mach number of $M_\infty = 0.3$ and Reynolds numbers $R_e = 1.10^5$ to $7.83\ 10^6$, at two instants in time. (reproduced by permission of AIAA from K. Ishii *et al*, 1985)

flow fields at an incident Mach number of 0.3 and for values of Reynolds number ranging from subcritical (10^5) to supercritical (7.8 10^6). These numerical results can be compared with flow visualizations at a Reynolds number of 3.7 10^5 obtained with shadowgraph techniques at the Institut de Mécanique des Fluides de Lille, France, by Dyment (1982) and Rodriguez (1984) at a Mach number of 0.45. Successive pictures taken at 80 μs intervals and covering different phases of the vortex shedding cycle are shown in Figure 2.1.2.

The numerical treatment of boundary conditions can have a non-negligible impact on the computed flow field, since the physical flow properties do not allow us to impose conditions at the downstream boundary. An extrapolation is applied from inside points towards the boundary, situated at a finite distance from the cylinder. Although reasonable, this approximation introduces a perturbation on the downstream flow field, and it is expected that the effect of this perturbation are negligible. Computations at this level of approximation allow a very detailed analysis of the flow mechanism, but are still expensive to run, requiring 7–9 h on a supercomputer Hitachi S810 (Ishii *et al.*, 1985) for each Reynolds number.

The predicted drag coefficients agree well with the data in the subcritical and critical regions, as seen in Figure 2.1.3, while they are underpredicted in the supercritical regions at higher Reynolds numbers. This could be due to the absence of turbulence models in these computations or (and) to numerical errors.

Compressibility effects   With increasing Mach number and intensity of the acoustic waves the interaction and the coupling between compressibility effects, vortex shedding and separation on the cylinder becomes more pronounced. This appears on the calculated flow shown in Figure 2.1.4, where the velocity distribution at incident Mach number $M_\infty = 0.8$, at two instants in time, can be compared with a calculation at $M_\infty = 0.95$ for the same Reynolds number of 5.10^6. The Mach number and density contours show the trace of the oblique shock waves and the cylinder wake.

The influence of the compressibility on the flow around the cylinder is summarized in the beautiful series of pictures shown in Figure 2.1.5 taken at the Institut de Mécanique des Fluides de Lille, France, at different Mach numbers and at a Reynolds number close to 10^5. A strong shock is gradually generated downstream of the cylinder and a steady wake of increasing length appears for Mach numbers from 0.70 to 0.90, with a periodic vortex shedding downstream of the shock. Above a certain value of M_∞, lambda shocks appear and when they join, no disturbances can travel upstream, preventing the coupling between the wake and the vortex street. This can lead to a stationary regime such as observed in certain circumstances at $M_\infty = 0.98$. The flow visualizations show another important phenomenon, namely the appearance of more than one flow regime at certain values of Mach and Reynolds

36

$$H = 40\text{mm}$$

$$M = 0.45$$

$$U = 153\,m/s$$

$$\frac{UH}{\nu} = 3.7\ 10^5$$

$$\Delta t = 80\,\mu s$$

Shadowgraphs $0.3\,\mu s$

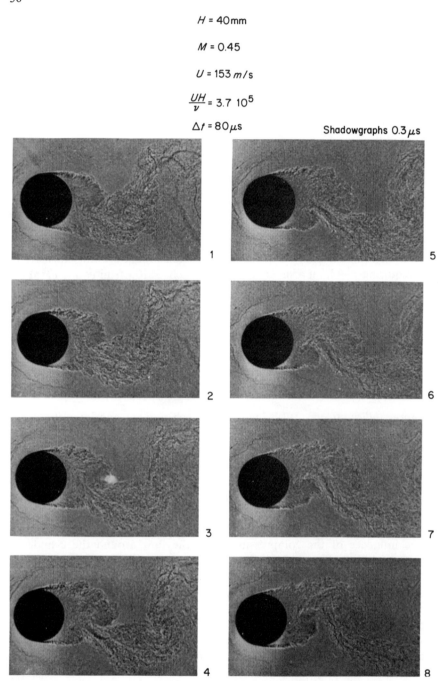

Figure 2.1.2 Shadowgraph of the flow around a cylinder at a Mach number of 0.45 and a Reynolds number of 3.710^5. The time interval between pictures in 80 μs. (Courtesy A. Dyment and M. Pianko, Institut de Mécanique des Fluides de Lille, France)

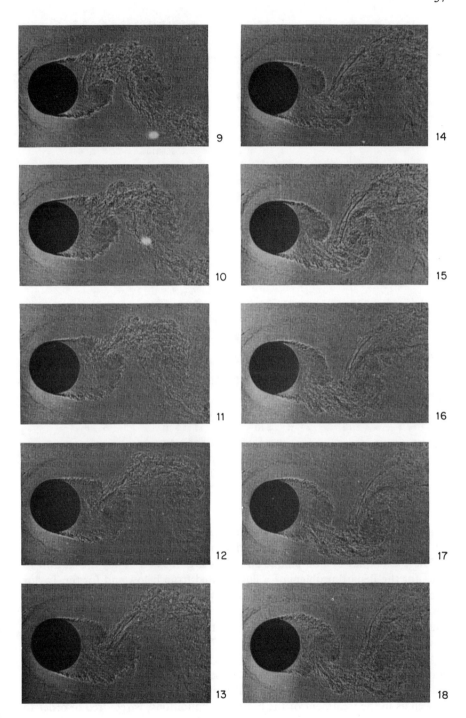

38

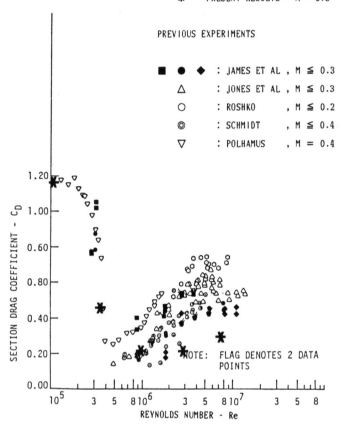

numbers. Two unsteady flow configurations can be distinguished at $M_\infty = 0.8$ while at $M_\infty = 0.98$ both unsteady and steady flow regimes can occur.

These examples of non-unique solutions are not new. They are known to exist for the Bénard problem of a fluid heated from below and for the Taylor problem of the flow between concentric cylinders, of which the inner one is rotating. It is interesting to observe here that the non-uniqueness of the stationary Navier–Stokes equations have been proved theoretically for these flow cases (see, for instance, Temam, 1977).

The non-uniqueness properties of the viscous flows, connected to the spontaneously generated unsteadiness, pose considerable problems for numerical simulation. Very high accuracy, at the level of the discretization schemes as well as in the treatment of the boundary conditions, is required in order to be able to recover numerically multiple solutions, when they exist.

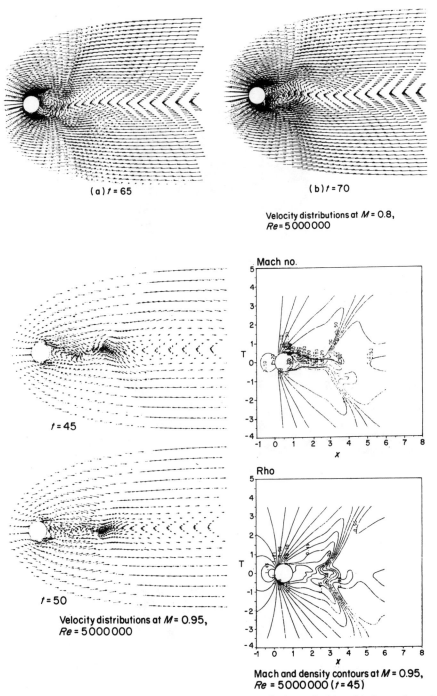

(a) $t = 65$

(b) $t = 70$

Velocity distributions at $M = 0.8$,
$Re = 5\,000\,000$

$t = 45$

$t = 50$

Velocity distributions at $M = 0.95$,
$Re = 5\,000\,000$

Mach no.

Rho

Mach and density contours at $M = 0.95$,
$Re = 5\,000\,000$ $(t = 45)$

Figure 2.1.4 Computed flow field around a circular cylinder at incident Mach number of $M_\infty = 0.8$ and $M_\infty = 0.95$, Reynolds numbers $Re = 5.10^6$, at two instants in time. For $t = 45$, Mach number and density contours at $M_\infty = 0.95$ (reproduced by permission of AIAA from Ishii and Kuwahara, 1984)

40

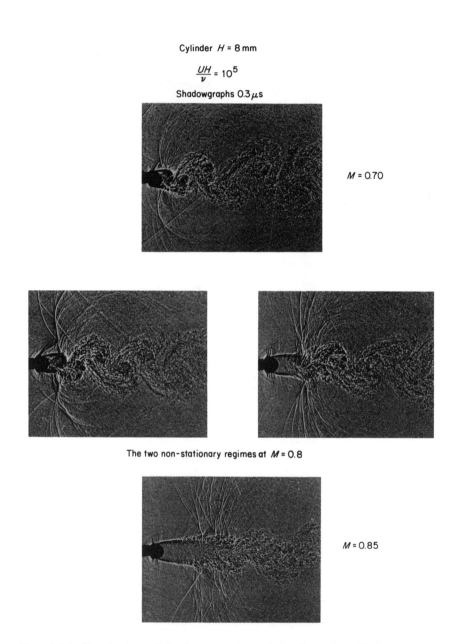

Figure 2.1.5 Visualizations of the flow around a cylinder for various Mach numbers at a Reynolds number of 10^5. (Courtesy A. Dyment and M. Pianko, Institut de Mécanique des Fluides de Lille, France)

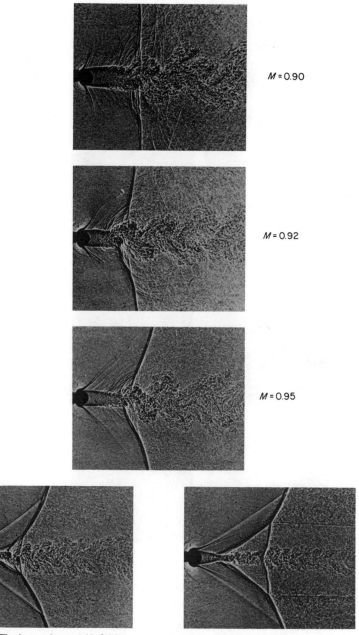

$M = 0.90$

$M = 0.92$

$M = 0.95$

The two regimes at $M = 0.98$; one non-stationary, the other stationary

42

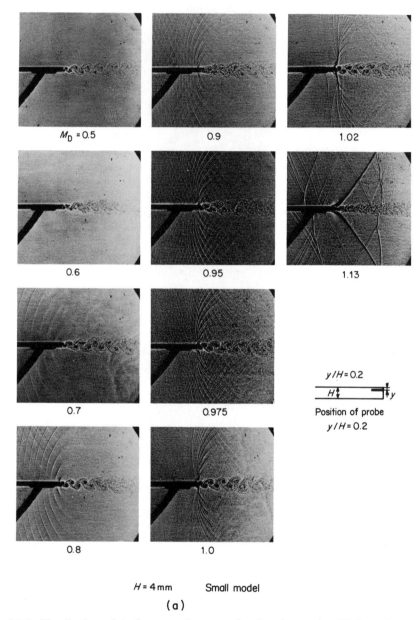

M_D = 0.5   0.9   1.02

0.6   0.95   1.13

y/H = 0.2
H
y
Position of probe
y/H = 0.2

0.7   0.975

0.8   1.0

H = 4 mm   Small model

(a)

Figure 2.1.6 Visualizations of the flow around a rectangular obstacle at various Mach numbers. (a) Acoustic wave pattern and vortex shedding with shock interaction at higher Mach numbers;

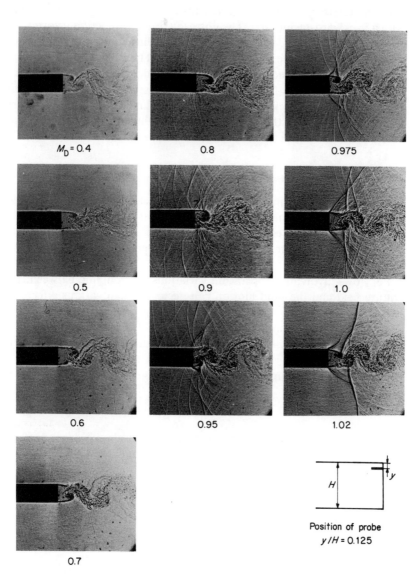

$M_D = 0.4$ 0.8   0.975

0.5   0.9   1.0

0.6   0.95   1.02

0.7

Position of probe
$y/H = 0.125$

$H = 20\,\text{mm}$ Large model

(b)

(b) Detailed view of the near-base flow and vortex-shock interaction. (Courtesy A. Dyment and M. Pianko, Institut de Mécanique des Fluides de Lille, France)

44

Compressible flow along a backward-facing step Another excellent test case
for compressible, viscous computations is the flow along a rectangular body
with a backward-facing step. The unsteady flow, with periodic vortex shedding
and progressively increasing interaction with the acoustic waves for increasing
Mach numbers, can be seen in Figure 2.1.6. These spectacular pictures show
the formation of the acoustic waves generated by the boundary layers along
the side walls ($M_\infty = 0.8$) and their progressive accumulation until their
coalescence into a shock at supersonic Mach numbers. Figure 2.1.6(b) shows a
more detailed view of the near base flow and the interaction between the shed
vortices and the shock waves at higher Mach numbers.

2.1.2 Incompressible fluid model

The Navier–Stokes equations simplify considerably for incompressible fluids
for which the specific mass may be considered as constant. This leads generally
to a separation of the energy equation from the other conservation laws if the
flow remains isothermal. This is the case for many applications which do not
involve heat transfer.

For flows involving temperature variations the coupling between the temperature field and the fluid motion can occur through various effects, such as
variations of viscosity or heat conductivity with temperature; the influence of
external forces as a function of temperature (for example, buoyancy forces
in atmospheric flows); and electrically, mechanically or chemically generated
heat sources.

In the case of incompressible flows the mass conservation equation reduces
to

$$\vec{\nabla} \cdot \vec{v} = 0 \qquad (2.1.28)$$

which appears as a kind of constraint to the general time-dependent equation
of motion, written here in non-conservative form:

$$\frac{\partial \vec{v}}{\partial t} + (\vec{v} \cdot \vec{\nabla})\vec{v} = -\frac{1}{\rho}\,\vec{\nabla}p + \nu\Delta\vec{v} + \vec{f_e} \qquad (2.1.29)$$

For incompressible flows, an alternative formulation can be obtained through
the Helmholtz vorticity equation (1.3.14):

$$\frac{\partial \vec{\zeta}}{\partial t} + (\vec{v} \cdot \vec{\nabla})\vec{\zeta} = (\vec{\zeta} \cdot \vec{\nabla})\vec{v} + \vec{\nabla}p \times \vec{\nabla}\frac{1}{\rho} + \vec{\nabla} \times \vec{f_e} + \nu\Delta\vec{\zeta} \qquad (2.1.30)$$

If no density stratification is to be considered the contribution of the pressure
term disappears completely from the vorticity equation. Moreover, for plane
two-dimensional flows the first term of the right-hand side vanishes. The
equation for the temperature field can be obtained through application of
equation (1.5.15), where the divergence-free condition for the velocity field is
introduced.

The system of equations for incompressible flow presents a particular situation in which one of the five unknowns, namely the pressure, does not appear under a time-dependence form due to the non-evolutionary character of the continuity equation. This actually creates a difficult situation for the numerical schemes and special techniques have to be adapted in order to treat the continuity equation. For more details we refer the reader to the corresponding sections of Volume 2.

An equation for the pressure can be obtained by taking the divergence of the momentum equation (2.1.29) and introducing the divergence-free velocity condition (2.1.28), leading to

$$\frac{1}{\rho}\,\Delta p = -\,\vec{\nabla}\cdot(\vec{v}\cdot\vec{\nabla})\vec{v} + \vec{\nabla}\cdot\vec{f_{e}} \tag{2.1.31}$$

which can be considered as a Poisson equation for the pressure for a given velocity field. Note that the right-hand side contains only products of first-order velocity derivatives because of the incompressibility condition (2.1.28). Indeed, in tensor notations the velocity term in the right-hand side is equal to $(\partial_j v_i)\cdot(\partial_i v_j)$.

For laminar, incompressible, isothermal flows no additional input is necessary to solve the system of flow equations besides the value of the fluid constants ρ and μ. Therefore it can be considered that the domain of laminar flows can be completely described for any set of initial and boundary conditions by computation, without having to resort to additional empirical information. Today, this phase can be considered to be close to realization, even for three-dimensional flow situations at reasonable computer times.

Examples

Direct simulation for large-scale coherent structures The numerical simulation of vortex shedding behind bluff bodies is of importance in view of applications such as atmospheric flows around buildings, vehicle aerodynamics or combustor flows. An impressive example of computation of the vortex shedding created by the flow around a square cylinder has been reported by Davis *et al* (1984). Figure 2.1.7 compares the calculated flow field with a visualization under similar conditions in a wind tunnel at a Reynolds number of $Re = 550$. Figure 2.1.8 is an illustration of a similar computation by these authors of an unstable mixing layer compared with a visualization under the same conditions.

These results emphasize the stage achieved nowadays in the numerical computation of complex flow fields via the resolution of Navier–Stokes equations. Although, as can be seen from the above results, many aspects of the flow can be reproduced, they are still to be considered as first approximations, since these computations are two-dimensional and do not contain the effect of the small-scale turbulence.

The system of Navier–Stokes equations is indeed valid for the laminar flow

46

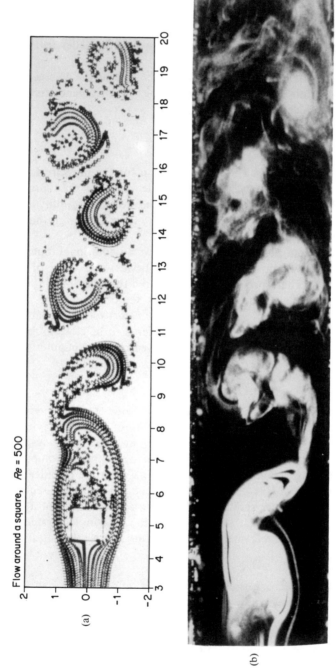

Figure 2.1.7 (a) Computed flow around a square cylinder at $Re = 500$; (b) comparison with a visualization at $Re = 550$. (Reproduced by permission of the American Institute of Physics from Davis *et al*, 1984.)

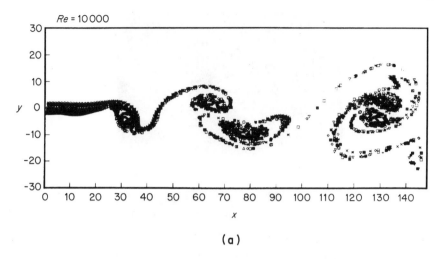

(a)

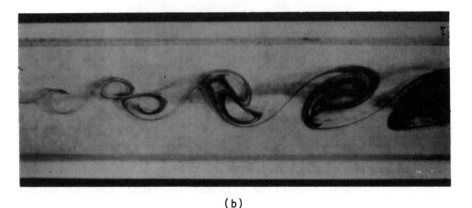

(b)

Figure 2.1.8 (a) Computation of the coherent structures in a mixing layer at $Re = 10\,000$; (b) comparison with visualizations. (Courtesy R. W. Davis, National Bureau of Standards)

of a viscous, Newtonian fluid. In reality, the flow will remain laminar up to a certain critical value of the Reynolds number $V \cdot L/\nu$, where V and L are representative values of velocity and length scales for the considered flow system. Above this critical value the flow becomes turbulent and is characterized by the appearance of fluctuations of all the variables (velocity, pressure, density, temperature, etc.) around mean values. These fluctuations are of a statistical nature and hence cannot be described in a deterministic way. However, they could be computed numerically in direct simulations of turbulence, such as the 'large eddy simulation' approach, whereby only the small-scale turbulent fluctuations are modelled and the larger-scale fluctuations are computed directly. The reader can find a review of the state of the art

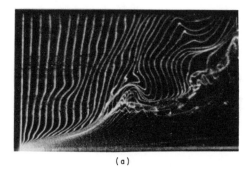

(a)

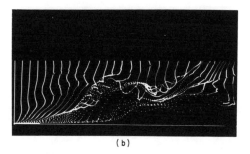

(b)

Figure 2.1.9 Time history of particles (hydrogen bubbles) generated along a line normal to the wall in a turbulent boundary layer, showing a coherent, bursting structure. (a) Experiments; (b) computations. (Reproduced by permission of AIAA from Moin, 1984)

of direct numerical simulation of turbulence in Rogallo and Moin (1984) and Moin (1984). Although this approach requires considerable computer resources, it has already led to very encouraging results.

A typical example of the direct numerical simulation of a turbulent bursting structure in a wall boundary layer as obtained by Moin and Kim (1982) is shown in Figure 2.1.9 compared with an experimental observation of the same structure. At present these approaches are still far from being applicable for practical calculations in industrial environments, due to the considerable requirements they put on computational resources. There is no doubt, however, that these methods will become increasingly important in the future, since they require the lowest possible amount of external information in addition to the basic Navier–Stokes equations.

Meanwhile there is a need to resort to a lower level of approximation, whereby the equations are averaged out, in time, over the turbulent fluctuations. This leads to the so-called *Reynolds-averaged Navier–Stokes equations*, which require, in addition, empirical or at least semi-empirical information on the turbulence structure and its relation to the averaged flow.

References

Chapman, D. R. (1979). 'Computational aerodynamics, development and outlook,' *AIAA Journal*, **17**, 1293–1313.

Davis, R. W., Moore, E. F., and Purtell, L. P. (1984). 'A numerical–experimental study of confined flow around rectangular cylinder.' *Physics of Fluids*, **27**, 46–59.

Dyment, A. (1982). 'Vortices following two dimensional separation.' In Hornung and Müller eds, *Vortex Motion*, Braunschweig: Vieweg & Sohn, pp. 18–30.

Ishii, K., and Kuwahara, K. (1984). 'Computation of compressible flow around a circular cylinder,' *AIAA Paper 84-1631*, AIAA 17th Fluid Dynamics, Plasma Dynamics and Lasers Conference.

Ishii, K., Kuwahara, K., Ogawa, S., Chyu, W. J., and Kawamura, T. (1985). 'Computation of flow around a circular cylinder in a supercritical regime.' *AIAA Paper 85-1660*, AIAA 18th Fluid Dynamics and Plasmadynamics and Lasers Conference.

Kline, S. J. *et al.* (1982). *Proceedings of the 1980–81 AFOSR/Stanford Conference on Complex Turbulent Flows*, Vols I–III, Stanford University, Stanford, California.

Kutler, P. (1983). 'A perspective of theoretical and applied computational fluid dynamics', *AIAA Paper 83-0037*, AIAA 21st Aerospace Sciences Meeting.

Le Balleur, J. C. (1983). 'Progres dans le calcul de l'interaction fluide parfait—fluide visqueux'. *AGARD Conference Proceedings CP 351* on *Viscous Effects in Turbomachines*.

Moin, P. (1984). 'Probing turbulence via large eddy simulation'. *AIAA Paper 84-0174*, AIAA 22nd Areospace Sciences Meeting.

Moin, P. and Kim, J. (1982). 'Numerical investigation of turbulent channel flows.' *Journal Fluid Mechanics*, **118**, 341–77.

Rodriguez, O. (1984). 'The circular cylinder in subsonic and transonic flow.' *AIAA Journal*, **22**, 1713–18.

Rogallo, R. S., and Moin, P. (1984). 'Numerical simulation of turbulent flows.' *Annual Review of Fluid Mechanics*, **16**, 99–137.

Temam, R. (1977). *Navier–Stokes Equations*, Amsterdam: North-Holland.

Vinokur, M. (1974). 'Conservation equations of gas dynamics in curvilinear coordinate systems,' *Journal Computational Physics*, **14**, 105–25.

Viviand, H. (1974). 'Conservative forms of gas dynamic equations.' *La Recherche Aerospatiale*, **1**, 65.

Thompson, J. F. (Ed.) (1982). *Numerical Grid Generation*, Amsterdam: North-Holland; New York: Elsevier.

2.2 THE REYNOLDS-AVERAGED NAVIER–STOKES EQUATIONS

The turbulent averaging process is introduced in order to obtain the laws of motion for the 'mean', time-averaged, turbulent quantities. This time averaging is to be defined in such a way as to remove the influence of the turbulent fluctuations while not destroying the time dependence associated with other time-dependent phenomena with time scales distinct from those of turbulence.

Turbulent averaged quantities

For any quantity A the separation

$$A = \bar{A} + A' \tag{2.2.1}$$

50

is introduced with

$$\bar{A}(\vec{x}, t) = \frac{1}{T} \int_{-T/2}^{T/2} A(\vec{x}, t + \tau)\, d\tau \tag{2.2.2}$$

where T is to be chosen large enough compared with the same time scale of the turbulence but still small compared with those of all other unsteady phenomena.

Obviously, this might not be always possible: if unsteady phenomena occur with time scales of the same order as those of the turbulent fluctuations the Reynolds-averaged equations will not allow us to model these phenomena. However, it can be considered that most of the unsteady phenomena in fluid dynamics have frequency ranges outside the frequency range of turbulence, Chapman (1979).

For compressible flows the averaging process leads to products of fluctuations between density and other variables such as velocity or internal energy. In order to avoid their explicit occurrence a *density-weighted* average can be introduced, through

$$\tilde{A} = \frac{\overline{\rho A}}{\bar{\rho}} \tag{2.2.3}$$

with

$$A = \tilde{A} + A'' \tag{2.2.4}$$

and

$$\overline{\rho A''} = 0 \tag{2.2.5}$$

This way of defining mean turbulent variables will remove all additional products of density fluctuations with other fluctuating quantities. This is easily seen by performing the averaging process defined by equation (2.2.3) on the continuity equation, leading to

$$\frac{\partial}{\partial t} \bar{\rho} + \vec{\nabla} \cdot (\bar{\rho} \vec{\tilde{v}}) = 0 \tag{2.2.6}$$

A more complete discussion can be found in Cebeci and Smith (1974).

Applied to the momentum equations, we obtain the following equation for the turbulent mean momentum, in the absence of body forces:

$$\frac{\partial}{\partial t} (\bar{\rho} \vec{\tilde{v}}) + \vec{\nabla} \cdot (\bar{\rho} \vec{\tilde{v}} \otimes \vec{\tilde{v}} + \bar{p} \bar{\bar{I}} - \bar{\bar{\tau}}^{\,v} - \bar{\bar{\tau}}^{\,R}) = 0 \tag{2.2.7}$$

where the Reynolds stresses, $\bar{\bar{\tau}}^{\,R}$, defined by

$$\bar{\bar{\tau}}^{\,R} = -\overline{\rho \vec{v}'' \otimes \vec{v}''} \tag{2.2.8a}$$

are added to the *averaged* viscous shear stresses $\bar{\bar{\tau}}^{\,v}$. In Cartesian co-ordinates we have

$$\tau_{ij}^{R} = - \overline{\rho v_i'' v_j''} \qquad (2.2.8b)$$

The relations between the Reynolds stresses and the mean flow quantities are unknown. Therefore the application of the Reynolds-averaged equations to the computation of turbulent flows requires the introduction of some modelling of these unknown relations, based on theoretical considerations coupled with unavoidable empirical information. Several of the most widely used of these turbulence models will be presented in Volume 2 when dealing with the Navier–Stokes equations.

In a similar way, the turbulent averaged energy conservation equation can be obtained under different forms according to the definition taken for the averaged total energy, and again we refer the reader to Volume 2 for a more detailed discussion.

Practical examples

Unsteady oscillatory flow in an axisymmetric inlet The spontaneously generated instability of the flow in an engine inlet (called 'inlet buzz') has been computed by Newsome (1983) with the Reynolds-averaged Navier–Stokes equations and a Cebeci and Smith algebraic turbulence model. Figure 2.2.1 shows the sequence of Mach contours at different times during the third buzz cycle at low mass flow (subcritical) regime for an incident Mach number of two. The oscillations develop as a consequence of a shear layer instability due to separated boundary layers, which amplify small pressure disturbances in a closed feedback loop of reflected expansion and compression waves. Strong shock wave interactions and unsteady boundary layer separations are marked phenomena of this complex flow pattern. We can observe during the oscillation cycle the backward displacement of the bow shock, which is forced towards the tip of the centrebody as a result of the interaction with a reflected compression wave generated from the separated flow downstream of the shock. During this phase ($t = 15.9$–17.1 ms) a region of reverse flow extends between the base of the bow shock and the cowl lip, with a shear layer dividing the two regions. The bow shock remains in its position at the centrebody tip for a period of time corresponding to the propagation and reflection of an expansion wave. Then the inlet begins again to ingest mass, and the shock moves forward to the cowl lip. During this phase ($t = 21.5$–23.9 ms) several regions of separated flow can be observed alternatively on the centrebody and the cowl.

Since the shear layer separation is the essential mechanism which amplifies the small disturbances this flow computation is very sensitive to the turbulence model. Too large values of the eddy viscosity prevent the occurrence of the instability and, clearly, more sophisticated turbulence models are required for this very complex flow, Newsome (1983).

52

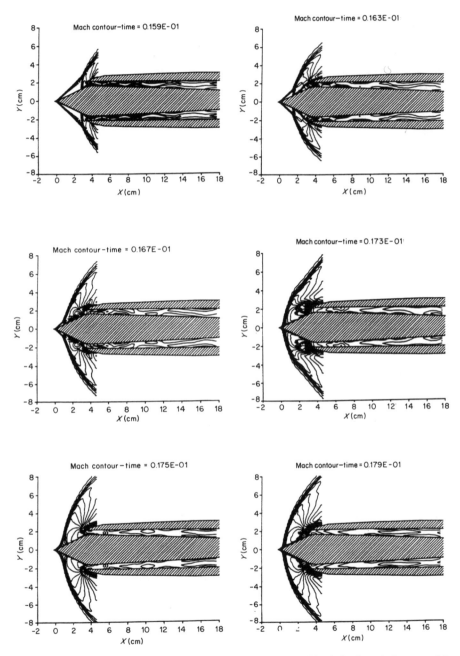

Figure 2.2.1 Reynolds-averaged Navier–Stokes computation, with algebraic turbulance model of the unsteady flow field in an axisymmetric inlet. Mach contours are shown at different instants during the inlet buzz cycle. (Reproduced by permission of AIAA from Newsome, 1983)

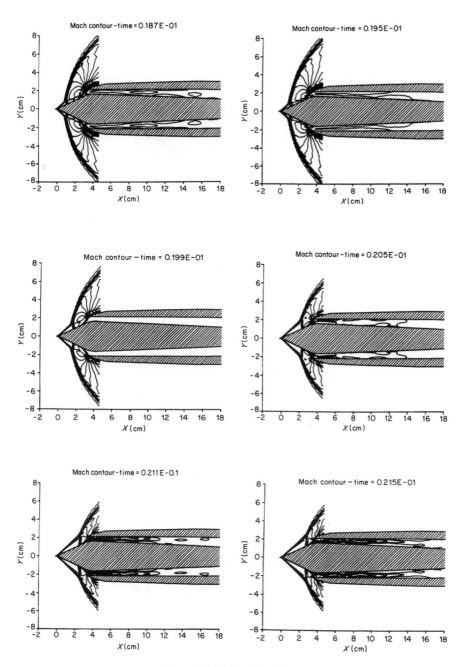

Figure 2.2.1 (*Continued*)

54

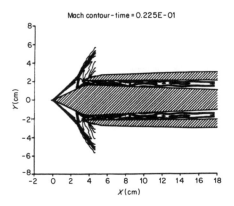

Mach contour - time = 0.225E - 01

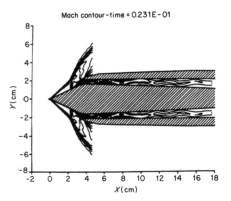

Mach contour - time = 0.231E - 01

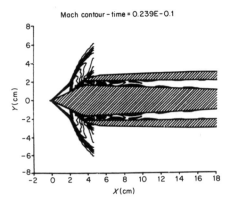

Mach contour - time = 0.239E - 0.1

Figure 2.2.1 (*Continued*)

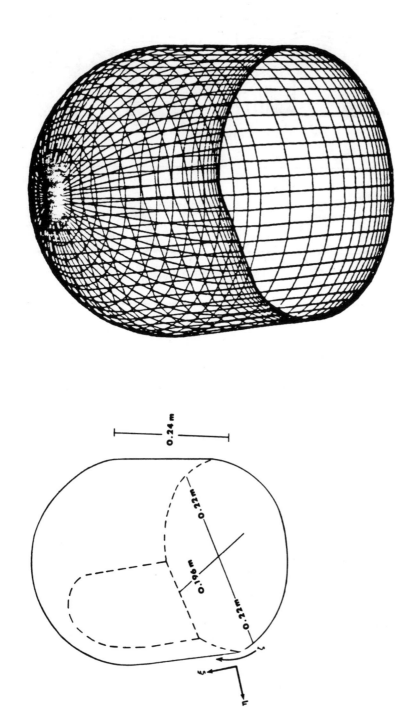

Figure 2.2.2 Geometry and surface mesh of a turret, resting on a flat surface. (Reproduced by permission of AIAA from Purohit *et al.*, 1982)

56

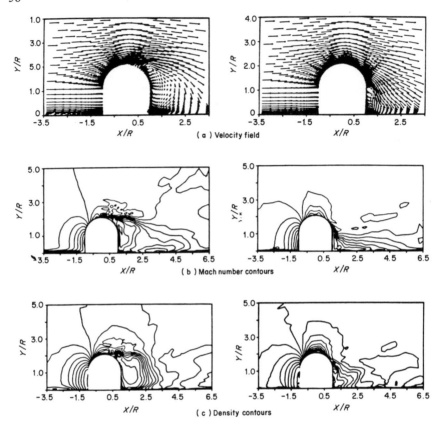

Figure 2.2.3 Three-dimensional flow field around the turret of Figure 2.2.2, Reynolds-averaged Navier–Stokes computation with algebraic turbulence model. Flow in a vertical cross section at two instants in time (Reproduced by permission of AIAA from Purohit *et al.*, 1982)

Flow around a three-dimensional obstacle The fully three-dimensional, separated flow around a surface-mounted bluff obstacle (turret) has been computed by Purohit *et al.* (1982) with a Cebeci and Smith algebraic turbulence model. Figure 2.2.2 shows the geometry of the turret and the surface mesh system, while some flow patterns are shown in Figures 2.2.3 and 2.2.4 in a vertical and horizontal plane cross-section at an incident Mach number of 0.55 and a Reynolds number of 10^7 per metre and at two instants in time. A clear appearance of vortex shedding, similar to the flow around a cylinder, can be observed in the *z-x* horizontal plane (Figure 2.2.4). The figures display the velocity field, Mach number and density contours at two instants in time, showing the naturally occurring unsteady flow field for stationary incident flow conditions.

As already shown in previous examples, large separated flow regions behind bluff bodies tend to generate a time-dependent flow field, associated with

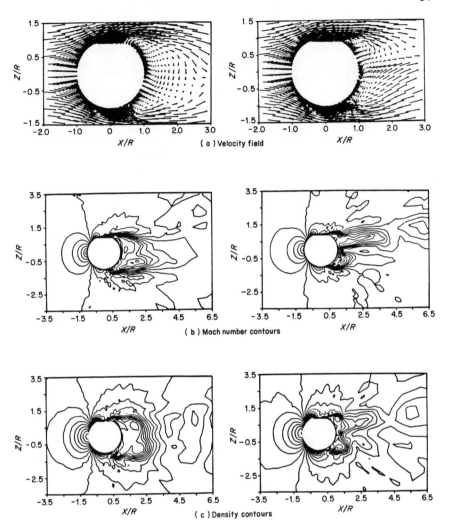

Figure 2.2.4 Three-dimensional flow field around the turret of Figure 2.2.2, Reynolds-averaged Navier–Stokes computation with algebraic turbulence model. Flow in a horizontal cross section at two instants in time. (Reproduced by permission of AIAA from Purohit *et al.*, 1982)

vortex shedding and fluctuating separation lines. In this three-dimensional flow the unsteady separation and vortex formation in the vertical cross-sections, as well as in the horizontal planes, can be observed in Figure 2.2.4. Figure 2.2.5 shows an instantaneous pattern of limiting streamlines on the developed turret surface, illustrating the complex system of separated regions and saddle-point singularities. The points marked S are saddle-point singularities, while the points N are nodes of attachment of the flow. Observe also the asymmetric nature of the separation lines. Although this computation

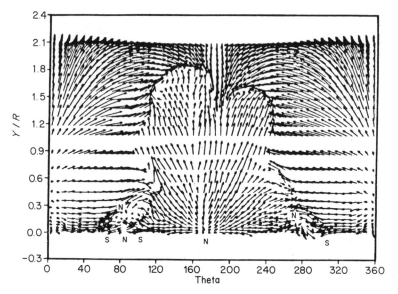

Figure 2.2.5 Three-dimensional flow field around the turret of Figure 2.2.2, Reynolds-averaged Navier–Stokes computation with algebraic turbulence model. Instantaneous picture of the surface flow and limiting streamline pattern on the turret surface, showing saddle points (S) and reattachment nodes (N). (Courtesy Dr. J. Shang (AFWAL. Wright-Patterson AFB. Ohio, USA))

provides more information and flow details than experimental observations might give, it is still essential to compare it with experimental data in order to assess the validity of the turbulence models.

Flow in an internal combustion engine The computation performed by Schock *et al.* (1984) aims at analysing the effects of compression ratio, rotational speed and valve-setting angle on the formation and destruction of vortices within an axisymmetric piston–cylinder configuration. A two-equation model for turbulence is applied, since the time evolution of the turbulent intensity is of primary importance to the combustion process. Figures 2.2.6 and 2.2.7 show the mean velocity field and turbulent profiles at different instants within the cycle for two valve-setting angles of zero and 45°. The formation of valve and cylinder vortices during the intake stroke can be seen. The valve vortex breaks down into two new vortices which merge with the cylinder vortex, followed by their dissipation. The creation and decrease of turbulent intensity is also shown at different times. Turbulence is generated at the shear layers of the air jet drawn into the cylinder and stimulated by the value vortex breakdown.

This effect is very sensitive to the valve-seat angle and the generated turbulence is higher for a valve seat of 45° when compared with the corresponding situation at 0° seat angle. The calculation of Schock *et al.* (1984) is the first to have shown the relation between the intensity of turbulence

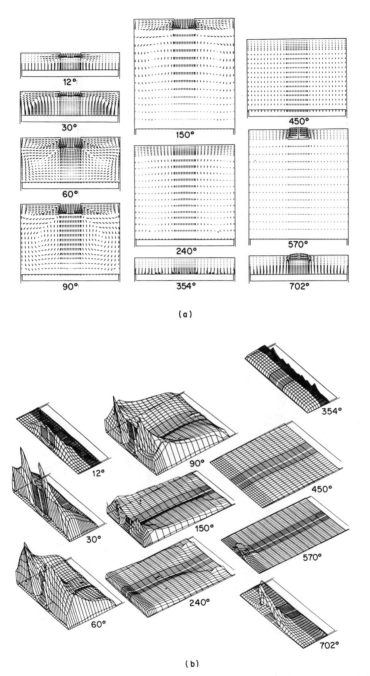

(a)

(b)

Figure 2.2.6 Flow field in the cylinder of an internal combustion engine for an air intake angle (value setting) of 0, at various instants during the cycle. (a) Mean velocities; (b) normalized rms turbulence intensities. (Courtesy H. J. Schock (NASA Lewis Research Center))

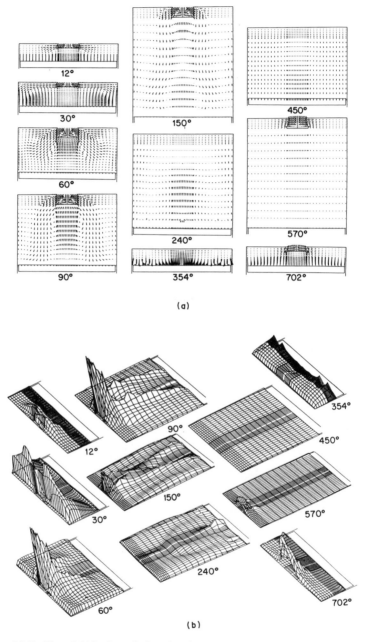

(a)

(b)

Figure 2.2.7 Flow field in the cylinder of an internal combustion engine for an air intake angle (value setting angle) of 45°, at various instants during the cycle. (a) Mean velocities; (b) normalized rms turbulence intensities. (Courtesy H. J. Schock (NASA Lewis Research Center))

and the valve geometry. This is of great importance, since the turbulence intensity determines the rates of fuel–air mixing and of combustion.

Flow in a turbine blade row   The three-dimensional flow in turbine blade rows is characterized by a strong leading edge horseshoe vortex, generated by the incident end wall boundary-layer velocity profile. Figure 2.2.8 shows the oil traces of the end-wall flow with the traces of the vortex, confirmed by a visualization in the vertical plane, through a laser light sheet technique superimposed on the same picture. A viscous calculation of this flow (for a similar turbine blade row) has been performed by Hah (1984) with the stationary Reynolds-averaged Navier–Stokes equations and a two-equation turbulence model, corrected for curvature effects via an algebraic Reynolds stress formulation.

The horseshoe vortex flow generates a saddle-point singularity where two separation lines, s_1, s_2, and two re-attachment lines, a_1, a_2, meet. These lines divide the end-wall flow field into distinct regions of three-dimensional flows. This can be seen on the computed results of Figure 2.2.9(a) (at 1% span from the wall) for an inlet angle of $32.2°$. The experimental position of the saddle point, measured at the wall, is also shown. The details of the separation regions appear, as well as the trace s_1-s_2 of the horseshoe vortex. Figure 2.2.9(b) shows a similar computation with a fully developed and thicker inlet

Figure 2.2.8 Visualization of the leading-edge horseshoe vortex for a turbine blade row. A vertical, laser sheet cross section is superimposed, showing the vortical structure. (Courtesy C. Sieverding, Von Karman Institute for Fluid Dymanics, Belgium)

62

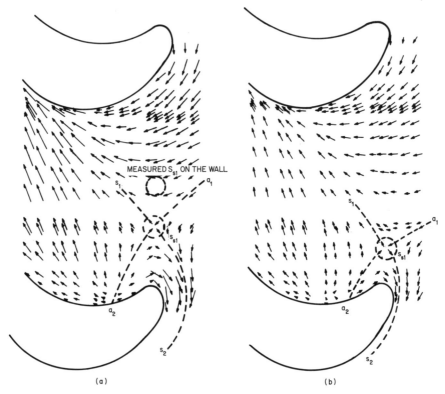

Figure 2.2.9 (a) Computed velocity field in a turbine blade row at 1% span from the end wall for an inlet angle of 32.2°; (b) Computed velocity field at 1% span from the end wall for a fully developed inlet boundary layer. (Reproduced by permission of the American Society of Mechanical Engineers from Hah, 1984.)

boundary layer. The effect of the thicker boundary layer on the location of the saddle point can be noticed, showing a displacement towards the pressure side of the blade and a more severe crossflow.

References

Chapman, D. R. (1979). 'Computational aerodynamics, development and outlook.' *AIAA Journal*, **17**, 1293–1313.

Cebeci, T., and Smith, A. M. O. (1974). *Analysis of Turbulent Boundary Layers*, New York: Academic Press.

Hah, C. (1984). 'A Navier–Stokes analysis of three-dimensional turbulent flow inside turbine blade rows at design and off-design conditions.' *Trans. ASME, Journal of Engineering for Gas Turbines and Power*, **106**, 421–9.

Newsome, R. W. Jr (1983). 'Numerical simulation of near-critical and unsteady subcritical inlet flow fields.' *AIAA Paper 83-0175*, AIAA 21st Aerospace Sciences Meeting.

Purohit, S. C., Shang, J. G., and Hankey, W. L., Jr (1982). 'Numerical simulation of

flow around a three-dimensional turret.' *AIAA Paper-82-1020*, AIAA/ASME 3rd Joint Thermophysics, Fluids, Plasma and Heat Transfer Conference.
Schock, H. J., Sosoka, D. J., and Ramos, J. I. (1984). 'Formation and destruction of vortices in a motored four-stroke piston–cylinder configuration.' *AIAA Journal*, **22**, 948–9.

2.3 THE THIN SHEAR LAYER (TSL) APPROXIMATION

At high Reynolds numbers wall shear layers, wakes or free shear layers will be of limited size, and if the extension of the viscous region remains limited during the flow evolution then the dominating influence of the shear stresses will come essentially from the gradients transverse to the main flow direction.

If we consider an arbitrary curvilinear system of co-ordinates with ξ^1 and ξ^2 along the surface and $\xi^3 = n$, directed towards the normal, then the thin *shear layer (TSL) approximation* (Figure 2.3.1) of the Navier–Stokes equations consists of neglecting all ξ^1 and ξ^2 derivatives occurring in the turbulent and viscous shear stress terms (Pulliam and Steger, 1978; Steger, 1978). This approximation is also supported by the fact that, generally, at high Reynolds numbers (typically $Re > 10^4$) the mesh is made dense only in the direction normal to the shear layer, and therefore the neglected terms are computed with a lower accuracy that the normal derivatives.

The general form of the conservation equations (2.1.5) remains unchanged in the TSL approximation:

$$\frac{\partial U}{\partial t} + \vec{\nabla} \cdot \vec{F} = Q \tag{2.3.1}$$

but the flux $\vec{F}$ is simplified such that, in the contributions of the viscous terms,

$$\vec{\nabla} \cdot \bar{\bar{\tau}}^{\mathrm{T}} \quad \text{and} \quad \vec{\nabla} \cdot (\bar{\bar{\tau}}^{\mathrm{T}} \cdot \vec{v}) \tag{2.3.2}$$

all derivatives with respect to ξ^1 and ξ^2 are neglected: that is, only the normal derivatives are maintained.

This leaves as the remaining contribution of the ith component of the shear stress gradient, with n the normal direction:

$$(\vec{\nabla} \cdot \bar{\bar{\tau}})^i \approx \frac{\partial \tau^{\mathrm{in}}}{\partial n} \equiv \frac{\partial}{\partial n} (\vec{\tau}^n)^i \tag{2.3.3}$$

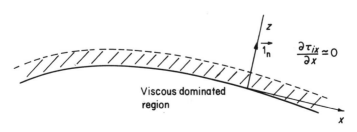

Figure 2.3.1 Thin shear layer approximation

in which the derivatives with respect to ξ^1 and ξ^2 are also neglected in the calculation of $\bar{\bar{\tau}}$. In equation (2.3.3) a shear stress vector $\vec{\tau}^n$ is introduced with components τ^{in}.

This approximation is actually close to a boundary-layer approximation, since viscous terms which are neglected in the boundary layer approximation are also neglected here. However, the momentum equation in the directions normal to the shear layer is retained, instead of the constant pressure rule over the boundary layer thickness along a normal to the wall. Therefore the transition from viscous dominated regions to the inviscid region outside the wall layer is integrally part of the calculation, and we have here a form of 'higher order' boundary-layer approximation. The classical boundary-layer approximation is obtained when the momentum equation in the direction normal to the wall is replaced by

$$\frac{\partial p}{\partial n} = 0 \tag{2.3.4}$$

The TSL approximation amounts to neglecting the viscous diffusion in the direction parallel to the shear surface and keeping only the contributions from the diffusion in the normal direction.

In Cartesian co-ordinates, with x, y co-ordinates in the tangent plane and z normal to the surface, the shear stresses are approximated by

$$\tau_{ij} \approx \mu\left(\delta_{3i}\frac{\partial v_j}{\partial z} + \delta_{3j}\frac{\partial v_i}{\partial z}\right) - \frac{2}{3}\mu\delta_{ij}\left(\frac{\partial v_z}{\partial z}\right) \tag{2.3.5}$$

and, in particular,

$$\tau_{i3} \approx \mu\left(\frac{\delta_{i3}}{3}\frac{\partial v_z}{\partial z} + \frac{\partial v_i}{\partial z}\right) \tag{2.3.6}$$

where $i = 3$ corresponds to the variable z.

The explicit form of the shear stress components in the TSL approximation are obtained from equation (2.3.5), writing u, v, w for the components v_x, v_y, v_z respectively:

$$\tau_{xx} = \tau_{yy} = -\frac{2}{3}\mu\frac{\partial w}{\partial z}$$

$$\tau_{xy} = 0$$

$$\tau_{xz} = \mu\frac{\partial u}{\partial z} \tag{2.3.7}$$

$$\tau_{yz} = \mu\frac{\partial v}{\partial z}$$

$$\tau_{zz} = \frac{4}{3}\mu\frac{\partial w}{\partial z}$$

Although, with the exception of τ_{xy}, the shear stress components do not vanish, the only contributions to the shear stress gradients are, according to equation (2.3.3),

$$\vec{\nabla} \cdot \bar{\bar{\tau}} \approx \left(\frac{\partial}{\partial z}\,\tau_{xz}\right)\vec{1}_x + \left(\frac{\partial}{\partial z}\,\tau_{yz}\right)\vec{1}_y + \left(\frac{\partial}{\partial z}\,\tau_{zz}\right)\vec{1}_z \qquad (2.3.8)$$

The flux components f, g, h, defined by equation (2.1.7), simplify to

$$f = \begin{vmatrix} \rho u \\ \rho u^2 + p \\ \rho uv \\ \rho uw \\ \rho uH \end{vmatrix} \qquad g = \begin{vmatrix} \rho v \\ \rho uv \\ \rho v^2 + p \\ \rho vw \\ \rho vH \end{vmatrix}$$

$$h = \begin{vmatrix} \rho w \\ \rho uw - \mu \dfrac{\partial u}{\partial z} \\ \rho vw - \mu \dfrac{\partial v}{\partial z} \\ \rho w^2 + p - \dfrac{4}{3}\mu\dfrac{\partial w}{\partial z} \\ \rho wH - (\tau_{zx}u + \tau_{zy}v + \tau_{zz}w) - k\dfrac{\partial T}{\partial z} \end{vmatrix} \qquad (2.3.9)$$

As can be observed, the x and y components of the flux vectors reduce to their inviscid form, and the viscous terms are entirely concentrated in the normal flux component h.

The x-momentum equation becomes, in the TSL approximation,

$$\frac{\partial}{\partial t}(\rho u) + \frac{\partial}{\partial x}(\rho u^2 + p) + \frac{\partial}{\partial y}(\rho uv) + \frac{\partial}{\partial z}(\rho uw) = \frac{\partial}{\partial z}\left(\mu\,\frac{\partial u}{\partial z}\right) \quad (2.3.10)$$

and a similar equation for the y-component:

$$\frac{\partial}{\partial t}(\rho v) + \frac{\partial}{\partial x}(\rho uv) + \frac{\partial}{\partial y}(\rho v^2 + p) + \frac{\partial}{\partial z}(\rho vw) = \frac{\partial}{\partial z}\left(\mu\,\frac{\partial v}{\partial z}\right) \quad (2.3.11)$$

The normal projection, in the z-direction, of the momentum equation becomes

$$\frac{\partial}{\partial t}(\rho w) + \frac{\partial}{\partial x}(\rho uw) + \frac{\partial}{\partial y}(\rho vw) + \frac{\partial}{\partial z}(\rho w^2 + p) = \frac{4}{3}\frac{\partial}{\partial z}\left(\mu\,\frac{\partial w}{\partial z}\right) \quad (2.3.12)$$

Practical examples

Viscous flow along an airfoil The TSL approximation, coupled to an algebraic turbulence model (and a laminar–turbulent transition point fixed at 11% of the chord) has been applied by Pulliam and Steger (1985) to the flow along an airfoil. Figure 2.3.2 shows the computed transonic pressure distribution, plotted as a pressure coefficient $C_p = (p - p_\infty)/q$, where q is the upstream dynamic pressure, compared to experimental data, for the RAE 2822 supercritical airfoil, at a Reynolds number of $5.7\ 10^6$ and incident conditions of $M_\infty = 0.676$ and $\alpha_\infty = 1.93°$. In the same figure we can observe a comparison between the computed and observed boundary layer growth, expressed by the chordwise evolution of the displacement thickness δ^* and of the momentum thickness θ. This example is typical of the flow situations for which the TSL approximation is fully valid.

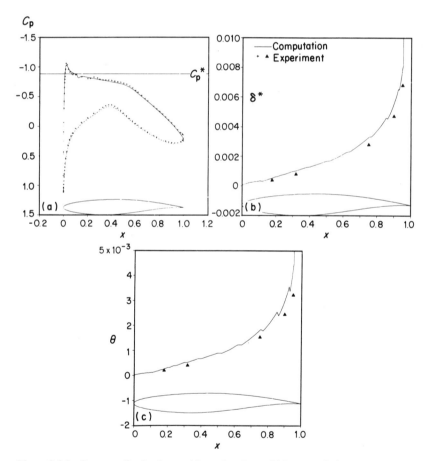

Figure 2.3.2 Pressure distribution and boundary layer thickness variations as computed with a thin shear layer approximation for the RAE 2822 supercritical airfoil. Incident conditions are $M_\infty = 0.676$, $\alpha_\infty = 1.93°$ and $Re = 5.7\ 10^6$. δ^*: displacement thickness; θ: momentum thickness; Δ experimental data. (Reproduced by permission of AIAA from Pulliam and Steger, 1985)

The supercritical airfoils are characterized by a shock-free transition from supersonic to subsonic flow conditions at well-defined incident Mach numbers and flow angles. Actually, the development of supercritical profiles is one of the most spectacular outcomes of transonic computational fluid dynamics, since its use on civil aircraft allows a significant reduction of wing drag and hence in fuel consumption. Most recent civil aircraft are therefore designed with supercritical wings.

Figure 2.3.3 shows a computation of the same airfoil but at off-design incident conditions of $M_\infty = 0.73$, $\alpha_\infty = 2.79°$ and $R_e = 6.5\ 10^6$. The shock is correctly captured and the considerable increase in boundary layer thickness arising from the shock boundary layer interaction can clearly be seen.

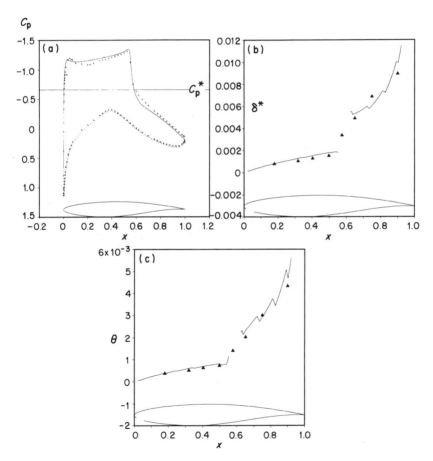

Figure 2.3.3 Pressure distribution and boundary layer thickness variations as computed with a thin shear layer approximation for the RAE 2822 supercritical airfoil. Incident conditions are $M_\infty = 0.73$; $\alpha_\infty = 2.79$ and $Re = 6.5\ 10^6$. δ^* Displacement thickness; θ momentum thickness; Δ experimental data. (Reproduced by permission of AIAA from Pulliam and Steger, 1985)

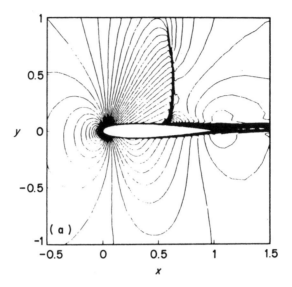

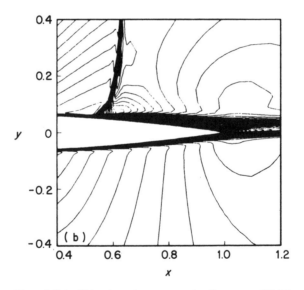

Figure 2.3.4 Thin shear layer approximation on a NACA
0012 airfoil at $M_\infty = 0.8$, $\alpha_\infty = 2$, $Re = 1.10^6$. Mach number
contours show the thickening of the viscous region after
interaction with the shock. (a) Complete flow picture; (b)
detailed flow picture showing the interaction region and the
wake. (Reproduced by permission of AIAA from Nakahashi
and Deiwert, 1985)

The comparison with the data shows that some improvement is still necessary in this strong interaction region, probably at the level of the turbulence modelling or at that of the TSL approximation itself, since a small region of separated flow appears at the base of the shock. A similar computation, performed on a NACA 0012 symmetrical airfoil (Nakahashi and Deiwert, 1985), shows on the Mach number contours the thickening of the viscous region after the interaction with the shocks (Figure 2.3.4).

Aircraft afterbody flows The strongly interacting flow behind an aircraft afterbody has been computed by Deiwert *et al.* (1984) with a TSL approximation. The shock structure in the exhaust plume and its interaction with the viscous base region can be seen in Figure 2.3.5, where a Schlieren picture of the flow field is also shown. The lines corresponding to the shocks, shear layers and contact discontinuities taken from the Schlieren photograph have been drawn on the computed isoline plots, and show very good agreement with computations.

In the near-base region large separated flows can sometimes be observed (for instance, when the jet diameter is smaller than the base diameter). In this case, afterbody shear layer-wake interactions induce a strong recirculation, and the computation of this effect seems to be at the limits of the validity of the TSL approximation (Deiwert *et al.*, 1984). This can be considered as a general statement: large separated, viscous dominated regions should be computed not with the TSL approximation but with the full Navier–Stokes equations.

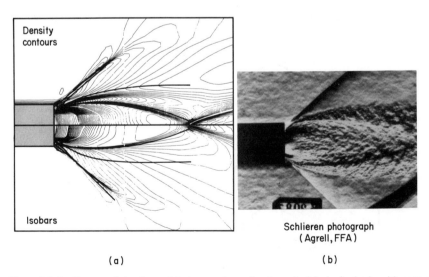

(a)   (b)

Figure 2.3.5 Computed density and isobar contours for the cylindrical afterbody with centred jet. The jet pressure $p_j = 6p_\infty$, $M_\infty = 2.0$ and the jet Mach number is $M_j = 2.5$. (a) Computed results; (b) experimental Schlieren photograph taken under the same conditions. (Reproduced by permission of AIAA from Deiwert *et al.*, 1984)

70

References

Deiwert, G. S., Andrews, A. E., and Nakahasi, K. (1984). 'Theoretical analysis of aircraft afterbody flow'. *AIAA Paper 84-1524*, AIAA 17th Fluid Dynamics, Plasma Dynamics and Lasers Conference.
Nakahasi, K., and Deiwert, G. S. (1985). 'A self-adaptive grid method with application to airfoil flow.' *AIAA Paper 85-1525, Proceedings AIAA 7th Computational Fluid Dynamics Conference.*
Pulliam, T. H., and Steger, J. L. (1978). 'Implicit finite difference simulations of three-dimensional flows.' *AIAA Paper 78-10*, 16th Aerospace Sciences Meeting; see also *AIAA Journal,* **18**, 1980, 159–67.
Pulliam, T. H., and Steger, J. L. (1985). 'Recent improvements in efficiency, accuracy and convergence for implicit approximate factorization algorithms.' *AIAA Paper 85-0360*, AIAA 23rd Aerospace Sciences Meeting.
Steger, J. L. (1978). 'Implicit finite difference simulation of flows about arbitrary geometries.' *AIAA Journal,* **16**, 679–86.

2.4 THE PARABOLIZED NAVIER–STOKES APPROXIMATION

The parabolized Navier–Stokes (PNS) approximation is based on considerations similar to the TSL approximation but applies only to the *steady-state formulation* of the Navier–Stokes equations. This approximation is directed towards flow situations with a predominant main flow direction, as would be the case in a channel flow, whereby the cross-flow components are of a lower order of magnitude. In addition, along the solid boundaries the viscous regions are assumed to be dominated by the normal gradients and, hence, the streamwise diffusion of momentum and energy can be neglected.

If x is the streamwise co-ordinate, the x-derivatives in the shear stress terms are all neglected compared with the derivatives in the two transverse directions

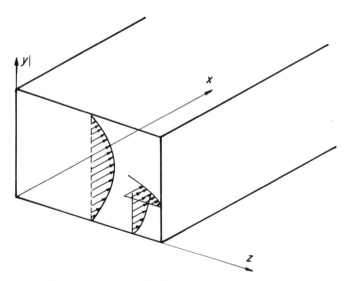

Figure 2.4.1 Parabolized Navier–Stokes approximation

y and *z*. A similar approximation is introduced into the energy diffusion terms. This approximation is therefore valid as long as the mainstream flow of velocity *u* is dominant, that is, as long as the positive *x*-direction corresponds to the forward flow direction. This will no longer be the case if there is a region of reverse flow of the streamwise velocity component. In this case, the streamwise derivatives of *u* will become of the same order as the transverse derivatives and the whole approximation breaks down.

In Cartesian co-ordinates the shear stress terms reduce to the following form for the x-component:

$$(\vec{\nabla} \cdot \bar{\bar{\tau}})_x = \frac{\partial}{\partial x} \tau_{xx} + \frac{\partial}{\partial y} \tau_{xy} + \frac{\partial}{\partial z} \tau_{xz}$$

$$\approx \frac{\partial}{\partial y} \tau_{xy} + \frac{\partial}{\partial z} \tau_{xz} \qquad (2.4.1)$$

neglecting the streamwise derivative. The same approximation is introduced into the computation of $\bar{\bar{\tau}}$. Hence

$$\tau_{xx} = 2\mu \frac{\partial u}{\partial x} - \frac{2}{3} \mu \left(\frac{\partial u}{\partial x} + \frac{\partial v}{\partial y} + \frac{\partial w}{\partial z} \right)$$

$$\approx -\frac{2}{3} \mu \left(\frac{\partial v}{\partial y} + \frac{\partial w}{\partial z} \right) \qquad (2.4.2)$$

$$\tau_{xy} \approx \mu \frac{\partial u}{\partial y} \qquad (2.4.3)$$

$$\tau_{xz} \approx \mu \frac{\partial u}{\partial z} \qquad (2.4.4)$$

resulting in

$$(\vec{\nabla} \cdot \bar{\bar{\tau}})_x \approx \frac{\partial}{\partial y} \left(\mu \frac{\partial u}{\partial y} \right) + \frac{\partial}{\partial z} \left(\mu \frac{\partial u}{\partial z} \right) \qquad (2.4.5)$$

The mainstream momentum equation reduces to

$$\frac{\partial}{\partial x} (\rho u^2 + p) + \frac{\partial}{\partial y} (\rho uv) + \frac{\partial}{\partial z} (\rho uw) = \frac{\partial}{\partial y} \left(\mu \frac{\partial u}{\partial y} \right) + \frac{\partial}{\partial z} \left(\mu \frac{\partial u}{\partial z} \right) \qquad (2.4.6)$$

Without these approximations the right-hand side of equation (2.4.6) would contain a term of the form $\partial_x(\mu \partial_x u)$, making the whole equation elliptic in the (x, y, z) space. However, with the present approximation there is no second-order derivative in *x* and the equation is therefore parabolic in *x*. This variable can be considered as playing the role of a pseudo-time, and hence the 'parabolized' *x*-momentum equation can be integrated by advancing in this direction, solving in each *x*-plane an elliptic problem in *y*-*z*.

The transverse momentum equations are obtained in a similar way. For instance, the y-component becomes

$$\frac{\partial}{\partial x}(\rho uv) + \frac{\partial}{\partial y}(\rho v^2 + p) + \frac{\partial}{\partial z}(\rho vw)$$

$$= \frac{4}{3}\frac{\partial}{\partial y}\left(\mu\frac{\partial v}{\partial y}\right) + \frac{\partial}{\partial z}\left(\mu\frac{\partial v}{\partial z}\right) - \frac{2}{3}\frac{\partial}{\partial y}\left(\mu\frac{\partial w}{\partial z}\right) + \frac{\partial}{\partial z}\left(\mu\frac{\partial w}{\partial y}\right) \tag{2.4.7}$$

where the approximations

$$(\vec{\nabla}\cdot\bar{\bar{\tau}})_y = \frac{\partial}{\partial x}\tau_{xy} + \frac{\partial}{\partial y}\tau_{yy} + \frac{\partial}{\partial z}\tau_{yz}$$

$$\approx \frac{\partial\tau_{yy}}{\partial y} + \frac{\partial\tau_{yz}}{\partial z} \tag{2.4.8}$$

with

$$\tau_{yy} \approx 2\mu\frac{\partial v}{\partial y} - \frac{2}{3}\mu\left(\frac{\partial v}{\partial y} + \frac{\partial w}{\partial z}\right) \tag{2.4.9}$$

$$\tau_{yz} = \mu\left(\frac{\partial v}{\partial z} + \frac{\partial w}{\partial y}\right) \tag{2.4.10}$$

have been introduced. Similarly, the 'parabolized' energy equation becomes

$$\frac{\partial}{\partial x}(\rho uH) + \frac{\partial}{\partial y}(\rho vH) + \frac{\partial}{\partial z}(\rho wH)$$

$$= \frac{\partial}{\partial y}(\bar{\bar{\tau}}\cdot\vec{v})_y + \frac{\partial}{\partial z}(\bar{\bar{\tau}}\cdot\vec{v})_z + \frac{\partial}{\partial y}\left(k\frac{\partial T}{\partial y}\right) + \frac{\partial}{\partial z}\left(k\frac{\partial T}{\partial z}\right) \tag{2.4.11}$$

where the same approximations are introduced in $(\bar{\bar{\tau}}\cdot\vec{v})$. Note that the continuity equation remains unchanged.

The above set of equations can be solved in various ways, and more details will be given in Volume 2. One approach, often applied, consists of solving equations (2.4.6) and (2.4.7) for u, v and a similar equation for w as a Poisson equation in each (y, z) plane, the streamwise derivatives of the pressure being obtained from a separate calculation. In particular, an initial guess for the pressure terms is introduced into the momentum equations, leading to an approximate velocity field which will generally not satisfy the continuity equation. Therefore an additional correction to the pressure is introduced and related to a corresponding velocity field correction, such as to satisfy the mass conservation equation. These methods are known as 'pressure correction methods', and will be discussed in Volume 2.

Practical example

Viscous flow in curved ducts   The subsonic, viscous flow in curved ducts has been computed by Towne (1984) with a parabolized Navier–Stokes approximation, both for laminar and turbulent flows. A comparison of the laminar and turbulent flows in the same S-shaped duct of Figure 2.4.2 can be seen from Figures 2.4.3–2.5.1, together with experimental data. These remarkable results show the validity of the parabolized Navier–Stokes model for this family of flows where no streamwise separation occurs. The circular S-shaped duct, shown in Figure 2.4.2, is formed by two circular arc bends, with $R = 336$ mm and $D = 48$ mm, corresponding to a ratio $R/D = 7$. The laminar flow case corresponds to a Reynolds number of $Re = 790$, while for the turbulent case $Re = 48\,000$. Figure 2.4.3 compares the laminar and turbulent computed streamwise velocity contours at four stations. The effects of the secondary flows on the streamwise velocities can be followed. The top boundary layer thickens as the flow progresses downstream due to the accumulation of secondary vorticity. This can also be seen in Figure 2.4.4, where the velocity profiles are shown and compared to experimental data. The agreement is excellent, particularly for the computation with the finer mesh. The turbulent flow behaves in a similar way to the laminar flow, with the difference of thinner boundary layers.

Figure 2.5.1 shows the secondary velocity field in two sections, one near the inflection plane and another near the exit plane. In the first half of the S-duct a secondary flow pattern sets in, typical of curved ducts. In the second half the secondary vorticity is of opposite sign and attenuates the crossflow vortices set up in the first part. Near the exit plane the sign of the crossflow near the walls has been reversed. The turbulent flow shows the same properties, but the secondary vortices and the crossflow are of lower magnitude.

Observe, however, in this example the considerable quantitative difference between laminar and turbulent flows, which stresses the importance of turbulence modelling.

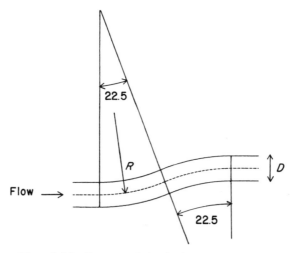

Figure 2.4.2 Geometry of circular S-duct configuration

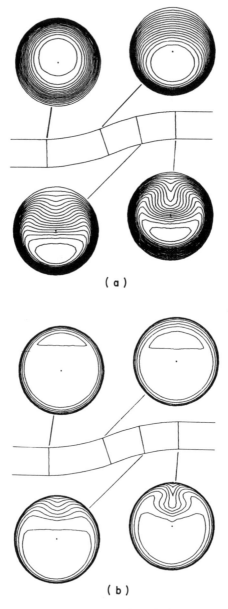

Figure 2.4.3 Computed streamwise velocity contours at four sections of the S-duct of Figure 2.4.2. (a) Laminar flow; (b) turbulent flow. (Reproduced by permission of AIAA from Towne, 1984.)

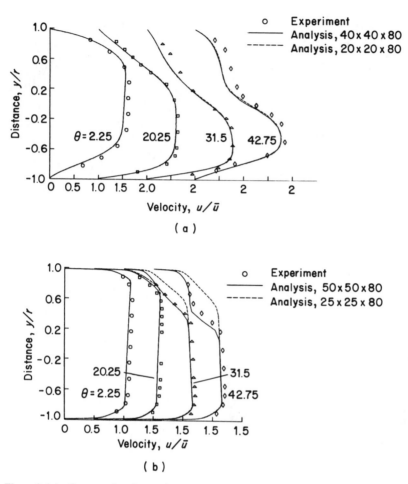

Figure 2.4.4 Computed and experimental streamwise velocity profiles in the symmetry planes of four stations of the S-duct of Figure 2.4.2. (a) Laminar flow; (b) turbulent flow. (Reproduced by permission of AIAA from Towne, 1984.)

Reference

Towne, C. E. (1984) 'Computation of viscous flow in curved ducts and comparison with experimental data.' *AIAA Paper 84-0531*, AIAA 22nd Aerospace Sciences Meeting.

2.5 THE BOUNDARY LAYER APPROXIMATION

It was the great achievement of Prandtl to recognize that at high Reynolds numbers the viscous regions remain of limited extension δ (of the order of $\delta/L \approx \sqrt{(\nu/UL)}$ for a body of length L) along the surfaces of solid bodies immersed in or limiting the flow. Hence in all cases where these viscous regions

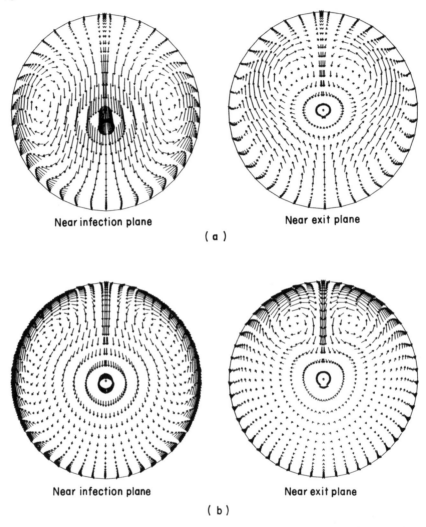

Near infection plane   Near exit plane

(a)

Near infection plane   Near exit plane

(b)

Figure 2.5.1 Computed secondary flow velocity field in two planes of the S-duct of Figure 2.4.2. (a) Laminar flow; (b) turbulent flow. (Reproduced by permission of AIAA from Towne, 1984.)

remain close to the body surfaces (that is, in the absence of separation) the calculation of the pressure field may be separated from that of the viscous velocity field. A detailed discussion of the condition for the derivation of the boundary layer equations can be found in Batchelor (1970), Schlichting (1971) and Cebeci and Bradshaw (1984).

The boundary layer equations can be considered as derived from the TSL equations of Section 2.3 by introducing additional assumptions concerning the velocity component w in the direction z, normal to the wall (Figure 2.5.2).

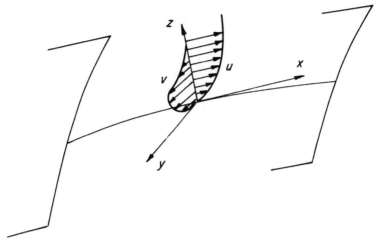

Figure 2.5.2 Boundary Layer velocity profiles

Referring to equation (2.3.12), it is assumed that the flow in the direction normal to the wall has a negligible effect on the flow in the direction parallel to the surface, since the normal velocity component w is much smaller than u and v almost everywhere in the boundary layer (the three components go to zero at the wall). Hence equation (2.3.12) reduces to the condition on the normal gradient of the pressure:

$$\frac{\partial p}{\partial z} \equiv \frac{\partial p}{\partial n} \cong 0 \tag{2.5.1}$$

As a consequence, the pressure $p(x, y, z)$ inside the viscous boundary layer may be taken to be equal to the pressure outside this layer and therefore equal to the value of the pressure $p_e(x, y)$, obtained from an inviscid computation. The pressure $p_e(x, y)$ is the value taken by the inviscid pressure field at the edge of the boundary layer of the surface point (x, y).

Hence the boundary layer equations are obtained from the streamwise and crossflow momentum equations of the TSL approximations (2.3.10) and (2.3.11), with the replacement of $p(x, y, z)$ by $p_e(x, y)$:

$$\frac{\partial}{\partial t} (\rho u) + \frac{\partial}{\partial x} (\rho u^2) + \frac{\partial}{\partial y} (\rho u v) + \frac{\partial}{\partial z} (\rho u w) = -\frac{\partial p_e}{\partial x} + \frac{\partial}{\partial z} \left(\mu \frac{\partial u}{\partial z} \right) \tag{2.5.2}$$

$$\frac{\partial}{\partial t} (\rho v) + \frac{\partial}{\partial x} (\rho u v) + \frac{\partial}{\partial y} (\rho v^2) + \frac{\partial}{\partial z} (\rho v w) = -\frac{\partial p_e}{\partial y} + \frac{\partial}{\partial z} \left(\mu \frac{\partial v}{\partial z} \right) \tag{2.5.3}$$

The inviscid pressure gradient, which is obtained from an inviscid calculation prior to the resolution of the boundary layer equations, acts as an external force on the viscous region. The same inviscid computation also provides the velocities $u_e(x, y)$ and $v_e(x, y)$ at the edge of the boundary layer, connected to the pressure field p_e by the inviscid equation

$$\rho \frac{\partial \vec{v}_e}{\partial t} + \rho (\vec{v}_e \cdot \vec{\nabla}) \vec{v}_e = - \vec{\nabla} p_e \qquad (2.5.4)$$

where $\vec{v}_e$ is the velocity vector of components u_e and v_e.

Equations (2.5.2) and (2.5.3) are to be solved with the additional boundary conditions

$$\begin{aligned} u &= u_e \\ v &= v_e \end{aligned} \qquad (2.5.5)$$

at the edge of the boundary layer. The system of equations obtained in this way has only the velocities as unknowns, and this represents a significant simplification of the Navier–Stokes equations. Therefore the boundary layer equations are much easier to solve, being basically close to standard parabolic second-order partial differential equations, and many excellent numerical methods have been developed (Kline, 1968; Cebeci and Bradshaw, 1977).

The inviscid region is limited by the edge of the boundary layer, which is initially unknown, since the computational process has to start by the calculation of the pressure field. In the classical boundary layer approximation the limits of the inviscid region are taken on the surface, which is justified for small boundary layer thicknesses. This leads to a complete separation of the pressure field and the velocity field, since the pressure in the remaining momentum equations (2.5.2) and (2.5.3) is equal to the values of the inviscid pressure field at the wall and is known at the moment these equations are to be solved.

When the influence of the boundary layers on the inviscid flow field is considered as non-negligible this interaction can be taken into account in an iterative way by recalculating the inviscid pressure field with the limits of the inviscid region located at the edge of the boundary layer obtained at the previous iteration. This procedure is applied for thick boundary layers up to small separated regions and is known as the *viscid–inviscid interaction* approximation (see Le Balleur, 1983, for a review of the subject).

Practical examples

In practice the boundary layer approximation is often used for non-separated flow conditions at high Reynolds numbers. This corresponds, in the aeronautical domain, to flight conditions close to design or cruise. In these cases the boundary layer approximation provides a valuable and economical way of estimating lift and drag of various aircraft components. The following examples have been obtained by Tinoco and Chen (1984).

Wing pressure distribution on B737-300 aircraft   The pressure distribution at different spanwise sections of the wing of a Boeing B737-300 aircraft (with

CFM 56-3 engines) is shown in Figure 2.5.3. These results are obtained from a three-dimensional potential flow computation of the wing–body combination, including the engine nacelle, followed by boundary layer computations, in an iterative way. We can observe the shock jump in the pressure distribution, with increasing magnitude from the base to the tip of the wing. The cross-section close to the engine shows an attenuated shock intensity, a consequence of the interaction between the flow around the nacelle and the wing.

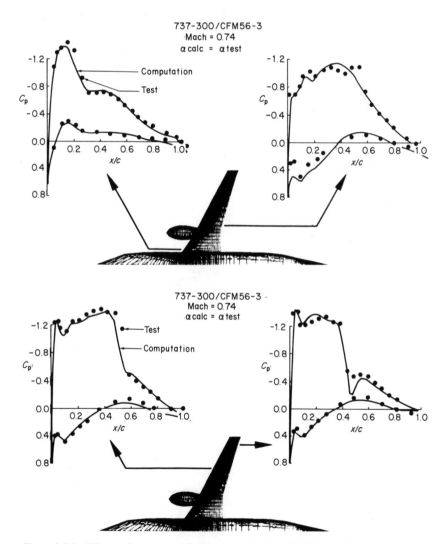

Figure 2.5.3 Wing surface pressure distribution on Boeing 737-300 aircraft computed with an inviscid model followed by a boundary layer computation and compared with experimental data. (Reproduced by permission of AIAA from Tinoco and Chen, 1984.)

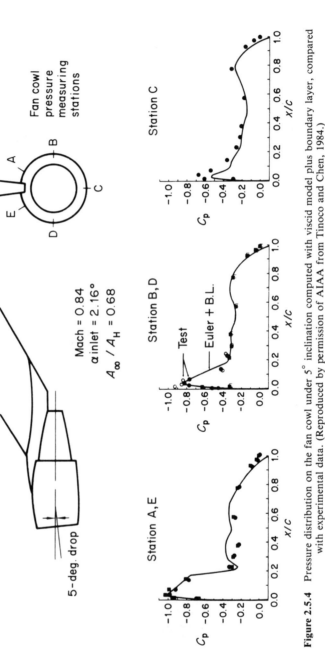

Figure 2.5.4 Pressure distribution on the fan cowl under 5° inclination computed with viscid model plus boundary layer, compared with experimental data. (Reproduced by permission of AIAA from Tinoco and Chen, 1984.)

Turbofan nacelle Figure 2.5.4 shows the pressure distribution at five positions around the fan cowl of a powered nacelle at an angle of inclination (droop angle) of $5°$. Flight Mach number is $M = 0.84$ and generates a transonic flow along the upper surface. The inviscid flow field is obtained from a resolution of the full Euler equations and is subsequently coupled iteratively to a three-dimensional boundary layer computation. The latter takes into account laminar–turbulent transition and empirical information on shock boundary layer interactions. Both these examples indicate a good overall agreement with experimental wind tunnel data.

A more detailed agreement, in difficult conditions such as these three-dimensional flow fields around complex geometries, requires computations at higher levels of approximation. The present examples are probably close to the best that can be achieved in such cases with the rather simple boundary layer approximations. This is confirmed by the author's indications that the data on shock–boundary layer interactions introduced in the computations are essential to obtain the level of agreement with experiments shown in Figures 2.5.3 and 2.5.4.

References

Batchelor, G. K. (1970). *An Introduction to Fluid Dynamics*, Cambridge: Cambridge University Press.
Cebeci, T., and Bradshaw, P. (1977). *Momentum Transfer in Boundary Layers*, Washington, DC: Hemisphere.
Cebeci, T., and Bradshaw, P. (1984). *Physical and Computational Aspects of Convective Heat Transfer*, New York: Springer Verlag.
Le Balleur, J. C. (1983). 'Progres dans le calcul de l'interaction fluide parfait–fluide visqueux.' *AGARD Conference Proceedings CP-351* on *Viscous Effects in Turbomachines*.
Kline, S. J. (1968). *Proceedings of the 1968 AFOSR-HTTM-Stanford Conference on Complex Turbulent Flows*, Vols. I and II, University of Stanford, Thermosciences Division, Stanford, California.
Schlichting, H. (1971). *Boundary Layer Theory*, New York: McGraw-Hill.
Tinoco, E. N., and Chen, A. W. (1984). 'Transonic CFD applications to engine/airframe integration.' *AIAA Paper 84-0381*, AIAA 22nd Aerospace Sciences Meeting.

2.6 THE DISTRIBUTED LOSS MODEL

The distributed loss model is an approximation applied essentially in internal and channel flows, particularly in the fields of turbomachinery, river hydraulics and oceanography. This model is defined by the assumption that the effect of the shear stresses on the motion is equivalent to a *distributed* friction force, defined by semi-empirical data.

Considering the internal flow in a multistage turbomachine, the background of this approximation can easily be understood. Due to its three-dimensional, unsteady, viscous nature, the description of the full three-dimensional flow through the various stages of the machine, including effects such as presence of end walls, tip-clearance flows and relative motions between rotating and non-rotating blade rows, is a considerable computing task which cannot be achieved at present within reasonable computing times and costs. Since such a calculation would, however, be the only way of obtaining, with an adequate turbulence model, a reliable estimation of the loss mechanisms and loss distributions, empirical information with regard to these losses will have to be introduced instead. In turbomachinery flow calculations these losses are defined on an overall or averaged basis as the stagnation pressure drop between the inlet and the outlet of a blade row, without taking into account details of the physical mechanism or the exact location of the producing regions. Therefore at this level of approximation there is no point in considering the detailed structure of the shear stresses in a flow description where the loss sources are introduced as empirical functions of certain flow parameters. Hence their effect on the flow can be attributed to a distributed friction force. The resulting approximation is then basically of an inviscid nature but not isentropic, since the entropy variation along the path of a fluid particle will be connected to the energy dissipation along this path.

Obviously, a certain number of three-dimensional flow details will be lost in such an approximation, especially all flow aspects which can be attributed to, or are strongly influenced by, viscous effects. In particular, the flow in the regions close to blade surfaces or end walls will not be described correctly in this model unless a boundary layer type of approximation is superimposed on the overall flow obtained by the distributed loss model. However, it is assumed that the main effects of the averaged flow can be correctly described by this approximation, at least in the main regions of the flow. A similar approximation is introduced in river hydraulics, where the effects of the wall friction are represented by an empirical resistance force.

The distributed loss model therefore consists of replacing the shear stress terms by an external friction force, function of velocity or other flow variables, but not directly expressed as second-order derivatives of the velocity field.

The general flow equations in a relative frame of reference rotating with a unique angular velocity $\vec{\omega}$ are summarized in Table 1.2, with the assumption that all quantities have been averaged for the turbulent fluctuations according to Section 2.2.

The momentum equation is considered in Crocco's form (1.5.25), without external forces:

$$\frac{\partial \vec{w}}{\partial t} - \vec{w} \times \vec{\zeta} = T\vec{\nabla}s - \vec{\nabla}I + \frac{1}{\rho}\,\vec{\nabla}\cdot\bar{\bar{\tau}} \tag{2.6.1}$$

The energy equation in conservation form (1.5.19) without heat sources $(q_H = 0)$, is

$$\frac{\partial}{\partial t}\left[\rho\left(e + \frac{\vec{w}^2}{2} - \frac{\vec{u}^2}{2}\right)\right] + \vec{\nabla}\cdot(\rho\vec{w}I) = \vec{\nabla}\cdot(k\vec{\nabla}T) + \vec{\nabla}\cdot(\bar{\bar{\tau}}\cdot\vec{w}) \tag{2.6.2}$$

The additional (but not independent) relation for the entropy variation

(1.5.18) plays an important role:

$$\rho T \frac{ds}{dt} = \vec{\nabla} \cdot (k \vec{\nabla} T) + \varepsilon_v \tag{2.6.3}$$

where the dissipation term ε_v is defined

$$\varepsilon_v = \frac{1}{2\mu_e} (\bar{\bar{\tau}} \otimes \bar{\bar{\tau}}^{\mathrm{T}}) = (\bar{\bar{\tau}} \cdot \vec{\nabla}) \cdot \vec{w} \tag{2.6.4}$$

The basic assumption of the distributed loss model lies in the consideration of the shear stress term in the momentum equation as a distributed friction force $\vec{F}_f$, responsible for the overall entropy increase in the flow:

$$\rho \vec{F}_f = \vec{\nabla} \cdot \bar{\bar{\tau}} \tag{2.6.5}$$

By a definition of the distributed loss model we have

$$T \frac{ds}{dt} = -\vec{w} \cdot \vec{F}_f \tag{2.6.6}$$

expressing that the work of the distributed friction forces $\vec{F}_f$ can only be transformed into non-reversible heat and hence serves only to increase the entropy. This can actually be considered as the definition of a friction force in the sense that it expresses the difference between a friction force and any other forces able to produce useful work.

The entropy equation can be transformed to

$$\rho T \frac{ds}{dt} = \vec{\nabla} \cdot (k \vec{\nabla} T) + (\bar{\bar{\tau}} \cdot \vec{\nabla}) \cdot \vec{w} = \vec{\nabla} \cdot (k \vec{\nabla} T) + \vec{\nabla} \cdot (\bar{\bar{\tau}} \cdot \vec{w}) - \vec{w} \cdot (\vec{\nabla} \cdot \bar{\bar{\tau}}) \tag{2.6.7}$$

Comparing these last two relations, it appears that the approximations of the distributed loss model imply that the contribution to the energy (and entropy) equation of the energy diffusion due to the viscous and turbulent stresses is compensated by the heat diffusion. Hence this model is defined by the approximation

$$\vec{\nabla} \cdot (k \vec{\nabla} T) + \vec{\nabla} \cdot (\bar{\bar{\tau}} \cdot \vec{w}) \approx 0 \tag{2.6.8}$$

A more severe assumption would be to neglect heat diffusion and consider the second term of equation (2.6.8) to vanish.

Obviously, this cannot be generally valid in regions with strong shear stress gradients and therefore the present model is not able to describe accurately the flow in the wall regions. The friction force $\vec{F}_f$ will be considered as an *external force* defined by equation (2.6.6) and not by equation (2.6.5). In addition, $\vec{F}_f$ is assumed to be oriented in the direction opposite to the local velocity vector $\vec{w}$. Hence for positive values of F_f,

$$\vec{F}_f = -F_f \cdot \vec{1}_w = -F_f \frac{\vec{w}}{w} \tag{2.6.9}$$

where $\vec{l}_w$ is the unit vector in the direction of the velocity vector $\vec{w}$ and equation (2.6.6) becomes

$$T \frac{\mathrm{d}s}{\mathrm{d}t} = + wF_f \qquad (2.6.10)$$

Assumption (2.6.9) for the direction of the friction force $\vec{F}_f$ is in agreement with the shear layer approximation and boundary layer theory whereby the dominant contribution to the gradient of the shear stresses is given by its normal component (see equation (2.3.3)):

$$\rho \vec{F}_f = \vec{\nabla} \cdot \vec{\tau} \approx \frac{\partial}{\partial n}\left(\mu \frac{\partial \vec{w}}{\partial n}\right) \approx \frac{\partial}{\partial n}\left(\mu \frac{\partial |\vec{w}|}{\partial n}\right) \cdot \vec{l}_w \qquad (2.6.11)$$

if n is the direction normal to the wall. Since the second derivative of w is generally negative in the boundary layer regions the vector $\vec{F}_f$ will indeed be opposite to the main velocity direction, as is to be expected from a friction force.

An important consequence of the assumptions made in the distributed loss model concerns the energy equation which reduces to the simplified form:

$$\frac{\partial}{\partial t}\left[\rho\left(e + \frac{\vec{w}^2}{2} - \frac{\vec{u}^2}{2}\right)\right] + \vec{\nabla} \cdot (\rho \vec{w}I) = 0 \qquad (2.6.12)$$

or, with equation (1.5.21):

$$\rho \frac{\mathrm{d}I}{\mathrm{d}t} = \frac{\partial p}{\partial t} \qquad (2.6.13)$$

Therefore the general form of the equations defining the present model are

$$\frac{\partial \rho}{\partial t} + \vec{\nabla} \cdot (\rho \vec{w}) = 0 \qquad (2.6.14a)$$

$$\frac{\partial \vec{w}}{\partial t} - \vec{w} \times \vec{\zeta} = T\vec{\nabla}s - \vec{\nabla}I + \vec{F}_f \qquad (2.6.14b)$$

$$\rho \frac{\mathrm{d}I}{\mathrm{d}t} = \frac{\partial p}{\partial t} \qquad (2.6.14c)$$

$$T \frac{\mathrm{d}s}{\mathrm{d}t} = wF_f \qquad (2.6.14d)$$

Since the details of the loss mechanism (that is, of the shear stresses) are not considered, these equations are to be taken as describing an inviscid model but with an entropy-producing term. The boundary conditions for the velocity field are therefore the inviscid conditions of vanishing normal velocity components at the walls, with a non-vanishing tangential velocity along these boundaries.

Steady-state formulation

This model is generally further simplified by the assumption of steady relative flows. In the turbomachinery environment this will be justified as long as the inlet flow conditions can be considered as uniform in the direction of blade rotation. This implies that the presence and effects of the wakes shed from the upstream blades are neglected. Recent detailed data on rotor stator interaction (Dring *et al.*, 1982) show that the unsteady pressure generated by upstream wakes can be locally very significant, although the time-averaged flow is in good agreement with steady-state calculations. Therefore we can consider the steady-state model for the relative flow as a valid approximation for the time-averaged flow at constant angular velocity. This leads to the following simplified models, where the equation of continuity has been introduced in the energy and entropy equations:

$$\vec{\nabla} \cdot (\rho \vec{w}) = 0 \tag{2.6.15a}$$

$$-\vec{w} \times \vec{\zeta} = T\vec{\nabla}s - \vec{\nabla}I + \vec{F_f} \tag{2.6.15b}$$

$$\vec{w} \cdot \vec{\nabla}I = 0 \tag{2.6.15c}$$

$$T\vec{w} \cdot \vec{\nabla}s = -\vec{w} \cdot \vec{F_f} = wF_f \tag{2.6.15d}$$

In particular, the energy equation reduces to the constancy of the rothalpy I along a flow path. It is to be noted also that the last equation (2.6.15d), which can be deduced from the momentum and energy conservation laws, is used as a definition for the friction force F_f.

Systems (2.6.14) or (2.6.15) contain six equations for five unknowns, $\vec{w}, s, I$. Therefore when use is made of the entropy equation one of the momentum equations can be dropped out of the system.

The model described above is used extensively in the field of turbomachinery flow calculations with the introduction of empirical data for the loss coefficients without the computation of the shear stresses, and a more detailed presentation can be found in Hirsch (1984) and Hirsch and Deconinck (1985).

Practical example

Meridional through-flow in a multistage compressor The distributed loss model is most frequently applied to multi-stage turbomachinery flows in connection with a quasi-three-dimensional approximation, which describes the properties of tangentially averaged flow variables. Figure 2.6.1 (from Hirsch and Warzee, 1976) represents flow variables averaged over the blade spacing, as a function of radius, at different stations of the meridional cross-section of a two-stage axial fan compared with experimental data. Observe that the accuracy of the computed flow variations is strongly dependent on the quantity of the empirical input as expressed by the friction force.

86

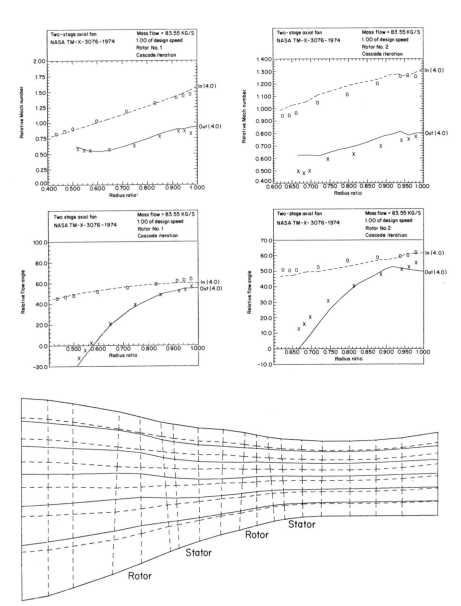

Figure 2.6.1 Meridional flow streamlines through a two-stage axial fan computed with the distributed loss model. Relative flow angle and Mach numbers are shown after the first and second rotor and compared with experimental data. (From Hirsch and Warzee, 1976)

References

Dring, R. P., Joslyn, H. D., and Hardin, L. W. (1982). 'An investigation of Compressor rotor aerodynamics.' *Trans. ASME, Journal of Engineering for Power*, **104**, 84–96.

Hirsch, Ch. (1984). 'Computational models for turbomachinery flows.' *Naval Postgraduate School Report NPS-67-84-022*, Monterey, California.

Hirsch, Ch., and Deconinck, H. (1985). 'Through flow models in turbomachines: stream surface and passage averaged representations.' In Ucer, S. A., Stow, P., and Hirsch, Ch. (eds), *Thermodynamics and Fluid Mechanics of Turbomachinery*, Vol. I, NATO ASI Series, Dordrecht: Martinus Nijhoff.

Hirsch, Ch., and Warzee, G. (1976). 'A finite element method for through-flow calculations in turbomachines.' *Trans. ASME, Journal of Fluids Engineering*, **98**, 403–21.

2.7 THE INVISCID FLOW MODEL—EULER EQUATIONS

The most general flow configuration for a non-viscous, non-heat-conducting fluid is described by the set of Euler equations, obtained from the Navier–Stokes equations (2.1.1) by neglecting all shear stresses and heat conduction terms. As is known from Prandtl's boundary layer analysis, this is a valid approximation for flows at high Reynolds numbers outside viscous regions developing in the vicinity of solid surfaces.

This approximation introduces a drastic change in the mathematical formulation with respect to all the previous models containing viscosity terms, since the system of partial differential equations describing the inviscid flow model reduces from second order to first order. This is of paramount importance, since it will determine the numerical and physical approach to the computation of these flows. Also, the number of allowable boundary conditions is modified by passing from the second-order viscous equations to the first-order inviscid system.

The time-dependent Euler equations, in conservation form and in an absolute frame of reference, for the conservative variables U defined by equation (2.1.2):

$$\frac{\partial U}{\partial t} + \vec{\nabla} \cdot \vec{F} = Q \qquad (2.7.1)$$

form a system of first-order partial differential equations hyperbolic in time (as will be shown later), where the flux vector $\vec{F}$ has the Cartesian components f, g, h given by equations (2.1.7) without the shear stresses terms:

$$f = \begin{vmatrix} \rho u \\ \rho u^2 + p \\ \rho u v \\ \rho u w \\ \rho u H \end{vmatrix} \qquad (2.7.2a)$$

$$g = \begin{vmatrix} \rho v \\ \rho uv \\ \rho v^2 + p \\ \rho vw \\ \rho vH \end{vmatrix} \qquad (2.7.2b)$$

$$h = \begin{vmatrix} \rho w \\ \rho uw \\ \rho vw \\ \rho w^2 + p \\ \rho wH \end{vmatrix} \qquad (2.7.2c)$$

and the source term Q is given by equation (2.1.4). Generally, heat sources q_H will not be considered since heat conduction effects are neglected in the system of Euler equations.

It is important to note the properties of the entropy variations in an inviscid flow. From equation (1.5.18), and in the absence of heat sources, the entropy equation for continuous flow variations reduces to

$$T\left(\frac{\partial s}{\partial t} + \vec{v} \cdot \vec{\nabla}s\right) = 0 \qquad (2.7.3)$$

expressing that entropy is constant along a flow path. Hence the Euler equations describe isentropic flows in the absence of discontinuities.

The value of the entropy can, however, vary from one flow path to another. This is best seen from Crocco's form of the momentum equation (1.5.24), when applied to a stationary, inviscid flow in the absence of external forces. We obtain

$$-\vec{v} \times \vec{\zeta} = T\vec{\nabla}s - \vec{\nabla}H \qquad (2.7.4)$$

In an intrinsic co-ordinate system $\vec{1}_l$, $\vec{1}_n$, $\vec{1}_b$, where $\vec{1}_l$ is directed along the velocity and $\vec{1}_b$ is the binormal unit vector, equation (2.7.4) becomes, projected in the normal direction n, for uniform total enthalpy,

$$w\zeta_b = T\frac{\partial s}{\partial n} \qquad (2.7.5)$$

This relation shows that entropy variations in the direction normal to the local velocity direction is connected to vorticity. Hence entropy variations will generate vorticity and, inversely, vorticity will create entropy gradients.

As is known, the set of Euler equations also allows discontinuous solutions in certain cases, namely vortex sheets, contact discontinuities or shock waves occurring in supersonic flows. The properties of these discontinuous solutions can only be obtained from the integral form of the conservation equations, since the gradients of the fluxes are not defined at discontinuity surfaces.

2.7.1 The properties of discontinuous solutions

For a discontinuity surface Σ, moving with velocity $\vec{C}$, the integral conservation laws are applied to the infinitesimal volume V of Figure 2.7.1. Referring to equations (1.1.2) and (1.1.8), the integral form of the Euler equations takes the following form in the absence of any source terms:

$$\frac{\partial}{\partial t} \int_V U \, d\Omega + \oint_S \vec{F} \cdot d\vec{S} = 0 \qquad (2.7.6)$$

The time derivative of the volume integral has to take into account the motion of the surface Σ, and hence of the control volume V, through

$$\frac{\partial}{\partial t} \int_V U \, d\Omega = \int_V \frac{\partial U}{\partial t} \, d\Omega + \int_V U \frac{\partial}{\partial t} (d\Omega)$$

$$= \int_V \frac{\partial U}{\partial t} \, d\Omega - \oint_S U\vec{C} \cdot d\vec{S} \qquad (2.7.7)$$

expressing the conservation of the volume V in the translation with velocity $\vec{C}$. The flux term in equation (2.7.6) can be rewritten for vanishing volumes $V(\Delta \to 0)$ as

$$\oint_S \vec{F} \cdot d\vec{S} = \int_\Sigma (\vec{F}_2 - \vec{F}_1) \cdot d\vec{\Sigma} \equiv \int_\Sigma [\vec{F} \cdot \vec{1}_n] d\Sigma \qquad (2.7.8)$$

where $d\vec{\Sigma}$ is normal to the discontinuity surface Σ and where the notation

$$[A] \equiv A_2 - A_1 \qquad (2.7.9)$$

denotes the jump in the variable A when crossing the discontinuity.

Combining equations (2.7.7) and (2.7.8) in equation (2.7.6) we obtain, for vanishing volumes V,

$$\int_\Sigma ([\vec{F}] - \vec{C}[U]) \cdot d\vec{\Sigma} = 0 \qquad (2.7.10)$$

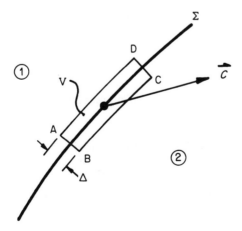

Figure 2.7.1 Control volume ABCD around a discontinuity surface Σ moving with velocity $\vec{C}$

leading to the local form of the conservation laws over a discontinuity, called the *Rankine–Hugoniot* relations:

$$[\vec{F}] \cdot \vec{I}_n - \vec{C}[U] \cdot \vec{I}_n = 0 \qquad (2.7.11)$$

If $\Sigma(\vec{x}, t) = 0$ is the discontinuity surface, then we have

$$\frac{\mathrm{d}\Sigma}{\mathrm{d}t} \equiv \frac{\partial\Sigma}{\partial t} + \vec{C} \cdot \vec{\nabla}\Sigma = 0 \qquad (2.7.12)$$

With the unit vector along the normal $\vec{I}_n$ defined by

$$\vec{I}_n = \frac{\vec{\nabla}\Sigma}{|\vec{\nabla}\Sigma|} \qquad (2.7.13)$$

equation (2.7.11) takes the form

$$[\vec{F}] \cdot \vec{\nabla}\Sigma + \frac{\partial\Sigma}{\partial t}[U] = 0 \qquad (2.7.14)$$

Various forms of discontinuities are physically possible; for example, *shocks* where all flow variables undergo a discontinuous variation, and *contact discontinuities* and *vortex sheets* (also called slip lines) across which no mass transfer takes place but where density, as well as the tangential velocity, may be discontinuous although pressure and normal velocity remain continuous.

The properties of these discontinuous solutions can best be seen from a reference system moving with the discontinuity. In this system the discontinuity surface is stationary ($\vec{C} = 0$) and the Rankine–Hugoniot relations for the Euler equations become

$$[\rho\vec{v} \cdot \vec{I}_n] = 0 \qquad (2.7.15a)$$
$$[\vec{v}]\,\rho\vec{v} \cdot \vec{I}_n + [p]\,\vec{I}_n = 0 \qquad (2.7.15b)$$
$$\rho\vec{v} \cdot \vec{I}_n[H] = 0 \qquad (2.7.15c)$$

This system admits solutions with the following properties.

Contact discontinuities

These are defined by the condition of no mass flow through the discontinuity:

$$v_{n1} = v_{n2} = 0 \qquad (2.7.16a)$$

and, following equation (2.7.15b), by continuity of pressure

$$[p] = 0 \qquad (2.7.16b)$$

allowing non-zero values for the jump in specific mass, as seen from equation (2.7.15a):

$$[\rho] \neq 0 \qquad (2.7.16c)$$

In addition, tangential velocity can be continuous:

$$[v_t] = 0 \qquad (2.7.16d)$$

Vortex sheets or slip lines

These are also defined by the condition of no mass flow through the discontinuity:

$$v_{n1} = v_{n2} = 0 \qquad (2.7.17a)$$

and, following equation (2.7.15b), by continuity of pressure

$$[p] = 0 \qquad (2.7.17b)$$

allowing non-zero values for the jump in tangential velocity, as seen from equation (2.7.15b):

$$[v_t] \neq 0 \qquad (2.7.17c)$$

coupled to a jump in density

$$[\rho] \neq 0 \qquad (2.7.17d)$$

Shock surfaces

Shocks are solutions of the Rankine–Hugoniot relations with non-zero mass flow through the discontinuity. Consequently pressure and normal velocity undergo discontinuous variations while the tangential velocity remains continuous. Hence shocks satisfy the following properties:

$$[\rho] \neq 0$$
$$[p] \neq 0 \qquad (2.7.18a)$$
$$[v_n] \neq 0$$

and

$$[v_t] = 0 \qquad (2.7.18b)$$

Note that since the stagnation pressure p_0 is not constant across the shock the inviscid shock relations imply a *discontinuous entropy variation through the shock*. This variation has to be positive, corresponding to compression shocks and excluding hereby expansion shocks, for physical reasons connected with the second principle of thermodynamics, Shapiro (1953); Zucrow and Hoffman (1976).

It has to be added that expansion shocks, whereby the entropy jump is negative, are valid solutions of the inviscid equations since, in the absence of heat transfer, they describe reversible flow variations. Hence there is no mechanism allowing for distinguishing between discontinuities with entropy

increase (positive entropy jump) or entropy decrease (negative entropy variation). An additional condition, called the *entropy condition*, has therefore to
be added to the inviscid equation in order to exclude these non-physical
solutions, Lax (1973). This is necessary for all inviscid flow models, and a
more detailed discussion of the entropy condition is presented in Volume 2.

The mathematical formulation of the second principle of thermodynamics
can be expressed, for an adiabatic flow without heat conduction or heat
sources $q_H = 0$, following equation (1.5.18):

$$\rho T \left(\frac{\partial s}{\partial t} + \vec{v} \cdot \vec{\nabla}s \right) = \varepsilon_v \tag{2.7.19}$$

Since ε_v is the viscous dissipation and always positive this equation states that
any solution of the Euler equations which has a physical sense as a limit, for
vanishing viscosity of real fluid flow phenomena, has to satisfy the following
entropy inequality:

$$\rho T \left(\frac{\partial s}{\partial t} + \vec{v} \cdot \vec{\nabla}s \right) \geq 0 \tag{2.7.20}$$

In addition, a non-uniform discontinuity, such as a shock with varying
intensity, will generate a non-uniform entropy field in the direction normal to
the velocity. Equation (2.7.5) then shows that as a consequence, vorticity will
be generated downstream of the shock. Hence even for irrotational flow
conditions upstream of the shock a rotational flow will be created by a
non-uniform shock intensity.

Compressible vorticity equation

An interesting relation is obtained for the vorticity transport in inviscid flow
conditions. The vorticity equation (1.3.14) becomes, in the absence of external
forces,

$$\frac{d}{dt} \vec{\zeta} = (\vec{\zeta} \cdot \vec{\nabla})\vec{v} - \vec{\zeta}(\vec{\nabla} \cdot \vec{v}) + \vec{\nabla}p \times \vec{\nabla} \frac{1}{\rho} \tag{2.7.21}$$

The last term will vanish if the isentropic relations are taken into account.
Indeed, the condition of isentropicity implies a unique relation between p and
ρ of the term $p = p(\rho)$. In particular, for a perfect gas we have, from equation
(2.1.17),

$$p = k \cdot \rho^\gamma \tag{2.7.22}$$

With the continuity equation the second term can be replaced by the material
derivative of the density, leading to

$$\frac{d}{dt} \vec{\zeta} = (\vec{\zeta} \cdot \vec{\nabla})\vec{v} + \frac{\vec{\zeta}}{\rho} \frac{d\rho}{dt} \tag{2.7.23}$$

or

$$\frac{\mathrm{d}}{\mathrm{d}t}\left(\frac{\vec{\xi}}{\rho}\right) = \left(\frac{\vec{\xi}\cdot\vec{\nabla}}{\rho}\right)\vec{v} \qquad (2.7.24)$$

This relation shows that in a two-dimensional flow, $\vec{\xi}/\rho$ is conserved along a path line.

Practical examples

Two-dimensional flows on airfoil sections The flow over the NACA 0012 airfoil is a widely used test case for two-dimensional flow computations, and a large number of results are available. The following can be considered as most accurate numerical solutions of the Euler equations, having been selected after a very careful investigation of a large number of computations by an AGARD Working Group, reporting in *AGARD AR 211* (1985). A series of computed flow distributions at increasing incident Mach numbers are shown in Figures 2.7.2 and 2.7.3 for the NACA 0012 profile, from $M_\infty = 0.85$ to $M_\infty = 1.2$ at zero degree incidence. These computations can now be obtained with high accuracy and low computational cost, and may be considered as excellent approximations to the 'exact' inviscid flow. However, comparison with experimental data (to be found in Harris, 1981) indicate the influence of viscosity on the overall flow properties. At subsonic velocities (Figure 2.7.4(a)) these effects are small, while in the presence of shock waves (Figure 2.7.4(b)) the shock boundary layer interaction clearly has a significant local effect which can only be described by inclusion of viscous effects.

Two-dimensional air intake The inviscid flow at supersonic speeds in an air intake represents a very complex flow system, and Figure 2.7.5 shows the results of a computation based on the inviscid Euler equations, where grid and isoMach lines are shown at incident Mach numbers of $M_\infty = 1.8$, 2 and 3. The shock structures are well captured and their gradual evolution with increasing Mach numbers can be analysed from these calculations.

Although a similar test case, computed with a Navier–Stokes model and shown in Figure 2.2.1, indicated an instability cycle (called inlet buzz) as a result of the the interaction between the shocks and the turbulent wall boundary layers, the determination of the flow properties obtained from an inviscid analysis is nevertheless essential for an understanding of the influence of viscous interactions. It is indeed imperative to dispose of an accurate numerical solution of Euler flows, before introducing viscous terms in the inviscid model, in order to make certain that the phenomena observed resulting from the introduction of viscous terms are not obscured by numerical errors, such as numerical viscosity.

Leading-edge vortex flow The flow field around a delta wing at supersonic conditions is a challenging case for computations with the Euler equations,

94

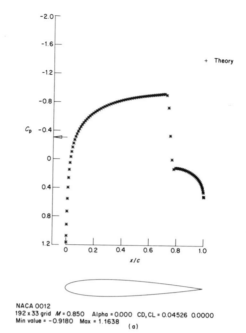

NACA 0012
192 x 33 grid M = 0.850   Alpha = 0.000 CD,CL = 0.04526 0.0000
Min value = -0.9180 Max = 1.1638
(a)

Mach Contours Min value = 0.0825 Max = 1.3587
NACA 0012
192 x 33 grid M = 0.850 Alpha = 0.000 CD,CL = 0.04526 0.0000
(b)

Figure 2.7.2 Pressure and Mach number distribution on a NACA 0012 profile at $M_\infty = 0.85$ and $0°$ incidence. (a) Surface pressure distribution; (b) isoMach contours. (Courtesy M. Salas, NASA Langley Research Center, USA)

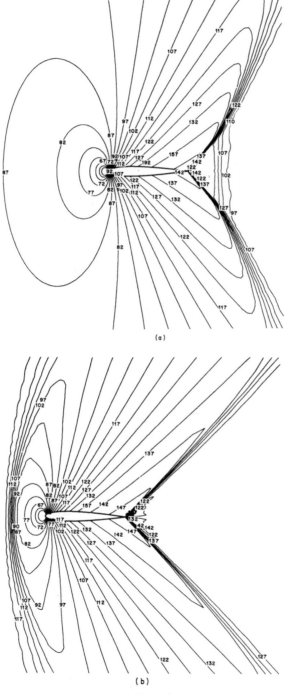

Figure 2.7.3   IsoMach contours for the NACA 0012 profile at incident Mach numbers $M_\infty = 0.95$ and $M_\infty = 1.2$. (From AGARD AR 211 Report, 1985)

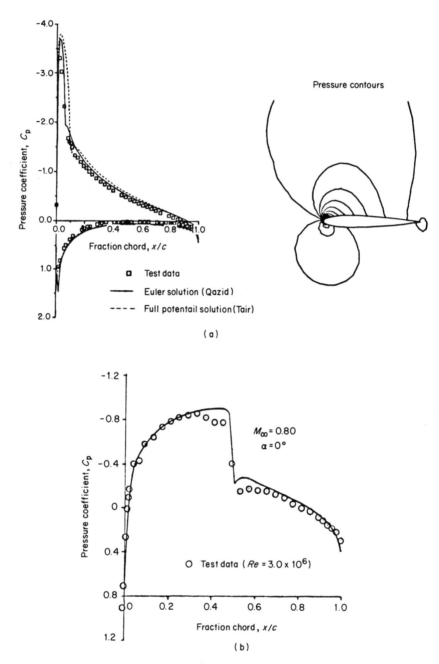

Figure 2.7.4 Comparison of inviscid surface pressure distribution with experimental data on the NACA 0012 airfoil. (a) $M_\infty = 0.50$, $7°$ incidence (note the comparison with the potential model); (b) $M_\infty = 0.80$, $0°$ incidence. (Courtesy A. Verhoff, McDonnell Aircraft Co., USA)

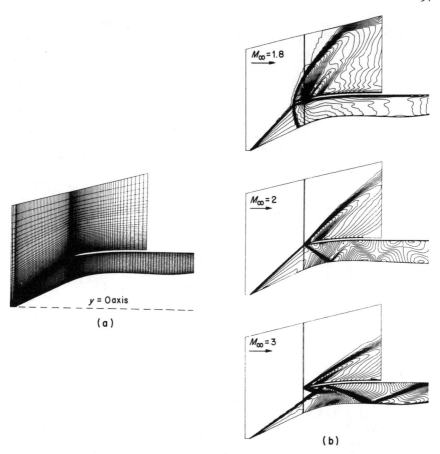

Figure 2.7.5 Inviscid computation of the flow in an air intake at supersonic Mach numbers. (a) Grid geometry; (b) isoMach lines at incident Mach numbers of $M_\infty = 1.8$, 2 and 3. (Reproduced by permission of AIAA, from Borrel and Montagné, 1985.)

since the vortex generated at the leading edge has to be captured by the calculations, although the incident flow is irrotational. In the inviscid Euler model the leading-edge vortex therefore appears as a vortex sheet discontinuity. A comparison of this flow field when computed with Euler and Navier–Stokes models provides indications of the validity range of the Euler models as a function of the required accuracy and the flow conditions.

In the examples of Figure 2.7.6 we can observe the validity of the inviscid Euler model for these low-incidence, high Mach number conditions, although the flow is subsonic in the leading-edge region. It appears that the location and intensity of the leading-edge vortex is predicted with good accuracy by the inviscid computation, although flow details close to the wing surface are obviously not to be relied upon in the inviscid model. It particular, a small secondary separated region appears in the Navier–Stokes computations as the

98

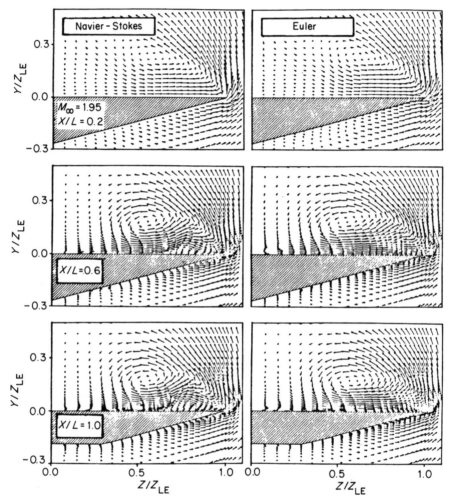

Figure 2.7.6 Comparison of computations with the Euler and Navier–Stokes flow models of the flow along a delta wing at incident Mach number of $M_\infty = 1.95$. Crossplane velocity fields are shown at three spanwise sections. (Reproduced by premission of AIAA from Rizzetta and Shang, 1984.)

vortex structure moves downstream, which is not seen by the Euler calculation. This secondary vortex results from the no-slip condition coupled to the inviscid recompression towards the leading edge, and is fully confirmed by experimental observations.

Three-dimensional inviscid flow in a steam turbine stage   Turbomachinery flows can be extremely complicated, particularly when the interaction of successive blade rows is to be taken into account. Figure 2.7.7 shows the computed three-dimensional inviscid flow in a complete steam turbine stage.

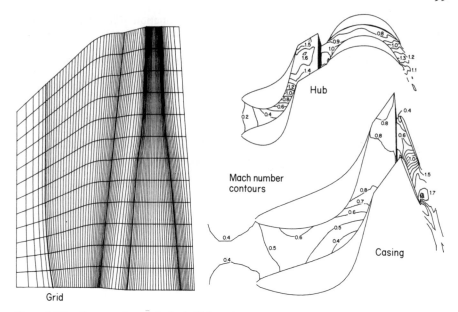

Figure 2.7.7 Computation of the inviscid flow through the last stage of a steam turbine. The grid shown as well as the Mach number contours for the hub and casing sections. Reproduced with permission from Denton, 1985. The original version of this material was first published in AGARD LS-140, published in May 1985 by the Advisory Group for Aerospace Research and Development, North Atlantic Treaty Organization (AGARD/NATO)

The flow is simultaneously computed in the stationary and the rotating systems and the boundary between the rotating and non-rotating blade rows is artificially assumed to have uniform flow conditions defined by pitchwise averaged flow quantities. This allows a realistic computation whereby the boundary conditions are imposed at the rotor exit only, while the pressure between the two blade rows results from the computation. This is a rather complex flow, with tip Mach numbers up to 1.9, and the computation allows the determination of essential aspects of the influence of three-dimensional geometry and of inviscid interaction between the two blade rows.

References

Borrel, M., and Montagné, J. L. (1985). 'Numerical study of a non-centred scheme with application to aerodynamics.' *AIAA Paper 85-1479, Proc. AIAA 7th Computational Fluid Dynamics Conference*, pp. 88–97.

Denton, J. (1985). 'The calculation of fully three-dimensional flow through any type of turbomachine blade row.' In *AGARD LS 140* on *3-D Computation Techniques Applied to Internal Flows in Propulsion Systems*.

Harris, C. D. (1981). 'Two-dimensional aerodynamic characteristics of the NACA 0012 airfoil in the Langley 8-foot transonic pressure tunnel.' *NASA TM 81927*, April 1981.

Lax, P. D. (1973). *Hyperbolic Systems of Conservation Laws and the Mathematical Theory of Shock Waves*, Philadelphia: SIAM Publications.

100

Rizzetta, D. P., and Shang, J. S. (1984). 'Numerical simulation of leading edge vortex flows.' *AIAA Paper 84-1544*, AIAA 17th Fluid Dynamics, Plasma Dynamics, and Lasers Conference.

Shapiro, A. H. (1953). *The Dynamics and Thermodynamics of Compressible Fluid Flow*, New York: Ronald Press.

Zucrow, M. J., and Hoffman, J. D. (1976). *Gas Dynamics*, New York: John Wiley.

2.8 STEADY INVISCID ROTATIONAL FLOWS—CLEBSCH REPRESENTATION

An alternative, general representation of inviscid, rotational flows can be defined, leading in many cases to a most economical description in terms of the number of flow variables. The absolute vorticity of a frictionless flow in a rotating, relative frame of reference is generated by the gradients of entropy and total energy. This is best seen from equations (2.6.14), which become, in the absence of friction forces,

$$\frac{\partial \rho}{\partial t} + \vec{\nabla} \cdot (\rho \vec{w}) = 0$$

$$\frac{\partial \vec{w}}{\partial t} - \vec{w} \times \vec{\zeta} = T\vec{\nabla}s - \vec{\nabla}I$$

$$\frac{\mathrm{d}I}{\mathrm{d}t} = \frac{1}{\rho}\frac{\partial p}{\partial t}$$

$$\frac{\mathrm{d}s}{\mathrm{d}t} = 0$$

$$(2.8.1)$$

Consequently, a form of Clebsch representation can be defined for the *absolute* velocity field $\vec{v}$:

$$\vec{v} = \vec{\nabla}\phi + \psi_1\vec{\nabla}s + \psi_2\vec{\nabla}I \qquad (2.8.2)$$

In this expression ϕ is a potential function describing the irrotational flow components and ψ_1 and ψ_2 are two additional functions describing the magnitude of the rotational parts of the flow. Indeed, the absolute vorticity is

$$\vec{\zeta} = \vec{\nabla} \times \vec{v} = \vec{\nabla}\psi_1 \times \vec{\nabla}s + \vec{\nabla}\psi_2 \times \vec{\nabla}I \qquad (2.8.3)$$

Although the Clebsch representation (2.8.2) can be applied for time-dependent flows (see, for instance, Serrin 1959), its usefulness for practical applications appears mainly with stationary flows. Therefore introducing representation (2.8.2) into the stationary momentum equation and taking into account the stationary energy and entropy equations lead to

$$(\vec{w} \cdot \vec{\nabla}\psi_1)\vec{\nabla}s + (\vec{w} \cdot \vec{\nabla}\psi_2)\vec{\nabla}I = -\vec{\nabla}I + T\vec{\nabla}s \qquad (2.8.4)$$

For arbitrary and *independent* entropy and rothalpy gradients we obtain the two equations for ψ_1 and ψ_2:

$$(\vec{w} \cdot \vec{\nabla})\psi_1 = +T \qquad (2.8.5a)$$

$$(\vec{w} \cdot \vec{\nabla})\psi_2 = -1 \qquad (2.8.5b)$$

The equations are purely convective, and require only the initial values of ψ_1 and ψ_2 in the inlet surface of the flow region. The equation for the third function ϕ is obtained from the continuity equation, taking into account the relation $\vec{v} = \vec{u} + \vec{w}$ between the relative and absolute velocities:

$$\vec{\nabla} \cdot (\rho \vec{\nabla}\phi) = (\vec{u} \cdot \vec{\nabla})\rho - \vec{\nabla} \cdot (\rho\psi_1 \vec{\nabla}s + \rho\psi_2 \vec{\nabla}I) \qquad (2.8.6)$$

Generally, the specific mass can be expressed as a function of two thermodynamic variables (e.g. $\rho = \rho(s, h)$ or $\rho = \rho(s, I)$) and the three equations (2.8.5) and (2.8.6) have to be closed by the equations for the transport of energy and entropy:

$$(\vec{w} \cdot \vec{\nabla})s = 0 \qquad (2.8.7)$$

$$(\vec{w} \cdot \vec{\nabla})I = 0 \qquad (2.8.8)$$

In this representation we still have five unknown functions to solve, namely the three velocity components, s and I.

A simplified representation can be obtained if a unique relation between s and I exists in the inlet field, that is,

$$s = s(I) \qquad (2.8.9)$$

or

$$I = I(s) \qquad (2.8.10)$$

In this case we have

$$\vec{\nabla}I - T\vec{\nabla}s = \left(\frac{dI}{ds} - T\right)\vec{\nabla}s = \left(1 - T\frac{ds}{dI}\right)\vec{\nabla}I \qquad (2.8.11)$$

since relations (2.8.9) and (2.8.10) are valid in the whole flow field due to equations (2.8.7) and (2.8.8).

Consequently, one of the 'rotational functions' ψ_1 and ψ_2 is not necessary, and the three-dimensional rotational flow field can be described by

$$\vec{v} = \vec{\nabla}\phi + \psi_1 \vec{\nabla}s \qquad (2.8.12)$$

or, equivalently, by

$$\vec{v} = \vec{\nabla}\phi + \psi_2 \vec{\nabla}I \qquad (2.8.13)$$

In the first representation the function ψ_1 satisfies the equation

$$(\vec{w} \cdot \vec{\nabla})\psi_1 = -\frac{dI}{ds} + T \qquad (2.8.14)$$

and in the second case we have

$$(\vec{w} \cdot \vec{\nabla})\psi_2 = -1 + T\frac{ds}{dI} \qquad (2.8.15)$$

It is easily seen that assumptions (2.8.9) or (2.8.10) are equivalent to the conditions

$$(\vec{\zeta} \cdot \vec{\nabla})s = (\vec{\zeta} \cdot \vec{\nabla})I = 0 \qquad (2.8.16)$$

which state that the vorticity vector lies in the surfaces of constant entropy and total energy, as can be seen by taking the scalar product of equation (2.8.3) with $\vec{\nabla}s$ and $\vec{\nabla}I$ and applying equations (2.8.9) and (2.8.10). This condition has been shown by Cazal (1966) to be necessary and sufficient for representations (2.8.12) or (2.8.13) to be valid.

We dispose here of a very economical representation of a rotational flow since the three-dimensional velocity field is described by two scalar functions, and the complete flow description requires only three equations for ϕ, ψ_1 and s, since $I = I(s)$ is known from the inlet conditions. It is shown by Lacor and Hirsch (1982) that these conditions are satisfied if either H or s are constant in the inlet flow field or, more generally, if the inlet velocity field is uniform in at least one of the directions transverse to the inlet velocity.

An alternative representation to equations (2.8.12) and (2.8.13) can be defined by

$$\vec{v} = \vec{\nabla}\phi + s\,\vec{\nabla}\psi_1 \qquad (2.8.12a)$$

or

$$\vec{v} = \vec{\nabla}\phi + I\vec{\nabla}\psi_2 \qquad (2.8.13a)$$

In the first representation the function ψ_1 satisfies the equation

$$(\vec{w} \cdot \vec{\nabla})\psi_1 = \frac{\mathrm{d}I}{\mathrm{d}s} - T \qquad (2.8.14a)$$

and in the second case

$$(\vec{w} \cdot \vec{\nabla})\psi_2 = 1 - T\frac{\mathrm{d}s}{\mathrm{d}I} \qquad (2.8.15a)$$

The Clebsch representation with reduced variables

Restrictions (2.8.9) or (2.8.10) on the inlet flow, necessary to obtain the reduced representations (2.8.12) or (2.8.13), are not required for non-rotating flows or for incompressible flows, as is shown next. Indeed, in these two cases the number of independent unknowns can be reduced, through the introduction of scaled quantities (indicated by a subscript s):

$$\vec{w}_s = f_1\,\vec{w} \qquad (2.8.17)$$

$$\rho_s = f_2\,\rho \qquad (2.8.18)$$

The continuity equation becomes

$$\vec{\nabla} \cdot (\rho_s\vec{w}_s) = \rho(\vec{w} \cdot \vec{\nabla})f_1 f_2 \qquad (2.8.19)$$

If the scaling functions f_1, f_2 depend only on quantities purely convected by the flow, such as entropy or rothalpy, then

$$(\vec{w} \cdot \vec{\nabla})f_1 f_2 = 0 \qquad (2.8.20)$$

and the scaled density and velocity fields will satisfy the continuity equation

$$\vec{\nabla} \cdot (\rho_s \vec{w}_s) = 0 \qquad (2.8.21)$$

The scaled vorticity is defined by

$$\vec{\zeta}_s = \vec{\nabla} \times \vec{v}_s \qquad (2.8.22)$$

and if f_1 is chosen as a function of entropy only, the momentum equation for the scaled flow variables can be written as (Hirsch and Lacor, 1983)

$$\vec{w}_s \times \vec{\zeta}_s = \vec{\nabla}(If_1^2) - \left(I + \frac{Tf_1}{2(df_1/ds)} - \frac{\vec{v}^2}{2} + \frac{\vec{u} \cdot \vec{v}}{2}\right)\vec{\nabla}f_1^2 \qquad (2.8.23)$$

A particular choice for f_1 and f_2, in the case of a perfect gas, is given by Yih (1965):

$$f_1 = e^{-s/2c_p} \qquad (2.8.24)$$

and

$$f_2 = e^{s/c_p} = \frac{1}{f_1^2} \qquad (2.8.25)$$

This leads to the following momentum equation, when introduced into equation (2.8.23), taking into account the relation (1.5.22) between I and H:

$$\vec{w}_s \times \vec{\zeta}_s = \vec{\nabla}(Ie^{-s/c_p}) + \tfrac{1}{2}(\vec{u} \cdot \vec{v})\vec{\nabla}e^{-s/c_p} \qquad (2.8.26)$$

Representation (2.8.2) for the velocity field can be reduced to the simplified forms (2.8.12) or (2.8.13) if only one gradient appears in the right-hand side of the above equation. This will be the case for a non-rotating system where $u = 0$. As shown by Hawthorne (1974), there is no choice of f_1 which could remove one of the two gradients appearing in this equation for a general rotating flow unless some restrictions are imposed on the inlet flow field, as discussed in the previous section.

Non-rotating system

For a non-rotating flow, $\vec{u} = 0$ and the momentum equation (2.8.26) becomes, with equation (2.1.27),

$$\vec{v}_s \times \vec{\zeta}_s = \vec{\nabla}(He^{-s/c_p}) = \frac{H_A}{p_{0_A}^{(\gamma-1)/\gamma}}\vec{\nabla}p_0^{(\gamma-1)/\gamma} \qquad (2.8.27)$$

for a perfect gas of specific heat ratio γ, where p_0 is the stagnation pressure and where the subscript A indicates references values. Since the right-hand side of this equation contains only one gradient, the following representation for

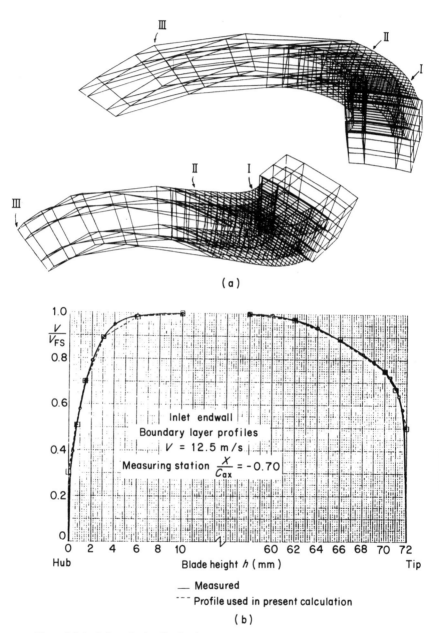

Figure 2.8.1 Inlet velocity distribution to the turbine passage. (a) Turbine passage geometry with part of the mesh; (b) inlet velocity distribution at hub and shroud walls

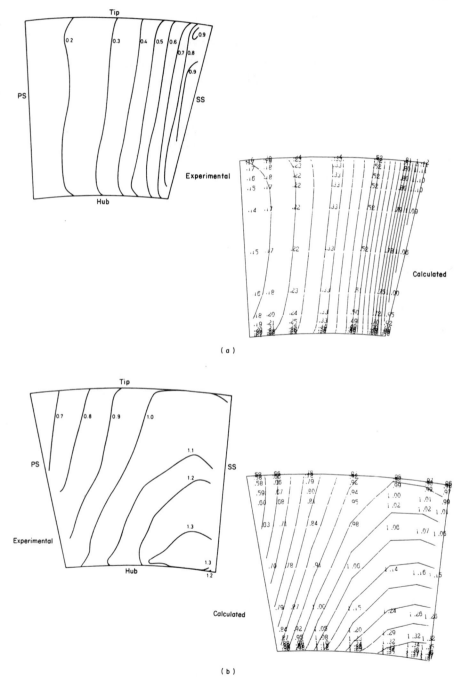

Figure 2.8.2 Isopressure contours at cross-sections I and II compared with experimental data from Sieverding (1982). (a) Section I; (b) section II. (From Lacor and Hirsch, 1984; Hirsch and Lacor, 1985)

the reduced velocity can be applied:

$$\vec{v}_s = \vec{\nabla}\phi + \psi\vec{\nabla}p_0 \qquad (2.8.28)$$

The two scalar functions ϕ and ψ satisfy the following equations, derived from equations (2.8.21) and (2.8.27):

$$\vec{\nabla} \cdot (\rho_s \vec{\nabla}\phi) = -\vec{\nabla} \cdot (\rho_s \psi \vec{\nabla}p_0) \qquad (2.8.29)$$

$$(\vec{v}_s \cdot \vec{\nabla})\psi \quad = -\frac{1}{\rho_{0_A}}\left(\frac{p_0}{p_{0_A}}\right)^{-1/\gamma} \qquad (2.8.30)$$

together with

$$(\vec{v}_s \cdot \vec{\nabla})p_0 = 0 \qquad (2.8.31)$$

$$(\vec{v}_s \cdot \vec{\nabla})H = 0 \qquad (2.8.32)$$

This is a completely general formulation for steady, non-rotating, inviscid rotational flows allowing a description with the aid of two scalar functions for the three scalar velocity components. Obviously choice (2.8.24) for the scaling function f_1 is not unique and many other choices are open. For instance, f_1 could also be chosen as a function of H only.

Incompressible fluid

For incompressible flows the scaling is actually not necessary since equation (1.4.7) reduces, with the continuity equations for incompressible flows,

$$\vec{\nabla} \cdot \vec{w} = 0 \qquad (2.8.33)$$

to the momentum equation

$$\vec{w} \times \vec{\zeta} = \frac{1}{\rho}\vec{\nabla}p^* \qquad (2.8.34)$$

where

$$p^* = p + \tfrac{1}{2}\rho(\vec{w}^2 - \vec{u}^2) \qquad (2.8.35)$$

is often called the *rotary stagnation pressure*. Obviously,

$$(\vec{w} \cdot \vec{\nabla})p^* = 0 \qquad (2.8.36)$$

and the following representation can be defined:

$$\vec{v} = \vec{\nabla}\phi + \psi\vec{\nabla}p^* \qquad (2.8.37)$$

The equations

$$\Delta\phi = -\vec{\nabla} \cdot (\psi\vec{\nabla}p^*) \qquad (2.8.38)$$

$$(\vec{w} \cdot \vec{\nabla})\psi = -\frac{1}{\rho} \qquad (2.8.39)$$

together with equation (2.8.36) form the complete set of equations for the three-dimensional, steady, inviscid, incompressible flow in a steadily rotating system. This flow system is therefore completely determined by the knowledge of the three scalar functions ϕ, ψ and p^*.

The above representations of three-dimensional stationary, rotational, inviscid flows have been applied by Lacor and Hirsch (1980, 1982), Chang and Adamczyck (1983) and Chaviaropoulos *et al.* (1986) to various internal flow problems. Ecer and Akay (1983) have applied representation (2.8.12) to two-dimensional, transonic potential flows in order to add non-isentropic contributions in the presence of shock discontinuities.

Practical example

Three-dimensional flow in a turbine passage   In turbine passages the influence of viscous boundary layers remains limited, and the above model provides an acceptable prediction of the secondary flows generated from non-uniform velocity profiles at inlet. Figure 2.8.1 displays the inlet velocity distribution to the turbine geometry shown in the same figure. The rotation of the isopressure surfaces due to the secondary flow can be seen from Figure 2.8.2 in two cross-sections, indicated I and II. The comparison with the experimental data shows a good correspondence with the calculated pressure field.

References

Cazal, P. (1966). 'Principes variationnels en fluide compressible et en magnéto-hydrodynamique.' *J. De Mécanique*, **5**, 149–66.

Chang, S. C., and Adamczyk, J. J. (1983). 'A semi-direct solver for compressible three-dimensional rotational flow.' *AIAA Paper 83-1909, Proceedings 6th AIAA Computational Fluid Dynamics Conference*, pp. 163–82.

Chaviaropoulos, P., Giannakoglou, K., and Papailiou, K. D. (1986). 'Numerical computation of three-dimensional rotation inviscid subsonic flows, using the decomposition of the flow field into a potential and a rotational part.' American Society of Mechanical Engineers, *ASME Paper 86-GT-169.*

Ecer, A., and Akay, H. V. (1983). 'A finite element formulation for steady transonic Euler equations.' *AIAA Journal*, **21**, 343–50.

Hawthorne, J. (1974) 'Secondary vorticity in stratified compressible fluids in rotating systems'. University of Cambridge, *Report CUED/A-Turbo/TR63.*

Hirsch, Ch., and Lacor, C. (1983). 'Three dimensional, inviscid, rotational flow calculations in turbomachines.' *Proc. 15th International Conference on Combustion Engines (CIMAC)*, pp. 287–313.

Hirsch, Ch., and Lacor, C. (1985). 'Computation of three dimensional, inviscid, rotational flows.' In *AGARD-LS 140* on *3D Computation Techniques Applied to Internal Flows in Propulsion Systems.*

Lacor, C., and Hirsch, Ch. (1980). 'Non-viscous, three-dimensional, rotational flow calculations in blade passages.' *Proc. 4th GAMM-Conference on Numerical Methods in Fluid Mechanics*, New York: Springer Verlag, pp. 150–61.

Lacor, C., and Hirsch, Ch. (1982). 'Rotational flow calculations in three dimensional blade passages.' American Society of Mechanical Engineers, *ASME Paper 82-GT-316.*

Lacor, C., and Hirsch, Ch. (1984). 'Three dimensional, inviscid, rotational flow calculations in turbomachines.' *Proc. of the Institute of Mechanical Engineers Conference on Computational Methods in Turbomachinery*, London: I. Mech. E. Conf. Publications.

Serrin, J. (1959). 'Mathematical principles of classical fluid mechanics.' In Flugge, S. (ed.), *Handbuch der Physik*, Vol. VIII/I, Berlin: Springer Verlag.

Sieverding, C. H. (1982). In *Numerical Methods for Flows in Turbomachinery*, Von Karman Institute Lecture Series 1982–5.

Yih, C. S. (1965). *Dynamics of Non-Homogeneous Fluids*, New York: Macmillan.

2.9 THE POTENTIAL FLOW MODEL

The most impressive simplification of the mathematical description of a flow system is obtained with the approximation of a non-viscous, *irrotational* flow. From

$$\vec{\zeta} = \vec{\nabla} \times \vec{v} = 0 \tag{2.9.1}$$

the three-dimensional velocity field can be described by a single scalar *potential* function, ϕ, defined by

$$\vec{v} = \vec{\nabla}\phi \tag{2.9.2}$$

reducing the knowledge of the three velocity components to the determination of a single potential function ϕ.

As seen from the preceding section, if the initial conditions are compatible with a uniform entropy, then for continuous flows, equation (2.8.1) implies that the entropy is constant over the whole flow field. Hence for isentropic flows the momentum equation becomes

$$\frac{\partial}{\partial t}(\vec{\nabla}\phi) + \vec{\nabla}H = 0 \tag{2.9.3}$$

or

$$\frac{\partial \phi}{\partial t} + H = \text{constant} = H_0 \tag{2.9.4}$$

the constant H_0 having the same value along all the streamlines.

This equation shows that the energy equation is no longer independent of the momentum equation, and therefore the flow will be completely determined by initial and boundary conditions, on the one hand, and by the knowledge of the single function ϕ on the other. This is a very considerable simplification indeed.

The equation for the potential function is obtained from the continuity equation, taking into account the isentropic conditions to express the density as a function of velocity and hence of the gradient of the potential function. We obtain the basic potential equation in conservation form:

$$\frac{\partial \rho}{\partial t} + \vec{\nabla} \cdot (\rho\,\vec{\nabla}\phi) = 0 \tag{2.9.5}$$

and the relation between density and potential function is obtained by introducing the definition of stagnation enthalpy as a function of velocity and static enthalpy h, for a perfect gas:

$$\frac{\rho}{\rho_A} = \left(\frac{h}{h_A}\right)^{1/(\gamma-1)} = \left[\left(H_0 - \frac{\vec{v}^2}{2} - \frac{\partial\phi}{\partial t}\right)/h_A\right]^{1/(\gamma-1)} \qquad (2.9.6)$$

The subscript A refers to an arbitrary reference state; for instance, the stagnation conditions $\rho_A = \rho_0$ and $h_A = H_0$.

Steady potential flows

A further simplification is obtained for *steady* potential flows. With $H = H_0 =$ constant the potential equation reduces to

$$\vec{\nabla} \cdot (\rho\,\vec{\nabla}\phi) = 0 \qquad (2.9.7)$$

with the density given by equation (2.9.6), where h_A can be chosen equal to H_0. Hence we have

$$\frac{\rho}{\rho_0} = \left(1 - \frac{(\vec{\nabla}\phi)^2}{2H_0}\right)^{1/(\gamma-1)} \qquad (2.9.8)$$

where ρ_0 is the stagnation density, constant throughout the whole flow field.

Both for steady and unsteady flows the boundary condition along a solid boundary is the condition of vanishing relative velocity between flow and solid boundary in the direction n normal to the solid wall:

$$v_n = \frac{\partial\phi}{\partial n} = \vec{u}_w \cdot \vec{l}_n \qquad (2.9.9)$$

where $\vec{u}_w$ is the velocity of the solid boundary with respect to the system of reference being considered.

2.9.1 Irrotational flow with circulation—Kutta–Joukowski condition

Although the local vorticity in the flow is zero for a potential flow in non-simply connected domains it may happen that the circulation around a closed curve C becomes non-zero. This is essentially the case for lifting airfoils. To achieve a non-zero lift on the body a circulation Γ around the airfoil is imposed. This circulation is represented by a free vortex singularity, although it originates from a vorticity production physically generated in the boundary layer. It follows that the value of Γ cannot be determined from irrotational theory and is an externally given value for a potential flow. It is also to be remembered that, with the addition of the free vortex singularity Γ, an infinity of different potential flows can be obtained for the same incident flow conditions, each of these solutions having another value of Γ. However, for aerodynamically shaped bodies such as airfoil profiles a fairly good approximation of the circulation, and hence the lift, may be obtained by the

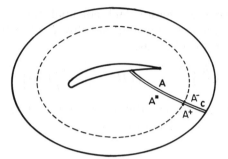

Figure 2.9.1 Potential jump over the cut C associated with a circulation

Kutta—Joukowski condition, provided that no boundary layer separation occurs in the physical flow. The Kutta–Joukowski condition states that the value of the circulation which approximates best to the real (viscous) unseparated flow is obtained if the stagnation point at the downstream end of the body is located at the trailing edge.

The non-zero circulation around the body requires the introduction of an artificial boundary or cut (C), emanating from the body to the farfield boundary (Figure 2.9.1) and over which a jump in the potential function is allowed. Calculating the circulation around the body for an arbitrary curve starting at a point A^- on the cut and ending in the corresponding point A^+ on the opposite side yields

$$\Gamma = \oint_{A^-A^+} \vec{v}\,d\vec{l} = \oint_{A^-A^+} \vec{\nabla}\phi \cdot d\vec{l} = \phi(A^+) - \phi(A^-) \qquad (2.9.10)$$

which is non-zero due to the circulation Γ. Since the potential jump is constant in each point of the cut the circulation for any closed curve not surrounding the body remains zero even when crossing the cut. Hence the flow remains irrotational. To satisfy mass conservation over the cut it is sufficient to require continuity of the normal derivatives of the potential, since all flow variables, particularly the density ρ, depend only on velocity $\vec{v}$ and inlet stagnation conditions. Thus the cut can be interpreted as a periodic boundary with conditions, for any point on the cut C,

$$\phi(A^+) = \phi(A^-) + \Gamma$$
$$\frac{\partial \phi}{\partial n}(A^+) = \frac{\partial \phi}{\partial n}(A^-) \qquad (2.9.11)$$

where n is the direction normal to the cut.

2.9.2 The limitations of the potential flow model for transonic flows

If we consider the steady-state potential model for continuous flows the constancy of entropy and total enthalpy, coupled to irrotationality, form a set of conditions fully consistent with the system of Euler equations. Hence the model defined by

$$s = s_0 = \text{constant}$$
$$H = H_0 = \text{constant} \tag{2.9.12}$$

and $\vec{v} = \vec{\nabla}\phi$ or $\vec{\nabla} \times \vec{v} = 0$, where ϕ is solution of the mass conservation equation, ensures that the momentum and energy conservation laws are also satisfied. Therefore it can be considered that an inviscid continuous flow, with initial conditions satisfying the above conditions (2.9.12), will be exactly described by the potential flow model.

However, in the presence of discontinuities such as shock waves this will no longer be the case since, as shown in Section 2.7, the Rankine–Hugoniot relations lead to an entropy increase through a shock. If the shock intensity is uniform then the entropy will remain uniform downstream of the shock but at a value other than the initial constant value. In this case, according to equation (2.7.5) the flow remains irrotational. However, if the shock intensity is not constant, which is most likely to occur in practice (for instance, for curved shocks), then equation (2.7.5) shows that the flow is no longer irrotational and hence the mere existence of a potential downstream of the discontinuity cannot be justified rigorously. Therefore the *potential flow model in the presence of shock discontinuities cannot be made fully compatible with the system of Euler equations*, since the potential model implies a constant entropy and has therefore no mechanisms to generate entropy variations over discontinuities. As will be seen next, the potential model allows shock discontinuities but with *isentropic* jump relations and, hence, the isentropic potential shocks will not satisfy all the Rankine–Hugoniot relations (2.7.14) or (2.7.15).

Potential shock relations

The irrotationality condition (2.9.2) can be integrated over the line segment $\overline{AB}$ of Figure 2.7.1, leading to

$$\int_{AB} \vec{v} \cdot d\vec{l} = \phi_B - \phi_A \tag{2.9.13}$$

where $d\vec{l}$ is the line element tangent to AB. Taking the limit for the distance going to zero, we obtain the jump condition

$$[\phi] = 0 \tag{2.9.14}$$

where the square brackets indicate the variation $[\phi] = \phi_B - \phi_A$ over the discontinuity. This relation states that the potential function remains continuous over a shock discontinuity.

Applying the same integration procedure over the closed contour ABCD of Figure 2.7.1 and expressing that no circulation is generated at the discontinuity surface Σ, we obtain

$$\oint_{ABCD} \vec{\nabla}\phi \cdot d\vec{l} = 0 \tag{2.9.15}$$

For the limits AB and CD tending to zero equation (2.9.15), together with equation (2.9.14), becomes, where $\vec{I}_l$ is the unit vector along AD,

$$[\vec{\nabla}\phi \cdot \vec{I}_l] = [\vec{v} \cdot \vec{I}_l] = 0 \tag{2.9.16}$$

expressing the continuity of the tangential velocity component at the discontinuity surface Σ. This is in agreement with the second equation (2.7.15b) projected in the tangential direction $\vec{I}_l$.

Finally, the potential equation (2.9.7), expressing mass conservation, can be treated as in Section 2.7, leading to

$$[\rho\,\vec{\nabla}\phi \cdot \vec{I}_n] = 0 \tag{2.9.17}$$

which is the first of the Rankine–Hugoniot relations (2.7.15a) for a steady shock.

The third Rankine–Hugoniot relation (2.7.15c) expresses the conservation of energy and hence is satisfied by the potential model, which assumes that the total energy is constant throughout the whole flow field. However, since the entropy is considered as constant everywhere in the potential model, and hence also over the discontinuity, this conflicts with equations (2.7.15) and, in particular, the normal projection of the momentum conservation equation (2.7.15b):

$$[\rho v_n \cdot v_n + p] = 0$$

will *not* be satisfied by the potential flows. Hence the isentropic potential model satisfies conservation of mass and energy but does not satisfy momentum conservation over a shock discontinuity. Actually, the difference in the momentum $[\rho v_n^2 + p]$ is used to estimate the drag due to the shock wave (Steger and Baldwin, 1972; see also Yu *et al.*, 1983, for a recent discussion on the numerical aspects connected with the inviscid drag computation).

Comparison between isentropic and Rankine–Hugoniot shock relations

The differences between the exact shock relations given by the Rankine–Hugoniot equations (2.7.15) and the isentropic shock relations implied by the potential flow model can best be illustrated for a one-dimensional normal shock (Figure 2.9.2). The exact normal shock relations (2.7.15) applied to the case of Figure 2.9.2 can be found in many textbooks (see, for instance, Shapiro, 1953; Zucrow and Hoffman, 1976).

The entropy increase over the normal shock, defined by equation (2.1.17),

$$\frac{\Delta s}{c_v} \equiv \frac{s_2 - s_1}{c_v} = \ln \frac{p_2/\rho_2^{\gamma}}{p_1/\rho_1^{\gamma}} \tag{2.9.18}$$

can be expanded as a function of the upstream Mach number M_1 as follows (Zucrow and Hoffman, 1976):

$$\frac{\Delta s}{c_v} = \frac{2}{3}\frac{\gamma(\gamma - 1)}{(\gamma + 1)^2}\,(M_1^2 - 1)^3 + 0[(M_1^2 - 1)^4] \tag{2.9.19}$$

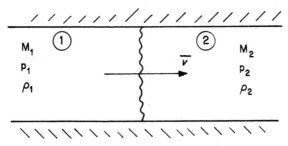

Figure 2.9.2 Normal shock configuration

Hence for sufficiently low supersonic upstream velocities the entropy increase will remain low and the irreversibility of the shock can be ignored.

A quantitative estimate of the error introduced by the isentropic shocks can be obtained from the computation of the stagnation pressure variations over the shock. From equation (2.1.25) and the energy conservation over the shock (equation (2.7.15c)) we have

$$\frac{\Delta s}{r} = -\ln\left(\frac{p_{02}}{p_{01}}\right) \tag{2.9.20}$$

with

$$\frac{p_{02}}{p_{02}} = \frac{\{[(\gamma+1)/2]\,M_1^2/[1+(\gamma-1)/2\cdot M_1^2]\}^{\gamma/(\gamma-1)}}{[2\gamma/(\gamma+1)\cdot M_1^2 - (\gamma-1)/(\gamma+1)]^{1/(\gamma-1)}} \tag{2.9.21}$$

for the exact shocks while for an isentropic shock the stagnation pressure remains constant. The total pressure loss $(p_{02}/p_{01} - 1)$ is plotted in Figure 2.9.3 as a function of M_1. For $M_1 \le 1.25$ the total pressure loss is lower than 2%. This is an acceptable error for technical applications using an inviscid model, since for higher incoming Mach numbers the level of accuracy of the inviscid model might become questionable due to the possible occurrence of strong shock–boundary layer interactions. Hence for Mach numbers upstream of the shock below a value of approximately 1.25 it might be acceptable to consider a constant stagnation pressure and hence constant entropy even in the presence of shocks.

Another view of the error introduced in the shock intensity by the isentropic assumption of potential theory is obtained by comparing the values of the Mach numbers downstream of the shock. The Rankine–Hugoniot relation for mass conservation (2.7.15a), leads to

$$\frac{M_2}{[1+(\gamma-1)/2\cdot M_2^2]^{(\gamma+1)/2(\gamma-1)}} = \frac{M_1}{[1+(\gamma-1)/2\cdot M_1^2]^{(\gamma+1)/2(\gamma-1)}}\frac{p_{01}}{p_{02}} \tag{2.9.22}$$

where (p_{02}/p_{01}) is given by equation (2.9.21).

The isentropic values of the Mach number downstream of the shock, $(M_2)_{is}$, which are obtained from a potential flow calculation are derived from the same

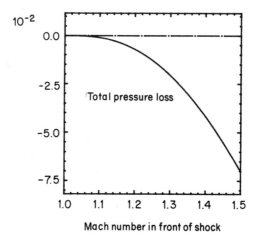

Figure 2.9.3 Stagnation pressure loss as a function of upstream Mach number for a normal shock

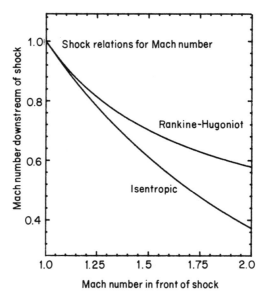

Figure 2.9.4 Comparison between Rankine–Hugoniot and isentropic Mach numbers downstream of the shock as a function of the incoming Mach number for a uniform normal shock

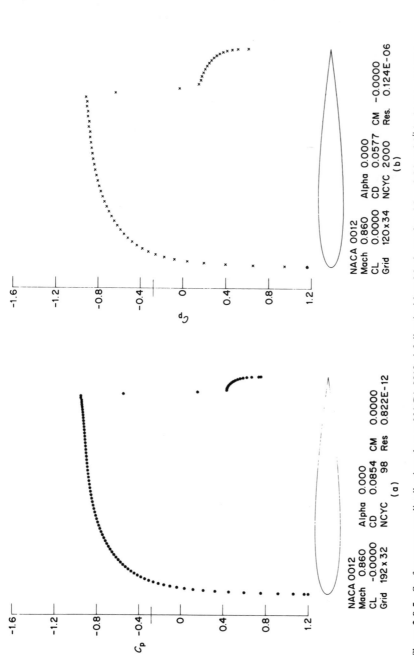

Figure 2.9.5 Surface pressure distribution along a NACA 0012 airfoil at incident Mach number $M_\infty = 0.86$ and $0°$ incidence. (a) Computed with a potential flow model; (b) computed with an Euler flow model. (Courtesy M. Salas, NASA Langley Research Center, USA)

equation (2.9.22) when the isentropic shock relation $(p_{01}/p_{02}) = 1$ is introduced. These relations are plotted in Figure 2.9.4. For $M_1 = 1.25$ an error of

$$\Delta M_2 = \frac{(M_2)_{is} - M_2}{M_1} \approx 0.03$$

is made, and it is to the user to decide, as a function of the overall level of accuracy required, if this is acceptable or not.

However, next to the error in the shock intensity, the isentropicity of the potential model also leads to shock positions which can be incorrectly located when compared with inviscid computations from Euler models. This can be seen in Figures 2.9.5 and 2.9.6 for a flow on a NACA 0012 airfoil at incident Mach number of $M_\infty = 0.86$ with $0°$ incidence and at $M_\infty = 0.8$ with $0.5°$ incidence.

This large error in shock position and hence on lift is a severe problem for potential flow models in the presence of shocks. An approximate cure to this problem can be achieved by the introduction of non-isentropic corrections to the basic potential model. An example (Klopfer and Nixon, 1983) illustrates the effect of these corrections for a computation on the same NACA 0012 airfoil at $M_\infty = 0.8$ and $1.25°$ incidence (Figure 2.9.7). More details will be given in the corresponding chapter of Volume 2, dealing with the comput-

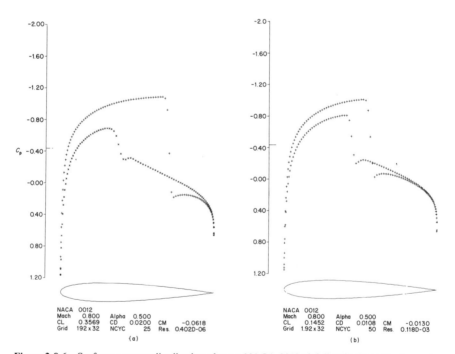

Figure 2.9.6 Surface pressure distribution along a NACA 0012 airfoil at incident Mach number $M_\infty = 0.80$ and $0.5°$ incidence. (a) Computed with a potential flow model; (b) computed with an Euler flow model. (Courtesy A. Jameson. Princeton University, USA)

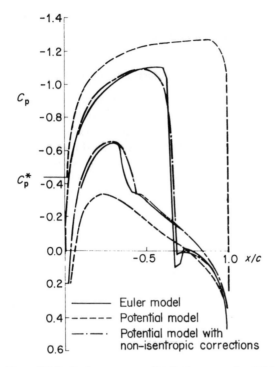

Figure 2.9.7 Surface pressure distribution around a NACA 0012 airfoil at $M_\infty = 0.80$ and $1.25°$ incidence, obtained by a potential model plus non-isentropic corrections. (Reproduced by permission of AIAA from Klopfer and Nixon, 1983.)

ational techniques for potential flows. However, the potential model remains an excellent approximation for inviscid flows, fully equivalent to the Euler model for irrotational conditions in the absence of shock discontinuities. This is illustrated by the following examples.

Supercritical airfoils

The development of supercritical airfoils, defined as having a shock-free transition from supersonic to subsonic surface velocities, is one of the most spectacular outcomes of the early developments of computational fluid dynamics. These airfoils are now of general use on civil aircrafts, allowing important savings on fuel costs due to the absence of the pressure drag produced by a shock. They are also being applied in axial flow compressors under the name of controlled diffusion blades.

Figure 2.9.8 shows the surface pressure distribution on the so-called Korn airfoil (Bauer *et al.*, 1975) and the isoMach lines for the design conditions of $M_\infty = 0.75$ and $0°$ incidence. The design conditions for shock-free transition represent a singular point, since the smallest perturbation will lead to the

118

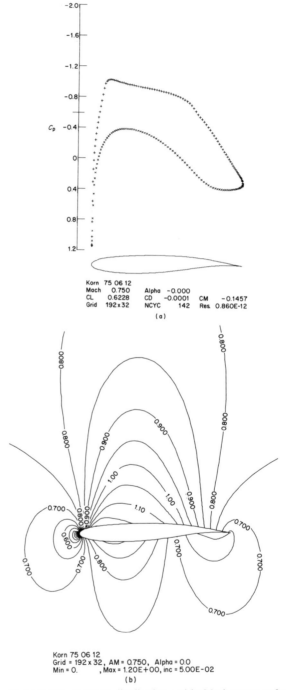

Figure 2.9.8 Pressure distribution and isoMach contours for the Korn supercritical airfoil at design conditions. $M_\infty = 0.75$ and $0°$ incidence. (Courtesy C. Gumbert and J. South, NASA Langley Research Center, USA)

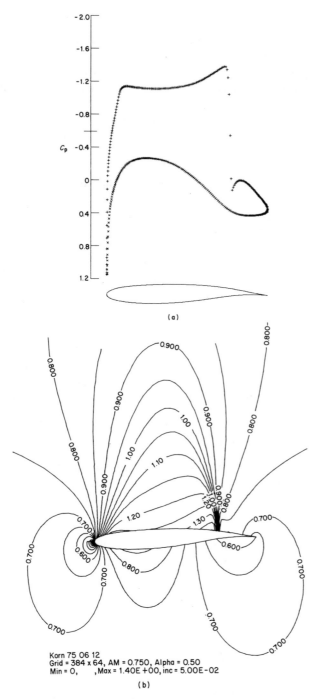

(a)

Korn 75 06 12
Grid = 384 x 64, AM = 0.750, Alpha = 0.50
Min = 0,   , Max = 1.40E +00, inc = 5.00E-02

(b)

Figure 2.9.9 Pressure distribution and isoMach contours for the Korn supercritical airfoil at off-design conditions, $M_\infty = 0.75$ and $0.5°$ incidence. (Courtesy C. Gumbert and J. South, NASA Langley Research Center, USA)

appearance of shocks. This is best illustrated by a calculation on the same airfoil with 0.5° incidence instead of the design value of 0°, the results of which are displayed in Figure 2.9.9. The strong shock can be noted as well as the sharp increase in pressure drag coefficient (from 0 to 0.108). These calculations were performed by C. Gumbert of NASA Langley Research Center with a very fine mesh of 284 × 64 points.

Subsonic potential flows

In the subsonic range the potential model has the same validity as the Euler model for uniform inflow conditions on a body since the flow remains irrotational in this case. An example is shown in Figure 2.9.10, where the inviscid potential flow through a compressor cascade of controlled diffusion blades is compared with experimental data, demonstrating excellent correspondence. Flow inlet conditions correspond to an incident Mach number of 0.22 and an inlet flow angle of 28°. The coarse mesh used in this computation by Schulz *et al.* (1984) is also shown, together with the isoMach contours.

2.9.3 The non-uniqueness of transonic potential models

The isentropic restriction of potential models in transonic flows has still another limitation, which can be very severe in certain cases of external flows as well as for internal transonic flows in channels, nozzles and cascades.

Non-uniqueness in internal flows

The limitations in transonic internal flows are linked with the existence of an infinite number of equally valid solutions for the same *isentropic outlet physical variables* such as back pressure or outlet Mach number.* This is best illustrated on a one-dimensional channel or nozzle problem. Considering a nozzle with varying cross-section $S(x)$, it is well known that, according to the level of the outlet pressure and for given inlet conditions, the flow can be either supersonic or subsonic. The mass flow per unit area

$$\left(\frac{\dot{m}}{S}\right) = \rho u \qquad (2.9.23)$$

can be expressed as a function of Mach number by

$$\rho u = \rho_0 \sqrt{(\gamma r T_0)} \frac{M}{[1 + (\gamma - 1)/2 \cdot M^2]^{(\gamma + 1)/2(\gamma - 1)}}$$

$$\equiv \rho_0 \sqrt{(\gamma r T_0)} F(M) \qquad (2.9.24)$$

---

* The non-uniqueness discussed here is the one remaining when the necessary precautions (entropy conditions) have been taken to remove the possible expansion shocks.

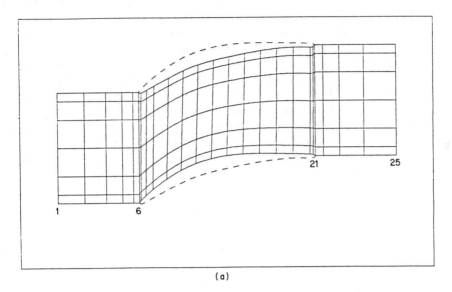

(a)

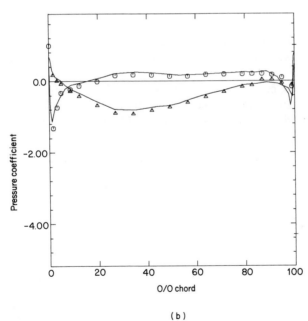

(b)

Figure 2.9.10 Mesh and surface pressures for the subsonic flow through a compressor cascade of controlled diffusion blades (a) Mesh and computational domain; (b) pressure distribution compared with experimental data (Reproduced by permission of the Department of the Navy from Schulz *et al.*, 1984.)

which is based on the isentropic assumption. A study of this function (Figure 2.9.11) shows that it attains an absolute maximum for $M = 1$, corresponding to the maximum mass flow the nozzle can allow for the given stagnation conditions p_0, T_0, the choking mass flow. For lower values of $F(M) < F(1)$ there are two solutions, of which one is subsonic and the other supersonic. Actually, the isentropic jump relation (equation (2.9.22)) with $p_{02} = p_{01}$ is given precisely by $F(M_1) = F(M_2)$. For subsonic inlet conditions the flow will accelerate to the throat and, if the outlet pressure is sufficiently high, will decelerate in the divergent part. When the outlet pressure is reduced sonic conditions will be reached in the throat and, depending on the outlet pressure, the flow can become supersonic in the divergent section (branch (A) of Figure 2.9.12) or remain subsonic and decelerate (branch (B) of Figure 2.9.12). In the first case a shock can occur which will bring the flow conditions from the supersonic branch (A) to the subsonic branch (B).

Due to the isentropic restriction all the different shock positions between the isentropic branches (A) and (B) will have the same outlet conditions M_B and corresponding pressure p_B. Therefore fixing the value of M_B (or p_B) will not determine uniquely the position of the shock of the transonic solution. This is a consequence of the fact that in the isentropic world there is no mechanism which connects the shock position to the outlet conditions. In the inviscid world governed by the Euler equations and the Rankine–Hugoniot shock relations, the two shocks of Figure 2.9.12 have different intensities and hence different entropy variations. Therefore they will correspond to different outlet pressures (or Mach numbers) and hence the physical outlet conditions uniquely determine the shock position. This is not the case for the isentropic potential model, and the same outlet value will allow an infinity of solutions with different shock positions and intensities defined by the jump between the two

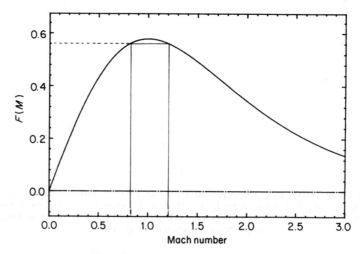

Figure 2.9.11 Mass flow per unit area

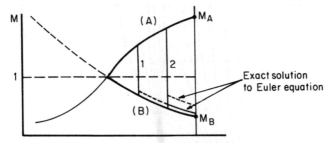

Figure 2.9.12 Shock transitions for Euler and potential flows

branches (A) and (B) of Figure 2.9.12. For instance, potential shocks 1 and 2 correspond to the same outlet Mach number M_B.

This non-uniqueness of the isentropic potential flow for given physical boundary conditions can be removed by a condition on the potential difference between inlet and outlet (Deconinck and Hirsch, 1983). Considering a model problem defined by a channel of constant section with uniform supersonic inlet velocity, a trivial solution is given by the uniform supersonic flow in the entire channel with Mach number equal to the inflow value. However, as seen in Figure 2.9.12, the same mass flow can also be passed with subsonic velocity.

More insight into the distinction between the different solutions is obtained by considering the variations of the potential ϕ as a function of distance, instead of the physical variables which are determined by the gradients of ϕ. Since the velocities are constant, the variations of ϕ with distance are linear with different slopes for the supersonic and subsonic branches (Figure 2.9.13). At the shock location the slope of the potential function changes and a unique exit value, ϕ_E, is obtained for each shock position. Therefore the value of the potential difference between inlet and exit will uniquely determine the shock position.

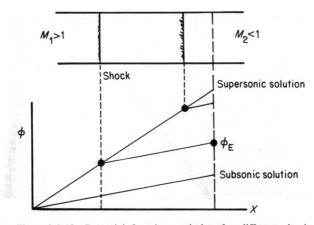

Figure 2.9.13 Potential function variation for different shock positions

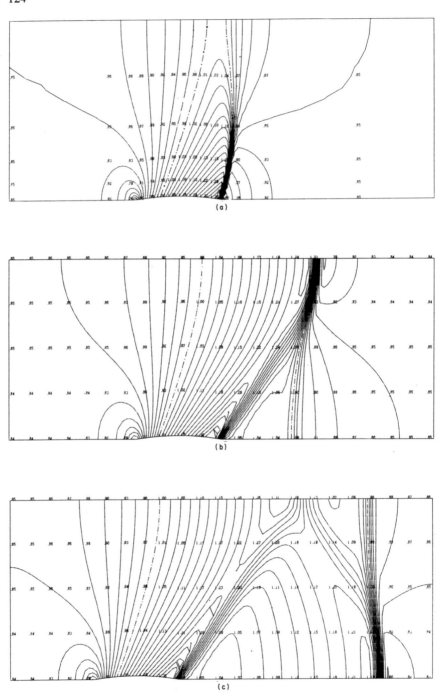

Figure 2.9.14 Isomach lines for choked channel flow. Potential solutions with different shock positions. (Reproduced with permission from Deconinck, 1983.)

If a streamline is followed from inlet to outlet the potential difference $(\phi_{out} - \phi_{in})$ fixes the circulation along the flow path according to equation (2.9.13), and therefore the circulation along the streamline completely determines the shock location.

Figure 2.9.14 shows the variety of solutions obtained in a channel flow for the same inlet and outlet physical conditions (from Deconinck, 1983). The three flow configurations shown in this figure have the same inlet and outlet isentropic Mach numbers of $M_\infty = 0.85$, but represent three different, but equally valid, potential flow solutions. However, compared with the exact inviscid Euler flow, the isentropic potential flow with the same shock position will not have the same exit pressure, or if the two flows have the same outlet parameters they will lead to different shock locations and intensities.

Non-uniqueness in external flows

A different non-uniqueness of the isentropic potential equation has been observed for external flows along airfoils. In a series of extremely careful computations Steinhoff and Jameson (1982) discovered multiple-flow configurations with different shock positions, for the same physical boundary conditions of incident flow angles and Mach numbers. Those unexpected and surprising results were further analysed by Salas *et al.* (1983, 1985).

This non-uniqueness has also been observed by Glowinski *et al.* (1984) with a finite element potential method based on a completely different numerical approach. Therefore there seems to be no doubt that the numerically observed non-uniqueness indeed corresponds to multiple solutions of the isentropic potential equation. For certain regions of incident Mach numbers, three or more different isentropic solutions, corresponding to different values of the circulation around the airfoil, were obtained for the same incident angle.

Figure 2.9.15 illustrates the computed non-unique solutions around a two-dimensional symmetrical NACA 0012 airfoil. A symmetrical (zero lift) and a non-symmetrical solution are found at zero incidence. In addition, as can be seen from the lift incidence curve in Figure 2.9.16, within the range of Mach numbers where non-unique solutions are found three different flow configurations are possible for the same incidence angle. Also in Figure 2.9.16 results from a full Euler model are shown, indicating no presence of non-unique, non-physical solutions. Note that in the non-uniqueness region none of the solutions has a physical significance, since the lift incidence angle curve has not the correct, physical slope.

The authors selected a given potential solution by fixing the circulation and iteratively allowing the incidence angle to adapt in order to satisfy the Kutta–Joukowski condition at the trailing edge. In this way the shock conditions are uniquely determined, in agreement with the approach for internal flows mentioned above. Note that within the same conditions non-isentropic computations of the same flows, by solutions of the Euler equations (or even by a non-conservative potential model) did produce unique solutions

126

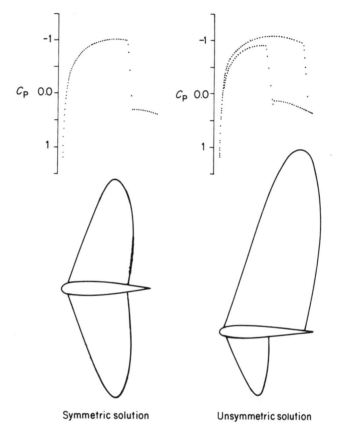

Mach = 0.832 Alph = 0.0

Symmetric solution   Unsymmetric solution

Figure 2.9.15 Non-unique solutions of the conservative potential
flow models for a symmetrical NACA 0012 airfoil. (Reproduced by
permission of AIAA from Steinhoff and Jameson, 1982)

for all incident flow conditions. More systematic computations (Salas and
Gumbert 1985) seem to indicate that a non-uniqueness region of incidence
angles exists for all supercritical values of Mach numbers for a given airfoil.
This raises an uncertainty with respect to transonic isentropic potential flow
models.

2.9.4 The small-disturbance approximation of the potential equation

In steady or unsteady transonic flow around wings and airfoils with thickness
to chord ratios of a few per cent we can generally consider that the flow is
predominantly directed along the chordwise direction, taken as the x-direc-
tion. In this case the velocities in the transverse direction can be neglected and
the potential equation reduces to the so-called small-disturbance potential

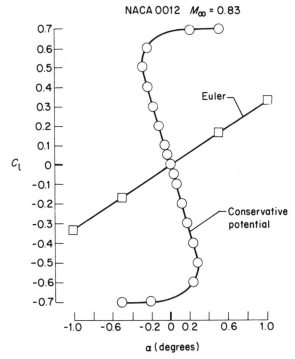

Figure 2.9.16 Computed variation of lift coefficient with incidence angle in the region of non-unique solutions for the potential flow around a NACA 0012 airfoil at $M_\infty = 0.83$. For comparison, the results from an Euler calculation are indicated. (Courtesy M. Salas, NASA Langley Research Center, USA)

equation:

$$(1 - M_\infty^2)\phi_{xx} + \phi_{yy} + \phi_{zz} = \frac{1}{a^2} (\phi_{tt} + 2\phi_x\phi_{xt}) \qquad (2.9.25)$$

Historically, the steady-state, two-dimensional form of this equation was used by Murman and Cole (1971) to obtain the first numerical solution for a transonic flow around an airfoil with shocks.

2.9.5 Linearized potential flows—singularity methods

If the flow can be considered as incompressible the potential equation becomes a linear Laplace equation for which many standard solution techniques exist. One of these, based on the linearity of the equation, is the singularity method, whereby a linear superposition of known elementary flow fields such as vortex and source singularities are defined. The unknown coefficients of this linear superposition are obtained by stipulating that the resultant velocity field satisfies the condition of vanishing normal velocity along solid body surfaces (in the absence of wall suction or blowing).

The three-dimensional extension of the singularity method (the panel method) has been widely used in the aeronautical industry in order to compute the three-dimensional flow field around complex configurations. The method is still in use and, although extensions to handle compressibility and transonic regimes can be developed, these methods are best replaced, for high-speed flows, by higher approximations such as the non-linear potential model and the Euler equations for the inviscid flow description. We will therefore omit any detailed discussion of this approach as the reader will find this information in the specialized literature.

2.10 SUMMARY

Different flow models, involving various degrees of approximation, have been defined and illustrated by a variety of examples. With the exception of laminar flows, which can be resolved by the Navier–Stokes model with the addition of empirical information on the dependence of viscosity and heat conductivity coefficients, all other models are limited by either empirical knowledge about

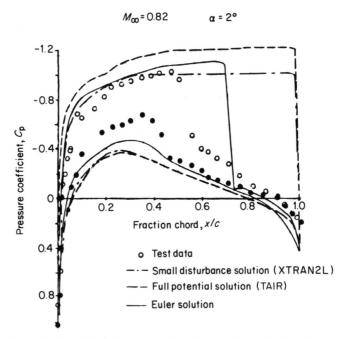

Figure 2.10.1 Comparison of calculations with small disturbance, potential Euler models for the NACA 0012 airfoil at $M_\infty = 0.82$ and 2° incidence and experimental data. (Courtesy A. Verhoff, McDonnell Aircraft Co., USA)

turbulence, as for the Reynolds-averaged Navier–Stokes equations, or by some approximations. Thin shear layer models are valid if no severe viscous separated regions exist and, similarly, the parabolized Navier–Stokes models for stationary formulations are limited by the presence of streamwise separation.

Inviscid flow models provide a valid approximation far from solid walls or when the influence of boundary layers can be neglected and, although the isentropic potential flow model is of questionable accuracy in transonic flows with shocks, it remains a valid and economical model for subsonic and

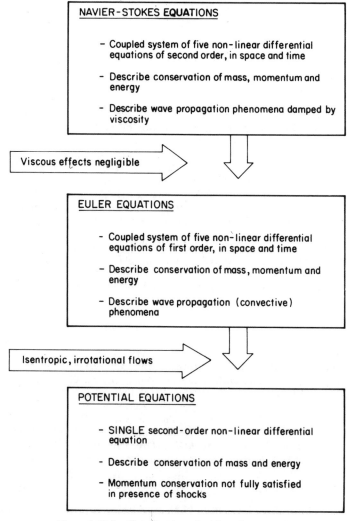

Figure 2.10.2 Classification of various flow models

130

shock-free supercritical flows. In addition, the introduction of non-isentropic corrections will lead to shock predictions close to those obtained with Euler models.

This should be kept in mind in the selection of a flow model, and the limits of validity have to be established for each family of applications by comparison with experimental data or with computations from a higher-level model. For instance, Figure 2.10.1 shows that for the flow on a NACA 0012 airfoil at $2°$ incidence viscous corrections are required to fit the experimental data, while this requirement is not so severe for the examples shown in Figures 2.7.4, 2.7.9 or 2.9.10. Figure 2.10.2 summarizes the interrelation between the various models in decreasing order of complexity.

References

Barton, J. T., and Pulliam, T. H. (1984). 'Airfoil computation at high angles of attack, inviscid and viscous phenomena.' *AIAA Paper 84-0524*, AIAA 22nd Aerospace Sciences Meeting.

Bauer, F., Garabedian, P., Korn, D., and Jameson, A. (1975). *Supercritical Wing Sections II*, Lecture Notes in Physics, Vol. 108, New York: Springer Verlag.

Deconinck, H. (1983). 'The numerical computation of transonic potential flows.' PhD Thesis, Vrije Universiteit Brussel, Dept of Fluid Mechanics, Brussels, Belgium.

Deconinck, H., and Hirsch, Ch. (1983). 'Boundary conditions for the potential equation in transonic internal flow calculations.' *Paper ASME-83-GT-135*, 26th International ASME Gas Turbine Conference.

Flores, J., Barton, J., Holst, T., and Pulliam, T. (1985). 'Comparison of the full potential and Euler formulations for computing transonic airfoil flows.' *Proc. 9th Int. Conf. on Numerical Methods in Fluid Dynamics*, Lecture Notes in Physics, Vol. 218, Berlin: Springer Verlag.

Fujii, K., and Obayashi, S. (1985). 'Evaluation of Euler and Navier–Stokes solutions for leading edge and shock-induced separation.' *AIAA Paper 85-1563*, AIAA 18th Fluid Dynamics, Plasmadynamics and Lasers Conference.

Klopfer, G. H., and Nixon, D. (1983). 'Non isentropic potential formulation for transonic flows.' *AIAA Paper 83-0375*, AIAA 21st Aerospace Sciences Meeting, Reno.

Murman, E. M., and Cole, J. D. (1971). 'Calculation of plane steady transonic flows'. *AIAA Journal*, **9**, 114–21.

Salas, M. D., Jameson, A., and Melnik, R. E. (1983). 'A comparative study of the non-uniqueness problem of the potential equation.' *AIAA Paper 83-1888, Proc. AIAA 6th Computational Fluid Dynamics Conf.*, pp. 48–60.

Salas, M. D., and Gumbert, C. R. (1985). 'Breakdown of the conservative potential equation.' *AIAA Paper 85-0367*, AIAA 23rd Aerospace Sciences Meeting.

Schulz, H. D., Neuhoff, F., Hirsch, Ch., and Shreeve, R. P (1984). 'Application of finite element code Q3DFLO-81 to turbomachinery flow fields.' *Naval Postgraduate School Technical Report NPS67-84-005PR*, December 1987.

Shapiro, A. H. (1953). '*The dynamics and thermodynamics of compressible fluid flow*'. New York: Ronald Press.

Steger, J. L., and Baldwin, B. S. (1972). 'Shock waves and drag in the numerical calculation of isentropic transonic flow.' *NASA TN-D-6997*.

Steinhoff, J., and Jameson, A. (1982). 'Multiple solutions of the transonic potential flow equation. *AIAA Journal*, **20**, 1521–5.

Yu, N. J., Chen, H. C., Samant, A. A., and Rubbert, P. E. (1983). 'Inviscid drag calculations for transonic flows.' *AIAA Paper 83-1928, Proc. AIAA 6th Computational Fluid Dynamics Conference*, pp. 283–92.

Zucrow, M. J., and Hoffman, J. D. (1976). *Gas Dynamics*, New York: John Wiley.

PROBLEMS

Problem 2.1

By developing explicitly the shear stress gradient and the momentum terms derive equations (2.1.7).

Problem 2.2

By using the definition of the shear stress tensor (equation (1.3.2)) work out the full, explicit form of the Navier–Stokes equations for non-constant viscosity coefficients, as a function of velocity components, in Cartesian co-ordinates. Show also that in the case of constant viscosity the equations reduce to the projections of equation (1.3.9).

Hint: Applying equation (1.3.2) we have, in condensed derivative notation,

$$\tau_{xx} = \tfrac{4}{3}\mu(\partial_x u) - \tfrac{2}{3}\mu(\partial_y v + \partial_z w)$$
$$\tau_{xy} = \mu(\partial_x v + \partial_y u)$$
$$\tau_{xz} = \mu(\partial_x w + \partial_z u)$$

and the x-projection of the momentum equation becomes

$$\frac{\partial}{\partial t}(\rho u) + \frac{\partial}{\partial x}(\rho u^2 + p) + \frac{\partial}{\partial y}(\rho uv) + \frac{\partial}{\partial z}(\rho uw) = \frac{\partial}{\partial x}\tau_{xx} + \frac{\partial}{\partial y}\tau_{xy} + \frac{\partial}{\partial z}\tau_{xz}$$

Problem 2.3

Derive the energy conservation equation for a three-dimensional incompressible flow in the presence of gravity forces.

Hint: Apply equation (1.3.13) to the momentum equation (2.1.29) and multiply scalarly by $\vec{v}$. Introducing the total energy $H = p/\rho + \vec{v}^2/2 + gz$, where z is the vertical co-ordinate, proof the Bernoulli equation:

$$\frac{\partial}{\partial t}\frac{\vec{v}^2}{2} + (\vec{v}\cdot\vec{\nabla})H = \nu\vec{v}\cdot\Delta\vec{v}$$

Problem 2.4

Proof equations (2.3.5)–(2.3.7)

Problem 2.5

Obtain the momentum equation for the z-component of the velocity, w, in the parabolized Navier–Stokes approximation, following the lines leading to equation (2.4.7) for the y-component v.

Problem 2.6

Obtain a Poisson equation for the pressure by taking the divergence of the 'parabolized' momentum equations (2.4.6) and (2.4.7) and the similar equation for w, taking the compressibility into account.

Problem 2.7

Write the equations for the distributed loss model (Section 2.6) in conservative, time-dependent form.

Problem 2.8

Proof equation (2.7.7) by expressing that the volume $d\Omega$ remains unchanged in the translation with speed $\vec{C}$.

Hint: Referring to Figure 2.7.1, the two sides AD and BC of the control volume ABCD have the same displacement velocity, hence $d(d\Omega)/dt = 0$.

Problem 2.9

Proof equation (2.8.23).

Hint: Calculate explicitly the scaled vorticity ζ_s; replace the entropy gradient by $\vec{\nabla}s = ds/df_1 \cdot \vec{\nabla}f_1$ and take into account that fact that $\vec{w} \cdot \vec{\nabla}f_1 = 0$ because of equation (2.8.7). Show also, for choice (2.8.24), that $f_1 \cdot ds/df_1 = -2c_p$.

Problem 2.10

Proof equations (2.8.27) and (2.8.30).

Problem 2.11

Show that a non-rotating system, the representation $\vec{v} = \vec{\nabla}\phi + p_0 \cdot \vec{\nabla}\psi$ is equivalent to representation (2.8.28).

Hint: Obtain the relation

$$\vec{v}_s \cdot \vec{\nabla}\psi = \frac{1}{\rho_{0A}} \left(\frac{p_0}{p_{0A}}\right)^{-1/\gamma} = \frac{1}{\rho_0}$$

Problem 2.12

By working out explicitly the gradients of specific mass as a function of the velocities, show that the potential equation can be written in the quasi-linear form as a function of the Mach numbers $M_i = v_i/c$:

$$(\delta_{ij} - M_i M_j) \frac{\partial^2 \phi}{\partial x_i \partial x_j} = \frac{1}{c^2} \left[\frac{\partial^2 \phi}{\partial t^2} + \frac{\partial}{\partial t}(\vec{\nabla}\phi)^2\right]$$

with a summation on the Cartesian subscripts $i, j = 1, 2, 3$ or x, y, z. Show that in two dimensions the potential equation reduces to

$$\left(1 - \frac{u^2}{c^2}\right)\frac{\partial^2 \phi}{\partial x^2} - \frac{2uv}{c^2}\frac{\partial^2 \phi}{\partial x \partial y} + \left(1 - \frac{v^2}{c^2}\right)\frac{\partial^2 \phi}{\partial y^2} = 0$$

Hint: Apply the isentropic laws and the energy equation (2.9.4) to derive

$$dh = c^2 \, d\rho/\rho$$

$$\frac{c^2}{\rho}\frac{\partial \rho}{\partial t} = -\frac{\partial^2 \phi}{\partial t^2} - \frac{\partial}{\partial t}\frac{(\vec{\nabla}\phi)^2}{2}$$

$$\frac{c^2}{\rho}\vec{\nabla}\rho = -\frac{\partial(\vec{\nabla}\phi)}{\partial t} - \frac{\vec{\nabla}(\vec{\nabla}\phi)^2}{2}$$

where c is the speed of sound, and substitute into equation (2.9.5).

Chapter 3

The mathematical nature of the flow equations and their boundary conditions

3.1 INTRODUCTION

The mathematical models of the various approximations defined in the previous chapters can all be classified as first-order or, at most, second-order systems of quasi-linear partial differential equations. Their mathematical properties are directly connected to the physical properties of the flow. As discussed in Chapter 1, any flow configuration is the outcome of a balance between the effects of convective fluxes, diffusive fluxes and the external or internal sources. The various approximation levels which have been defined can be considered as resulting from *a priori* estimates of the relative influence and balance between the contributions of these various fluxes and sources.

From a mathematical point of view, the diffusive fluxes appear through second-order derivative terms as a consequence of the generalized Fick's law, equation (1.1.6), which expresses the essence of the molecular diffusion phenomenon as a tendency to smooth out gradients. The convective fluxes, on the other hand, appear as first-order derivative terms and express the transport properties of a flow system. Therefore each of these contributions will influence the mathematical nature of the equations, particularly the competition between the elliptic, parabolic and hyperbolic character of the systems of equations describing the approximation level being considered.

Before presenting a more rigorous mathematical description it might be useful to illustrate these connections in a more straightforward way. Consider one projection of the Navier–Stokes equations, say the x-component of the momentum equation, in a Cartesian system and a laminar, incompressible flow under the form

$$\rho\frac{\partial u}{\partial t} + \rho(\vec{v}\cdot\vec{\nabla})u = -\frac{\partial p}{\partial x} + \mu\Delta u \qquad (3.1.1)$$

If all variables are non-dimensionalized through a reference length L for the space co-ordinate, a time scale T for the time co-ordinate, a velocity scale V for the velocity field and (ρV^2) for the pressure, we obtain, keeping the same notation for all variables now considered as *non-dimensionalized*.

$$\frac{VT}{L}\frac{\partial u}{\partial t} + (\vec{v} \cdot \vec{\nabla})u = -\frac{\partial p}{\partial x} + \frac{1}{Re}\Delta u \qquad (3.1.2)$$

where Re is the Reynolds number, defined by

$$Re = \frac{\rho VL}{\mu} = \frac{VL}{\nu} \qquad (3.1.3)$$

For very small values of the Reynolds number, that is, for strongly viscous dominated flows, the convection terms can be neglected with respect to the viscous terms, and we obtain the Stokes equation

$$-\frac{V^2 T}{\nu}\frac{\partial u}{\partial t} + \Delta u = Re\left(\frac{\partial p}{\partial x}\right) \qquad (3.1.4)$$

This equation is purely of an elliptic type in the steady-state case for a fixed pressure gradient, but parabolic in the unsteady case due to the Laplace operator on the left-hand side. Actually, the Laplace equation (or the Poisson equation) can be considered as the standard form of an elliptic equation describing an isotropic *diffusion* in all space directions.

At the other end, at very high Reynolds numbers and outside the boundary layers, the viscous terms have a negligible influence on the flow field, which is then dominated by the non-viscous transport terms describing the effect of the convective fluxes. Hence the equation reduces to the Euler equation

$$\frac{\partial u}{\partial t} + (\vec{v} \cdot \vec{\nabla})u = -\frac{1}{\rho}\frac{\partial p}{\partial x} \qquad (3.1.5)$$

which in a one-dimensional space takes the form

$$\frac{\partial u}{\partial t} + u\frac{\partial u}{\partial x} = -\frac{1}{\rho}\frac{\partial p}{\partial x} \qquad (3.1.6)$$

and is a basic hyperbolic equation in space and time describing a *propagation* phenomenon.

This distinction is of paramount importance, since the numerical discretization and solution methods will have to take into account the differences between phenomena as distinctive and far apart in their physical behaviour as diffusion and propagation. The former property is essentially independent of the flow direction acting in all directions and in the whole space domain, while the latter is essentially direction dominated and acts in specific regions of space defined by the wave-propagation directions.

Between these two extremes the parabolic type of equations (in space and time) for a time-dependent, diffusion-dominated system

$$\left(\frac{VT}{L}Re\right)\frac{\partial u}{\partial t} = \Delta u \qquad (3.1.7)$$

represents an intermediate situation between hyperbolic and elliptic. This equation, which reduces to a pure diffusive process in the steady state, describes a diffusion effect propagating in all space directions but damped in time. Hence the system of time-dependent Navier–Stokes equation is essentially parabolic in time and space, although the continuity equation has a hyperbolic structure. Therefore they are considered as parabolic–hyperbolic. For the same reason, the steady-state form of the Navier–Stokes equations leads to elliptic–hyperbolic properties.

3.2 THE CONCEPT OF CHARACTERISTIC SURFACES AND WAVE-LIKE SOLUTIONS

The systems of partial differential equations (PDE) describing the various levels of approximation discussed in Chapter 2 are quasi-linear and, at most, second order. It can be shown, however, that any second-order equation, or system of equations, can be transformed into a first-order system. Although this transformation is not unique and could lead to an artificially degenerate system, it will be considered that an appropriate transformation has been defined such that the system of first order represents correctly the second-order equations.

The classification of the flow equations is connected to the mathematical concept of characteristics, which can be defined as families of surfaces or hypersurfaces in a general three-dimensional unsteady flow, along which certain properties remain constant or certain derivatives can become discontinuous. The discussion of these properties can be found in many textbooks, and we refer to Courant and Hilbert (1962) for a mathematical presentation. We will give here the preference to a more 'physical' presentation of the structure of PDEs and of the associated concept of characteristic surfaces.

A system of quasi-linear partial differential equations of the first order will be called hyperbolic if its homogeneous part admits wave-like solutions. This implies that an hyperbolic set of equations will be associated with propagating waves and that the behaviour and properties of the physical system described by these equations will be dominated by wave-like phenomena. On the other hand, if the equations admit solutions corresponding to damped waves the system will be called parabolic, and if it does not admit wave-like solutions the equations are said to be elliptic. In this case the behaviour of the physical system being considered is dominated by diffusion phenomena.

3.2.1 Partial differential equation of second order

These different concepts are best introduced through the classical example of the quasi-linear partial differential equation of second order:

$$a \frac{\partial^2 \phi}{\partial x^2} + 2b \frac{\partial^2 \phi}{\partial x \partial y} + c \frac{\partial^2 \phi}{\partial y^2} = 0 \qquad (3.2.1)$$

where a, b and c can depend on the co-ordinates x and y, the function ϕ and its first derivatives. This equation can be written as a system of first-order equations, after introduction of the variables u and v defined by

$$u = \frac{\partial \phi}{\partial x} \qquad v = \frac{\partial \phi}{\partial y} \qquad (3.2.2)$$

Equation (3.2.1) is then equivalent to the following system:

$$a \frac{\partial u}{\partial x} + 2b \frac{\partial u}{\partial y} + c \frac{\partial v}{\partial y} = 0$$

$$\frac{\partial v}{\partial x} - \frac{\partial u}{\partial y} = 0 \qquad (3.2.3)$$

which can be written in matrix form:

$$\begin{vmatrix} a & 0 \\ 0 & 1 \end{vmatrix} \frac{\partial}{\partial x} \begin{vmatrix} u \\ v \end{vmatrix} + \begin{vmatrix} 2b & c \\ -1 & 0 \end{vmatrix} \frac{\partial}{\partial y} \begin{vmatrix} u \\ v \end{vmatrix} = 0 \qquad (3.2.4)$$

Introducing the vector U and the matrices A^1 and A^2,

$$U = \begin{vmatrix} u \\ v \end{vmatrix} \qquad A^1 = \begin{vmatrix} a & 0 \\ 0 & 1 \end{vmatrix} \qquad A^2 = \begin{vmatrix} 2b & c \\ -1 & 0 \end{vmatrix} \qquad (3.2.5)$$

equation (3.2.4) is written as

$$A^1 \frac{\partial U}{\partial x} + A^2 \frac{\partial U}{\partial y} = 0 \qquad (3.2.6)$$

A simple plane wave solution, propagating in the direction $\vec{n}$, is sought of the form

$$U = \hat{U} e^{I(\vec{n} \cdot \vec{x})} = \hat{U} e^{I(n_x x + n_y y)} \qquad (3.2.7)$$

where $I = \sqrt{-1}$.

Equation (3.2.6) will have solutions of form (3.2.7) if the homogeneous system

$$(A^1 n_x + A^2 n_y)\hat{U} = 0 \qquad (3.2.8)$$

admits non-trivial solutions. This will be the case when the determinant of the matrix $(A^1 n_x + A^2 n_y)$ vanishes, that is if

$$\det | A^1 n_x + A^2 n_y | = 0 \qquad (3.2.9a)$$

or

$$\begin{vmatrix} a n_x + 2b n_y & c n_y \\ -n_y & n_x \end{vmatrix} = 0 \qquad (3.2.9b)$$

Hence from the roots of

$$a \left(\frac{n_x}{n_y}\right)^2 + 2b \left(\frac{n_x}{n_y}\right) + c = 0 \qquad (3.2.10)$$

the well-known conditions defining the type of the second-order quasi-linear partial differential equation (3.2.1) are obtained.

Solution (3.2.7) will represent a true wave if n_y is real for all real values of n_x. Therefore if $(b^2 - ac)$ is positive there are two wave-like solutions, and the equation is hyperbolic, while for $(b^2 - ac) < 0$ the two solutions are complex conjugate and the equation is elliptic. When $(b^2 - ac) = 0$ the two solutions are reduced to one single direction $n_x/n_y = b/2a$ and the equation is parabolic.

3.2.2 Wave front or characteristic surfaces

A more general representation of *non-linear* wave propagation consists of defining a wavefront surface, which separates the points already influenced by the propagating disturbance from the points not yet reached by the wave. If $S(x, y) = S_0$ (where S_0 is a constant) is such a surface (also called the phase of the wave) a solution of the form

$$U = \hat{U} e^{IS(x,y)}$$ (3.2.11)

represents a general wave.

A system of equations is hyperbolic if equation (3.2.11) is a solution for real values of $S(x, y)$. Hence introducing this solution into equation (3.2.6) leads to the condition

$$\det | A^1 S_x + A^2 S_y | = 0$$ (3.2.12)

where the notation S_x and S_y are used to denote the partial derivatives of S with respect to x and y. This condition is identical to equation (3.2.9) if the normal $\vec{n}$ to the surface $S(x, y)$ is interpreted as the propagation vector in representation (3.2.7). That is, if

$$\vec{n} = \vec{\nabla} S$$ (3.2.13)

The surfaces $S(x, y)$, which satisfy equation (3.2.12) for real values of S, are called *characteristic* surfaces, and the directions $\vec{n}$ obtained from equation (3.2.10) are the normals to the characteristic surfaces.

If equation (3.2.13) is introduced into the wave form (3.2.7) a general non-linear wave representation is defined, equivalent to solution (3.2.11), as

$$U = \hat{U} e^{I(\vec{x} \cdot \vec{\nabla} S)} = \hat{U} e^{I(xS_x + yS_y)}$$ (3.2.14)

By definition, certain properties are transported along the surface $S(x, y)$ and the vectors tangent to the characteristic surface are obtained by expressing that along the wave front:

$$dS = \vec{\nabla} S \cdot d\vec{x} = \frac{\partial S}{\partial x} dx + \frac{\partial S}{\partial y} dy = 0$$ (3.2.15)

Hence the direction of the characteristic surface (a line in two dimensions) is given by

$$\frac{dy}{dx} = -\frac{S_x}{S_y} = -\frac{n_x}{n_y}$$ (3.2.16)

Example 3.2.1   Stationary potential equation

An interesting example is provided by the stationary potential flow equation in two dimensions x, y (see Problem 2.12), where c_0 designates here the speed of sound:

$$\left(1 - \frac{u^2}{c_0^2}\right)\frac{\partial^2\phi}{\partial x^2} - \frac{2uv}{c_0^2}\frac{\partial^2\phi}{\partial x\partial y} + \left(1 - \frac{v^2}{c_0^2}\right)\frac{\partial^2\phi}{\partial y^2} = 0 \qquad (E3.2.1)$$

With

$$a = 1 - \frac{u^2}{c_0^2}; \quad b = -\frac{uv}{c_0^2}; \quad c = 1 - \frac{v^2}{c_0^2} \qquad (E3.2.2)$$

we can write the potential equation under the form (3.2.1). In this particular case the discriminant $(b^2 - ac)$ becomes, introducing the Mach number M,

$$b^2 - ac = \frac{u^2 + v^2}{c_0^2} - 1 = M^2 - 1 \qquad (E3.2.3)$$

and hence the stationary potential equation is *elliptic for subsonic flows* and *hyperbolic for supersonic flows*. Along the sonic line $M = 1$ the equation is parabolic. This mixed nature of the potential equation has been a great challenge in the numerical computation of transonic flows since the transition line between the subsonic and the supersonic regions is part of the solution. An additional complication arises from the presence of shock waves which are discontinuities of the potential derivatives and which can arise in the supersonic regions. The particular problems of transonic potential flow with shocks and their numerical treatment will be discussed in Volume 2.

The small-disturbance potential equation   If the vertical velocity component is negligible (for instance, for a flow along a thin body) the stationary potential equation reduces to the form

$$(1 - M_\infty^2)\frac{\partial^2\phi}{\partial x^2} + \frac{\partial^2\phi}{\partial y^2} = 0 \qquad (E3.2.4)$$

where M_∞ is the upstream Mach number (see equation (2.9.25)).
The solutions of equation (3.2.10) are

$$\frac{n_y}{n_x} = \pm \sqrt{(M_\infty^2 - 1)} \qquad (E3.2.5)$$

defining the normals to the two characteristics for supersonic flows. Their directions are obtained from equation (3.2.16) as

$$\frac{dy}{dx} = \pm 1/\sqrt{(M_\infty^2 - 1)} = \pm \tan \mu \qquad (E3.2.6)$$

Referring to Figure E3.2.1 it can be seen that these characteristics are identical to the Mach lines at an angle μ to the direction of the velocity, with

$$\sin \mu = 1/M_\infty \qquad (E3.2.7)$$

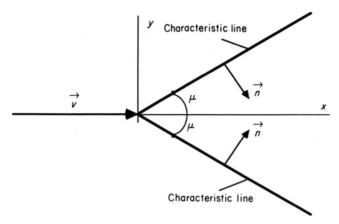

Figure E3.2.1 Characteristics for two-dimensional potential equation

3.2.3 General definition

Let us consider a system of n first-order partial differential equations for the n unknown functions u^j in the m-dimensional space x^k $(k = 1, ..., m)$, eventually including time, written in the conservation form with a summation convention of repeated super- or subscripts:

$$\frac{\partial}{\partial x^k} F_i^k = Q_i \qquad k = 1, ..., m \qquad i = 1, ..., n \qquad (3.2.17)$$

The analysis of the properties of this system relies on the quasi-linear form, obtained after introduction of the Jacobian matrices A^k, where

$$A_{ij}^k = \frac{\partial F_i^k}{\partial u^j} \qquad (3.2.18)$$

is the Jacobian matrix element of the flux F_i^k with respect to the variable u^j. System (3.2.17) takes the quasi-linear form:

$$A_{ij}^k \cdot \frac{\partial u^j}{\partial x^k} = Q_i \qquad i, j = 1, ... n \qquad k = 1, ..., m \qquad (3.2.19)$$

This can be condensed into the matrix form:

$$A^k \frac{\partial U}{\partial x^k} = Q \qquad k = 1, ..., m \qquad (3.2.20)$$

where the $(n \times 1)$ vector column U contains the u^j unknowns, A^k are $(n \times n)$ matrices and Q is a column vector of the non-homogeneous source terms. The matrices A^k and Q can depend on x^k and U but not on the derivatives of U.

A plane wave solution of the form (3.2.7), or of the more general form (3.2.11), will exist if the homogeneous system

$$[A^k n_k] \hat{U} = 0 \qquad (3.2.21a)$$

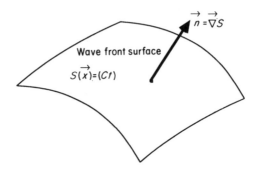

Figure 3.2.1   Wave front surface and associated normal

or

$$\left[A^k \frac{\partial S}{\partial x^k} \right] \hat{U} = 0 \qquad k = 1, \ldots, m \qquad (3.2.21b)$$

has non-trivial solutions. Defining the normal $\vec{n}$ to the wavefront surface by $\vec{n} = \vec{\nabla} S$ (Figure 3.2.1), this will be the case if the determinant of the system vanishes, that is, if

$$\det | A^k n_k | = 0 \qquad (3.2.22)$$

Equation (3.2.22) then defines a condition on the normal $n_k = \partial S/\partial x^k$ to the surface S. This equation can have, at most, n solutions, that is, there are, at most, n characteristic surfaces. For each of these normals $\vec{n}^{(\alpha)}$, system (3.2.21) has a non-trivial solution.

   The system is said to be *hyperbolic* if all the n characteristic normals $\vec{\nabla} S^{(\alpha)} = \vec{n}^{(\alpha)}$ are real and if the solutions of the n associated systems of equations (3.2.21) are linearly independent. If all the characteristics are complex, the system is said to be *elliptic* and if some are real and others complex the system is considered as *hybrid*. If the matrix $(A^k n_k)$ is not of rank n (that is, there are less than n *real* characteristics) then the system is said to be *parabolic*. This will occur, for instance, when at least one of the variables, say u^1 has derivatives with respect to one co-ordinate, say x^1, missing. This implies that $A^1_{i1} = 0$ for all equations i.

Example 3.2.2   System of two first-order equations in two dimensions

The above-mentioned properties can be illustrated in a two-dimensional space x, y with the system

$$a \frac{\partial u}{\partial x} + c \frac{\partial v}{\partial y} = f_1$$

$$b \frac{\partial v}{\partial x} + d \frac{\partial u}{\partial y} = f_2$$

$$(E3.2.8)$$

or in matrix form:

$$\begin{vmatrix} a & 0 \\ 0 & b \end{vmatrix} \cdot \frac{\partial}{\partial x} \begin{vmatrix} u \\ v \end{vmatrix} + \begin{vmatrix} 0 & c \\ d & 0 \end{vmatrix} \cdot \frac{\partial}{\partial y} \begin{vmatrix} u \\ v \end{vmatrix} = \begin{vmatrix} f_1 \\ f_2 \end{vmatrix} \qquad \text{(E3.2.9)}$$

Hence with $x^1 = x$, $x^2 = y$:

$$A^1 = \begin{vmatrix} a & 0 \\ 0 & b \end{vmatrix} \qquad A^2 = \begin{vmatrix} 0 & c \\ d & 0 \end{vmatrix} \qquad \text{(E3.2.10)}$$

The determinant equation (3.2.22) becomes, after division by n_y (assumed to be different from zero),

$$\begin{vmatrix} \dfrac{an_x}{n_y} & c \\[3mm] d & \dfrac{bn_x}{n_y} \end{vmatrix} = 0 \qquad \text{(E3.2.11)}$$

leading to the conditions for the characteristic normals:

$$\left| \frac{n_x}{n_y} \right|^2 = \frac{cd}{ab} \qquad \text{(E3.2.12)}$$

If $cb/ab > 0$, the system is hyperbolic: for instance, $a = b = 1$; $c = d = 1$ with vanishing right-hand side, leading to the well-known wave equation

$$\frac{\partial^2 u}{\partial x^2} - \frac{\partial^2 u}{\partial y^2} = 0 \qquad \text{(E3.2.13)}$$

If $cd/ab < 0$, the system is elliptic; for instance, $a = b = 1$; $c = -d = -1$ and vanishing right-hand side, leading to the Laplace equation, which is the standard form of elliptic equations and describes diffusion phenomena.

Finally, if $b = 0$ there is only one characteristic normal $n_x = 0$ and the system is parabolic. For instance, with $a = 1$, $b = 0$, $c = -d = -1$ and $f_1 = 0$, $f_2 = v$ we obtain the standard form for a parabolic equation:

$$\frac{\partial u}{\partial x} = \frac{\partial^2 u}{\partial y^2} \qquad \text{(E3.2.14)}$$

Example 3.2.3   Stationary shallow-water equations

The stationary shallow-water equations describe the spatial distribution of the height h of the free water surface in a stream with velocity components u and v. They can be written in the following form (where g is the earth's gravity

142

acceleration):

$$u\frac{\partial h}{\partial x} + v\frac{\partial h}{\partial y} + h\frac{\partial u}{\partial x} + h\frac{\partial v}{\partial y} = 0$$

$$u\frac{\partial u}{\partial x} + v\frac{\partial u}{\partial y} + g\frac{\partial h}{\partial x} = 0 \qquad \text{(E3.2.15)}$$

$$u\frac{\partial v}{\partial x} + v\frac{\partial v}{\partial y} + g\frac{\partial h}{\partial y} = 0$$

Introducing the vector

$$U = \begin{vmatrix} h \\ u \\ v \end{vmatrix} \qquad \text{(E3.2.16)}$$

system (E3.2.15) is written in the matrix form (3.2.19):

$$\begin{vmatrix} u & h & 0 \\ g & u & 0 \\ 0 & 0 & u \end{vmatrix}\frac{\partial}{\partial x}\begin{vmatrix} h \\ u \\ v \end{vmatrix} + \begin{vmatrix} v & 0 & h \\ 0 & v & 0 \\ g & 0 & v \end{vmatrix}\frac{\partial}{\partial y}\begin{vmatrix} h \\ u \\ v \end{vmatrix} = 0 \qquad \text{(E3.2.17)}$$

or

$$A^1\frac{\partial U}{\partial x} + A^2\frac{\partial U}{\partial y} = 0 \qquad \text{(E3.2.18)}$$

The three characteristic normals $\vec{n}$ are obtained as the solutions of equation (3.2.22), with $\lambda = n_x/n_y$:

$$\begin{vmatrix} u\lambda + v & h\lambda & h \\ g\lambda & u\lambda + v & 0 \\ g & 0 & u\lambda + v \end{vmatrix} = 0 \qquad \text{(E3.2.19)}$$

Working out determinant (E3.2.19) leads to the solution

$$\lambda^{(1)} = -\frac{v}{u} \qquad \text{(E3.2.20)}$$

and the two solutions of the quadratic equation

$$(u^2 - gh)\lambda^2 + 2\lambda uv + (v^2 - gh) = 0 \qquad \text{(E3.2.21)}$$

$$\lambda^{(2),(3)} = \frac{-uv \pm \sqrt{(u^2 + v^2 - gh)}}{u^2 - gh} \qquad \text{(E3.2.22)}$$

It is seen that $\sqrt{(gh)}$ plays the role of a sonic, critical velocity and the system is hyperbolic for supercritical velocities $\vec{v}^2 = u^2 + v^2 > gh$. Otherwise, the system is hybrid, since the first solution $\lambda^{(1)}$ is always real.

Note that the characteristic surface associated with the solution $\lambda^{(1)}$ is the streamline, since the vector $\vec{n}^{(1)}$ has components proportional to $n_x^{(1)} = -v$ and $n_y^{(2)} = u$ and therefore $\vec{n}^{(1)} \cdot \vec{v} = 0$.

3.2.4 Domain of dependence—zone of influence

The propagation property of hyperbolic problems has important consequences with regard to the way the information is transmitted through the flow region. Considering Figure 3.2.2, where Γ is a boundary line distinct from a characteristic, the solution U along a segment AB of Γ will propagate in the flow domain along the characteristics issued from AB.

For a two-dimensional problem in the variables x, y determined by a second-order equation such as (3.2.1) there are two characteristics if the problem is hyperbolic. Hence the two characteristics out of A and B limit the region PAB, which determines the solution at point P. The region PAB is called the *region of dependence* of point P, since the characteristics out of any point C outside AB will never reach point P. On the other hand, the region downstream of P, and located between the characteristics, defines the zone where the solution is influenced by the function value in P. This region is called the *zone of influence* of P.

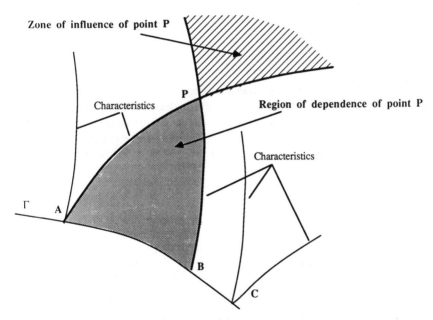

Figure 3.2.2 Region of dependence and zone of influence of point P for a hyperbolic problem with two characteristics per point

Parabolic problems

For parabolic problems the two characteristics are identical (Figure 3.2.3) and the region of dependence of point P reduces to the segment BP. The zone of influence of P, on the other hand, is the whole region right of the characteristic BP.

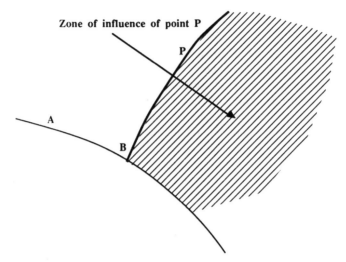

Zone of influence of point P

P

A

B

Figure 3.2.3 Region of dependence and zone of influence of point P for a parabolic problem with one characteristic per point

Elliptic problems

In this case there are no real characteristics and the solution in a point P depends on all the surrounding points, since the physical problem is of the diffusive type. Inversely, the whole boundary ACB surrounding P is influenced by point P (Figure 3.2.4). Hence we can consider that the dependence region is identical to the zone of influence, both of them being equal to the whole of the flow domain.

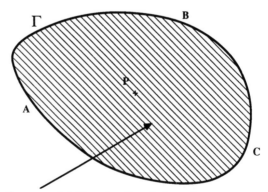

Γ

B

P

A

C

P depending on and influencing the whole region

Figure 3.2.4 Region surrounding P in an elliptic problem

3.3 Alternative definition—Compatibility relations

An alternative definition of characteristic surfaces and hyperbolicity can be obtained from the fact that wave front surfaces carry certain properties and that a complete description of the physical system is obtained when all these properties are known. This implies that the original system of equations, if hyperbolic, can be reformulated as differential relations written along the wave front or characteristic surfaces only. Hence the following definition can be given: a *characteristic surface* $S(x^1, ..., x^m) = 0$ will exist if the first-order system of equations (3.2.20) can be transformed through a linear combination of the form, where l^i are n arbitrary coefficients:

$$l^i A_{ij}^k \frac{\partial u^j}{\partial x^k} = l^i Q_i \qquad i, j = 1, ..., n \qquad k = 1, ..., m \qquad (3.3.1)$$

into an equivalent system containing only derivatives along the surface S.

Along the surface $S(x^1, ..., x^m) = 0$ one of the co-ordinates can be eliminated, for instance x^m, by expressing

$$dS = \frac{\partial S}{\partial x^k} dx^k = 0 \qquad (3.3.2)$$

or, along the surface S,

$$\frac{\partial x^m}{\partial x^k}\bigg|_S = -\frac{\partial S/\partial x^k}{\partial S/\partial x^m} = -\frac{n_k}{n_m} \qquad (3.3.3)$$

where the components of the normal vector $\vec{n} = \vec{\nabla} S$ are introduced. Hence we can define derivatives $\overline{\partial}/\partial x^k$ along the surface S in the following way. For any variable u^j the partial derivative $\overline{\partial}/\partial x^k$ along the surface S is given by

$$\frac{\overline{\partial}}{\partial x^k} = \frac{\partial}{\partial x^k} + \left(\frac{\partial x^m}{\partial x^k}\right) \frac{\partial}{\partial x^m} = \frac{\partial}{\partial x^k} - \left(\frac{n_k}{n_m}\right) \frac{\partial}{\partial x^m} \qquad k = 1, ..., m \qquad (3.3.4)$$

Note that for the variable x^m the surface derivative is zero, that is, $\overline{\partial}/\partial x^m = 0$. Introducing this relation into the linear combination (3.3.1) leads to

$$l^i A_{ij}^k \left[\frac{\overline{\partial}}{\partial x^k} + \left(\frac{n_k}{n_m}\right) \cdot \frac{\partial}{\partial x^m}\right] u^j = l^i Q_i \qquad (3.3.5)$$

The summation over k extends from $k = 1$ to $k = m$. A characteristic surface will exist for any u^j if the system is reduced to the form

$$l^i A_{ij}^k \frac{\overline{\partial}}{\partial x^k} u^j = l^i Q_i \qquad i, j = 1, ..., n \qquad k = 1, ..., m \qquad (3.3.6)$$

This is satisfied if the surface S obeys the relations, for any u^j,

$$l^i A_{ij}^k n_k = 0 \qquad (3.3.7a)$$

or,

$$l^i A_{ij}^k \frac{\partial S}{\partial x^k} = 0 \qquad i, j = 1, ..., n \qquad k = 1, ..., m \qquad (3.3.7b)$$

The conditions for this homogeneous system, in the l^i unknowns, to be compatible is the vanishing of the determinant of the coefficients, leading to condition (3.2.22). For each solution $\vec{n}^{(\alpha)}$ of equation (3.2.22), system (3.3.7) has a non-trivial solution for the coefficients $l^{i(\alpha)}$, up to an arbitrary scale factor. The $l^{i(\alpha)}$ coefficients can be grouped into a $(n \times 1)$ line vector $\vec{l}^{(\alpha)}$.

The system is said to be *hyperbolic* if all the n characteristic normals $\partial_k S^{(\alpha)}$ or $\vec{n}^{(\alpha)}$ are real and if the n vectors $\vec{l}^{(\alpha)}$, $(\alpha = 1, ..., n)$ solutions of the n systems of equations (3.3.7) are linearly independent.

3.3.1 Compatibility relations

The reduced form (3.3.6) expresses that the basic equations can be combined to a form containing only derivatives confined to a $(n-1)$ dimensional space. That is, the system of equations, if hyperbolic, can be considered as describing phenomena occurring on hypersurfaces $S^{(\alpha)}$. Indeed, defining a set of n vectors $\vec{Z}_j$ in the m-dimensional space, with components $Z_j^k : j = 1, ..., n; k = 1, ..., m$ by the relations

$$Z_j^k = l^i A_{ij}^k \tag{3.3.8}$$

equation (3.3.1) can be rewritten as

$$Z_j^k \frac{\partial u^j}{\partial x^k} = l^i Q_i \tag{3.3.9}$$

The vectors $\vec{Z}$ define n *characteristic directions* of which $(n-1)$ are independent.

The operators $(Z_j^k \partial_k)$ are the derivatives in the direction of the vector $\vec{Z}_j$. Hence, defining

$$d_j \equiv Z_j^k \frac{\partial}{\partial x^k} = \vec{Z}_j \cdot \vec{\nabla} \tag{3.3.10}$$

in the m-dimensional space, the transformed equation (3.3.1) can be written as a sum of derivatives along the vectors $\vec{Z}_j$:

$$d_j u^j = \vec{Z}_j \cdot \vec{\nabla} u^j = (l^i Q_i) \tag{3.3.11}$$

Equations (3.3.11) are known as the *compatibility relations,* and represent an alternative formulation to system (3.2.19).

Condition (3.3.7) expresses that all the $\vec{Z}_j$ vectors lie in the characteristic surface whose normal is $\vec{n}$. Indeed, equation (3.3.7) becomes, with the introduction of $\vec{Z}_j$,

$$\vec{Z}_j \cdot \vec{n} = Z_j^k n_k = 0 \qquad \text{for all } j = 1, ..., n \tag{3.3.12}$$

These phenomena correspond to propagating wavefronts, as seen earlier, and it can be shown, see, for instance, Whitham (1974), that the characteristic surfaces can also contain discontinuities of the normal derivatives $\partial u^j / \partial n$, satisfying

$$A^k_{ij}n_k\left[\frac{\partial u^j}{\partial n}\right]=0 \qquad (3.3.13)$$

where

$$\left[\frac{\partial u^j}{\partial n}\right]=\left(\frac{\partial u^j}{\partial n}\right)_+ -\left(\frac{\partial u^j}{\partial n}\right)_- \qquad (3.3.14)$$

is the jump, over the surface S, of the normal derivatives of the solution u^j. This relation can also be written

$$n_k\left[\frac{\partial F^k_i}{\partial n}\right]=0 \qquad (3.3.15)$$

where $\partial/\partial n$ is the normal derivative to the surface S.

Example 3.3.1   Small-disturbance potential equation (E3.2.4)

The two vectors l associated to the two characteristic normals (E3.2.5) are obtained from system (3.3.7), which is written here as

$$(l^1, l^2)\begin{vmatrix}(1-M^2_\infty)n_x & n_y \\ -n_y & n_x\end{vmatrix}=0 \qquad (E3.3.1)$$

where the ratio $\lambda \equiv n_y/n_x$ is defined as the solution (E3.2.5). Choosing $l^1 = 1$, system (E3.3.1) has the solution

$$\begin{aligned}l^1 &= 1 \\ l^2 &= \sqrt{(M^2_\infty - 1)}\end{aligned} \qquad (E3.3.2)$$

The two characteristic directions $\vec{Z}_j$, defined by equation (3.3.8), become here

$$\begin{aligned}\vec{Z}_1 &\equiv (1-M^2_\infty, \lambda) \\ \vec{Z}_2 &\equiv (-\lambda, 1)\end{aligned} \qquad (E3.3.3)$$

Observe that these two directions are parallel to each other, since $1 - M^2_\infty = \lambda^2$ and that their common direction is the characteristic line of Figure E3.2.1, making the angle μ with the x-direction, since $\lambda = \cos \mu/\sin \mu$. This can also be seen from a direct verification of equation (3.3.12), with the vector of the characteristic normal defined by the components $\vec{n} \equiv (1, \lambda)$, which indicates that the Z-directions are orthogonal to the normals n.

The compatibility relations (3.3.9) or (3.3.11) become here

$$\left[(1-M^2_\infty)\frac{\partial}{\partial x}+\lambda\frac{\partial}{\partial y}\right]u + \left[-\lambda\frac{\partial}{\partial x}+\frac{\partial}{\partial y}\right]v = 0 \qquad (E3.3.4)$$

or

$$\cot \mu\left[\cos \mu\frac{\partial}{\partial x}\mp \sin \mu\frac{\partial}{\partial y}\right]u \pm \left[\cos \mu\frac{\partial}{\partial x}\mp \sin \mu\frac{\partial}{\partial y}\right]v = 0 \qquad (E3.3.5)$$

For constant Mach angles μ the compatibility relation expresses the property that the velocity component along one characteristic ($u \cos \mu \pm v \sin \mu$) is conserved along the other characteristic.

3.4. Time-like variables

More insight into the significance of characteristics is obtained if one space variable, say x^m, is singled out and the corresponding Jacobian matrix A^m is taken as the unit matrix. This will always be possible if the matrix is positive definite, since we can multiply the original system of equations by $A^{m^{(-1)}}$. We will call this variable *time-like* and take $x^m = t$, with the conditions

$$A^m_{ij} = \delta_{ij} \text{ with } x^m \equiv t \tag{3.4.1}$$

System (3.2.20) is written as

$$\frac{\partial U}{\partial t} + A^k \frac{\partial U}{\partial x^k} = Q \qquad k = 1, \ldots, m-1 \tag{3.4.2}$$

Note that $(m-1)$ is now the number of *space-like* variables. The characteristic condition, equation (3.2.22), becomes with $n_m = n_t$

$$\det | n_t + A^k n_k | = 0 \qquad k = 1, \ldots, m-1 \tag{3.4.3}$$

Equation (3.4.3) is therefore an eigenvalue problem where the characteristic normals are obtained as the eigenvalues of the matrix

$$K_{ij} = A^k_{ij} n_k \qquad k = 1, \ldots, m-1 \tag{3.4.4}$$

satisfying

$$\det | K - \lambda I | = 0 \tag{3.4.5}$$

If the n eigenvalues $\lambda_{(\alpha)}$ are real there are n characteristic surfaces with

$$n_t^{(\alpha)} = -\lambda_{(\alpha)} \tag{3.4.6}$$

The corresponding vectors $\vec{l}$ are obtained from equation (3.3.7), written as

$$l^i(\delta_{ij} n_t + A^k_{ij} n_k) = 0 \qquad k = 1, \ldots, m-1 \tag{3.4.7}$$

or

$$l^i K_{ij} = \lambda_{(\alpha)} l^i \delta_{ij} \qquad i, j = 1, \ldots, n \qquad (\alpha) = 1, \ldots, n \tag{3.4.8}$$

Hence the vectors $\vec{l}^{(\alpha)}$ of components $l^{i(\alpha)}$ are the left eigenvectors of the matrix K corresponding to the eigenvalue $\lambda_{(\alpha)}$. If the n eigenvectors $\vec{l}^{(\alpha)}$ are linearly independent, the system will be hyperbolic.

If the n eigenvectors $\vec{l}^{(\alpha)}$ are grouped in a matrix L^{-1}, where each row contains the components of an eigenvector $\vec{l}^{(\alpha)}$, that is,

$$(L^{-1})^{i\alpha} = l^{i(\alpha)} \tag{3.4.9}$$

we obtain from the eigenvector equation (3.4.8) that the matrix L diagonalizes the matrix K:

$$L^{-1}KL = \Lambda \qquad (3.4.10)$$

where Λ is the diagonal matrix containing the eigenvalues $\lambda_{(\alpha)}$:

$$\Lambda = \begin{vmatrix} \lambda_{(1)} & & & \\ & \lambda_{(2)} & & \\ & & \ddots & \\ 0 & & & \lambda_{(n)} \end{vmatrix} \qquad (3.4.11)$$

It is also interesting to note that, within the same assumptions, the intensities of the propagating disturbances are, following equation (3.2.21), the right eigenvectors $\vec{r}^{(\alpha)}$ of K, since equations (3.2.21) or (3.3.13) can be written, with equations (3.4.1) and (3.4.5),

$$K_{ij}r^j = \lambda_{(\alpha)}r^j\delta_{ij} \qquad i, j = 1, ..., n \qquad (3.4.12)$$

3.4.1 Plane wave solutions with time-like variable

With the introduction ot the time-like variables the plane wave solutions (3.2.7) can be written in a more conventional way:

$$U = \hat{U}\, e^{I(\vec{\varkappa} \cdot \vec{x} - \omega t)} \qquad (3.4.13)$$

The vector $\vec{\varkappa}$ is called the *wave number* vector, and its magnitude is the number of periods or wavelengths over a distance 2π in the direction of the vector $\vec{\varkappa}$. Since this simple wave is a solution of the convection equation

$$\frac{\partial U}{\partial t} + (\vec{a} \cdot \vec{\nabla})U = 0 \qquad (3.4.14)$$

where

$$\vec{a} = \frac{\omega}{\varkappa^2}\, \vec{\varkappa} \qquad (3.4.15)$$

it is clear that the *direction of $\vec{\varkappa}$ is the propagation direction of the wave* with a *phase velocity*, given by

$$a = \frac{\omega}{\varkappa} \qquad (3.4.16)$$

We also have the following relation between frequency ν, wavelength λ and the variables $\varkappa$ and ω:

$$\lambda = \frac{2\pi}{\varkappa} \qquad \omega = 2\pi\nu \qquad \lambda\nu = a \qquad (3.4.17)$$

typical of plane waves.

Wave-like solutions to the homogeneous system of equations (3.4.2), with $Q_i = 0$, will exist if the components of equation (3.4.13)

$$u^j = \hat{u}^j\, e^{I(\vec{\varkappa} \cdot \vec{x} - \omega t)} \qquad (3.4.18)$$

150

are solutions of the equation, with $x^m = t$:

$$(-\omega\delta_{ij} + \varkappa_k A_{ij}^k)\hat{u}^j = 0 \qquad (3.4.19)$$

The condition of compatibility is that the determinant of the system vanishes:

$$\det | -\omega\delta_{ij} + \varkappa_k A_{ij}^k | = 0 \qquad (3.4.20)$$

which leads to condition (3.4.3) with $n_t = -\omega$ and $n_k = \varkappa_k$.

Hence for each wave number vector $\vec{\varkappa}$ a perturbation in the *surface normal to* $\vec{\varkappa}$ propagates in the direction of $\vec{\varkappa}$ with phase velocity $\vec{a}$ and frequency ω equal to the eigenvalue of the matrix $K = A^k \varkappa_k$. If we group the $(m-1)$ matrices A^k in a vector $\vec{A}$ of dimensions $(m-1)$, $\vec{A}$ $(A^i,\dots A^{m-1})$, we can write the matrix K as a scalar product:

$$K = \vec{A} \cdot \vec{\varkappa} \qquad (3.4.21)$$

The propagation speed associated with the characteristic frequency of the perturbation $\omega_{(\alpha)}$ is obtained as the eigenvalue of the matrix

$$\hat{K} = \vec{A} \cdot \frac{\vec{\varkappa}}{\varkappa} = \vec{A} \cdot \vec{I}_x \qquad (3.4.22)$$

where $\vec{I}_x$ is the unit vector in the direction $\vec{\varkappa}$. Hence the *characteristic speeds* $a_{(\alpha)}$ are obtained as solutions of the eigenvalue problem:

$$\det | -a_{(\alpha)ij} + \hat{K}_{ij} | = 0 \qquad (3.4.23)$$

For an elliptic system the eigenvalues are complex and the solution takes the form, for an eigenvalue $\omega_1 = \xi + I\eta$:

$$U_1 = \hat{U}_1 e^{-I\xi t} e^{I\vec{x} \cdot \vec{x}} e^{+\eta t} \qquad (3.4.24)$$

Since the coefficients of the Jacobian matrices A^k are considered to be real, each eigenvalue $\omega_1 = \xi + I\eta$ is associated with a complex conjugate eigenvalue $\omega_2 = \xi - I\eta$, leading to a solution of the form

$$U_2 = \hat{U}_2 e^{-I\xi t} e^{I\vec{x} \cdot \vec{x}} e^{-\eta t} \qquad (3.4.25)$$

Hence according to the sign of the imaginary part of the eigenvalue one of the two solutions U_1 or U_2 will be damped in time while the other will be amplified.

Example 3.4.1   Time-dependent shallow-water equations in one dimension

The one-dimensional form of the time-dependent shallow-water equations can be written as

$$\frac{\partial h}{\partial t} + u\frac{\partial h}{\partial x} + h\frac{\partial u}{\partial x} = 0$$

$$\frac{\partial u}{\partial t} + u\frac{\partial u}{\partial x} + g\frac{\partial h}{\partial x} = 0 \qquad (E3.4.1)$$

In matrix form, we have

$$\frac{\partial}{\partial t}\begin{vmatrix} h \\ u \end{vmatrix} + \begin{vmatrix} u & h \\ g & u \end{vmatrix}\frac{\partial}{\partial x}\begin{vmatrix} h \\ u \end{vmatrix} = 0 \qquad \text{(E3.4.2)}$$

The two characteristic velocities $a_{1,2}$ are obtained from equation (3.4.23) as solutions of

$$\begin{vmatrix} -a+u & h \\ g & -a+u \end{vmatrix} = 0 \qquad \text{(E3.4.3)}$$

or

$$a_{1,2} = u \pm \sqrt{(gh)} \qquad \text{(E3.4.4)}$$

Since these eigenvalues are always real, the system is always hyperbolic in (x, t).

3.4.2 Non-linear wave solutions and time-like variable

A general solution of forms (3.2.11) or (3.2.14), with condition (3.4.1) can be written

$$U = \hat{U}\,e^{I(\vec{x}\cdot\vec{\nabla}S + tS_t)} \qquad \text{(3.4.26)}$$

For this wave to be a solution of the homogeneous system

$$\frac{\partial U}{\partial t} + A^k\frac{\partial U}{\partial x^k} = 0 \qquad k = 1,\ldots,m-1 \qquad \text{(3.4.27)}$$

S has to satisfy the equation

$$\det\left|\frac{\partial S}{\partial t} + A^k\frac{\partial S}{\partial x^k}\right| = 0 \qquad \text{(3.4.28)}$$

which is identical to the characteristic condition equation (3.4.3). Hence the frequency ω of the wave is defined by

$$\omega = -\frac{\partial S}{\partial t} = -n_t \qquad \text{(3.4.29)}$$

and the wave number vector $\vec{x}$ is defined by

$$\vec{x} = \vec{\nabla}S = \vec{n} \qquad \text{(3.4.30)}$$

In this notation $\vec{\nabla}S$ is the normal to the intersection of the characteristic or wavefront surface $S(\vec{x}, t)$ with the hypersurfaces $t = $ constant. Hence the normals are defined in the $(m-1)$ dimensional space of the space-like variables. On the other hand, the normals $\vec{n}$ defined in Section 3.2.2, equation (3.2.13), are normals to the wavefront surfaces in the m-dimensional space $x^1,\ldots,x^m$. Observe also that the wave-number $\vec{x}$, in the space $x^k (k = 1, m-1)$ is normal to the characteristic subsurfaces $S(\vec{x}, t)$ at constant t.

The n eigenvalues of the matrix $K_{ij} = A^{k}_{ij} \varkappa_k$ define the n *dispersion relations* for the frequencies $\omega_{(\alpha)}$:

$$\omega_{(\alpha)} = \lambda_{(\alpha)}(\vec{\varkappa}(\vec{x}, t)) \qquad \alpha = 1, ..., n \qquad (3.4.31)$$

Note that a non-linear wave can be only written under form (3.4.13) with

$$\vec{\varkappa} = \vec{\varkappa}(\vec{x}, t)$$
$$\omega = \omega(\vec{\varkappa}(\vec{x}, t)) \qquad (3.4.32)$$

and the same *local* definition of wave number and frequency, under certain conditions.

Indeed, when introduced into equations (3.4.14) the above non-linear solution gives the following contributions:

$$\frac{\partial U}{\partial t} = IU\left[-\omega - t\frac{\partial \omega}{\partial t} + \vec{x} \cdot \frac{\partial \vec{\varkappa}}{\partial t}\right] = IU\left[-\omega + (\vec{x} - t\vec{\nabla}_x\omega) \cdot \frac{\partial \vec{\varkappa}}{\partial t}\right]$$

$$\frac{\partial U}{\partial x^k} = IU\left[\varkappa_k - t\frac{\partial \omega}{\partial x^k} + \vec{x} \cdot \frac{\partial \vec{\varkappa}}{\partial x^k}\right] = IU\left[\varkappa_k + (\vec{x} - t\vec{\nabla}_x\omega) \cdot \frac{\partial \vec{\varkappa}}{\partial x^k}\right] \qquad (3.4.33)$$

The derivative of the frequency with respect to the wave number component $\varkappa_j$ is the j-component of the *group velocity* of the wave

$$v_j^{(G)} = \frac{\partial \omega}{\partial \varkappa_j} \qquad (3.4.34)$$

or, in condensed notation,

$$\vec{v}^{(G)} = \vec{\nabla}_x\omega \qquad (3.4.35)$$

The terms in parentheses in equation (3.4.33) will vanish for an observer moving with the group velocity, that is, for

$$\frac{\vec{x}}{t} = \vec{\nabla}_x\omega \qquad (3.4.36)$$

Note that the group velocity is the velocity at which the wave energy propagates. The reader will find an extensive discussion of non-linear waves in Whitman (1974).

3.5 INITIAL AND BOUNDARY CONDITIONS

The information necessary for the initial and boundary conditions to be imposed with a given system of differential equations in order to have a *well-posed problem* can be gained from the preceding considerations. The condition of being well posed (according to Hadamard) is established if the solution depends in a continuous way on the initial and boundary conditions. That is, a small perturbation of these conditions should give rise to a small variation of the solution at any point of the domain at a finite distance from the boundaries.

Two types of problems are considered with regard to the time-like variable $x^m = t$: an *initial value* or a *Cauchy* problem, where the solution is given in the subspace $x^m = t = 0$ as $U = U(\vec{x}, t = 0)$ and is to be determined at subsequent values of t. If the subspace $t = 0$ is bounded by some surface $\Omega(\vec{x})$ then additional conditions have to be imposed along that surface at all values of t, and this defines an *initial boundary value problem*.

A solution of the system of first-order partial differential equations can be written as a superposition of wave-like solutions of the type corresponding to the n-eigenvalues of the matrix K:

$$U = \sum_{\alpha=1}^{n} \hat{U}_\alpha e^{I(\vec{x} \cdot \vec{x} - \omega_{(\alpha)} t)} \tag{3.5.1}$$

where the summation extends over all the eigenvalues $\lambda_{(\alpha)}$, U being the column containing the unknowns u^j.

If N_r and N_c denote, respectively, the number of real and complex eigenvalues, considered to be of multiplicity one, with $n = N_c + N_r$, it is seen from equation (3.4.24) that the complex eigenvalues will generate amplified modes for $\eta > 0$. If such a mode is allowed the problem will not be well posed according to Hadamard. Therefore the number of initial and boundary conditions to be imposed have to be selected to make sure that such modes are neither generated nor allowed.

If the problem is hyperbolic, $N_c = 0$ and $N_r = n$, and since no amplified modes are generated, n initial conditions for the Cauchy problem have to be given in order to determine completely the solution. That is, as many conditions as unknowns have be given at $t = 0$.

On the other hand, if the problem is elliptic or hybrid there will be $N_c/2$ amplified modes and hence only $N_r + N_c/2$ conditions are allowed. Since this number is lower than n, the pure initial value or Cauchy problem is not well posed for non-hyperbolic problems, and only boundary value problems will be well posed in this case. The inverse is also true: a pure boundary value problem is ill posed for a hyperbolic problem. For an elliptic system $N_r = 0$, and the number of boundary conditions to be imposed at every point of the boundary is equal to half the order of the system. For instance, for a second-order hyperbolic equation two conditions will have to be fixed along the initial Cauchy line, while for a second-order elliptic equation one condition will have to be given along the boundaries.

For initial boundary value problems the n boundary conditions have to be distributed along the boundaries at all values of t, according to the direction of propagation of the corresponding waves. If a wave number $\vec{x}$ is taken in the direction of the interior normal vector $\vec{n}$, then the corresponding wave, whose phase velocity is obtained as an eigenvalue of the matrix $(A^k \cdot n_k)$, will propagate information inside the domain if this velocity is positive. Hence the number of conditions to be imposed for the hyperbolic initial boundary value problem at a given point of the boundary is equal to the number of positive eigenvalues of the matrix $A^k \cdot n_k$ at that point. The total number of conditions obviously remains equal to the total number of eigenvalues, that is, to the order of the system. For hybrid problems the conclusions are the same for the real characteristics, but $N_c/2$ additional conditions have to be imposed everywhere along the boundary. Note also that, next to the number of boundary conditions to impose, the nature of these conditions can also be important in order to avoid ill-posed conditions along the boundaries. This will be discussed more in detail in the following chapters.

Parabolic problems in t and $\vec{x}$ define initial boundary value problems. Hence the solution is to be defined at $t = 0$, thàt is, for an order n, n conditions have to be given at $t = 0$. Along the boundaries for all times, $n/2$ boundary conditions have to be imposed. This is the case for the standard form of parabolic equations $\partial_t u = L(u)$, where $L(u)$ is a second-order elliptic operator in space.

A more complex parabolic structure arises in boundary layer theory, where the equations are of the form $u_{yy} = L(u)$, where $L(u)$ is an hyperbolic first-order operator in the space x, y, z, with y the co-ordinate normal to the wall. This leads to complex mixed parabolic–hyperbolic phenomena in three-dimensional boundary layer calculations. (Some of these aspects are described in Krause, 1973 and Dwyer, 1981). The boundary conditions are of the initial value type for the hyperbolic components and of boundary value nature for the elliptic parts of the system.

The whole system of Navier-Stokes equations is essentially parabolic in time and space or parabolic–hyperbolic, while the steady-state part is elliptic–hyperbolic, due to the hyperbolic character of the continuity equation considered for a known velocity field. On the other hand, in the absence of viscosity and heat conduction effects, the system of time-dependent Euler equations is purely hyperbolic in space and time.

The various approximations to the Navier–Stokes equations discussed in this chapter have evidently different mathematical properties. For instance, in the TSL approximation or the boundary layer approximations the diffusive effect of viscosity is neglected in all directions except in those normal to the wall. Therefore the resulting equations remain parabolic in time and in the direction normal to the surface, while the behaviour of the system will be purely hyperbolic in the other two directions and time. The global property remains, however, parabolic, although the local behaviour of the system is modified compared with the full Navier–Stokes model. This leads to important consequences for the numerical simulation of three-dimensional boundary layers; see Dwyer (1981), for a recent review.

From a mathematical point of view, no general, global existence theorems for the non-stationary compressible Navier–Stokes equation with a defined set of boundary and initial conditions can be defined. Some partial, local existence theorems have been obtained, Temam (1977); Solonnikov and Kazlikhov (1981), for both the Cauchy problem, that is, given distributions of density, velocity and temperature at time $t = 0$, and for initial boundary value problems where the flow parameters are given at $t = 0$, boundary conditions are imposed at all times for velocity and temperature on the boundary of the flow domain.

These investigations do not lead at present to practical rules for the establishment of boundary conditions, and therefore case-by-case considerations have to be used as a function of the type of the equations and of physical properties of the system. In general, the elliptic time-independent problems will impose the values of the flow variables (Dirichlet conditions) or their derivative (Neumann conditions) on the boundaries of the flow domain. Due to physical considerations, for fluid conditions far from the molecular free motions (Knudsen numbers below 10^{-2}) the velocity should be continuous at the material boundaries. This leads to the well-known no-slip conditions for the velocity for the Navier–Stokes equations. For the temperature, one of the following three conditions can be used, T_w being the wall temperature:

$$T = T_w \qquad \text{fixed wall temperature (Dirichlet condition)}$$

$$k \frac{\partial T}{\partial n} = q \qquad \text{fixed heat flux (Von Neumann condition)}$$

$$k \frac{\partial T}{\partial n} = \alpha(T - T_w) \qquad \text{heat flux proportional to local heat transfer (mixed condition)}$$

For other elliptic equations, such as the subsonic potential or streamfunction equations, the choice will be made on the basis of the physical interpretation of these functions, and this will be discussed in the appropriate chapters.

Inviscid flow equations, being first order, allow only one condition on the velocity, namely that the velocity component normal to the wall is fixed by the mass transfer through that wall, while the tangential component will have to be determined from the computation and will generally be different from the non-slip value, since slip velocities are allowed. For free surfaces the physical conditions are chosen on the basis of continuity of the normal and tangential stresses and of the statement that the free boundary is a streamsurface.

References

Courant, R., and Hilbert, D. (1962). *Methods of Mathematical Physics*, New York: Interscience.

Dwyer, H. A. (1981). 'Some aspects of three dimensional laminar boundary layers.' *Annual Review of Fluid Mechanics*, **13**, 217–29.

Krause, E. (1973). 'Numerical treatment of boundary layer problems'. *Advances in Numerical Fluid Dynamics, AGARD-LS-64.*

Solonnikov, J. A., and Kazhikhov, A. V. (1981). 'Existence theorems for the equations of motion of a compressible viscous fluid.' *Annual Review of Fluid Mechanics,* **13**, 79–95.

Temam, R. (1977). *Navier–Stokes Equations*, Amsterdam: North-Holland.

Whitham, G. B. (1974). *Linear and Non-linear Waves*, New York: John Wiley.

PROBLEMS

Problem 3.1.

Consider the steady potential equation (E3.2.1) for supersonic flows, $M > 1$. From Example 3.2.1 it is known that the equation is hyperbolic. Obtain the two vectors $\vec{l}^{(\alpha)}$, $\alpha = 1, 2$ associated with the two characteristic normal directions $\vec{n}^{(\alpha)}$, solutions of equation (3.2.9).

Show that the two characteristics form an angle $\pm \mu$ with the velocity vector $\vec{v}$, with $\sin \mu = 1/M$. The angle μ is called the Mach angle. Write also the compatibility relations (3.3.11) after having defined the characteristic directions $\vec{Z}_j$ according to equation (3.3.8).

Hint: Define β as the angle of the velocity vector by $\cos \beta = u/|\vec{v}|$, $\sin \beta = v/|\vec{v}|$. Setting $n_x = 1$, show that $n_y = -\cotan(\beta \pm \mu)$. Selecting $l_1 = 1$, obtain $l_2 = -cn_y$ from equation (3.2.7). Obtain

$$\vec{Z}_1 = \begin{vmatrix} a \\ -2b + cn_y \end{vmatrix} \qquad \vec{Z}_2 = \begin{vmatrix} -cn_y \\ c \end{vmatrix}$$

and verify equation (3.3.12). Note also that $\vec{Z}_1$ and $\vec{Z}_2$ are in the same direction since they are both orthogonal to $\vec{n}$. Show by a direct calculation that the vector product of $\vec{Z}_1$ and $\vec{Z}_2$ is indeed zero.

Referring to the general form of equation (3.2.1), obtain the compatibility relation

$$a \frac{\partial u}{\partial x} + (cn_y - 2b) \frac{\partial u}{\partial y} - cn_y \frac{\partial v}{\partial x} + c \frac{\partial v}{\partial y} = 0$$

Problem 3.2.

Show that the solutions $\lambda_{(2)}$ and $\lambda_{(3)}$ for the shallow-water equations, Example 3.2.3, have the same properties as the two characteristic normals of the potential equation obtained in Problem 3.1.

Problem 3.3

Show that for the transformation leading to equation (3.4.2) $A_{ij}^m = \delta_{ij}$, the characteristic directions (3.3.8) become

$$Z_j^m = l^i \delta_{ij}$$
$$Z_j^k = l^i A_{ij}^k \qquad k = 1, ..., m-1$$

Form the $(n \times m)$ matrix Z_j^k and note that the last line (j being the column index) is formed by the vector $\vec{l}$. Show that the orthogonality condition (3.3.12) is equivalent to equation (3.4.8) and that we have, for a wave number vector $\vec{\varkappa}$,

$$Z_j^k \varkappa_k = \lambda l^i \delta_{ij} \qquad k = 1, ..., m-1$$

Problem 3.4

Referring to the one-dimensional shallow-water equations treated in Example 3.4.1, find the eigenvectors $\vec{l}^{(1)}$ and $\vec{l}^{(2)}$ as well as the characteristic vectors $\vec{Z}_1$ and $\vec{Z}_2$. Derive also the compatibility relations (3.3.11).

Hint: Show that the left eigenvectors are proportional to $\vec{l}[1, \pm \sqrt{(h/g)}]$. Since

$m - 1 = 1$, and $x_1 = x$, the characteristic vectors have the components

$$\vec{Z}_1 = \begin{vmatrix} u \pm \sqrt{(gh)} \\ 1 \end{vmatrix} \qquad \vec{Z}_2 = \pm \frac{\sqrt{h}}{g} \begin{vmatrix} u \pm \sqrt{(gh)} \\ 1 \end{vmatrix}$$

Show that we obtain

$$\frac{\partial h}{\partial t} + [u \pm \sqrt{(gh)}] \frac{\partial h}{\partial x} \pm \frac{\sqrt{h}}{g} \left\{ \frac{\partial u}{\partial t} + [u \mp \sqrt{(gh)}] \frac{\partial u}{\partial x} \right\} = 0$$

where the upper signs refer to the first compatibility relation and the lower signs refer to the second.

Problem 3.5

Show that the system of Cauchy–Riemann equations

$$\frac{\partial u}{\partial x} + \frac{\partial v}{\partial y} = 0$$

$$\frac{\partial v}{\partial x} - \frac{\partial u}{\partial y} = 0$$

is of an elliptic nature.

Problem 3.6

Consider the one-dimensional Euler equations:

$$\frac{\partial \rho}{\partial t} + u \frac{\partial \rho}{\partial x} + \rho \frac{\partial u}{\partial x} = 0$$

$$\frac{\partial u}{\partial t} + u \frac{\partial u}{\partial x} + \frac{1}{\rho} \frac{\partial p}{\partial x} = 0$$

$$\frac{\partial H}{\partial t} + u \frac{\partial H}{\partial x} = \frac{1}{\rho} \frac{\partial p}{\partial t}$$

Introduce the isentropic assumption, with c the speed of sound:

$$\frac{\partial p}{\partial \rho} = c^2$$

Replace the third equation by an equation on the pressure by applying the perfect gas laws and the definition of H. Obtain the equation

$$\frac{\partial p}{\partial t} + u \frac{\partial p}{\partial x} + \rho c^2 \frac{\partial u}{\partial x} = 0$$

Write the system in matrix form for the variable vector:

$$\begin{vmatrix} \rho \\ u \\ p \end{vmatrix}$$

Show that the system is hyperbolic and has the eigenvalues

$$a_{(1)} = u, \qquad a_{(2)} = u + c, \qquad a_{(3)} = u - c$$

Obtain the left and right eigenvectors.

158

Problem 3.7

Show that the one-dimensional Navier–Stokes equation without pressure gradient (known as the 'viscous' Burger's equation)

$$\frac{\partial u}{\partial t} + u \frac{\partial u}{\partial x} = \alpha \frac{\partial^2 u}{\partial x^2}$$

is parabolic in x, t.

Hint: Write the equation as a system, introducing $v = \partial u/\partial x$ as a second variable. Apply equation (3.2.22) and show that the matrix is not of rank 2.

Problem 3.8

Consider the system

$$\frac{\partial u}{\partial t} + \frac{1}{2}\frac{\partial u}{\partial x} + \frac{\partial v}{\partial x} = 0$$

$$\frac{\partial v}{\partial t} + \frac{\partial u}{\partial x} + \frac{1}{2}\frac{\partial v}{\partial x} = 0$$

(a) Write the system in matrix form (3.4.2) and obtain the matrix A:

$$\frac{\partial U}{\partial t} + A \frac{\partial U}{\partial x} = 0 \quad \text{with} \quad U = \begin{vmatrix} u \\ v \end{vmatrix}$$

(b) Find the eigenvalues of A and show that the system is hyperbolic.
(c) Derive the left and right eigenvectors and obtain the matrix L which diagonalizes A. Explain why the left and right eigenvectors are identical.
(d) Obtain the characteristic variables and the compatibility relations.

Hint: The eigenvalues of A are $\lambda_1 = 3/2$ and $\lambda_2 = -1/2$. The matrix L has the form

$$L = \frac{1}{\sqrt{2}}\begin{vmatrix} 1 & 1 \\ 1 & -1 \end{vmatrix}$$

The characteristic variables are $w_1 = (u+v)/\sqrt{2}$ and $w_2 = (u-v)/\sqrt{2}$. The compatibility relations are

$$\frac{\partial(u+v)}{\partial t} + \frac{3}{2}\frac{\partial(u+v)}{\partial x} = 0$$

$$\frac{\partial(u-v)}{\partial t} - \frac{1}{2}\frac{\partial(u-v)}{\partial x} = 0$$

Problem 3.9

Consider the stationary, invisied equations for an incompressible fluid

$$\frac{\partial u}{\partial x} + \frac{\partial v}{\partial y} = 0$$

$$u\frac{\partial u}{\partial x} + v\frac{\partial u}{\partial y} = -\frac{1}{\rho}\frac{\partial p}{\partial x}$$

$$u\frac{\partial v}{\partial x} + v\frac{\partial v}{\partial y} = -\frac{1}{\rho}\frac{\partial p}{\partial y}$$

Show that this system is hybrid, having one real and two complex eigenvalues.

Hint: Consider the variable vector

$$\mathbf{U} = \begin{pmatrix} u \\ v \\ p \end{pmatrix}$$

and write the system as

$$A \frac{\partial \mathbf{U}}{\partial x} + B \frac{\partial \mathbf{U}}{\partial y} = 0$$

The determinant $(A + B\lambda)$ has the eigenvalues $-v/u$ and $\pm I = \pm \sqrt{-1}$

PART II: BASIC DISCRETIZATION TECHNIQUES

The definition of a computational approach involves several steps leading from an initial mathematical model to a final numerical solution. The first step, discussed in the previous chapters, is the selection of a *level of approximation* to the physical problem to be solved, dependent on the accuracy required as much as on the computational power available. The second step is the choice of the *discretization method* of the mathematical formulation and involves two components, the *space discretization* and the *equation discretization*. The space discretization consists of setting up a mesh or a grid by which the continuum of space is replaced by a finite number of points where the numerical values of the variables will have to be determined. It is intuitively obvious that the accuracy of a numerical approximation will be directly dependent on the size of the mesh, that is, the better the discretized space approaches the continuum, the better the approximation of the numerical scheme. In other words, the error of a numerical simulation has to tend to zero when the mesh size tends to zero, and the rapidity of this variation will be characterized by the *order* of the numerical discretization of the equations.

On the other hand, for complex geometries the solution will also be dependent on the form of the mesh, since in these cases we will tend to develop meshes which are adapted to the geometrical complexities, as for flows along solid walls, and the mesh form and size will vary through the flow field. Therefore the generation of meshes for complex geometries is a problem whose importance increases with the space dimension, making this aspect a most important one in three-dimensional calculations along complex bodies such as aircrafts. In recent years methods have been, and are being, developed in order to generate meshes adapted to arbitrary geometries. An excellent review of available numerical mesh generation techniques can be found in Thompson (1984).

Once a mesh has been defined the equations can be discretized, leading to the transformation of the differential or integral equations to discrete algebraic operations involving the values of the unknowns at the mesh points. The basis of all numerical methods consists of this transformation of the physical equations into an algebraic, linear or non-linear, system of equations. For time-dependent problems an intermediate step is obtained, namely a *system of ordinary differential equations (ODEs) in time* which, through an integration scheme in time, will ultimately lead to an algebraic system for the unknowns at a given time level.

The introduction of the additional time variable allows the application of a whole class of schemes which form the body of the theories of numerical solutions of systems of ordinary differential equations. Since these schemes differ generally from those applied to solve algebraic systems obtained from time-independent space discretizations it is important to distinguish *time-dependent* from *time-independent* formulations. For physical time-dependent problems, such as those associated with transient flow behaviour or those connected to time-varying boundary conditions, there is obviously no alternative to the use of a time-dependent mathematical model whereby, in addition, time accuracy of the numerical solution is required.

However, with stationary problems an alternative exists, and the user can decide to work with a time-independent formulation, or apply a time-dependent model, and follow the numerical solution in time until the steady state is reached. This last family of methods is often called *time marching* or *pseudo-unsteady*, since the time accuracy is not required in order to reach the steady state in the smallest possible number of time steps. In this case the numerical schemes will be taken from the family of methods for the solution of systems of ODEs in time, while in the former the numerical solution techniques will have to rely on the methods for solving algebraic systems of equations (in space).

Although the discretization of the time derivatives ultimately leads to an algebraic system of equations for the unknowns at a given time step as a function of the variables at previous time steps the structure of these algebraic systems is generally much simpler than that obtained from time-independent formulations. We distinguish two families of methods, the *explicit* and the *implicit* methods. In explicit methods the matrix of the unknown variables at the new time is a diagonal matrix, while the right-hand side of the system is dependent only on the flow variables at the previous times. This leads therefore to a trivial matrix inversion and hence to a solution with a minimal number of arithmetic operations for each time step. However, this advantage is counterbalanced by the fact that *stability* and *convergence* conditions impose severe restrictions on the maximum admissible time step. While this might not be a limitation for physical unsteady problems it leads to the necessity of a large number of time steps in order to reach the steady-state solution corresponding to a physical time-independent problem.

In implicit methods the matrix to be inverted is not diagonal, since more than one set of variables are unknown at the same time level. In most cases, however, the structure of the matrix will be rather simple, such as block pentadiagonal, block tridiagonal or block bidiagonal, allowing simple algorithms for the solution of the system at each time step, although the number of operations required will be higher when compared with explicit methods. This is compensated by the fact that many implicit methods have, at least for linear problems, no limitation on the time step, and hence a smaller number of iterations will be needed to reach the steady state.

In time-independent formulations the space discretization will lead to a system of linear or non-linear algebraic equations. A large number of methods are available for the solution of linear algebraic systems and much research work is still being performed in this field, since this step is most important for the determination of the total computational work involved. This effort is measured in either the total computer time or the number of iterations necessary to obtain a given level of accuracy.

Two families of methods for solving algebraic systems can be distinguished: *direct* and *iterative* methods. The former can be defined as leading to the solution of a linear system in one step, while the latter will require many iterative steps. For non-linear problems all approaches will necessarily be iterative, being either direct-iterative or inserted within one of the basically linear iterative methods.

It is important to observe that the distinction between the two large classes of methods, namely the pseudo-unsteady approach and the time-independent formulation, is not as great as one might suspect from the definitions given above. Indeed, as will be discussed later, a bridge can be defined between the two formulations, and it will be shown in Chapter 12 that any iterative method for the solution of a linear, or non-linear, algebraic system can be written in a pseudo-unsteady formulation of the same problem, whereby the iteration number plays the role of an artificial time index.

Convergence acceleration techniques linked with iterative methods have recently been developed and have led to dramatic improvements in convergence rates. They are known as *preconditioning* and *multigrid* methods.

Finally, these different steps can be strongly interacting on each other. The solution techniques for the algebraic system can be greatly influenced by the type of discretization chosen as well as by the characteristics of the physical properties of the flow system. This will appear clearly in various examples in the following chapters and in particular for the discretization of hyperbolic systems such as the Euler equations. Table II.1 gives in condensed form an overview of the various approaches that can be taken. To each of the various options represented in this table corresponds an extremely large number of possible choices. Starting from the discretization technique, finite difference or finite element methods, each of these possibilities still allows an unquantifiable number of variants. In addition, as will also appear from the Chapters 11 and 12, the number of possibilities for either the resolution of the system of ODE in time or for the solution of the algebraic system of equations is considerable. This explains the large variety of methods available, or still to be developed, in order to solve numerically a given physical flow problem.

The various schemes and resolution techniques are not necessarily equivalent, either in their accuracy or in their performance characteristics, expressed as the central processor time required to obtain the solution at the desired level

Table II.1. The structure of numerical schemes

Physical problem	Mathematical model	Numerical scheme	Solution technique	Limitations
Time dependent	Time-dependent equations	Time and space discretization	ODE in time ⟨Explicit / Implicit⟩	Δt limitations / Δx
		Time accuracy required		Δt limited by accuracy
Time independent	*Steady-state equations* Linear ⟨ Non-linear⟩	Discretization of equations in space	Resolution of algebraic system of equations ⟨Direct method / Iterative method⟩	Diagonal dominance criteria / Convergence rates Acceleration techniques
		Discretization of equations in space ⊕ Linearization procedure	⊕ Iterative solution of non-linearity	Acceleration of convergence rates
	Pseudo-unsteady formulation	Time integration towards steady state	ODE systems ⟨Explicit / Implicit⟩	Δt limitations
	Time-dependent equations	Time accuracy not required		Generally no Δt limitations for linear stability

Table II.2.   Structure of a numerical simulation

166

of accuracy. Therefore in order to evaluate a selected scheme in terms of accuracy, stability, convergence properties and operation count we can rely on various techniques which will be summarized in Chapters 7–10.

In the following three chapters the three most important tools for the space and time discretization of differential operators will be presented. The *finite difference (FD) method* (Chapter 4) is probably the most popular technique today, while the *finite element (FE) method* (Chapter 5) is gaining increasing popularity for a variety of problems of fluid mechanics. The *finite volume (FV) method* (Chapter 6) by which the integral form of the conservation laws are discretized can be treated as an independent method due to its large flexibility on arbitrary meshes. For all three methods we will summarize basic properties and methodology in order to allow the reader to choose a discretization method and adapt it to his problem. As will be clear from the following, a quasi-unlimited number of options are available and a large subset of these can be chosen with equivalent properties of convergence and accuracy.

In summary, the following steps have to be defined in the process of setting up a numerical scheme:

(1) *Selection* of a discretization method of the equations. This implies selection between finite difference, finite element or finite volume methods as well as selection of the order of accuracy of the spatial and, eventually, time discretization.
(2) *Selection* of a resolution method for the system of ordinary differential equations in time, for the algebraic system of equations and for the iterative treatment of eventual non-linearities.
(3) *Analysis* of the selected numerical algorithm. This step concerns the analysis of the 'qualities' of the scheme in terms of stability and convergence properties as well as investigation of the errors generated.

Step (1) will be discussed in Chapters 4–6, step (3) in Chapters 7–10 and step (2) in Chapters 11 and 12.

Table II.2 gives a synoptic overview of the above–mentioned components.

Chapter 4

The Finite Difference Method

The finite difference method is based on the properties of Taylor expansions and on the straightforward application of the definition of derivatives. It is perhaps the simplest method to apply, particularly on uniform meshes, but it requires a high degree of regularity of the mesh. In particular, the mesh must be set up in a structured way, whereby the mesh points, in an n-dimensional space, are located at the intersections of n family of rectilinear or curved lines. These curves appear as a form of numerical coordinate lines and each point must lie on one, and only one, line of each family.

Finite difference formulas for first and higher order derivatives can be defined in a general manner and some of their properties are introduced for a one-dimensional space in section 4.1 to 4.3, under the assumption of a uniform mesh. Section 4.4 deals with two-dimensional extensions and section 4.5 introduces some applications to non-uniform meshes.

4.1 THE BASICS OF FINITE DIFFERENCE METHODS

The finite difference approximation is the oldest of the methods applied to obtain numerical solutions of differential equations, and the first application is considered to have been developed by Euler in 1768. The idea of finite difference methods is actually quite simple, since it corresponds to an estimation of a derivative by the ratio of two differences according to the definition of the derivative.

For a function $u(x)$ the derivative at point x is defined by

$$u_x \equiv \left(\frac{\partial u}{\partial x}\right) = \lim_{\Delta x \to 0} \frac{u(x + \Delta x) - u(x)}{\Delta x} \tag{4.1.1}$$

If Δx is small but finite the expression on the right-hand side is an approximation to the exact value of u_x. The approximation will be improved by reducing Δx, but for any finite value of Δx an error (the *truncation* error) is introduced which goes to zero for Δx tending to zero. The power of Δx with which this error tends to zero is called *the order of the difference approximation*, and can be obtained from a Taylor series development of $u(x + \Delta x)$ around point x. Actually, the whole concept of finite difference approximations is based on the properties of Taylor expansions. Developing $u(x + \Delta x)$

167

we obtain

$$u(x + \Delta x) = u(x) + \Delta x\, u_x(x) + \frac{\Delta x^2}{2}\, u_{xx}(x) + \cdots \tag{4.1.2}$$

and therefore, to the highest order in Δx,

$$\frac{u(x + \Delta x) - u(x)}{\Delta x} = u_x(x) + \frac{\Delta x}{2}\, u_{xx}(x) + \cdots \tag{4.1.3}$$

This approximation for $u_x(x)$ is said to be first order in Δx, and we write

$$\frac{u(x + \Delta x) - u(x)}{\Delta x} = u_x(x) + 0(\Delta x) \tag{4.1.4}$$

indicating that the truncation error $0(\Delta x)$ goes to zero like the first power in Δx.

A very large number of finite difference approximations can be obtained for the derivatives of functions and a general procedure will be described in the following based on formal difference operators and their manipulation.

4.1.1 The properties of difference formulas

Let us consider a one-dimensional space, the x-axis, where a space discretization has been performed such that the continuum is replaced by N discrete mesh points x_i, $i = 1, ..., N$ (Figure 4.1.1). We will indicate by u_i the values of the function $u(x)$ at the points x_i (that is, $u_i = u(x_i)$) and consider that the spacing between the discrete points is constant and equal to Δx. Without loss of generality we can consider that $x_i = i\,\Delta x$, and this point will also be referred to as 'point x_i' or 'point i'.

The following finite difference approximations can be defined for the first derivative $(u_x)_i \equiv (\partial u / \partial x)_{x = x_i}$.

$$(u_x)_i \equiv \left(\frac{\partial u}{\partial x}\right)_{x = x_i} = \frac{u_{i+1} - u_i}{\Delta x} + 0(\Delta x) \tag{4.1.5}$$

$$(u_x)_i \equiv \left(\frac{\partial u}{\partial x}\right)_i = \frac{u_i - u_{i-1}}{\Delta x} + 0(\Delta x) \tag{4.1.6}$$

With respect to the point $x = x_i$ the first formula is called a *forward* difference, while the second is a *backward* difference, both being first-order approximations to $(u_x)_i$. Both are considered as *one-sided difference formulas*.

A second-order approximation is obtained from the *central* difference:

$$(u_x)_i = \frac{u_{i+1} - u_{i-1}}{2\Delta x} + 0(\Delta x^2) \tag{4.1.7}$$

as is easily verified by a Taylor expansion of u_{i+1} around point x_i. These three approximations are represented geometrically in Figure 4.1.1.

The forward difference formula for $(u_x)_i$ can be considered as a central

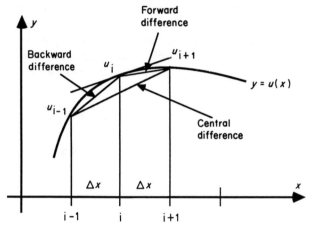

Figure 4.1.1 Geometrical interpretation of difference formulas for first-order derivatives

difference with respect to the point

$$x_{i+1/2} = \frac{x_i + x_{i+1}}{2} \qquad (4.1.8)$$

leading to a second-order approximation for the derivative $(u_x)_{i+1/2}$ in this point. This is an important property which is often used in computations due to its compact character. The same formula (4.1.5) is either a first-order forward difference for $(u_x)_i$ or a second-order central approximation for $(u_x)_{i+1/2}$ but involving only the same two mesh points i and $(i+1)$. We therefore have

$$(u_x)_{i+1/2} \equiv \left(\frac{\partial u}{\partial x}\right)_{i+1/2} = \frac{u_{i+1} - u_i}{\Delta x} + 0(\Delta x^2) \qquad (4.1.9)$$

and similarly at $(i - 1/2)$:

$$(u_x)_{i-1/2} = \frac{u_i - u_{i-1}}{\Delta x} + 0(\Delta x^2) \qquad (4.1.10)$$

Compared with formulas (4.1.5) and (4.1.6) for $(u_x)_i$ we have gained an order of accuracy by considering the same expressions as approximations for the mid-points $(i + 1/2)$ or $(i - 1/2)$, respectively.

4.1.2 Difference formulas with an arbitrary number of points

Actually, difference formulas for the first derivative $(u_x)_i$ can be constructed involving any number of adjacent points, with the order of the approximation increasing with the number of points. In any numerical scheme a balance will have to be defined between the order of accuracy and the number of points simultaneously involved in the computation. The bandwidth of the algebraic system which has finally to be solved n order to obtain the solution u_i is generally proportional to the number of simultaneous points involved. For instance, a one-sided, second-order difference formula for $(u_x)_i$, containing only the upstream points $i - 2, i - 1, i$, can be obtained by an expression of the form

$$(u_x)_i = \frac{au_i + bu_{i-1} + cu_{i-2}}{\Delta x} + 0(\Delta x^2) \qquad (4.1.11)$$

The coefficients (a, b, c) are found from a Taylor expansion of u_{i-2} and u_{i-1} around u_i. Writing

$$u_{i-2} = u_i - 2\Delta x(u_x)_i + 2\Delta x^2(u_{xx})_i - \frac{(2\Delta x)^3}{6}(u_{xxx})_i + \cdots \qquad (4.1.12)$$

$$u_{i-1} = u_i - \Delta x(u_x)_i + \frac{\Delta x^2}{2}(u_{xx})_i - \frac{\Delta x^3}{6}(u_{xxx})_i + \cdots \qquad (4.1.13)$$

and multiplying the first equation by c, the second by b and adding au_i leads to

$$cu_{i-2} + bu_{i-1} + au_i$$

$$= (a + b + c)u_i - \Delta x(2c + b)(u_x)_i + \frac{\Delta x^2}{2}(4c + b)(u_{xx})_i + 0(\Delta x^3) \qquad (4.1.14)$$

Hence identifying with equation (4.1.11) we obtain the three conditions

$$\begin{aligned} a + b + c &= 0 \\ (2c + b) &= -1 \\ 4c + b &= 0 \end{aligned} \qquad (4.1.15)$$

and the second-order accurate one-sided formula:

$$(u_x)_i = \frac{3u_i - 4u_{i-1} + u_{i-2}}{2\Delta x} + 0(\Delta x^2) \qquad (4.1.16)$$

This is a general procedure for obtaining finite difference formulas with an arbitrary number of points and an adapted order of accuracy. In general, a first-order derivative at mesh point i can be made of order of accuracy p by an explicit formula such as equation (4.1.11), involving $(p + 1)$ points. For instance, a formula involving the forward points $i + 2, i + 1, i$ is

$$(u_x)_i = \frac{-3u_i + 4u_{i+1} - u_{i+2}}{2\Delta x} + 0(\Delta x^2) \qquad (4.1.17)$$

Higher orders of accuracy can also be obtained with a reduced number of mesh points at the cost of introducing *implicit formulas* (see Section 4.3).

Finite difference approximations of higher-order derivatives can be obtained by repeated application of first-order formulas. For instance, a second-order approximation to the second derivative $(u_{xx})_i$ is obtained by

$$(u_{xx})_i \equiv \left(\frac{\partial^2 u}{\partial x^2}\right)_i = \frac{(u_x)_{i+1} - (u_x)_i}{\Delta x}$$

$$= \frac{u_{i+1} - 2u_i + u_{i-1}}{\Delta x^2} + 0(\Delta x^2)$$

(4.1.18)

where backward approximations for $(u_x)_{i+1}$ and $(u_x)_i$ are selected.

The symmetrical, central difference formula is of second-order accuracy, as can be seen from a Taylor expansion. We obtain indeed.

$$\frac{u_{i+1} - 2u_i + u_{i-1}}{\Delta x^2} = (u_{xx})_i + \frac{\Delta x^2}{12}\left(\frac{\partial^4 u}{\partial x^4}\right) + \cdots$$

(4.1.19)

As with equation (4.1.11), we can define formulas with an arbitrary number of points around point i by a combination of Taylor series developments. For instance, an expression such as equation (4.1.11) for the second derivative u_{xx} will lead to the conditions

$$a + b + c = 0$$
$$2c + b = 0$$
$$4c + b = 2$$

(4.1.20)

and the one-sided, backward formula for the second derivative:

$$(u_{xx})_i = \frac{u_i - 2u_{i-1} + u_{i-2}}{\Delta x^2} + \Delta x \cdot u_{xxx} + \cdots$$

(4.1.21)

This one-sided formula is only first-order accurate at point i. Note also that this same formula is a second-order accurate approximation to the second derivative *at point (i − 1)*, as can be seen by a comparison with the central formula (4.1.18).

The above procedure, with undetermined coefficients, can be put into a systematic framework in order to obtain finite difference approximations to all derivatives with a preselected order of accuracy. In order to achieve this a formalization of the relations between differentials and difference approximations is to be defined via the introduction of appropriate difference operators.

4.2   GENERAL METHODS FOR FINITE DIFFERENCE FORMULAS

General procedures developed in order to generate finite difference formulas to any order of accuracy and a general theory can be found in Hildebrand (1956). This approach is based on the definition of the following difference operators.

Displacement operator E:

$$Eu_i = u_{i+1}$$

(4.2.1a)

Forward difference operator δ^+:

$$\delta^+ u_i = u_{i+1} - u_i \tag{4.2.1b}$$

Backward difference operator δ^-:

$$\delta^- u_i = u_i - u_{i-1} \tag{4.2.1c}$$

Central difference operator δ:

$$\delta u_i = u_{i+1/2} - u_{i-1/2} \tag{4.2.1d}$$

Central difference operator $\bar{\delta}$:

$$\bar{\delta} u_i = \tfrac{1}{2}(u_{i+1} - u_{i-1}) \tag{4.2.1e}$$

Averaging operator μ:

$$\mu u_i = \tfrac{1}{2}(u_{+1/2} + u_{-1/2}) \tag{4.2.1f}$$

Differential operator D:

$$Du = u_x \equiv \frac{\partial u}{\partial x} \tag{4.2.1g}$$

From these definitions some obvious relations can be defined between these operators; for example,

$$\delta^+ = E - 1 \tag{4.2.2}$$

$$\delta^- = 1 - E^{-1} \tag{4.2.3}$$

where the inverse displacement operator E^{-1} is introduced, defined by

$$E^{-1} u_i = u_{i-1} \tag{4.2.4}$$

This leads to the following relations:

$$\delta^- = E^{-1} \delta^+ \tag{4.2.5}$$

and

$$\delta^+ \delta^- = \delta^- \delta^+ = \delta^+ - \delta^- = \delta^2 \tag{4.2.6}$$

With the general definition, n being positive or negative,

$$E^n u_i = u_{i+n} \tag{4.2.7}$$

we also have

$$\delta = E^{1/2} - E^{-1/2} \tag{4.2.8}$$

$$\mu = \tfrac{1}{2}(E^{1/2} + E^{-1/2}) \tag{4.2.9}$$

and

$$\delta = \tfrac{1}{2}(E - E^{-1}) \tag{4.2.10}$$

Any of the above difference operators taken to a given power, n, is interpreted as n repeated actions of this operator. For instance,

$$\delta^{+2} = \delta^+ \delta^+ = E^2 - 2E + 1 \qquad (4.2.11)$$

$$\delta^{+3} = (E-1)^3 = E^3 - 3E^2 + 3E - 1 \qquad (4.2.12)$$

4.2.1 Generation of difference formulas for first derivatives

The key to the operator technique for generating finite difference formulas lies in the relation between the derivative operator D and the finite displacement operator E. This relation is obtained from the Taylor expansion

$$u(x + \Delta x) = u(x) + \Delta x\, u_x(x) + \frac{\Delta x^2}{2!}\, u_{xx}(x) + \frac{\Delta x^3}{3!}\, u_{xxx}(x) + \cdots \qquad (4.2.13)$$

or, in operator form,

$$Eu(x) = \left(1 + \Delta x D + \frac{(\Delta x D)^2}{2!} + \frac{(\Delta x D)^3}{3!} + \cdots \right) u(x) \qquad (4.2.14)$$

This last relation can be written formally as

$$Eu(x) = e^{\Delta x D} u(x) \qquad (4.2.15)$$

and therefore we have symbolically

$$E = e^{\Delta x D} \qquad (4.2.16)$$

This relation has to be interpreted as giving identical results when acting on the exponential function e^{ax} and on any polynomial of degree n. In the latter case the expansion on the right-hand side has only n terms and therefore *all the expressions to be defined in the following are exact up to* n *terms for polynomials of degree* n. The basic operation is then to use equation (4.2.16) in the inverse way, leading to

$$\Delta x D = \ln E \qquad (4.2.17)$$

Forward differences

Formulas for forward differences are obtained by introducing relation (4.2.2) between E and the forward operator δ^+. We obtain, after a formal development of the ln function,

$$\Delta x D = \ln E = \ln(1 + \delta^+)$$

$$= \delta^+ - \frac{\delta^{+2}}{2} + \frac{\delta^{+3}}{3} - \frac{\delta^{+4}}{4} + \cdots \qquad (4.2.18)$$

The order of accuracy of the approximation increases with the number of terms kept in the right-hand side. The first neglected term gives the *truncation error*. For instance, keeping the first term only leads to the first-order formula (4.1.5) and a truncation error equal to $(\Delta x / 2u_{xx})$. If the first two terms are considered we obtain the second-order formula (4.1.17) with the truncation error $\Delta x^3/3 u_{xxx}$:

$$(u_x)_i \equiv Du_i = \frac{-3u_i + 4u_{i+1} - u_{i+2}}{2\Delta x} + \frac{\Delta x^2}{3} u_{xxx} \qquad (4.2.19)$$

Hence this relation leads to the definition of various forward finite difference formulas for the first derivative with an increasing order of accuracy. Since the forward difference operator can be written as $\delta^+ = \Delta x u_x + 0(\Delta x^2)$, the first neglected operator δ^{+n} is of order n, showing that the truncation error is $0(\Delta x^{n-1})$.

Backward Differences

Similarly, backward difference formulas with increasing order of accuracy can be obtained by application of relation (4.2.3):

$$\Delta x D = \ln E = -\ln(1 - \delta^-)$$

$$= \delta^- + \frac{\delta^{-2}}{2} + \frac{\delta^{-3}}{3} + \frac{\delta^{-4}}{4} + \cdots \qquad (4.2.20)$$

To second-order accuracy we have

$$(u_x)_i = Du_i = \frac{3u_i - 4u_{i-1} + u_{i-2}}{2\Delta x} + \frac{\Delta x^2}{3} u_{xxx} \qquad (4.2.21)$$

Central differences

Central difference formulas are obtained from equation (4.2.8):

$$\delta u_i = u_{i+1/2} - u_{i-1/2} = (E^{1/2} - E^{-1/2})u_i$$

and therefore

$$\delta = e^{\Delta x D/2} - e^{-\Delta x D/2} = 2 \sinh\left(\frac{\Delta x D}{2}\right) \qquad (4.2.22)$$

which, through inversion, leads to

$$\Delta x D = 2 \sinh^{-1} \delta/2 = 2\left[\frac{\delta}{2} - \frac{1}{2\cdot 3}\left(\frac{\delta}{2}\right)^3 + \frac{1\cdot 3}{2\cdot 4\cdot 5}\left(\frac{\delta}{2}\right)^5\right.$$

$$\left. - \frac{1\cdot 3\cdot 5}{2\cdot 4\cdot 6\cdot 7}\left(\frac{\delta}{2}\right)^7 + \cdots\right) \qquad (4.2.23)$$

$$= \delta - \frac{\delta^3}{24} + \frac{3\delta^5}{640} - \frac{5\delta^7}{7168} + \cdots$$

This formula generates a family of central difference approximations to the first-order derivative $(u_x)_i$ based on the values of the function u at half-integer mesh point locations. By keeping only the first term we obtain, with second-order accuracy,

$$(u_x)_i = \frac{u_{i+1/2} - u_{i-1/2}}{\Delta x} - \frac{\Delta x^2}{24} u_{xxx} \tag{4.2.24}$$

Keeping the first two terms we obtain a fourth-order accurate approximation:

$$(u_x)_i = \frac{-u_{i+3/2} + 27u_{i+1/2} - 27u_{i-1/2} + u_{i-3/2}}{24\Delta x} + \frac{3}{640} \Delta x^4 \left(\frac{\partial^5 u}{\partial x^5}\right) \tag{4.2.25}$$

In order to derive central differences involving only integer mesh points we could apply the above procedure to the operator $\bar{\delta}$. From equation (4.2.10) we have

$$\bar{\delta} = \tfrac{1}{2}(E - E^{-1}) = \tfrac{1}{2}(e^{\Delta xD} - e^{-\Delta xD}) = \sinh(\Delta xD) \tag{4.2.26}$$

and therefore, as a function of $\bar{\delta}$,

$$\begin{aligned} \Delta xD &= \sinh^{-1}\bar{\delta} \\ &= \left(\bar{\delta} - \frac{\bar{\delta}^3}{6} + \frac{3}{2 \cdot 4 \cdot 5} \bar{\delta}^5 + \cdots\right) \end{aligned} \tag{4.2.27}$$

This formula can be used to replace equation (4.2.23) for the central difference at x_i. However, although the first term is the second-order central difference approximation (4.1.7) the next one leads to a fourth-order formula for $(u_x)_i$, involving the four points $i-3, i-1, i+1, i+3$. This is of no interest for numerical computations, since we would expect a fourth-order formula for $(u_x)_i$ to involve the points $i-2, i-1, i+1, i+2$. This can be obtained from the identity

$$\mu^2 = 1 + \delta^2/4 \tag{4.2.28}$$

After multiplication of equation (4.2.23) by

$$1 = \mu\left(1 + \frac{\delta^2}{4}\right)^{-1/2} = \mu\left(1 - \frac{\delta^2}{8} + \frac{3\,\delta^4}{128} - \frac{5\,\delta^6}{1024} + \cdots\right) \tag{4.2.29}$$

we obtain the relation

$$\begin{aligned} \Delta xD &= \mu\left(\delta - \frac{1}{3!}\delta^3 + \frac{1^2 2^2}{5!}\delta^5 - \cdots\right) \\ &= \bar{\delta}\left(1 - \frac{\delta^2}{3!} + \frac{2^2}{5!}\delta^4 - \frac{2^2 \cdot 3^2}{7!}\delta^6 + \cdots\right) \end{aligned} \tag{4.2.30}$$

Hence we obtain the following second- and fourth-order accurate central difference approximations to the derivative $(u_x)_i$ with integer mesh point values:

$$(u_x)_i = \frac{u_{i+1} - u_{i-1}}{2\Delta x} - \frac{\Delta x^2}{6} u_{xxx} \tag{4.2.31}$$

and

$$(u_x)_i = \frac{-u_{i+2} + 8u_{i+1} - 8u_{i-1} + u_{i-2}}{12\Delta x} + \frac{\Delta x^4}{30}\left(\frac{\partial^5 u}{\partial x^5}\right) \tag{4.2.32}$$

4.2.2 Higher-order derivatives

Applying the operator technique an unlimited number of finite difference formulas can be applied to obtain second-order and higher-order derivatives. From equation (4.2.18) we have the one-sided, forward difference formula, see for instance Ames (1977),

$$\left(\frac{\partial^n u}{\partial x^n}\right)_i = D^n u_i = \frac{1}{\Delta x^n}\left[\ln(1 + \delta^+)\right]^n u_i$$

$$= \frac{1}{\Delta x^n}\left[\delta^{+n} - \frac{n}{2}\delta^{+(n+1)} + \frac{n(3n+5)}{24}\delta^{+(n+2)}\right.$$

$$\left. - \frac{n(n+2)(n+3)}{48}\delta^{+(n+3)} + \cdots\right]u_i \tag{4.2.33}$$

In terms of the backward difference operator δ^- we have

$$\left(\frac{\partial^n u}{\partial x^n}\right)_i = -\frac{1}{\Delta x^n}\left[\ln(1 - \delta^-)\right]^n u_i$$

$$= \frac{1}{\Delta x^n}\left(\delta^- + \frac{\delta^{-2}}{2} + \frac{\delta^{-3}}{3} + \cdots\right)^n u_i \tag{4.2.34}$$

$$= \frac{1}{\Delta x^n}\left[\delta^{-n} + \frac{n}{2}\delta^{-(n+1)} + \frac{n(3n+5)}{24}\delta^{-(n+2)}\right.$$

$$\left. + \frac{n(n+2)(n+3)}{48}\delta^{-(n+3)} + \cdots\right]u_i$$

Central difference formulas for higher-order derivatives can also be obtained through

$$D^n u_i = \left(\frac{2}{\Delta x}\sinh^{-1}\frac{\delta}{2}\right)^n u_i$$

$$= \frac{1}{\Delta x^n}\left[\delta - \frac{\delta^3}{24} + \frac{3\delta^5}{640} - \frac{5\delta^7}{7168} + \cdots\right]^n u_i$$

$$= \frac{1}{\Delta x^n}\delta^n\left[1 - \frac{n}{24}\delta^2 + \frac{n}{64}\left(\frac{22+5n}{90}\right)\delta^4\right. \tag{4.2.35}$$

$$\left. - \frac{n}{4^5}\left(\frac{5}{7} + \frac{n-1}{5} + \frac{(n-1)(n-2)}{3^5}\right)\delta^6 + \cdots\right]u_i$$

For *n* even, this equation generates difference formulas with the function values at the integer mesh point. For *n* uneven, the difference formulas involve points at half-integer mesh points. The reverse is true for the following difference equation.

In order to involve only points at integer values of *i* for *n* uneven, using equation (4.2.28), we define

$$D^n u_i = \frac{\mu}{[1 + (\delta^2/4)]^{1/2}} \left(\frac{2}{\Delta x} \sinh^{-1} \frac{\delta}{2} \right)^n u_i$$

$$= \mu \frac{\delta^n}{\Delta x^n} \left[1 - \frac{n+3}{24} \delta^2 + \frac{5n^2 + 52n + 135}{5760} \delta^4 + \cdots \right] u_i$$

(4.2.36)

Second-order derivative

For instance, second-order derivative formulas are

$$(u_{xx})_i = \frac{1}{\Delta x^2} \left(\delta^{+2} - \delta^{+3} + \frac{11}{12} \delta^{+4} - \frac{5}{6} \delta^{+5} + \cdots \right) u_i \qquad (4.2.37)$$

$$(u_{xx})_i = \frac{1}{\Delta x^2} \left(\delta^{-2} + \delta^{-3} + \frac{11}{12} \delta^{-4} + \frac{5}{6} \delta^{-5} + \cdots \right) u_i \qquad (4.2.38)$$

$$(u_{xx})_i = \frac{1}{\Delta x^2} \left(\delta^2 - \frac{\delta^4}{12} + \frac{\delta^6}{90} - \frac{\delta^8}{560} + \cdots \right) u_i \qquad (4.2.39)$$

$$(u_{xx})_i = \frac{\mu}{\Delta x^2} \left(\delta^2 - \frac{5\delta^4}{24} + \frac{259}{5760} \delta^6 + 0(\Delta x^8) \right) u_i \qquad (4.2.40)$$

These equations define four families of difference operators for the second derivative to various orders of accuracy. By maintaining only the first term we obtain the following difference formulas.

Forward difference: first-order accurate

$$(u_{xx})_i = \frac{1}{\Delta x^2} (u_{i+2} - 2u_{i+1} + u_i) - \Delta x \, u_{xxx} \qquad (4.2.41)$$

Backward difference: first-order accurate

$$(u_{xx})_i = \frac{1}{\Delta x^2} (u_i - 2u_{i-1} + u_{i-2}) + \Delta x \, u_{xxx} \qquad (4.2.42)$$

Central difference: integer points—second-order accurate

$$(u_{xx})_i = \frac{1}{\Delta x^2} (u_{i+1} - 2u_i + u_{i-1}) - \frac{\Delta x^2}{12} \left(\frac{\partial^4 u}{\partial x^4} \right) \qquad (4.2.43)$$

Central difference: half-integer mesh points—second-order accurate

$$(u_{xx})_i = \frac{1}{2\Delta x^2} (u_{i+3/2} - u_{i+1/2} - u_{i-1/2} + u_{i-3/2}) - \frac{5}{24} \Delta x^2 \left(\frac{\partial^4 u}{\partial x^4}\right) \quad (4.2.44)$$

With the exception of the last, these difference approximations for the second derivative involve three mesh points like the first derivatives. The one-sided difference formulas are only first-order accurate, while the central differences always lead to a higher order of accuracy.

By keeping the two first terms of the above formulas we obtain difference formulas with a higher order of accuracy:

Forward difference: second-order accurate

$$(u_{xx})_i = \frac{1}{\Delta x^2} (2u_i - 5u_{i+1} + 4u_{i+2} - u_{i+3}) + \frac{11}{12} \Delta x^2 \left(\frac{\partial^4 u}{\partial x^4}\right) \quad (4.2.45)$$

Backward difference: second-order accurate

$$(u_{xx})_i = \frac{1}{\Delta x^2} (2u_i - 5u_{i-1} + 4u_{i-2} - u_{i-3}) - \frac{11}{12} \Delta x^2 \left(\frac{\partial^4 u}{\partial x^4}\right) \quad (4.2.46)$$

Central difference: integer points—fourth-order accurate

$$(u_{xx})_i = \frac{1}{12\Delta x^2} (-u_{i+2} + 16u_{i+1} - 30u_i + 16u_{i-1} - u_{i-2})$$

$$+ \frac{\Delta x^4}{90} \left(\frac{\partial^6 u}{\partial x^6}\right) \quad (4.2.47)$$

Central difference: half-integer mesh points—fourth-order accurate

$$(u_{xx})_i = \frac{1}{48\Delta x^2} (-5u_{i+5/2} + 39u_{i+3/2} - 34u_{i+1/2} - 34u_{i-1/2}$$

$$+ 39u_{i-3/2} - 5u_{i-5/2}) + \frac{259}{5760} \Delta x^4 \left(\frac{\partial^6 u}{\partial x^6}\right) \quad (4.2.48)$$

The last formula is of little practical use, since it requires six mesh points to obtain a fourth-order accurate approximation to the second derivative at point i, while formula (4.2.47) requires only four mesh points.

A more complex operator, often occurring in second-order differential problems, is $\partial_x[k(x)\partial_x u]$. A central difference formula of second-order accuracy with three mesh points is given by

$$\frac{\partial}{\partial x}\left[k(x)\frac{\partial}{\partial x}\right] u_i = \frac{1}{\Delta x^2} \delta^+ (k_{i-1/2}\, \delta^-)\, u_i + 0(\Delta x^2) \quad (4.2.49)$$

which takes the explicit form

$$\frac{\partial}{\partial x}\left[k(x)\frac{\partial}{\partial x}\right]u_i = \frac{k_{+1/2}(u_{i+1}-u_i)}{\Delta x^2} - \frac{k_{i-1/2}(u_i-u_{i-1})}{\Delta x^2} + 0(\Delta x^2) \quad (4.2.50)$$

An equivalent formula is obtained by inverting the forward and backward operators:

$$\frac{\partial}{\partial x}\left[k(x)\frac{\partial}{\partial x}\right]u_i = \frac{1}{\Delta x^2}\,\delta^-(k_{i+1/2}\,\delta^+)\,u_i \quad (4.2.51)$$

leading to the same expression (4.2.50).

Third-order derivatives

Approximations for third-order derivatives are obtained from the above general expressions. To the lowest orders of accuracy we have the following difference formulas.

Forward difference

$$\left(\frac{\partial^3 u}{\partial x^3}\right)_i \equiv (u_{xxx})_i$$

$$= \frac{1}{\Delta x^3}(u_{i+3}-3u_{i+2}+3u_{i+1}-u_i) - \frac{\Delta x}{2}\left(\frac{\partial^4 u}{\partial x^4}\right) \quad (4.2.52)$$

or with second-order accuracy:

$$(u_{xxx})_i = \frac{1}{2\Delta x^3}(-3u_{i+4}+14u_{i+3}-24u_{i+2}+18u_{i+1}-5u_i)$$

$$+ \frac{21}{12}\,\Delta x^2\left(\frac{\partial^5 u}{\partial x^5}\right) \quad (4.2.53)$$

Backward difference

$$(u_{xxx})_i = \frac{1}{\Delta x^3}(u_i-3u_{i-1}+3u_{i-2}-u_{i-3}) + \frac{\Delta x}{2}\left(\frac{\partial^4 u}{\partial x^4}\right) \quad (4.2.54)$$

or with second-order accuracy:

$$(u_{xxx})_i = \frac{1}{2\Delta x^3}(5u_i-18u_{i-1}+24u_{i-2}-14u_{i-3}+3u_{i-4})$$

$$- \frac{21}{12}\,\Delta x^2\left(\frac{\partial^5 u}{\partial x^5}\right) \quad (4.2.55)$$

Central difference: half-integer points

$$(u_{xxx})_i = \frac{1}{\Delta x^3}(u_{i+3/2}-3u_{i+1/2}+3u_{i-1/2}-u_{i-3/2}) - \frac{\Delta x^2}{8}\left(\frac{\partial^5 u}{\partial x^5}\right) \quad (4.2.56)$$

This is a second-order accurate approximation to the third derivative, and a fourth-order accuracy is obtained from the following:

$$(u_{xxx})_i = \frac{1}{8\Delta x^3} \left(-u_{i+5/2} + 13u_{i+3/2} - 34u_{i+1/2} + 34u_{i-1/2} - 13u_{i-3/2} \right.$$

$$\left. + u_{i-5/2} \right) + \frac{37}{1920} \Delta x^4 \left(\frac{\partial^7 u}{\partial x^7} \right) \tag{4.2.57}$$

Central difference: integer mesh point

$$(u_{xxx})_i = \frac{1}{2\Delta x^3} \left(u_{i+2} - 2u_{i+1} + 2u_{i-1} - u_{i-2} \right) - \frac{1}{4} \Delta x^2 \left(\frac{\partial^5 u}{\partial x^5} \right) \tag{4.2.58}$$

or with fourth-order accuracy:

$$(u_{xxx})_i = \frac{1}{8\Delta x^3} \left(-u_{i+3} + 8u_{i+2} - 13u_{i+1} + 13u_{i-1} - 8u_{i-2} + u_{i-3} \right)$$

$$+ \frac{7}{120} \Delta x^4 \left(\frac{\partial^7 u}{\partial x^7} \right) \tag{4.2.59}$$

Fourth-order derivatives

To the lowest order of accuracy we have the following approximations.

Forward difference: first-order accurate

$$\left(\frac{\partial^4 u}{\partial x^4} \right)_i = \frac{1}{\Delta x^4} \left(u_{i+4} - 4u_{i+3} + 6u_{i+2} - 4u_{i+1} + u_i \right) - 2\Delta x \left(\frac{\partial^5 u}{\partial x^5} \right) \tag{4.2.60}$$

Backward difference: first-order accurate

$$\left(\frac{\partial^4 u}{\partial x^4} \right)_i = \frac{1}{\Delta x^4} \left(u_i - 4u_{i-1} + 6u_{i-2} - 4u_{i-3} + u_{i-4} \right) + 2\Delta x \left(\frac{\partial^5 u}{\partial x^5} \right) \tag{4.2.61}$$

Central difference: second-order accurate

$$\left(\frac{\partial^4 u}{\partial x^4} \right)_i = \frac{1}{\Delta x^4} \left(u_{i+2} - 4u_{i+1} + 6u_i - 4u_{i-1} + u_{i-2} \right) - \frac{\Delta x^2}{6} \left(\frac{\partial^6 u}{\partial x^6} \right) \tag{4.2.62}$$

Obtaining these formulas is left as an exercise to the reader (see Problems 4.13–4.15).

4.3 IMPLICIT FINITE DIFFERENCE FORMULAS

Implicit formulas are defined as expressions where derivatives at different mesh points appear simultaneously. Their essential advantage comes from the high order of accuracy which is generated when derivatives at different mesh points are related to each other. The price to be paid is that we generate an algebraic system for the approximated derivatives which cannot be written in an explicit way. The above expressions can be used to generate these high-order implicit formulas for the derivative operators as follows.

For instance, equation (4.2.30) gives, with a fourth-order accuracy,

$$\Delta x D = \mu \delta \left(1 - \frac{\delta^2}{6} \right) + 0(\Delta x^5) \tag{4.3.1}$$

or by a formal operation, to the same order of accuracy:

$$\Delta x D = \frac{\mu \delta}{1 + (\delta^2/6)} + 0(\Delta x^5) \tag{4.3.2}$$

This formula is a *rational fraction* or PADE *differencing approximation* (Kopal, 1961).

The interpretations of these two last formulas are quite distinct from each other. Equation (4.3.1), applied to u_i leads to the fourth-order formula (4.2.32), while equation (4.3.2) is to be interpreted *after* multiplication of both sides by the operator $(1 + \delta^2/6)$:

$$\left(1 + \frac{\delta^2}{6} \right) D = \frac{1}{6} \left[(u_x)_{i+1} + 4(u_x)_i + (u_x)_{i-1} \right]$$
$$\tag{4.3.3}$$
$$= \frac{u_{i+1} - u_{i-1}}{2 \Delta x} + 0(\Delta x^4)$$

The left-hand side has an implicit structure, and this formula has the important property of involving only three spatial points while being of the same fourth order as equation (4.2.32), which requires five mesh points. These schemes are called sometimes *Hermitian* schemes, and can also be obtained from a finite element formulation (see Chapter 5).

Similar procedures can be applied to generate other implicit formulas; for instance, equation (4.2.18) leads to

$$\Delta x D = \delta^+ - \frac{\delta^{+2}}{2} + 0(\Delta x^3) = \delta^+ \left(1 - \frac{\delta^+}{2} \right) + 0(\Delta x^3) = \frac{\delta^+}{1 + (\delta^+/2)} + 0(\Delta x^3)$$
$$\tag{4.3.4}$$

After multiplication by $1 + \delta^+/2$ we obtain the two-point implicit relation, of second-order accuracy,

$$\frac{1}{2}[(u_x)_i + (u_x)_{i+1}] = \frac{u_{i+1} - u_i}{\Delta x} + 0(\Delta x^2) \tag{4.3.5}$$

Formulas such as (4.3.3) or (4.3.5) do not permit the explicit determination of the numerical approximations to the derivatives $(u_x)_i$. Instead they have to be written for all the mesh points and solved simultaneously as an algebraic system of equations for the unknowns $(u_x)_i$, $i = 1, ..., N$.

For instance, the fourth-order implicit approximation (4.3.3) for $(u_x)_i$ will be obtained from the solution of the *tridiagonal* system:

$$\begin{vmatrix} \cdot & \cdot & & \\ 1 & 4 & 1 & \\ & 1 & 4 & 1 \\ & & 1 & 4 & 1 \\ & & & \cdot \end{vmatrix} \cdot \begin{vmatrix} \vdots \\ (u_x)_{i-1} \\ (u_x)_i \\ (u_x)_{i+1} \\ \cdot \end{vmatrix} = \frac{3}{\Delta x} \begin{vmatrix} \vdots \\ u_i - u_{i-2} \\ u_{i+1} - u_{i-1} \\ u_{i+2} - u_i \\ \cdot \end{vmatrix}$$

while equation (4.3.5) leads to a *bidiagonal* system:

$$\begin{vmatrix} \vdots & \cdot & & \\ 1 & 1 & & \\ & 1 & 1 & \\ & & 1 & 1 \\ & & & \cdot & \vdots \end{vmatrix} \begin{vmatrix} \vdots \\ (u_x)_{i-1} \\ (u_x)_i \\ (u_x)_{i+1} \\ \cdot \end{vmatrix} = \frac{2}{\Delta x} \begin{vmatrix} \vdots \\ u_i - u_{i-1} \\ u_{i+1} - u_i \\ u_{i+2} - u_{i+1} \\ \cdot \end{vmatrix}^0$$

As a consequence, the numerical value of $(u_x)_i$, obtained as a solution of the above systems, is influenced by *all* the mesh point values u_i. This explains why these formulas are of a higher order of accuracy than the corresponding explicit formulas involving the same number of mesh points. When applied to practical flow problems the function values and the derivatives are considered as unknowns. They are obtained as solutions of an algebraic system formed by adding the basic equations to be solved to the above implicit relations.

Along the same lines, we obtain implicit formulas for second-order derivatives with a higher order of accuracy and a number of mesh point values limited to two or three. From equation (4.2.39) we have, to fourth-order accuracy,

$$(u_{xx})_i = \frac{1}{\Delta x^2} \delta^2 \left(1 - \frac{\delta^2}{12}\right) u_i + 0(\Delta x^4)$$

$$= \frac{1}{\Delta x^2} \frac{\delta^2 u_i}{1 + (\delta^2/12)} + 0(\Delta x^4)$$

(4.3.6)

Multiplying formally by $[1 + (\delta^2/12)]$ we obtain the implicit, compact expression for the second-order derivative:

$$\left(1 + \frac{\delta^2}{12}\right)(u_{xx})_i = \frac{1}{\Delta x^2} \delta^2 u_i + 0(\Delta x^4)$$

(4.3.7a)

or

$$\frac{1}{12}[(u_{xx})_{i+1} + 10(u_{xx})_i + (u_{xx})_{i-1}] = \frac{1}{\Delta x^2}(u_{i+1} - 2u_i + u_{i-1}) + 0(\Delta x^4)$$

(4.3.7b)

Here again a tridiagonal system is to be solved in order to calculate $(u_{xx})_i$ from the mesh point values u_i.

A one-sided relation of second-order accuracy can be obtained from equation (4.2.37), leading to

$$(u_{xx})_i = \frac{1}{\Delta x^2} \delta^+ (1 - \delta^{+^2}) \, u_i + 0(\Delta x^2)$$

(4.3.8)

$$= \frac{1}{\Delta x^2} \frac{\delta^+ u_i}{1 + \delta^{+^2}} + 0(\Delta x^2)$$

After multiplication by $(1 + \delta^{+^2})$ we obtain

$$(u_{xx})_{i+2} - 2(u_{xx})_{i+1} + (u_{xx})_i = \frac{1}{\Delta x^2} (u_{i+1} - u_i) + 0(\Delta x^2) \qquad (4.3.9)$$

There is no way of obtaining an implicit relation for the second derivatives with only values at the two mesh points i and $i + 1$ without also involving first-derivative values (Hirsh, 1975).

4.3.1 General derivation of implicit finite difference formulas for first and second derivatives

Implicit finite difference relations for first and second derivatives have been derived by various methods and given a variety of names. Many can be found in Collatz (1966), under the names of the *Mehrstellen* or *Hermitian* methods by analogy with Hermitian finite elements. We have already mentioned Pade approximations, and recently a large number of applications to the solution of fluid-mechanical equations have been developed by Krause (1971); Hirsh (1975) as *compact* methods; Rubin and Graves (1975); Rubin and Khosla (1976) as *spline* methods; Adam (1975, 1977), Ciment and Leventhal (1975); and Leventhal (1980) as *(operator) compact implicit* (OCI) methods. However, following Peyret (1978) (see also Peyret and Taylor, 1982) all the implicit formulas can be derived in a systematic way from a Taylor series expansion.

With a limitation to three-point expressions the general form of an implicit finite difference relation between a function and its first two derivatives would be

$$a_+ u_{i+1} + a_0 u_i + a_- u_{i-1} + b_+ (u_x)_{i+1} + b_0 (u_x)_i + b_- (u_x)_{i-1}$$
$$+ c_+ (u_{xx})_{i+1} + c_0 (u_{xx})_i + c_- (u_{xx})_{i-1} = 0 \qquad (4.3.10)$$

Developing all the variables in a Taylor series about point i we have the following expansion for equal mesh spacing:

$$u_{i \pm 1} = u_i \pm \Delta x (u_x)_i + \frac{\Delta x^2}{2} (u_{xx})_i \pm \frac{\Delta x^3}{6} (u_{xxx})_i + \frac{\Delta x^4}{24} \left(\frac{\partial^4 u}{\partial x^4}\right)_i \pm \frac{\Delta x^5}{5!} \left(\frac{\partial^5 u}{\partial x^5}\right)_i$$

$$+ \frac{\Delta x^6}{6!} \left(\frac{\partial^6 u}{\partial x^6}\right)_i \pm \frac{\Delta x^7}{7!} \left(\frac{\partial^7 u}{\partial x^7}\right)_i + \frac{\Delta x^8}{8!} \left(\frac{\partial^8 u}{\partial x^8}\right)_i + \cdots \qquad (4.3.11)$$

$$(u_x)_{i\pm 1} = (u_x)_i \pm \Delta x (u_{xx})_i + \frac{\Delta x^2}{2}(u_{xxx})_i \pm \frac{\Delta x^3}{6}\left(\frac{\partial^4 u}{\partial x^4}\right)_i + \frac{\Delta x^4}{24}\left(\frac{\partial^5 u}{\partial x^5}\right)_i$$

$$\pm \frac{\Delta x^5}{5!}\left(\frac{\partial^6 u}{\partial x^6}\right)_i + \frac{\Delta x^6}{6!}\left(\frac{\partial^7 u}{\partial x^7}\right)_i \pm \frac{\Delta x^7}{7!}\left(\frac{\partial^8 u}{\partial x^8}\right)_i + \cdots \qquad (4.3.12)$$

$$(u_{xx})_{i\pm 1} = (u_{xx})_i \pm \Delta x (u_{xxx})_i + \frac{\Delta x^2}{2}\left(\frac{\partial^4 u}{\partial x^4}\right)_i \pm \frac{\Delta x^3}{6}\left(\frac{\partial^5 u}{\partial x^5}\right)_i + \frac{\Delta x^4}{24}\left(\frac{\partial^6 u}{\partial x^6}\right)_i$$

$$\pm \frac{\Delta x^5}{5!}\left(\frac{\partial^7 u}{\partial x^7}\right)_i + \frac{\Delta x^6}{6!}\left(\frac{\partial^8 u}{\partial x^8}\right)_i + \cdots \qquad (4.3.13)$$

When introduced into the implicit relation (4.3.10) we can request the coefficients up to the third-order derivative of the truncation error to vanish, in order to obtain at least second-order accuracy for the second derivatives. This leads to the conditions

$$a_+ + a_0 - a_- = 0$$

$$\Delta x(a_+ - a_-) + b_+ + b_0 + b_- = 0$$

$$\frac{\Delta x^2}{2}(a_+ + a_-) + \Delta x(b_+ - b_-) + c_+ + c_0 + c_- = 0 \qquad (4.3.14)$$

$$\frac{\Delta x^3}{6}(a_+ - a_-) + \frac{\Delta x^2}{2}(b_+ + b_-) + \Delta x(c_+ - c_-) = 0$$

from which we can choose to eliminate a_+, a_0, a_- and b_0, for instance (other choices are obviously possible, see Problem 4.6):

$$a_+ = \frac{1}{2\Delta x}\left[-5b_+ - b_- + \frac{2}{\Delta x}(2c_- - 4c_+ - c_0)\right]$$

$$a_0 = \frac{2}{\Delta x}\left[b_+ - b_- + \frac{1}{\Delta x}(c_+ + c_0 + c_-)\right]$$

$$a_- = \frac{1}{2\Delta x}\left[b_+ + 5b_- + \frac{2}{\Delta x}(2c_+ - 4c_- - c_0)\right] \qquad (4.3.15)$$

$$b_0 = 2(b_+ + b_-) + \frac{6}{\Delta x}(c_+ - c_-)$$

and the truncation error R reduces to

$$R = \frac{\Delta x^3}{24}\left[2(b_+ - b_-) + \frac{10}{\Delta x}(c_+ + c_-) - \frac{2}{\Delta x}c_0\right]\frac{\partial^4 u}{\partial x^4}$$

$$+ \frac{\Delta x^4}{120}\left[2(b_+ + b_-) + \frac{14}{\Delta x}(c_+ - c_-)\right]\frac{\partial^5 u}{\partial x^5}$$

$$+ \frac{\Delta x^5}{6!}\left[4(b_+ - b_-) + \frac{28}{\Delta x}(c_+ + c_-) - \frac{2}{\Delta x}c_0\right]\frac{\partial^6 u}{\partial x^6}$$

$$+ \frac{\Delta x^6}{7!} \left[4(b_+ + b_-) + \frac{36}{\Delta x}(c_+ - c_-) \right] \frac{\partial^7 u}{\partial x^7}$$

$$+ \frac{\Delta x^7}{8!} \left[6(b_+ - b_-) + \frac{54}{\Delta x}(c_+ + c_-) - \frac{2}{\Delta x}c_0 \right] \frac{\partial^8 u}{\partial x^8} \qquad (4.3.16)$$

Hence we have a four-parameter family of implicit relations (one parameter may always be set arbitrarily to one since equation (4.3.10) is homogeneous). These parameters can be selected on the basis of various conditions, according to the number of derivatives and mesh points we wish to maintain in the implicit relation or by imposing a minimum order of accuracy. For instance, the second-order relation (4.3.7) is obtained with $b_+ = b_- = b_0 = 0$ and by selecting $c_+ = c_- = 1$, $c_0 = 10$.

As can be seen from the expression of the truncation error, the highest order of accuracy that can be achieved is six. This is obtained by imposing the coefficients of the three first terms in R to vanish. This gives the relations

$$b_+ = \frac{1}{\Delta x}(8c_+ + c_-)$$

$$b_- = \frac{1}{\Delta x}(c_+ + 8c_-) \qquad (4.3.17)$$

$$c_0 = -4(c_+ + c_-)$$

Inserted into the above formulas a one-parameter family of implicit relations is obtained between the function u and its first two derivatives, with $\alpha = c_-/c_+$:

$$\frac{3}{2\Delta x^2}(13 + 3\alpha)u_{i+1} - \frac{24}{\Delta x^2}(1 + \alpha)u_i + \frac{3}{2\Delta x^2}(3 + 13\alpha)u_{i-1}$$

$$- \frac{1}{\Delta x}(8 + \alpha)(u_x)_{i+1} - \frac{8}{\Delta x}(1 - \alpha)(u_x)_i + \frac{1}{\Delta x}(1 + 8\alpha)(u_x)_{i-1}$$

$$+ (u_{xx})_{i+1} - 4(1 + \alpha)(u_{xx})_i + \alpha(u_{xx})_{i-1} = 0 \qquad (4.3.18)$$

with the truncation error

$$R = \frac{8\Delta x^5}{7!}(1 - \alpha)\frac{\partial^7 u}{\partial x^7} + \frac{\Delta x^6}{8!}(1 + \alpha)\frac{\partial^8 u}{\partial x^8} \qquad (4.3.19)$$

The unique, implicit relation of order six is obtained from $\alpha = 1$:

$$\frac{24}{\Delta x^2}(u_{i+1} - 2u_i + u_{i-1}) - \frac{9}{\Delta x}[(u_x)_{i+1} - (u_x)_{i-1}]$$

$$+ (u_{xx})_{i+1} - 8(u_{xx})_i + (u_{xx})_{i-1} = 0 \qquad (4.3.20)$$

with a truncation error

$$R = \frac{2}{8!}\Delta x^6 \left(\frac{\partial^8 u}{\partial x^8} \right) \qquad (4.3.21)$$

Implicit relations, with first derivatives only, are obtained from $c_+ = c_0 = c_- = 0$, and can therefore be, at most, fourth-order accurate. From equations (4.3.15) and (4.3.16) we obtain the one parameter family, with $\beta = b_-/b_+$:

$$\frac{1}{2\Delta x}(-5 - \beta)u_{i+1} + \frac{2}{\Delta x}(1 - \beta)u_i + \frac{1}{2\Delta x}(1 + 5\beta)u_{i-1}$$

$$+ (u_x)_{i+1} + 2(1 + \beta)(u_x)_i + \beta(u_x)_{i-1} = 0 \qquad (4.3.22)$$

with a truncation error

$$R = \frac{\Delta x^3}{12}(1 - \beta)\frac{\partial^4 u}{\partial x^4} + \frac{\Delta x^4}{60}(1 + \beta)\frac{\partial^5 u}{\partial x^5} \qquad (4.3.23)$$

For $\beta = 1$ we obtain the unique fourth-order relation (4.3.3). For other choices of β the formula is only third-order accurate.

Two-point implicit difference formulas

The most general two-point relation, with at least second-order accuracy for the second derivatives, is obtained from $a_- = b_- = c_- = 0$. We obtain the one-parameter family of relations from equation (4.3.14) with $\gamma = b_+/(\Delta x a_+)$

$$\frac{1}{\Delta x^2}(u_i - u_{i+1}) + \frac{1 + \gamma}{\Delta x}(u_x)_i - \frac{\gamma}{\Delta x}(u_x)_{i+1}$$

$$+ \frac{1}{6}[(1 + 3\gamma)(u_{xx})_{i+1} + (2 + 3\gamma)(u_{xx})_i] = 0 \qquad (4.3.24)$$

with the truncation error

$$R = \frac{\Delta x^2}{12}\left(\gamma + \frac{1}{2}\right)\frac{\partial^4 u}{\partial x^4} + \frac{\Delta x^3}{24}\left(\gamma + \frac{7}{15}\right)\frac{\partial^5 u}{\partial x^5} \qquad (4.3.25)$$

For $\gamma = -1/2$ we have the unique third-order accurate relation,

$$\frac{u_{i+1} - u_i}{\Delta x^2} - \frac{1}{2\Delta x}[(u_x)_{i+1} + (u_x)_i] + \frac{1}{12}[(u_{xx})_{i+1} - (u_{xx})_i] = 0 \quad (4.3.26)$$

with the truncation error

$$R = \frac{-\Delta x^3}{720}\left(\frac{\partial^5 u}{\partial x^5}\right) \qquad (4.3.27)$$

Many other formulas can be derived, according to the points and/or the derivatives we wish to isolate.

4.4 MULTI-DIMENSIONAL FINITE DIFFERENCE FORMULAS

The partial derivatives of functions of several variables can be approximated by the formulas derived in the previous section considering each variable

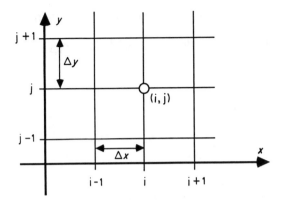

Figure 4.4.1 Two-dimensional Cartesian mesh

separately. We will represent in the following the operators for forward and backward differences by $\delta_x^{\pm}, \delta_y^{\pm}$, with the subscript to indicate the space co-ordinate.

In a two-dimensional space a rectangular mesh can be defined by the points of co-ordinates $x_i = x_0 + i\Delta x$ and $y_j = y_0 + j\Delta y$ (Figure 4.4.1). Defining $u_{ij} = u(x_i, y_j)$, all the above formulas can be applied on either variable x, y, acting separately on the i and j subscripts. For instance for the first partial derivative in the x-direction with first order accuracy

$$(u_x)_{ij} \equiv \left(\frac{\partial u}{\partial x}\right)_{ij} = \frac{u_{i+i,j} - u_{i,j}}{\Delta x} = \frac{1}{\Delta x}\delta_x^+ u_{ij} + 0(\Delta x) \qquad (4.4.1)$$

and similarly in the y-direction:

$$(u_y)_{ij} \equiv \left(\frac{\partial u}{\partial y}\right)_{ij} = \frac{u_{i,j+1} - u_{i,j}}{\Delta y} = \frac{1}{\Delta y}\delta_y^+ u_{i,j} + 0(\Delta y) \qquad (4.4.2)$$

Also, a second-order, central difference formula for the second derivative will be, referring to formula (4.2.43),

$$(u_{xx})_{ij} = \left(\frac{\partial^2 u}{\partial x^2}\right)_{ij} = \frac{u_{i+1,j} - 2u_{i,j} + u_{i-1,}}{\Delta x^2} - \frac{\Delta x^2}{12}\left(\frac{\partial^4 u}{\partial x^4}\right) \qquad (4.4.3)$$

and similar expressions can be derived for the y-derivatives. Besides the straightforward application of the various formulas presented in the previous section, additional forms can be defined by introducing an interaction between the two space directions; for instance, through a semi-implicit form on one of the two space co-ordinates.

4.4.1 Difference schemes for the Laplace operator

In order to illustrate this point let us consider the Laplace operator $\Delta u = u_{xx} + u_{yy}$ in two dimensions. Application of second-order central differencing

188

in both directions leads to the well-known *five-point difference operator*, $\Delta^{(1)}$:

$$\Delta u_{ij} = \frac{u_{i-1,j} - 2u_{ij} + u_{i+1,j}}{\Delta x^2} + \frac{u_{i,j-1} - 2u_{ij} + u_{i,j+1}}{\Delta y^2} + 0(\Delta x^2, \Delta y^2)$$

$$\equiv \Delta^{(1)}u_{ij} + 0(\Delta x^2, \Delta y^2) \tag{4.4.4}$$

or for a uniform mesh $\Delta x = \Delta y$:

$$\Delta u_{ij} = \frac{u_{i+1,j} + u_{i-1,j} + u_{i,j-1} + u_{i,j+1} - 4u_{ij}}{\Delta x^2} - \frac{\Delta x^2}{12}\left(\frac{\partial^4 u}{\partial x^4} + \frac{\partial^4 u}{\partial y^4}\right) \tag{4.4.5}$$

This formula is illustrated by the *computational molecule* of Figure 4.4.2. In operator form the five-point approximation $\Delta^{(1)}$, written as

$$\Delta^{(1)}u_{ij} = \left(\frac{\delta_x^2}{\Delta x^2} + \frac{\delta_y^2}{\Delta y^2}\right)u_{ij} \tag{4.4.6}$$

is the most widely applied difference scheme of second order for the Laplace operator. This is generalized to the operator $\vec{\nabla}(k\vec{\nabla}u)$, following equation (4.2.49), as the second-order difference operator:

$$\vec{\nabla}.(k\vec{\nabla}u)_{ij} = \frac{1}{\Delta x^2}(\delta_x^+ k_{i+1/2,j}\,\delta_x^-)u_{ij} + \frac{1}{\Delta y^2}(\delta_y^+ k_{i,j+1/2}\,\delta_y^-)u_{ij} + 0(\Delta x^2, \Delta y^2)$$

$$\tag{4.4.7}$$

Other combinations are possible whereby difference operators on the two space co-ordinates are mixed. For instance, the approximation

$$\Delta^{(2)} = \frac{1}{\Delta x^2}(\mu_y\,\delta_x)^2 + \frac{1}{\Delta y^2}(\mu_x\,\delta_y)^2 \tag{4.4.8}$$

is of the same order of accuracy as the schemes of Figure 4.4.2.

Worked out in detail, we have after introduction of the shift operators E_x

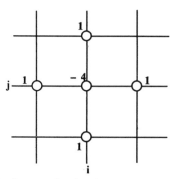

Figure 4.4.2 Computational molecule for the five-point Laplace operator (equation (4.4.5))

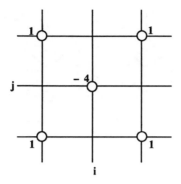

Figure 4.4.3 Five-point molecule for Laplace operator $\Delta^{(2)}u$ of equation (4.4.10) for $\Delta x = \Delta y$

and E_y following the definitions of Section 4.2,

$$\Delta^{(2)}u_{ij} \equiv \left[\left(\frac{1}{\Delta x}\mu_y\delta_x\right)^2 + \left(\frac{1}{\Delta y}\mu_x\delta_y\right)^2\right]u_{ij} = \frac{1}{4\Delta x^2}(E_y + 2 + E_y^{-1}).$$

(4.4.9)

$$(E_x - 2 + E_x^{-1})u_{ij} + \frac{1}{4\Delta y^2}(E_x + 2 + E_x^{-1})(E_y - 2 + E_y^{-1})u_{ij}$$

For $\Delta x = \Delta y$ we obtain

$$\Delta^{(2)}u_{ij} = \frac{1}{4\Delta x^2}(u_{i+1,j+1} + u_{i+1,j-1} + u_{i-1,j-1} + u_{i-1,j+1} - 4u_{ij})$$

(4.4.10)

which is represented in Figure 4.4.3.

Up to higher-order terms we can write

$$\Delta^{(2)}u_{ij} = \left(1 + \frac{\Delta y^2}{4}\frac{\partial^2}{\partial y^2}\right)\left(u_{xx} + \frac{\Delta x^2}{12}\frac{\partial^4 u}{\partial x^4}\right)_{ij} + \left(1 + \frac{\Delta x^2}{4}\frac{\partial^2}{\partial x^2}\right)\left(u_{yy} + \frac{\Delta y^2}{12}\frac{\partial^4 u}{\partial y^4}\right)_{ij}$$

(4.4.11)

$$= \Delta u_{ij} + \frac{1}{12}\Delta x^2\frac{\partial^4 u}{\partial x^4} + \frac{1}{12}\Delta y^2\frac{\partial^4 u}{\partial y^4} + \left(\frac{\Delta x^2 + \Delta y^2}{4}\right)\frac{\partial^4 u}{\partial x^2\partial y^2} + \cdots$$

defining the truncation error of the $\Delta^{(2)}$ difference operator.

However, this operator is not to be recommended for a Laplace equation, since the odd-numbered points are detached from the even-numbered ones. Referring to Figure 4.4.4., we observe that point (i, j) is coupled to the points marked by a square, while there is no connection to the even-numbered points marked by a circle. Hence a situation such as the one shown in Figure 4.4.4(b), where the solution oscillates between the two values a and b when passing from an even- to an odd numbered point, will satisfy the difference equation $\Delta^{(2)}u_{ij} = 0$. Clearly, this solution will not satisfy the difference equation (4.4.5).

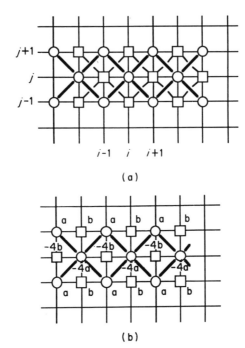

Figure 4.4.4 Generation of odd–even oscillations by the five-point scheme of equation (4.4.10), represented in Figure 4.4.3

We can define a family of nine-point schemes for the Laplace operator on a uniform Cartesian mesh $\Delta x = \Delta y$ by the combination

$$\Delta^{(3)} u_{ij} = (a\Delta^{(1)} + b\Delta^{(2)})u_{ij} \qquad \text{with } a + b = 1 \qquad (4.4.12)$$

With the operator notations of equations (4.4.2) and (4.4.7) we obtain (see also Problem 4.9)

$$\Delta^{(3)} u_{ij} = \frac{1}{\Delta x^2}\left[(\delta_x^2 + \delta_y^2) + \frac{b}{2}\,\delta_x^2\,\delta_y^2\right]u_{ij} = \Delta^{(1)} u_{ij} + \frac{b}{2}\,\delta_x^2\,\delta_y^2 u_{ij}$$

$$= \Delta u_{ij} + \frac{\Delta x^2}{12}\left[\frac{\partial^4 u}{\partial x^4} + \frac{\partial^4 u}{\partial y^4} + 6b\,\frac{\partial^4 u}{\partial x^2 \partial y^2}\right]$$

$$(4.4.13)$$

The particular choice of $b = 2/3$ leads to the well-known scheme of Figure 4.4.5, which is also obtained from a Galerkin, finite element discretization of the Laplace operator on the same mesh, using bilinear quadrilateral elements (See chapter 5).

With $b = 1/3$ we obtain the computational molecule of Figure 4.4.6, which is recommended by Dahlquist and Bjorck (1974), because the truncation error is equal to

$$-\frac{\Delta x^2}{12}\left(\frac{\partial^2}{\partial x^2} + \frac{\partial^2}{\partial y^2}\right)^2 u = -\frac{\Delta x^2}{12}\cdot\Delta^2 u$$

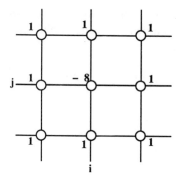

Figure 4.4.5 Nine-point molecule for the Laplace operator
(4.4.13) with $b = 2/3$

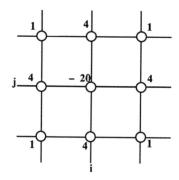

Figure 4.4.6 Nine-point molecule for the Laplace operator
(4.4.13) with $b = 1/3$

Hence the equation $\Delta u = \lambda u$ can be discretized with this nine-point operator $\Delta^{(3)} = \frac{2}{3} \Delta^{(1)} + \frac{1}{3} \Delta^{(2)}$ and will have a truncation error equal to

$$- \frac{\lambda^2 \Delta x^2}{12} u$$

Therefore the corrected difference scheme

$$\Delta^{(3)} u_{ij} = \left(\lambda + \frac{\lambda^2 \Delta x^2}{12} \right) u$$

will have a fourth-order truncation error.

4.4.2 Mixed derivatives

Mixed derivatives of any order can be discretized in much the same way by using for $D_x = \partial/\partial x$ and $D_y = \partial/\partial y$ the various formulas and their possible combinations described above. The simplest, *second-order central* formula for the mixed derivative is obtained from the application of equation (4.2.30) in both directions x and y:

$$u_{xy} \equiv \frac{\partial^2 u}{\partial x\, \partial y} = \frac{1}{\Delta x\, \Delta y}\, \mu_x\, \delta_x \left[\left(1 - \frac{\delta_x^2}{6} + 0(\Delta x^4) \right) \right] \mu_y\, \delta_y \left[\left(1 - \frac{\delta_y^2}{6} + 0(\Delta y^4) \right) \right] u_{ij}$$

$$(4.4.14)$$

and to second-order accuracy:

$$(u_{xy})_{ij} = \frac{1}{\Delta x\, \Delta y}\, (\mu_x\, \delta_x \mu_y\, \delta_y) u_{ij} + 0(\Delta x^2,\, \Delta y^2)$$

$$(4.4.15)$$

$$= \frac{u_{i+1,j+1} - u_{i+1,j-1} - u_{i-1,j+1} + u_{i-1,j-1}}{4\Delta x\, \Delta y} + 0(\Delta x^2)$$

which is illustrated by the molecule in Figure 4.4.7.

Other combinations are possible; for instance,

$$(u_{xy})_i = \frac{1}{\Delta x\, \Delta y}\, (\mu_x\, \delta_x\, \delta_y^+) u_{ij} + 0(\Delta x^2,\, \Delta y)$$

$$(4.4.16)$$

$$= \frac{1}{2\Delta x\, \Delta y}\, (u_{i+1,j+1} - u_{i-1,j+1} - u_{i+1,j} + u_{i-1,j}) + 0(\Delta x^2,\, \Delta y)$$

which is first order in Δy and second order in Δx. This formula is represented in Figure 4.4.8.

Permuting x and y, we obtain a formula which is second order in y and first order in x. A first order formula in both x and y is obtained from the application of δ_x^+ and δ_y^+, leading to

$$(u_{xy})_i = \frac{1}{\Delta x\, \Delta y}\, \delta_x^+\, \delta_y^+\, u_{ij} + 0(\Delta x,\, \Delta y)$$

$$(4.4.17)$$

$$= \frac{1}{\Delta x\, \Delta y}\, (u_{i+1,j+1} - u_{i+1,j} - u_{i,j+1} + u_{ij}) + 0(\Delta x,\, \Delta y)$$

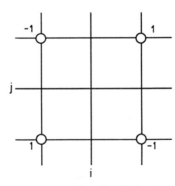

Figure 4.4.7 Computational molecule for the second-order accurate, mixed derivative formula (4.4.15)

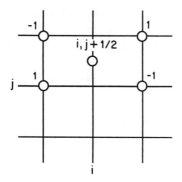

Figure 4.4.8 Mixed derivative formula (4.4.16)

Observe that the same formula (4.4.17) will give a second-order accurate estimation of the mixed derivative taken at the point $(i + 1/2, j + 1/2)$, that is,

$$(u_{xy})_{i+1/2,j+1/2} = (u_{i+1,j+1} - u_{i+1,j} - u_{i,j+1} + u_{ij}) + 0(\Delta x^2, \Delta y^2)$$

$$= \frac{1}{\Delta x \, \Delta y} \delta_x \, \delta_y u_{i+1/2,j+1/2} + 0(\Delta x^2, \Delta y^2) \qquad (4.4.18)$$

These formulas are represented in Figure 4.4.9.

In a similar way equation (4.4.16) is a second-order approximation to $(u_{xy})_{i,j+1/2}$.

Applying backward differences in both directions leads to

$$(u_{xy})_i = \frac{1}{\Delta x \, \Delta y} \delta_x^- \, \delta_y^- u_{ij} + 0(\Delta x, \Delta y)$$

$$= \frac{1}{\Delta x \, \Delta y} (u_{i-1,j-1} - u_{i-1,j} - u_{i,j-1} + u_{ij}) + 0(\Delta x, \Delta y) \qquad (4.4.19)$$

$$= \frac{1}{\Delta_x \, \Delta_y} \delta_x \, \delta_y u_{i-1/2,j-1/2} + 0(\Delta x, \Delta y)$$

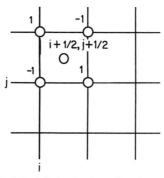

Figure 4.4.9 Mixed derivative formulas (4.4.17) and (4.4.18)

Since the truncation errors of equations (4.4.19) and (4.4.17) are equal but of opposite signs (see Problem 4.10), the sum of the two expressions will lead to a second-order formula for the mixed derivative $(u_{xy})_i$:

$$(u_{xy})_i = \frac{1}{2\Delta x\,\Delta y}\,[\delta_x^+\,\delta_y^+ + \delta_x^-\,\delta_y^-]u_{ij} + 0(\Delta x^2, \Delta y^2)$$

$$= \frac{1}{2\Delta x\,\Delta y}\,[u_{i+1,j+1} - u_{i+1,j} - u_{i,j+1} + u_{i-1,j-1} - u_{i-1,j} - u_{i,j-1}$$

$$+ 2u_{ij}) + 0(\Delta x^2, \Delta y^2) \tag{4.4.20}$$

This formula is represented in Figure 4.4.10(a) and, compared with the central approximation (4.4.15) shown in Figure 4.4.7, has a non-zero coefficient for u_{ij}. This might be advantageous in certain cases by enhancing the weight of the u_{ij} coefficients in the matrix equations obtained after discretization, that is, enhancing the diagonal dominance (see, for instance, O'Carroll, 1976).

An alternative to the last formulation is obtained by a combination of forward and backward differences, leading to the second-order approximation for $(u_{xy})_i$, shown in Figure 4.4.10(b):

$$(u_{xy})_i = \frac{1}{2\Delta x\,\Delta y}\,[\delta_x^+\,\delta_y^- + \delta_x^-\,\delta_y^+]u_{ij} + 0(\Delta x^2, \Delta y^2)$$

$$= \frac{1}{2\Delta x\,\Delta y}\,[u_{i+1,j} - u_{i+1,j-1} + u_{i,j+1} + u_{i,j-1} - u_{i-1,j+1} + u_{i-1,j} - 2u_{ij}]$$
$$\tag{4.4.21}$$

$$= \frac{1}{2\Delta x\,\Delta y}\,(\delta_x\,\delta_y u_{i+1/2,j-1/2} + \delta_x\,\delta_y u_{i-1/2,j+1/2}) + 0(\Delta x^2, \Delta y^2)$$

It can also be seen, by adding up the two last expressions, that we recover the fully central second-order approximation (4.4.15).

Hence we can define the most general, second-order mixed derivative

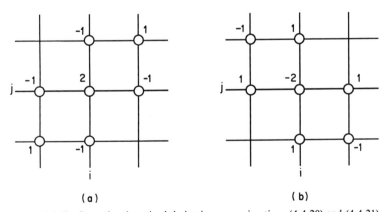

(a) (b)

Figure 4.4.10 Second-order mixed derivative approximations (4.4.20) and (4.4.21)

approximation by an arbitrary linear combination of formulas (4.4.20) and (4.4.21) (see also Mitchell and Griffiths, 1980):

$$(u_{xy})_i = \frac{1}{2\Delta_x\,\Delta_y}\,\delta_x\,\delta_y(au_{i+1/2,j+1/2} + au_{i-1/2,j-1/2}$$

$$+ bu_{i+1/2,j-1/2} + bu_{i-1/2,j+1/2}) + 0(\Delta x^2, \Delta y^2) \qquad (4.4.22)$$

with

$$a + b = 1 \qquad (4.4.23)$$

for consistency.

4.5 FINITE DIFFERENCE FORMULAS ON NON-UNIFORM CARTESIAN MESHES

For non-uniform or curvilinear meshes the discretization of the equation can be performed after a transformation from the physical space (x, y, z) to a Cartesian, computational space (ξ, η, ζ). The relations between the two spaces are defined through the co-ordinate transformation formulas such as $\xi = \xi(x, y, z)$ and similar formulas for η and ζ, and are considered as performing a mapping from the physical space to the computational space (ξ, η, ζ), Figure 4.5.1.

Therefore all the formulas derived above can be applied in the (ξ, η, ζ) space on the equations written in the curvilinear co-ordinates. These transformed equations contain metric coefficients which have to be discretized in a consistent way. They introduce the mesh size influence in the difference formulas. Hence this procedure extends, in a straightforward way, the use of finite difference techniques to arbitrary geometries and meshes, the only restriction of finite difference meshes being that *all mesh points have to be positioned on families of non-intersecting lines with one family for each space*

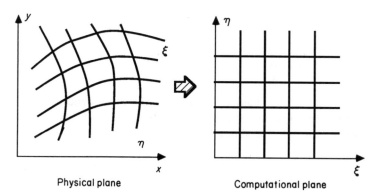

Physical plane   Computational plane

Figure 4.5.1 Mapping between an arbitrary co-ordinate system (ξ, η) in the physical plane and the computational plane

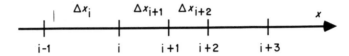

Figure 4.5.2 Arbitrary mesh point distribution in one-dimensional space x

co-ordinate. For instance, in Figure 4.5.1 the ξ lines may not intersect with each other since this would lead to two different values of ξ for the same point.

An insight into the influence of non-uniform mesh sizes can be obtained from the following approximations, derived from Taylor series expansions. On an arbitrary mesh point distribution x_i we have for the first derivatives the one-sided first-order formulas (Figure 4.5.2)

$$(u_x)_i = \frac{u_{i+1} - u_i}{\Delta x_{i+1}} - \frac{\Delta x_{i+1}}{2} u_{xx} \tag{4.5.1}$$

Backward difference

$$(u_x)_i = \frac{u_i - u_{i-1}}{\Delta x_i} + \frac{\Delta x_i}{2} u_{xx} \tag{4.5.2}$$

where the notation

$$\Delta x_i = x_i - x_{i-1} \tag{4.5.3}$$

is introduced

Central difference

Combining the above two formulas in order to eliminate the first-order truncation error leads to the second-order formula (also see Problem 4.11):

$$(u_x)_i = \frac{1}{\Delta x_i + \Delta x_{i+1}} \left[\frac{\Delta x_i}{\Delta x_{i+1}} (u_{i+1} - u_i) + \frac{\Delta x_{i+1}}{\Delta x_i} (u_i - u_{i-1}) \right]$$

$$- \frac{\Delta x_i \, \Delta x_{i+1}}{6} u_{xxx} \tag{4.5.4}$$

We can also obtain one-sided forward or backward second-order formulas, involving three mesh points by the standard techniques of Taylor expansions. For a forward formula we obtain

$$(u_x)_i = \left(\frac{\Delta x_{i+1} + \Delta x_{i+2}}{\Delta x_{i+2}} \cdot \frac{u_{i+1} - u_i}{\Delta x_{i+1}} - \frac{\Delta x_{i+1}}{\Delta x_{i+2}} \cdot \frac{u_{i+2} - u_i}{\Delta x_{i+1} + \Delta x_{i+2}} \right)$$

$$+ \frac{\Delta x_{i+1}(\Delta x_{i+1} + \Delta x_{i+2})}{6} u_{xxx} \tag{4.5.5}$$

Second derivatives

The three-point, central difference formula for the second derivative is obtained as

$$(u_{xx})_i = \left(\frac{u_{i+1} - u_i}{\Delta x_{i+1}} - \frac{u_i - u_{i-1}}{\Delta x_i}\right) \frac{2}{\Delta x_{i+1} + \Delta x_i}$$

$$+ \frac{1}{3}(\Delta x_{i+1} - \Delta x_i)u_{xxx} - \frac{\Delta x_{i+1}^3 + \Delta x_i^3}{12(\Delta x_{i+1} + \Delta x_i)}\left(\frac{\partial^4 u}{\partial x^4}\right) \quad (4.5.6)$$

It is important to observe here the presence of a truncation error proportional to the difference of two consecutive mesh lengths Δx_{i+1} and Δx_i. If the mesh size varies abruptly, for instance if $\Delta x_{i+1} \approx 2\Delta x_i$, then the above formula will be only first-order accurate. This is a general property of finite difference approximations on non-uniform meshes. If the mesh size does not vary smoothly a loss of accuracy is unavoidable. More details can be found in Hoffman (1982).

References

Adam, Y. (1975). 'A Hermitian finite difference method for the solution of parabolic equations.' *Comp. Math. Applications*, **1**, 393–406.

Adam, Y. (1977). 'Highly accurate compact implicit methods and boundary conditions.' *Journal Computational Physics*, **24**, 10–22.

Ames, W. F. (1977). *Numerical Methods for Partial Differential Equations*, 2nd edn, New York: Academic Press.

Ciment, M., and Leventhal, S. H. (1975). 'Higher order compact implicit schemes for the wave equations.' *Math. of Computation*, **29**, 885–944.

Collatz, L. (1966). *The Numerical Treatment of Differential Equations*, 3rd edn, Berlin: Springer Verlag.

Dahlquist, G., and Bjork, A. (1974). *Numerical Methods*, Englewood Cliffs, NJ: Prentice-Hall.

Hildebrand, F. B. (1956). *Introduction to Numerical Analysis*, New York: McGraw-Hill.

Hirsh, R. S. (1975). 'Higher order accurate difference solutions of fluid mechanics problems by a compact differencing scheme.' *J. Computational Physics*, **19**, 20–109.

Hirsh, R. S. (1983). 'Higher order approximations in fluid mechanics—compact to spectral.' *Von Karman Institute Lecture Series 1983–6*, Brussels, Belgium.

Hoffman, J. D. (1982). 'Relationship between the truncation errors of centered finite difference approximation on uniform and non-uniform meshes'. *Journal Comp. Physics*, **46**, pp. 469–77.

Kopal, Z. (1961). *Numerical Analysis*, New York: John Wiley.

Krause, E. (1971). Mehrstellen Verfahren zur Integration der Grenzschichtgleichungen.' *DLR Mitteilungen*, **71-13**, 109–40.

Leventhal, S. H. (1980). The operator compact implicit method for reservoir simulation.' *Journal Soc. Petroleum Engineers*, **20**, 120–8.

Mitchell, A. R., and Griffiths, D. F. (1980). *The Finite Difference Method in Partial Differential Equations*, New York: John Wiley.

O'Carroll, M. J. (1976). 'Diagonal dominance and SOR performance with skew nets.' *Int. Journal for Numerical Methods in Engineering*, **10**, 225–40.

198

Peyret, R. (1978). 'A Hermitian finite difference method for the solution of the Navier–Stokes equations.' *Proc. First Cong. On Num. Meth. in Laminar and Turbulent Flows*, Plymouth: Pentech Press, pp. 43–54.

Peyret, R., and Taylor, T. D. (1982). *Computational Methods for Fluid Flow*, New York: Springer Verlag.

Rubin, S. G., and Graves, R. A. (1975). 'Viscous flow solutions with a cubic spline approximation.' *Computer & Fluids*, **3**, 1–36.

Rubin, S. G., and Khosla, P. K. (1977). 'Polynomial interpolation method for viscous flow calculations.' *J. Computational Physics*, **24**, 217–46.

Thompson, J. F. (1984). 'A survey of grid generation techniques in computational fluid dynamics.' *AIAA Journal*, **22**, 1505–23.

PROBLEMS

Problem 4.1

Derive third-order accurate formulas for $(u_x)_i$, with forward and backward difference formulas, applying equations (4.2.18) and (4.2.20).

Problem 4.2

Apply a Taylor series expansion to a mixed backward formula, for the first derivative

$$(u_x)_i = au_{i-2} + bu_{i-1} + cu_i + du_{i+1}$$

Obtain the truncation error and show that this formula is second-order accurate.
Hint: obtain

$$(u_x)_i = \frac{1}{8\Delta x}(u_{i-2} - 7u_{i-1} + 3u_i + 3u_{i+1}) - \frac{\Delta x^2}{24} u_{xxx}$$

Problem 4.3

Apply a Taylor series expansion to the general form

$$(u_x)_i = au_{i+2} + bu_{i+1} + cu_i + du_{i-1} + eu_{i-2}$$

and obtain the central fourth-order accurate finite difference approximation to the first derivative $(u_x)_i$ at mesh point i. Repeat the same procedure to obtain an approximation to the second derivative $(u_{xx})_i$ with the same mesh points. Show that the formula is also fourth-order accurate. Calculate the truncation error for both cases.
Hint: Show that we have

$$(u_x)_i = \frac{(-u_{i+2} + 8u_{i+1} - 8u_{i-1} + u_{i-2})}{12\Delta x} + \frac{\Delta x^4}{30}\left(\frac{\partial^5 u}{\partial x^5}\right)$$

$$(u_{xx})_i = \frac{(-u_{i+2} + 16u_{i+1} - 30u_i + 16u_{i-1} - u_{i-2})}{12\Delta x^2} + \frac{1}{90}\Delta x^4\left(\frac{\partial^6 u}{\partial x^6}\right)$$

Problem 4.4

Repeat Problem 4.3 for the third and fourth derivatives and obtain

$$(u_{xxx})_i = \frac{u_{i+2} - 2u_{i+1} + 2u_{i-1} - u_{i-2}}{2\Delta x^3} - \frac{1}{4}\Delta x^2\left(\frac{\partial^5 u}{\partial x^5}\right)$$

$$\left(\frac{\partial^4 u}{\partial x^4}\right)_i = \frac{u_{i+2} - 4u_{i+1} + 6u_i - 4u_{i-1} + u_{i-2}}{\Delta x^4} - \frac{\Delta x^2}{6}\left(\frac{\partial^6 u}{\partial x^6}\right)$$

Problem 4.5

Evaluate numerically the first and second derivatives of $\cos x, \sin x, e^x$ at $x = 0$ with forward, backward and central differences of first- and second- or fourth-order each. Compare the error with the estimated truncation error. Take $\Delta x = 0.1$.

Problem 4.6

Derive a family of compact implicit finite difference formulas by eliminating the coefficients a_+, b_+, c_+ and a_0 from system (4.3.14). Derive the truncation error and obtain the formulas with the highest order of accuracy.

Problem 4.7

Find the highest-order implicit difference formula, involving second derivatives at only one point. Write this expression as an explicit relation for $(u_{xx})_i$ and derive the truncation error.
Hint: Select $c_+ = c_- = 0$, $c_0 = 1$. Obtain

$$(u_{xx})_i = -\frac{1}{2\Delta x}[(u_x)_{i+1} - (u_x)_{i-1}] + \frac{2}{\Delta x^2}(u_{i+1} - 2u_i + u_{i-1})$$

The truncation error is found to be

$$R = \frac{\Delta x^4}{360}\left(\frac{\partial^6 u}{\partial x^6}\right)$$

and the formula is fourth-order accurate.

Problem 4.8

Derive a family of implicit difference formulas involving no second derivatives.
Hint: Select $c_+ = c_0 = c_- = 0$ and set $\alpha = b_+/b_-$. Obtain the scheme

$$\alpha(u_x)_{i+1} + 2(1+\alpha)(u_x)_i + (u_x)_{i-1} - \frac{1}{2\Delta x}[(5\alpha+1)u_{i+1} + 4(1-\alpha)u_i - (5+\alpha)u_{i-1}] = 0$$

with the truncation error

$$R = \frac{\Delta x^3}{12}(\alpha - 1)\frac{\partial^4 u}{\partial x^4} + \frac{\Delta x^4}{60}(1+\alpha)\frac{\partial^5 u}{\partial x^5}$$

Problem 4.9

Define the computational molecule for the Laplace operator $\Delta^{(3)}$ on a uniform mesh, following formula (4.4.13) for an arbitrary b.
Hint: Show that $\delta_x^2 \delta_y^2$ is represented by the molecule

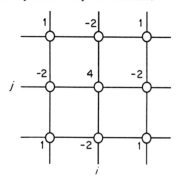

Figure P.4.9

Problem 4.10

Find the truncation errors of the mixed derivative formulas (4.4.14)–(4.4.22). In particular, proof the second-order accuracy of formulas (4.4.18) and (4.4.21).

Problem 4.11

Show that the arithmetic average of the one-sided formulas (4.5.1) and (4.5.2) is not second-order accurate on an arbitrary mesh.

Hint: Perform a Taylor series expansion and show that we obtain

$$(u_x)_i = \frac{1}{2} \left[\frac{u_{i+1} - u_i}{\Delta x_{i+1}} + \frac{u_i - u_{i-1}}{\Delta x_i} \right] - \frac{\Delta x_{i+1} - \Delta x_i}{4} (u_{xx})_i - \frac{\Delta x_{i+1}^2 + \Delta x_i^2}{12} (u_{xxx})_i$$

Problem 4.12

Obtain formulas (4.5.5) and (4.5.6).

Problem 4.13

Obtain formulas (4.2.45) to (4.2.48).

Problem 4.14

Obtain formulas (4.2.52) to (4.2.59).

Problem 4.15

Obtain formulas (4.2.60) to (4.2.62).

Chapter 5

The Finite Element Method

The finite element method originated from the field of structural analysis as a result of many years of research, mainly between 1940 and 1960. The concept of 'elements' can be traced back to the techniques used in stress calculations, whereby a structure was subdivided into small substructures of various shapes and re-assembled after each 'element' had been analysed. The development of this technique and its formal elaboration led to the introduction of what is now called the finite element method by Turner *et al.* (1956) in a paper dealing with the properties of a triangular element in plane stress problems. The expression 'finite elements' was introduced by Clough (1960).

After having been applied with great success to a variety of problems in linear and non-linear structural mechanics it soon appeared that the method could also be used to solve continuous field problems (Zienkiewicz and Cheung, 1965). From then on, the finite element method was used as a general approximation method for the numerical solution of physical problems described by field equations in continuous media, actually containing many of the finite difference schemes as special cases. Today, after its initial developments in an engineering framework, the finite element method has been put by mathematicians into a very elegant, rigorous, formal framework, with precise mathematical conditions for existence and convergence criteria and exactly derived error bounds.

Due to the particular character of finite element discretizations the appropriate mathematical background is functional analysis, and an excellent introduction to the mathematical formulation of the method can be found in Strang and Fix (1973) and Oden and Reddy (1976), and more advanced treatments in Oden (1972) and Ciarlet (1978). From the point of view of applications a general overview of the wide range of problems and a presentation of the engineering treatment of finite element methods can be found in Huebner (1975) and Zienkiewicz (1977). With regard to fluid flow problems a general introduction is given by Chung (1978) and more advanced developments are analysed in Temam (1977), Girault and Raviart (1979) and Thomasset (1981). A recent addition to the literature on the applications of finite elements to fluid flow problems is to be found in Baker (1983).

5.1 THE NATURE OF THE FINITE ELEMENT APPROXIMATION

The basic steps in a finite element approximation differ from the corresponding ones in a finite difference method essentially by the generality of their formulation. In the previous chapter we have seen that in order to set up a finite difference discretization scheme we must:

(1) Define a space discretization by which the mesh points are distributed along families of non-intersecting lines;
(2) Develop the unknown functions as a Taylor series expansion around the values at grid points; and
(3) Replace the differential equations by finite difference approximations of the derivatives.

The finite element method, on the other hand, defines for each of these three steps a more general formulation, as will become clear from the following.

5.1.1 Finite element definition of the space discretization

The space domain can be discretized by subdivision of the continuum into elements of arbitrary shape and size. Since any polygonal structure with rectilinear or curved sides can finally be reduced to triangular and quadrilateral figures the latter are the basis for the space subdivision. The only restriction is that the elements may not overlap and have to cover the complete computational domain. Figure 5.1.1 shows a schematic example while Figure 5.1.2 is a surface discretization used in a practical computation (Bristeau *et al.*, 1980) of the flow around an aircraft. Within each element a certain number of points are defined, which can be positioned along the straight (or curved) sides or inside the element. These nodes will be the points where the numerical value of the unknown functions, and eventually their derivatives, will have to be determined. The total number of unknowns at the nodes, function values and eventually their derivatives are called the *degrees of freedom* of the numerical problem, or *nodal values*.

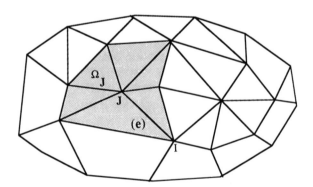

Figure 5.1.1 General finite element subdivision of a domain

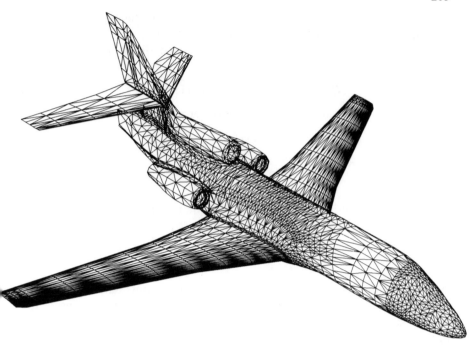

Figure 5.1.2 Surface finite element triangulation of an aircraft as applied in a potential flow computation by Bristeau *et al.* (1980). (Courtesy J. Periaux, Avions Marcel Dissault, France)

5.1.2 Finite element definition of interpolation functions

The field variables are approximated by linear combinations of known *basis* functions (also called *shape, interpolation* or *trial* functions). If $\tilde{u}$ is an approximate solution of $u(\vec{x})$ we write a series expansion of the form

$$\tilde{u}(\vec{x}) = \sum_I u_I N_I(\vec{x}) \qquad (5.1.1)$$

where the summation extends over all nodes I. Hence one interpolation function is attached to each nodal value or degree of freedom. These functions $N_I(\vec{x})$ can be quite general, with varying degrees of continuity at the inter-element boundaries.

Methods based on defining the interpolation functions on the *whole domain*, such as trigonometric functions leading to Fourier series, are used in *collocation* and *spectral* methods, where the functions $N_I(\vec{x})$ can be defined as orthogonal polynomials of Legendre, Chebyshev or similar types. Other possible choices are spline functions for $N_I(\vec{x})$, leading to *spline interpolation* methods. In these cases the coefficients u_I are obtained from the expansions in series of the base functions.

In standard finite element methods the interpolation functions are chosen to be *locally defined polynomials* within each element, being zero outside the considered element. In addition, the coefficients u_I of the expansion are the *unknown nodal values* of the dependent variables u. As a consequence the local interpolation functions satisfy the following conditions on each element (e), with I being a node of (e):

$$N_I^{(e)}(\vec{x}) = 0 \qquad \text{if } \vec{x} \text{ not in element } (e) \qquad (5.1.2)$$

and since u_I are the values of the unknowns at node number I:

$$\tilde{u}(x_I) = u_I \qquad (5.1.3)$$

we have for any point $\vec{x}_J$:

$$N_I^{(e)}(\vec{x}_J) = \delta_{IJ} \qquad (5.1.4)$$

An additional condition is provided by the requirement to represent exactly a constant function $u(\vec{x}) = \text{constant}$. Hence this requires

$$\sum_I N_I^{(e)}(\vec{x}) = 1 \qquad \text{for all } \vec{x} \in (e) \qquad (5.1.5)$$

The global function N_I is obtained by assembling the contributions $N_I^{(e)}$ of all the elements to which node I belongs. The above condition connects the various basis functions within an element and the allowed polynomials will be highly dependent on the number of nodes within each element.

5.1.3 Finite element definition of the equation discretization—integral formulation

This is the most essential and particular step of the finite element approximation since it requires the definition of an *integral* formulation of the physical problem equivalent to the field equations to be solved. Two possibilities are open for that purpose: either a *variational principle* can be found expressing the physical problem as the extremum of a functional or an integral formulation is obtained from the differential system through a *weak* formulation, also called the *method of weighted residuals*. Although many physical models can be expressed through a variational equation (for instance, the potential flow model) it is well known that it is not always possible to find a straightforward variational principle for all physical problems (for instance, for the Navier–Stokes equations). Therefore the weak formulation, or method of weighted residuals, is the most general technique which allows us to define in all cases an equivalent integral formulation. Actually, in situations where discontinuous solutions are possible (such as shock waves in transonic flows) the integral formulation is the only one which is physically meaningful, since the derivatives of the discontinuous flow variables are not defined.

Since weak formulations involve the definition of functionals, where restricted continuity properties are allowed on the functions concerned, it is important to define in a clear and precise way the functional spaces as well as the appropriate norms, in order to derive correct convergence properties and error bounds. It is not our intention here to enter into these mathematical aspects and the reader will find these detailed developments in the references mentioned.

5.2 THE FINITE ELEMENT INTERPOLATION FUNCTIONS

Conditions (5.1.2)–(5.1.5) define the properties of the local polynomial interpolation functions used in finite element approximations. Two families of elements are generally considered, according to their degree of inter-element continuity and the associated nodal values. If the nodal values are defined by the values of the unknown functions, then C^0 continuity at the inter-element boundary is sufficient for systems described by partial differential equations no higher than two. These elements and their associated shape functions are called *Lagrangian* elements.

If first-order partial derivatives of the unknown functions are to be considered as additional degrees of freedom the inter-element continuity up to the highest order of these derivatives will generally be imposed, and the elements satisfying these conditions are *Hermitian elements*. When the required continuity conditions are satisfied along every point of the inter-element boundary, the element is called *conforming*. This condition is sometimes relaxed, and elements whereby this continuity condition is imposed only at a limited number of points of the boundary are said to be *non-conforming*.

5.2.1 One-dimensional elements

Linear Lagrangian elements

The simplest element has a piecewise linear interpolation function and contains two nodes. Referring to Figure 5.2.1, the element between nodes i and $i - 1$ is denoted as element 1 and the adjacent element between i and $i + 1$ as element 2. They have node i in common and have respective lengths Δx_i and Δx_{i+1}.

Considering element 1, relation (5.1.4) gives two conditions, and we obtain for the shape functions at the nodes i and $i - 1$ of element 1 the linear form

$$N_i^{(1)} = \frac{(x - x_{i-1})}{\Delta x_i} \qquad N_{i-1}^{(1)} = \frac{(x_i - x)}{\Delta x_i} \tag{5.2.1}$$

For element 2 we have the following linear shape functions:

$$N_i^{(2)} = \frac{(x_{i+1} - x)}{\Delta x_{i+1}} \qquad N_{i+1}^{(2)} = \frac{(x - x_i)}{\Delta x_{i+1}} \tag{5.2.2}$$

The global shape function N_i, associated with node i and obtained by assembling $N_i^{(1)}$ and $N_i^{(2)}$, is illustrated in Figure 5.2.1. It is zero for $x \geqslant x_{i+1}$ and $x \leqslant x_{i-1}$.

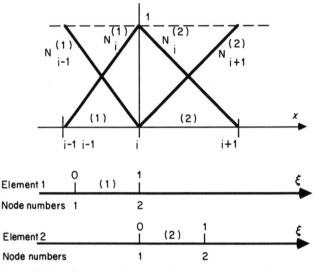

Figure 5.2.1 Linear one-dimensional elements and associated interpolation functions

If we define a *local co-ordinate* ξ within each element through the mapping

$$\xi = \frac{(x - x_{i-1})}{\Delta x_i} \tag{5.2.3}$$

the interpolation functions take the universal form

$$N_1(\xi) = 1 - \xi \qquad N_2(\xi) = \xi \tag{5.2.4}$$

where node 1 corresponds to mesh point $i - 1$ with $\xi = 0$ and node 2 to mesh point i with $\xi = 1$. Similarly, for element 2 the mapping between the subspace (x_i, x_{i+1}) and the subspace (0.1) of the ξ-space

$$\xi = \frac{(x - x_i)}{\Delta x_{i+1}} \tag{5.2.5}$$

leads to

$$N_i^{(2)} = 1 - \xi = N_1(\xi) \qquad N_{i+1}^{(2)} = \xi = N_2(\xi) \tag{5.2.6}$$

Hence this mapping allows the base functions to be defined through the universal forms given by equation (5.2.4), independently of the physical co-ordinates. They are determined by the nature of the element and the number of nodal points. The explicit form of the shape function in physical space is reconstructed from the knowledge of the mapping functions $\xi(x)$ between the element being considered and the normalized reference element $(0, 1)$:

$$N_i^{(1)}(x) = N_i^{(1)}(\xi(x)) \tag{5.2.7}$$

where the relation $\xi(x)$ is defined by equations (5.2.3) or (5.2.5).

The function $u(x)$ is approximated on element 1 by the linear representation

$$\tilde{u}(x) = u_{i-1}N^{(1)}_{i-1}(x) + u_i N^{(1)}_i(x) \tag{5.2.8}$$

or

$$\tilde{u}(x) = u_{i-1} + \frac{x - x_{i-1}}{\Delta x_i} \cdot (u_i - u_{i-1}) \qquad x_{i-1} \leqslant x \leqslant x_i \tag{5.2.9}$$

On element 2, $u(x)$ is approximated by

$$\tilde{u}(x) = u_i N^{(2)}_i(x) + u_{i+1}N^{(2)}_{i+1}(x)$$

$$\tag{5.2.10}$$

$$= u_i + \frac{x - x_i}{\Delta x_{i+1}}(u_{i+1} - u_i) \qquad x_i \leqslant x \leqslant x_{i+1}$$

The derivative of u is approximated in this finite element representation by

$$\frac{\partial u}{\partial x} = \sum_i u_i \frac{\partial N_i}{\partial x} \tag{5.2.11}$$

and, particularly at node i, we have within element 1

$$(u_x)_i = \left(\frac{\partial u}{\partial x}\right)^{(1)}_i = \frac{u_i - u_{i-1}}{\Delta x_i} \tag{5.2.12}$$

which corresponds to the first-order backward difference formula (4.5.2). In element 2 we have the approximation

$$(u_x)_i = \left(\frac{\partial u}{\partial x}\right)^{(2)}_i = \frac{u_{i+1} - u_i}{\Delta x_{i+1}} \tag{5.2.13}$$

which is identical to the forward difference formula (4.5.1). It should be observed that the derivatives of u are not continuous at the boundary between two elements, since, by definition, the interpolation functions have only C^0-continuity, as can be seen from Figure 5.2.1.

It is therefore customary, in finite element approximations, to define a 'local' approximation to the derivative $(u_x)_i$ by an average of the 'element' approximations (5.2.12) and (5.2.13). If a simple arithmetic average is taken the resulting formula is:

$$(u_x)_i = \frac{1}{2}\left[\left(\frac{\partial u}{\partial x}\right)^{(1)}_i + \left(\frac{\partial u}{\partial x}\right)^{(2)}_i\right] = \frac{1}{2}\left(\frac{u_{i+1} - u_i}{\Delta x_{i+1}} - \frac{u_i - u_{i-1}}{\Delta x_i}\right) \tag{5.2.14}$$

and has a truncation error equal to

$$\frac{\Delta x_i - \Delta x_{i+1}}{4}(u_{xx})_i$$

(see Problem 4.11). This approximation is only strictly second-order accurate on a regular mesh.

However, if each 'element' approximation is weighted by the relative length of the other element, that is, if we define

$$(u_x)_i = \frac{1}{\Delta x_i + \Delta x_{i+1}} \left[\Delta x_{i+1} \left(\frac{\partial u}{\partial x} \right)_i^{(1)} + \Delta x_i \left(\frac{\partial u}{\partial x} \right)_i^{(2)} \right] \qquad (5.2.15)$$

we obtain formula (4.5.4), which is second-order accurate on an arbitrary mesh.

Since any linear function can be represented exactly on the element, we can also express the linear mapping $\xi(x)$ or $x(\xi)$ as a function of the linear base functions (5.2.4). It is easily verified that equations (5.2.3) or (5.2.5) can be written as

$$x = \sum x_i N_i(\xi) \qquad (5.2.16)$$

where the sum extends over the two nodes of the element. This particular mapping is called an *isoparametric* mapping, and illustrates a general procedure which also applies to two- and three-dimensional elements.

Quadratic Lagrangian elements

An element with three nodes $(i - 1, i, i + 1)$ will require three conditions (5.1.4) and the base functions will be second-order polynomials, enabling the exact representation of quadratic functions on the element. Since a mapping from (x_{i-1}, x_i, x_{i+1}) to the ξ-space $(-1, 0, +1)$ can always be defined through a quadratic function $\xi = \xi(x)$, the interpolation functions can be directly defined in the normalized space ξ. Referring to Figure 5.2.2, we obtain

$$N_1 = -0.5\xi(1 - \xi) \qquad N_2 = (1 - \xi^2) \qquad N_3 = 0.5\xi(1 + \xi) \quad (5.2.17)$$

and we easily verify that condition (5.1.5) is satisfied.

The general mapping between x and ξ is quadratic, and, since any quadratic function on an element can be written as a linear combination of the basis functions $N_j (j = 1, 2, 3)$, the following isoparametric mapping can be defined:

$$x(\xi) = \sum_{j=1}^{3} x_j N_j(\xi) \qquad (5.2.18)$$

where the summation 1, 2, 3 corresponds to nodes $i - 1, i, i + 1$ and the basis functions N_j are given by equation (5.2.17). In evaluating the derivatives of the interpolation functions, $\partial N_j / \partial x$, we proceed as follows, by application of the derivative rules:

$$\frac{\partial N_j}{\partial x} = \frac{\partial N_j}{\partial \xi} \cdot \frac{\partial \xi(x)}{\partial x} = \frac{\partial N_j}{\partial \xi} \cdot \frac{1}{\partial x / \partial \xi}$$

$$= \frac{\partial N_j}{\partial \xi} \cdot \frac{1}{\sum_k x_k (\partial N_k / \partial \xi)} \qquad (5.2.19)$$

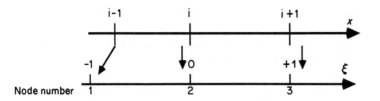

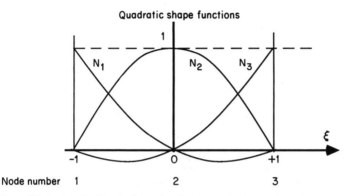

Quadratic shape functions

Figure 5.2.2 Quadratic one-dimensional elements and associated interpolation functions on normalized element $(-1, 0, 1)$

The denominator can be evaluated independently, since it depends only on the geometrical mapping. With functions (5.2.17) we have, referring to Figure 5.2.2,

$$\frac{\partial x(\xi)}{\partial \xi} = \xi(x_1 - 2x_2 + x_3) - \tfrac{1}{2}(x_1 - x_3)$$

$$= \xi(\Delta x_{i+1} - \Delta x_i) + \tfrac{1}{2}(\Delta x_{i+1} + \Delta x_i) \qquad (5.2.20)$$

For an equidistant mesh $\Delta x_{i+1} = \Delta x_i = \Delta x$, the right-hand side of equation (5.2.20) reduces to Δx.

Combining equations (5.2.19) and (5.2.20) gives the derivative of any of the three basis functions within the element $(i-1, i+1)$. For instance, for $j = i - 1$, we have

$$\frac{\partial N_{i-1}}{\partial x} = \frac{(2\xi - 1)}{2\xi(\Delta x_{i+1} - \Delta x_i) + (\Delta x_{i+1} + \Delta x_i)} \qquad (5.2.21)$$

leading to the nodal values

$$\left(\frac{\partial N_{i-1}}{\partial x}\right)_{i-1} = \frac{-3}{3\Delta x_i - \Delta x_{i+1}} \qquad \left(\frac{\partial N_{i-1}}{\partial x}\right)_{i} = -\frac{1}{\Delta x_{i+1} + \Delta x_i}$$

$$\left(\frac{\partial N_{i-1}}{\partial x}\right)_{i+1} = \frac{1}{3\,\Delta x_{i+1} - \Delta x_i} \qquad (5.2.22)$$

Similar relations are obtained for the other two base functions (see Problem 5.1).

These relations are applied to the computation of finite element approximations of first-order derivatives through

$$\left(\frac{\partial u}{\partial x}\right) = \sum_j u_j \left(\frac{\partial N_j}{\partial x}\right) = u_{i-1} \frac{\partial N_{i-1}}{\partial x} + u_i \frac{\partial N_i}{\partial x} + u_{i+1} \frac{\partial N_{i+1}}{\partial x} \qquad (5.2.23)$$

following the framework outlined with the linear elements (see also Problem 5.2).

Hermitian element with two nodes

Hermitian elements are characterized by the fact that the derivatives of the unknown functions are taken as additional degrees of freedom with continuity C^1 at the inter-element boundary. For an element with two nodes, four equations (5.1.4) have to be satisfied if the first derivatives are taken as nodal values, since four shape functions corresponding to the four degrees of freedom of the element have to be considered.

Referring to Figure 5.2.1 and denoting the first derivatives of the function u by u_x, conditions (5.1.4) require third-order polynomials for the shape functions. These are Hermitian cubic polynomials H^0, H^1, satisfying, in local co-ordinates,

$$H_i^0(\xi_j) = \delta_{ij} \qquad \frac{\mathrm{d}}{\mathrm{d}\xi} H_i^1(\xi_j) = \delta_{ij} \qquad (5.2.24)$$

and with the representation

$$\tilde{u} = u_1 H_1^0 + u_2 H_2^0 + (u_x)_1 H_1^1 + (u_x)_2 H_2^1 = \sum_{j=1}^{4} u_j N_j \qquad (5.2.25)$$

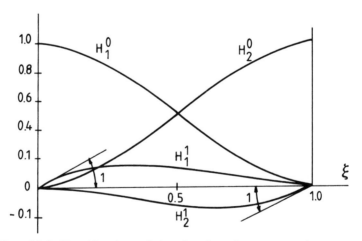

Figure 5.2.3 Hermitian interpolation functions for a two-node, one-dimensional element

where subscripts 1 and 2 refer to the nodes $i - 1$ and i of element 1. We obtain

$$H_1^0 = 1 - 3\xi^2 + 2\xi^3 \qquad H_2^0 = \xi^2(3 - 2\xi)$$
$$H_1^1 = (1 - \xi)^2 \qquad H_2^1 = \xi^2(\xi - 1) \qquad (5.2.26)$$

These Hermitian basis functions are represented in Figure 5.2.3.

5.2.2 Two-dimensional elements

The most currently used elements are widely described in the literature on finite elements, and their derivation and properties can be found in most textbooks on the subject (see, for instance, Huebner, 1975; Zienkiewicz, 1977; Chung, 1978). Therefore we will limit ourselves to a tabular presentation of various types of elements and associated base functions which have been or could be applied in finite element flow computations.

Since any two-dimensional polynomial figure, with straight or curved sides can always be subdivided into triangles and quadrilaterals there is actually no need to define other two-dimensional elements. Table 5.1 lists a selection of most widely used elements, their degrees of freedom, the associated expression of the interpolation function, generally in local co-ordinates, and some additional remarks and properties when necessary.

Table 5.1. Two-Dimensional Elements

Type of element	Interpolation function	Element geometry
TRIANGULAR **P1—Linear element**	*Piecewise polynomial of degree* 1 $N_i = ax + by + c$ $2a = (y_2 - y_3)/A$ $2b = (x_3 - x_2)/A$ $2c = (x_2 y_3 - x_3 y_2)/A$ $A = $ area of triangle 123	*Physical space*
Degrees of freedom u_1, u_2, u_3 C^0-continuity	$A = 1/2 \begin{vmatrix} x_1 & y_1 & 1 \\ x_2 & y_2 & 1 \\ x_3 & y_3 & 1 \end{vmatrix}$	
	Local co-ordinates: L_1, L_2, L_3 $L_1 = $ (Area P23)$/A$ $L_2 = $ (Area P13)$/A$ $L_3 = $ (Area P12)$/A$ $L_1 + L_2 + L_3 = 1$ $N_i = L_i$	*Local coordinates* L_1, L_2, L_3

Table 5.1. *(continued)*

Type of element	Interpolation function	Element geometry

Isoparametric transformation
$$\vec{x} = \vec{x}_1 L_1 + \vec{x}_2 L_2 + \vec{x}_3 L_3$$

Properties
Integration formula

$$\int_\Delta L_1^m L_2^n L_3^p \, d\Omega =$$

$$\frac{m! \, n! \, p!}{(m+n+p+2)!} \cdot 2A$$

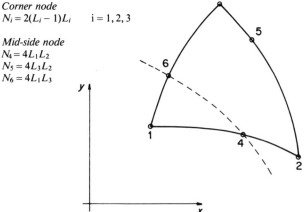

P2—Quadratic element

Degrees of freedom
$u_1, u_2, \ldots, u_6$

C^0-continuity

*Piecewise polynomials
of degree 2*
In local co-ordinates
Corner node
$N_i = 2(L_i - 1)L_i \qquad i = 1, 2, 3$

Mid-side node
$N_4 = 4L_1 L_2$
$N_5 = 4L_3 L_2$
$N_6 = 4L_1 L_3$

Physical space

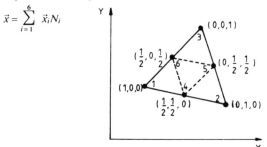

Local co-ordinates:
(L_1, L_2, L_3)

Isoparametric transformation

$$\vec{x} = \sum_{i=1}^{6} \vec{x}_i N_i$$

Table 5.1. (*continued*)

Type of element	Interpolation function	Element geometry

P1—Linear element
Lagrangian—
non-conforming

u_I, u_{II}, u_{III}

Local co-ordinates

$N_I = L_2 + L_3 - L_1$
$N_{II} = L_3 + L_1 - L_2$
$N_{III} = L_1 + L_2 - L_3$

Error estimate—semi-nom.
$\| u - u_h \|_{L^1_2} = 0(h)$
see Thomasset (1981)

$\int_\Delta N_I N_{II}\, dS = 0$ for $I \neq J$

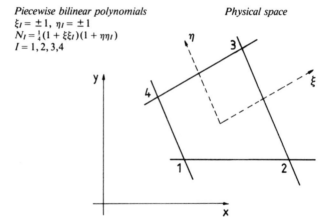

QUADRILATERAL ELEMENTS

Q1—Bilinear element
Lagrangian—conforming
u_1, u_2, u_3, u_4

Piecewise bilinear polynomials
$\xi_I = \pm 1,\ \eta_I = \pm 1$
$N_I = \frac{1}{4}(1 + \xi\xi_I)(1 + \eta\eta_I)$
$I = 1, 2, 3, 4$

Physical space

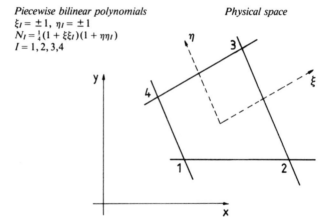

C^0-continuity

Local co-ordinates
ξ, η

Isoparametric transformation

$$\vec{x} = \sum_{I=1}^{4} \vec{x}_I N_I$$

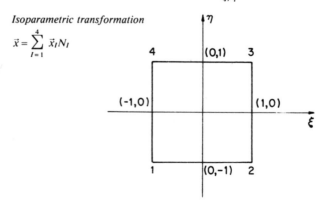

Table 5.1. (*continued*)

Type of element	Interpolation function	Element geometry

Q2—Biquadratic element
Lagrangian—conforming

Degrees of freedom
$u_1, \ldots, u_9$

C^0-continuity

Piecewise biquadratic polynomials
Define
$$\xi_i = i - 2 \qquad i = 1, 2, 3$$
$$\eta_j = j - 2 \qquad j = 1, 2, 3$$

$$K_1(\xi) = \tfrac{1}{2}\xi(\xi - 1)$$
$$K_2(\xi) = 1 - \xi^2$$
$$K_3(\xi) = \tfrac{1}{2}\xi(\xi + 1)$$

Physical space

For a node I with 'co-ordinates'
(i, j) we have
$$N_I(\xi, \eta) = K_i(\xi)K_j(\eta)$$

Example: Node 7
$$N_7 = K_1(\xi)K_3(\eta) = \tfrac{1}{4}\xi(\xi - 1)\eta(\eta + 1)$$

Node 9
$$N_9 = K_2(\xi)K_2(\eta) = (1 - \xi^2)(1 - \eta^2)$$

Local co-ordinates
ξ, η

Biquadratic Q^2 element
or
'serendipidity' element
Conforming
Degrees of freedom
$u_1, \ldots, u_8$

Piecewise biquadratic polynomials

Corner nodes $\xi_0 = \xi\xi_i$, $\eta_0 = \eta\eta_i$

$$N_I = \tfrac{1}{4}(1 + \xi_0)(1 + \eta_0)(\xi_0 + \eta_0 - 1)$$

Physical space

Table 5.1. (*continued*)

Type of element	Interpolation function	Element geometry

| | *Mid-side nodes* $N_I = \frac{1}{2}(1 - \xi^2)(1 + \eta_0)$ $\xi_i = 0$
 $N_I = \frac{1}{2}(1 - \eta^2)(1 + \xi_0)$ $\eta_i = 0$ | *Local co-ordiantes* |

5.2.3 Three-dimensional elements

Table 5.2 presents a summary of the most current elements used in three-dimensional flow computations.

Table 5.2. Three-dimensional elements

Type of element	Interpolation function	Element
TETRAHEDRALS	$N_I = L_I = ax + by + cz + d$	
P1—Linear	$L_I = $ Volume (P 234)	
Lagrangian—conforming	*For node 1*	
Degrees of freedom: u_1, u_2, u_3, u_4	$a = - \begin{vmatrix} 1 & y_2 & z_2 \\ 1 & y_3 & z_3 \\ 1 & y_4 & z_4 \end{vmatrix}$	
	$b = - \begin{vmatrix} x_2 & 1 & z_2 \\ x_3 & 1 & z_3 \\ x_4 & 1 & z_4 \end{vmatrix}$	
	$c = - \begin{vmatrix} x_2 & y_2 & 1 \\ x_3 & y_3 & 1 \\ x_4 & y_4 & 1 \end{vmatrix}$	
	$d = \begin{vmatrix} x_2 & y_2 & z_2 \\ x_3 & y_3 & z_3 \\ x_4 & y_4 & z_4 \end{vmatrix}$	
	Other L_I are obtained by cyclic permutation $V = $ volume tetrahedron (1234)	

Table 5.2. *(continued)*

Type of element	Interpolation function	Element

$$V = \begin{vmatrix} x_1 & y_1 & z_1 & 1 \\ x_2 & y_2 & z_2 & 1 \\ x_3 & y_3 & z_3 & 1 \\ x_4 & y_4 & z_4 & 1 \end{vmatrix}$$

Properties

$$\frac{1}{V} \int_V L_1^m L_2^n L_3^p L_4^q \, d\Omega = \frac{6m!n!p!q!}{(m+n+p+q+3)!}$$

P2—Quadratic
Lagrangian—conforming

Degrees of freedom
$u_1, \ldots, u_{10}$
C^0-continuity

Corner nodes: $N_I = (2L_I - 1)L_I$

Mid-side notes: $N_5 = 4L_1 L_3$

HEXAHEDRALS

Q1—Trilinear
Lagrangian—conforming

Degrees of freedom
$u_1, \ldots, u_8$

$N_I = \frac{1}{8}(1 + \xi\xi_I)(1 + \eta\eta_I)(1 + \zeta\zeta_I)$

$\xi_I = \pm 1$
$\eta_I = \pm 1$
$\zeta_I = \pm 1 \qquad I = 1, \ldots, 8$

5.3 INTEGRAL FORMULATION: THE METHOD OF WEIGHTED RESIDUALS OR WEAK FORMULATION

In order to illustrate the principles of this approach we will consider first the classical example of the two-dimensional quasi-harmonic equation,

$$\frac{\partial}{\partial x}\left(k \frac{\partial}{\partial x} u\right) + \frac{\partial}{\partial y}\left(k \frac{\partial}{\partial y} u\right) = q \tag{5.3.1}$$

written as $L(u) = q$, where L represents the differential operator.

If $\tilde{u}(x, y)$ is an approximation to the solution u, the quantity R, called the *residual*,

$$R(\tilde{u}) = L(\tilde{u}) - q$$
$$= \vec{\nabla} \cdot (k\vec{\nabla}\tilde{u}) - q \qquad (5.3.2)$$

is different from zero, otherwise $\tilde{u}$ would be the analytical solution. Any resolution algorithm will converge if it drives $R(\tilde{u})$ towards zero, although this value will never be reached in a finite number of operations. Hence the residual appears as a measure of the accuracy or of the error of the approximation $\tilde{u}$. Since this error canot be made to vanish simultaneously in all the points of the discretized domain a 'best' solution can be extracted by requiring that some weighted average of the residuals over the domain should be identically zero.

It $W(\vec{x})$ is some weight function with appropriate smoothness properties, the *method of weighted residuals*, or weak formulation, requires

$$\int_\Omega WR(\tilde{u}) \, d\Omega = 0 \qquad (5.3.3)$$

Applied to equation (5.3.2) this condition becomes

$$\int_\Omega W\vec{\nabla} \cdot (k\vec{\nabla}\tilde{u}) \, d\Omega = \int_\Omega qW \, d\Omega \qquad (5.3.4)$$

An essential step in this approach is an *integration by part of the second-order derivative terms*, according to Green's theorem:

$$\int_\Omega W\vec{\nabla} \cdot (k\vec{\nabla}\tilde{u}) \, d\Omega = - \int_\Omega k\vec{\nabla}\tilde{u} \cdot \vec{\nabla}W \, d\Omega + \oint_\Gamma k\frac{\partial \tilde{u}}{\partial n} W \, d\Gamma \qquad (5.3.5)$$

where the normal derivative along the boundary Γ of the domain appears in the right-hand side. Equation (5.3.4) becomes

$$- \int_\Omega k\vec{\nabla}\tilde{u} \cdot \vec{\nabla}W \, d\Omega + \oint_\Gamma k\frac{\partial \tilde{u}}{\partial n} W \, d\Gamma = \int_\Omega qW \, d\Omega \qquad (5.3.6a)$$

or, in condensed notation,

$$- (k\,\vec{\nabla}\tilde{u}, \vec{\nabla}W) + \left(k\frac{\partial \tilde{u}}{\partial n}, W \right)_\Gamma - (q, W) = 0 \qquad (5.3.6b)$$

where the inner functional product (f, g) is defined by

$$(f, g) = \int_\Omega fg \, d\Omega \qquad (5.3.7)$$

Equation (5.3.6) is the mathematical formulation of the weighted residual method, and is also called the *weak* formulation of the problem.

According to the choice of the weighting functions W (also called *test* functions) different methods are obtained. From a numerical point of view equation (5.3.6) is an algebraic equation, and therefore in any method as many weighting functions as unknown variables (degrees of freedom) will have to be defined within the chosen subspace of test functions. That is, a unique correspondence will have to be established between each nodal value and a corresponding weighting function in such a way that one equation of the type (5.3.6) is defined for each nodal value. Note also that if Ω_W is the subspace of the test functions then the weighted residual equation (5.3.3) expresses the condition that the projection of the residual in the subspace of the test functions is zero, that is, the residual is orthogonal to the subspace Ω_W.

5.3.1 The Galerkin method

The most widely method is the *Galerkin* method, in which the weighting functions are taken equal to the interpolation functions $N_I(x)$. This is also called the *Bubnow–Galerkin* method, to be distinguished from the *Petrov–Galerkin* method, in which the test functions are different from the interpolation functions N_I.

For each of the M degrees of freedom, with the finite element representation

$$\tilde{u}(\vec{x}, t) = \sum_I u_I(t) N_I(\vec{x}) \qquad (I = 1, \ldots, M) \qquad (5.3.8)$$

and the choice $W = N_J(\vec{x})$, in order to obtain the discretized equation for node J, we obtain from equation (5.3.6)

$$-\sum_I u_I \int_{\Omega_J} k \vec{\nabla} N_I \cdot \vec{\nabla} N_J \, d\Omega + \oint_\Gamma k \frac{\partial \tilde{u}}{\partial n} N_J \, d\Gamma = \int_{\Omega_J} q N_J \, d\Omega \quad (5.3.9)$$

where Ω_J is the subdomain of all elements containing node J and the summation over I covers all the nodes of Ω_J (see Figure 5.1.1). The matrix

$$K_{IJ} = \int_{\Omega_J} k \vec{\nabla} N_I \cdot \vec{\nabla} N_J \, d\Omega \equiv (k \vec{\nabla} N_I, \vec{\nabla} N_J) \qquad (5.3.10)$$

is called the *stiffness* matrix. For linear problems whereby k is independent of u this will depend only on the geometry of the mesh and the elements chosen.

Equation (5.3.9) can also be obtained from the Rayleigh–Ritz method for homogeneous boundary conditions. This is a general property, that is, if applied to a variational formulation it leads to the same system of numerical equations as the Galerkin-weighted residual method.

Example 5.3.1 One-dimensional equation

Consider the one-dimensional form of equation (5.3.1):

$$\frac{\partial}{\partial x} \left(k \frac{\partial u}{\partial x} \right) = q \qquad (E5.3.1)$$

and a Galerkin weak formulation with linear elements. Applying equation (5.3.9) we have explicitly, with the linear shape functions (5.2.4) and with

$$\Delta x_i = \Delta x_{i+1} = \Delta x,$$

$$\sum_{j=i-1}^{i+1} u_j \int_{i-1}^{i+1} k \frac{\partial N_j}{\partial x} \frac{\partial N_i}{\partial x} \, dx - \int_{i-1}^{i+1} q N_i(x) \, dx = 0 \qquad (\text{E5.3.2})$$

Performing the integrations, with the shape function derivatives equal to $\pm 1/\Delta x$, we obtain

$$k_{i+1/2} \frac{u_{i+1} - u_i}{\Delta x^2} - k_{i-1/2} \frac{u_i - u_{i-1}}{\Delta x^2} = \frac{(q_{i-1} + 4q_i + q_{i+1})}{6} \qquad (\text{E5.3.3})$$

where

$$k_{i+1/2} = \frac{1}{\Delta x} \int_i^{i+1} k \, dx \qquad (\text{E5.3.4})$$

and a similar expression for $k_{i-1/2}$. If a linear variation within each element is assumed for k, then

$$k_{i+1/2} = (k_i + k_{i+1})/2 \qquad (\text{E5.3.5})$$

It is interesting to note that the left-hand side of equation (E2.3.3) is identical to the central second-order finite difference formula (4.2.50). In the latter case the right-hand side would be equal to q_i while in the finite element Galerkin approach we obtain an average weighted over the three nodal points $i-1$, i, $i+1$. This is a typical property of the weighted residual Galerkin method.

Observe also that linear elements lead to second-order accurate discretizations. It is a general rule, on uniform meshes, that elements of order p lead to discretizations of order of accuracy $p+1$.

Example 5.3.2 *Laplace equation on a triangular uniform mesh*

A triangulation of a uniform Cartesian mesh ($\Delta x = \Delta y$) can be defined as in Figure 5.3.1. Node J is associated with the mesh co-ordinates (i, j) and the Laplace equation is considered with Dirichlet boundary conditions:

$$\Delta u = q$$
$$u = u_0 \quad \text{on } \Gamma \qquad (\text{E5.3.6})$$

The Galerkin equation (5.3.6) becomes

$$-\sum_I u_I \int_{\Omega_J} \left(\frac{\partial N_I}{\partial x} \cdot \frac{\partial N_J}{\partial x} + \frac{\partial N_I}{\partial y} \cdot \frac{\partial N_J}{\partial y} \right) dx \, dy = \int_{\Omega_J} q N_J \, dx \, dy \qquad (\text{E5.3.7})$$

There is no boundary integral, since the weight functions are taken to vanish on the boundaries. The integration domain Ω_J covers all the triangles containing node J, that is, triangles 1–6. The summation extends over all the nodes of these triangles.

With linear shape functions, according to Table 5.1 we have for $N_J = N_{i,j}$ in triangle 1

$$N_{ij}^{(1)} \equiv N_j^{(1)} = 1 - \frac{x - x_{ij}}{\Delta x}$$

$$N_{i+1,j}^{(1)} = 1 + \frac{x - x_{i+1,j}}{\Delta x} - \frac{y - y_{i+1,j}}{\Delta y} \qquad \text{(E5.3.8)}$$

$$N_{i+1,j+1}^{(1)} = 1 + \frac{y - y_{i+1,j+1}}{\Delta y}$$

Similarly, in triangles 2 and 3,

$$N_{ij}^{(2)} = 1 - \frac{y - y_{ij}}{\Delta y}$$

$$\text{(E5.3.9)}$$

$$N_{ij}^{(3)} = 1 + \frac{x - x_{ij}}{\Delta x} - \frac{y - y_{ij}}{\Delta y}$$

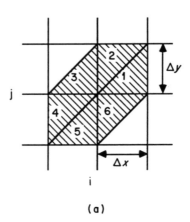

(a)

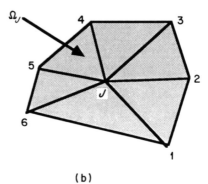

(b)

Figure 5.3.1 Finite element domain formed by six linear triangular elements. (a) Triangular elements on Cartesian mesh; (b) general triangular elements

The contributions from triangle 1 to the stiffness matrix K_{IJ} are obtained as follows, with the notation $K_{JJ} \equiv K_{ij}^{ij}$ and $\Delta x = \Delta y$:

$$K_{ij}^{i+1,\,j(1)} = \int_1 \left(\frac{\partial N_{ij}^{(1)}}{\partial x} \cdot \frac{\partial N_{i+1,\,j}^{(1)}}{\partial x} + \frac{\partial N_{ij}^{(1)}}{\partial y} \cdot \frac{\partial N_{i+1,\,j}^{(1)}}{\partial y} \right) dx\, dy \quad \text{(E5.3.10)}$$

$$= \int_1 \left(-\frac{1}{\Delta x} \right) \cdot \left(\frac{1}{\Delta x} \right) dx\, dy = -\tfrac{1}{2}$$

$$K_{ij}^{i+1,\,j+1(1)} = 0$$

$$\text{(E5.3.11)}$$

$$K_{ij}^{ij\,(1)} = \tfrac{1}{2}$$

Adding the contributions from all the triangles, we obtain

$$K_{ij}^{ij} = 4; \; K_{ij}^{i+1,\,j} = K_{ij}^{i-1,\,j} = K_{i,\,j+1}^{ij} = K_{ij}^{i,\,j-1} = -1$$
$$K_{ij}^{i+1,\,j+1} = K_{i-1,\,j-1}^{ij} = 0 \quad \text{(E5.3.12)}$$

and equation (E5.3.7) becomes, for a linear variation of the source term q within each triangle,

$$-4u_{ij} + (u_{i+1,\,j} + u_{i-1,\,j} + u_{i,\,j+1} + u_{i,\,j-1}) \quad \text{(E5.3.13)}$$

$$= \frac{\Delta x^2}{12} (6q_{ij} + q_{i+1,\,j} + q_{i-1,\,j} + q_{i+1,\,j+1} + q_{i,\,j+1} + q_{i-1,\,j-1} + q_{i,\,j-1})$$

Compared with the finite difference discretization of the Laplace operator it is seen that the left-hand side is identical to the five-point molecule of Figure 4.4.2. In a finite difference method the right-hand side would be equal to q_{ij}, while the finite element method generates an average of the points surrounding the node (i, j).

5.3.2 Finite element Galerkin method for a conservation law

Consider the conservation law of the form (2.1.5):

$$\frac{\partial U}{\partial t} + \vec{\nabla} \cdot \vec{F} = 0 \quad (5.3.11)$$

where $\vec{F}$ is the flux vector containing only *convective* contributions with the following initial boundary conditions on the domain Ω with boundary $\Gamma = \Gamma_0 \cup \Gamma_1$:

$$
\begin{array}{llll}
U(\vec{x}, 0) = U_0(\vec{x}) & \text{for} & t = 0 & \vec{x} \in \Omega \\
U(\vec{x}, t) = U_1(\vec{x}) & \text{for} & t \geqslant 0 & \vec{x} \in \Gamma_0 \\
\vec{F} \cdot \vec{1}_n \equiv F_n = g & \text{for} & t \geqslant 0 & \vec{x} \in \Gamma_1
\end{array}
\quad (5.3.12)
$$

Defining a weak formulation, with $W = 0$ on Γ_0,

$$\int_\Omega \frac{\partial U}{\partial t} W \, d\Omega + \int_\Omega (\vec{\nabla} \cdot \vec{F}) W \, d\Omega = \int_\Omega Q \, d\Omega \qquad (5.3.13)$$

followed by an integration by parts on the flux term leads to

$$\int_\Omega \frac{\partial U}{\partial t} W \, d\Omega - \int_\Omega (\vec{F} \cdot \vec{\nabla}) W \, d\Omega + \int_\Gamma W \vec{F} \cdot d\vec{S} = \int_\Omega Q \, d\Omega \qquad (5.3.14)$$

The finite element representation is defined by

$$U = \sum_I U_I(t) N_I(\vec{x}) \qquad (5.3.15)$$

and since the flux term $\vec{F}$ is generally a non-linear function of U it is preferable to define also a separate representation for the fluxes $\vec{F}$ as

$$\vec{F} = \sum \vec{F}_I N_I(\vec{x}) \qquad (5.3.16)$$

The discretized equation for node J is obtained via the Galerkin method, $W = N_J$, leading to

$$\sum_I \frac{dU_I}{dt} \int_{\Omega_J} N_I N_J \, d\Omega - \sum_I \vec{F}_I \int_{\Omega_J} N_I \cdot \vec{\nabla} N_J \, d\Omega + \int_{\Gamma_1} g N_J \, d\Gamma = \int_{\Omega_J} Q N_J \, d\Omega$$

$$(5.3.17)$$

where Ω_J is the subdomain of all elements containing node J and the summation over I covers all the nodes of Ω_J (Figure 5.1.1).

The matrix of the time-dependent term is called the *mass* matrix M_{IJ}:

$$M_{IJ} = \int_{\Omega_J} N_I N_J \, d\Omega \qquad (5.3.18)$$

and the stiffness matrix

$$\vec{K}_{IJ} = \int_{\Omega_J} N_I \vec{\nabla} N_J \, d\Omega \qquad (5.3.19)$$

is no longer symmetric. Hence equation (5.3.17) becomes

$$\sum_I M_{IJ} \frac{dU_I}{dt} - \sum_I \vec{F}_I \cdot \vec{K}_{IJ} = \int_{\Omega_J} Q N_J \, d\Omega - \int_{\Gamma_1} g N_J \, d\Gamma \qquad (5.3.20)$$

If the flux $\vec{F}$ contains in addition a diffusive term of the form $(- \varkappa \vec{\nabla} U)$ then the term $(\vec{\nabla} \cdot \varkappa \vec{\nabla} U)$ will be treated according to equation (5.3.9). In finite difference discretizations the time-dependent term will generally reduce to dU_J/dt, corresponding to a diagonal mass matrix, while the present formulation leads to an average over the various nodes in Ω_J. The presence of this mass matrix complicates the resolution of the system of ordinary differential equations (5.3.20) in time.

A rigorous way of diagonalizing the mass matrix is to introduce 'orthogonal' interpolation functions and to apply a Petrov–Galerkin method with these new functions, following Hirsch and Warzee (1978). A more currently applied approximation, called *mass lumping*, consists of replacing M_{IJ} by the sum over I at fixed J, this sum along the elements of a column being concentrated on the main diagonal. That is,

$$M_{IJ}^{(\text{lump})} = \left[\sum_I M_{IJ}\right] \delta_{IJ} \tag{5.3.21}$$

The modified equation (5.3.20), obtained in this way, is close to a finite volume formulation.

5.3.3 Subdomain collocation—finite volume method

The collocation methods (domain and point collocation) both use the residual equation (5.3.13) *without* partial integration on the weighting function W. If a subdomain Ω_J is attached to each nodal point J, with the corresponding weighting function defined by (see Figure 5.1.1)

$$\begin{aligned} W_J(\vec{x}) &= 0 \qquad \vec{x} \notin \Omega_J \\ W_J(\vec{x}) &= 1 \qquad \vec{x} \in \Omega_J \end{aligned} \tag{5.3.22}$$

then the residual equation (5.3.13) becomes

$$\int_{\Omega_J} \frac{\partial U}{\partial t} \, d\Omega + \int_{\Omega_J} \vec{\nabla} \cdot \vec{F} \, d\Omega = + \int_{\Omega_J} Q \, d\Omega \tag{5.3.23}$$

A more interesting form of this equation is obtained after application of Gauss's theorem on the flux term, leading to the conservation equation in integral form written for each subdomain Ω_J limited by the closed surface Γ_J:

$$\int_{\Omega_J} \frac{\partial U}{\partial t} \, d\Omega + \oint_{\Gamma_J} \vec{F} \cdot d\vec{S} = \int_{\Omega_J} Q \, d\Omega \tag{5.3.24}$$

This equation, which can be obtained directly from equation (5.3.14), is the basic equation for the *finite volume method* which takes as its starting point the physical conservation laws in integral form written for small volumes around every mesh point. The advantage of this method, especially in the absence of source terms, is that the fluxes are calculated only on two-dimensional surfaces instead of in the three-dimensional space. In this subspace the contour integrals can be discretized either by finite difference techniques or by finite element methods. For two-dimensional problems, where the flux integrals are one-dimensional, both approaches will lead to the same numerical discretization of these terms, at least for the lowest-order approximation. The subdomains Ω_J are allowed to overlap with the condition that each part of the surface Γ_J appears as part of an even number of different subdomains such that the overall integral conservation law holds for any combination of adjacent subdomains.

The finite volume method is a widely used approach, and will be presented in more detail in Chapter 6.

Example 5.3.3   Conservation law on linear triangles

We apply the Galerkin equation (5.3.17) with linear triangles for a node J not adjacent to the boundary Γ_1 of the domain and in the absence of source terms. Referring to Figure 5.3.1, the domain Ω_J contains six triangles and the nodes numbered 1–6 around node J. We will also lump the mass matrix, leading to the approximation (see Problem 5.10)

$$M_{IJ} = \frac{\Omega_J}{3}\,\delta_{IJ} \qquad (E5.3.14)$$

The flux terms can be written as

$$\int_{\Omega_J} \vec{F}\cdot\vec{\nabla}N_J\,d\Omega = \sum_I \vec{F}_I \int_{\Omega_J} N_I\cdot\vec{\nabla}N_J\,d\Omega = \sum_I f_I \int_{\Omega_J} N_I \frac{\partial N_J}{\partial x}\,d\Omega$$

$$+ \sum_I g_I \int_{\Omega_J} N_I \frac{\partial N_J}{\partial y}\,d\Omega \qquad (E5.3.15)$$

where f and g are the Cartesian components of $\vec{F}$.

With the relations in Table 5.1 we have for triangle $J12$

$$\frac{\partial N_J}{\partial x} = (y_1 - y_2)/2A_{12} \qquad \frac{\partial N_J}{\partial y} = (x_2 - x_1)/2A_{12} \quad (E5.3.16)$$

where A_{12} is the area of triangle $J12$. Since these derivatives are constants, the integrals in equation (E5.3.15) reduce to the integrals over N_I, which are equal to $A_{12}/3$. Hence the contribution to the flux term of the Galerkin formulation from triangle $J12$, becomes

$$\int_{J12} \vec{F}\cdot\vec{\nabla}N_J\,d\Omega = (y_1 - y_2)\tfrac{1}{6}(f_1 + f_2 + f_J) - \tfrac{1}{6}(x_1 - x_2)(g_1 + g_2 + g_J)$$

$$(E5.3.17)$$

Summing over all the triangles we obtain

$$-\int_{\Omega_J} \vec{F}\cdot\vec{\nabla}N_J\,d\Omega = \frac{1}{3}\sum_{\text{sides}} (f_{12}\,\Delta y_{12} - g_{12}\,\Delta x_{12}) \qquad (E5.3.18)$$

with

$$f_{12} = \tfrac{1}{2}(f_1 + f_2) \qquad g_{12} = \tfrac{1}{2}(g_1 + g_2)$$
$$\Delta y_{12} = y_2 - y_1 \qquad \Delta x_{12} = x_2 - x_1 \qquad (E5.3.19)$$

The contributions from f_J, g_J cancel out of the summation on all the triangles, since

$$\sum_{\text{sides}} \Delta y_{12} = \sum_{\text{sides}} \Delta x_{12} = 0 \qquad (E5.3.20)$$

because Γ_J is a closed contour.

Introducing this expression into the Galerkin equation (5.3.17) leads to the following discretized scheme:

$$\Omega_J \frac{dU_J}{dt} + \sum_{\text{sides}} (f_{12} \, \Delta y_{12} - g_{12} \, \Delta x_{12}) = 0 \qquad \text{(E5.3.21)}$$

which is nothing else than a finite volume discretization on the hexagonal contour Γ_J, as will be seen in Chapter 6.

Alternative formulations If the flux terms of equation (E5.3.21) are recombined, for instance by assembling terms such as

$$(f_1 + f_2)(y_2 - y_1) + (f_2 + f_3)(y_3 - y_2) = f_2(y_3 - y_1) + \cdots$$
$$= -y_2(f_3 - f_1) + \cdots$$

we obtain the alternative formulation

$$\Omega_J \frac{dU_J}{dt} + \frac{1}{2} \sum_I \left[f_I(y_{I+1} - y_{I-1}) - g_I(x_{I+1} - x_{I-1}) \right] = 0 \qquad \text{(E5.3.22)}$$

where the summation extends over all the nodes, with $y_0 = y_6$ and $y_7 = y_1$ (and similarly for the x co-ordinates). We can also write equation (E5.3.22) as

$$\Omega_J \frac{dU_J}{dt} + \frac{1}{2} \sum_I \left[(f_I - f_J)(y_{I+1} - y_{I-1}) - (g_I - g_J)(x_{I+1} - x_{I-1}) \right] = 0$$

$$\text{(E5.3.23)}$$

since all the contributions to f_J and g_J vanish, because of

$$\sum_I f_J(y_{I+1} - y_{I-1}) = f_J \sum_I (y_{I+1} - y_{I-1}) = 0 \qquad \text{(E5.3.24)}$$

The other formulation becomes

$$\Omega_J \frac{dU_J}{dt} - \frac{1}{2} \sum_I \left[y_I(f_{I+1} - f_{I-1}) - x_I(g_{I+1} - g_{I-1}) \right] = 0 \qquad \text{(E5.3.25)}$$

or alternatively

$$\Omega_J \frac{dU_J}{dt} + \frac{1}{2} \sum_I \left[(y_J - y_I)(f_{I+1} - f_{I-1}) - (x_J - x_I)(g_{I+1} - g_{I-1}) \right] = 0$$

$$\text{(E5.3.26)}$$

with the conventions $f_0 = f_6$ and $f_7 = f_4$ (and similarly for g). The above relations can also be derived from the results of Problem 5.11. Note that these relations are independent of the number of triangles inside Ω_J and therefore apply to any polygonal contour.

5.4 PRACTICAL COMPUTATIONAL TECHNIQUES

Two basic aspects are involved in finite element computations: the mapping from the physical co-ordinate space, where the element is defined, to the

226

computational space of the local co-ordinates and the numerical integration rules of the various integrals appearing in finite element formulations.

5.4.1 General mapping to local co-ordinates

The first topic is generally handled by the multi-dimensional generalization of the isoparametric transformation concept, whereby the mapping function from the $\vec{x}$-space to the $\vec{\xi}$-space makes use of the shape functions in local co-ordinates $N_i(\vec{\xi})$ through

$$\vec{x} = \sum_I \vec{x}_I N_I(\vec{\xi}) \qquad (5.4.1)$$

where the summation extends to all the nodes I of the element. This mapping performs a transformation on every element separately, and the whole space is mapped on an element-by-element basis. Of course, certain continuity and conformability conditions have to be satisfied, and Figure 5.4.1, from Zienkiewicz (1977), gives some rules for uniqueness of the mapping (5.4.1).

When the interpolation functions are of a degree higher than one, the mapping allows the representation of curved elements. For instance, with quadrilateral elements the curved side of a two-dimensional element will be represented in the computation by a parabolic curve.

Note that, in general, the nodes I used in the mapping defined above need

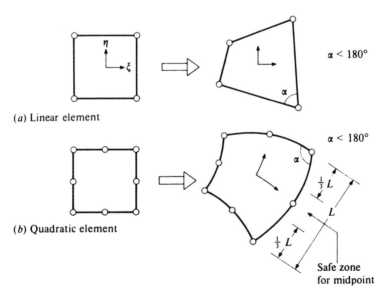

(a) Linear element

(b) Quadratic element

Figure 5.4.1 Rules for uniqueness of isoparametric mappings on linear and quadratic quadrilateral elements. (Reproduced by permission of McGraw-Hill Book Company from Zienkiewicz, 1977.)

not be identical in position or in number to the nodes used in the representation of the functions *u*. In this case the transformation is not called isoparametric but *super-parametric* or *sub-parametric* if the number of nodes used in the geometrical mapping is, respectively, higher or lower than the number used for the function representation. We refer the reader to the literature on finite elements for more details.

5.4.2 Numerical integration techniques

The various integrals appearing in finite element formulations can be evaluated using standard transformation rules based on the co-ordinate transformation (5.4.1). If $\bar{\bar{J}}$ is the Jacobian matrix of this transformation the gradient of any scalar function (for instance, one of the shape functions) defined in the physical space will be evaluated from the derivatives in the transformed space through

$$\vec{\nabla}_x N_I(\vec{x}) = \bar{\bar{J}}\, \vec{\nabla}_\xi N_I(\vec{\xi}) \tag{5.4.2}$$

The matrix elements of the Jacobian and the element volumes $d\Omega$ enter into the computation of the stiffness matrices (5.3.10) and are evaluated as follows:

$$\bar{\bar{J}}^{-1} = \begin{vmatrix} \dfrac{\partial x}{\partial \xi} & \dfrac{\partial y}{\partial \xi} \\[2mm] \dfrac{\partial x}{\partial \eta} & \dfrac{\partial y}{\partial \eta} \end{vmatrix} \quad \text{and} \quad \bar{\bar{J}} = \begin{vmatrix} \dfrac{\partial \xi}{\partial x} & \dfrac{\partial \eta}{\partial x} \\[2mm] \dfrac{\partial \xi}{\partial y} & \dfrac{\partial \eta}{\partial y} \end{vmatrix} \tag{5.4.3}$$

The elements of the Jacobian matrix are calculated from equation (5.4.1) by

$$\frac{\partial x}{\partial \xi} = \sum_I x_I \frac{\partial N_I}{\partial \xi} \qquad \frac{\partial y}{\partial \xi} = \sum_I y_I \frac{\partial N_I}{\partial \xi} \tag{5.4.4}$$

$$\frac{\partial x}{\partial \eta} = \sum_I x_I \frac{\partial N_I}{\partial \eta} \qquad \frac{\partial y}{\partial \eta} = \sum_I y_I \frac{\partial N_I}{\partial \eta} \tag{5.4.5}$$

The matrix elements of $\bar{\bar{J}}$ are calculated from

$$\bar{\bar{J}} = J \begin{vmatrix} \dfrac{\partial y}{\partial \eta} & -\dfrac{\partial y}{\partial \xi} \\[2mm] -\dfrac{\partial x}{\partial \eta} & \dfrac{\partial x}{\partial \xi} \end{vmatrix} \tag{5.4.6}$$

where $1/J$ is the determinant of $|\bar{\bar{J}}^{-1}|$.

The derivatives of N_I in physical space are obtained from equation (5.4.2). For instance,

$$\frac{\partial}{\partial x} N_I(x, y) = \frac{\partial N_I}{\partial \xi} \cdot \frac{\partial \xi}{\partial x} + \frac{\partial N_I}{\partial \eta} \cdot \frac{\partial \eta}{\partial x} = J\left(\frac{\partial y}{\partial \eta} \cdot \frac{\partial N_I}{\partial \xi} - \frac{\partial y}{\partial \xi} \cdot \frac{\partial N_I}{\partial \eta}\right) \tag{5.4.7}$$

With

$$d\Omega = \frac{1}{J} d\xi \, d\eta \tag{5.4.8}$$

the stiffness matrix becomes

$$K_{IJ} = \int_\Omega \vec{\nabla}_x N_I \cdot \vec{\nabla}_x N_J \, dx \, dy = \int \vec{\nabla}_\xi N_I (\bar{\bar{J}}^T \otimes \bar{\bar{J}}) \cdot \vec{\nabla}_\xi N_J \frac{1}{J} d\xi \, d\eta \tag{5.4.9}$$

$$= \int_\Omega g^{\alpha\beta} \partial_\alpha N_I \cdot \partial_\beta N_J \frac{1}{J} d\xi \, d\eta \qquad \alpha, \beta = 1, 2$$

where $g^{\alpha\beta}$ is the metric tensor:

$$g^{\alpha\beta} = (\bar{\bar{J}}^T \otimes \bar{\bar{J}})^{\alpha\beta} \tag{5.4.10}$$

and $\xi^1 = \xi$, $\xi^2 = \eta$.

Observe that formula (5.4.8) can be used to derive an approximation to the Jacobian of the element, when $d\xi$ and $d\eta$ are replaced by their extreme values. For a quadrilateral element of area A, with $\Delta\xi = \Delta\eta = 2$, we have $A = 4/J$.

In practical calculations the integrations are performed numerically using Gaussian quadrature:

$$\int_{-1}^{+1} f(\xi) \, d\xi = \sum_1^n H_i f(\xi_i) \tag{5.4.11}$$

where the summation extends to n well-defined points in the integration interval $(-1, +1)$ and where H_i are the corresponding weight coefficients. For most of the computational methods used in practice it will be unnecessary to take more than four Gauss points. Actually, in many instances two-point Gauss integration gives sufficient numerical accuracy, since with n points, a polynomial of order $(2n - 1)$ is integrated exactly.

The lowest Gauss abscissa and weight coefficients are given in Table 5.3. A two-dimensional integration of the form $\int f(\xi, \eta) \, d\xi \, d\eta$ is calculated with Gauss points in each direction separately, leading to

$$\int_{-1}^{+1} \int_{-1}^{+1} f(\xi, \eta) \, d\xi \, d\eta = \sum_{i=1}^n \sum_{j=1}^n H_i H_j f(\xi_i, \eta_j) \tag{5.4.12}$$

Table 5.3. Gauss point integration for quadrilaterals

n	ξ_i	H_i
1	0.0	2.0
2	± 0.5773502691896	1.0
3	± 0.7745966692415	0.5555555555556
	0.0	0.8888888888889
4	± 0.8611363115941	0.3478548451375
	± 0.3399810435849	0.6521451548625

The extension to three dimensions is straightforward. The above integration rule is valid on a rectangular domain or a cubic volume in three dimensions.

For triangular or tetrahedral regions in local co-ordinates the limits of integration involve the variables themselves, since in two dimensions we have, on a triangle with local co-ordinates L_i (see Table 5.1),

$$\int f(L_1, L_2, L_3) \, d\Omega = 2A \int_0^1 \int_0^{1-L_1} f(L_1, L_2, L_3) \, dL_1 \, dL_2 \quad (5.4.13)$$

Table 5.4. Gaussian integration formula from triangular elements. (Reproduced by permission of McGraw-Hill Book Company from Zienkiewicz, 1977)

Order	Fig.	Error	Points	Triangular co-ordinates	Weights
Linear		$R = O(h^2)$	a	$\frac{1}{3}, \frac{1}{3}, \frac{1}{3}$	1
Quadratic		$R = O(h^3)$	a b c	$\frac{1}{2}, \frac{1}{2}, 0$ $0, \frac{1}{2}, \frac{1}{2}$ $\frac{1}{2}, 0, \frac{1}{2}$	$\frac{1}{3}$ $\frac{1}{3}$ $\frac{1}{3}$
Cubic		$R = O(h^4)$	a b c d	$\frac{1}{3}, \frac{1}{3}, \frac{1}{3}$ $0.6, 0.2, 0.2$ $0.2, 0.6, 0.2$ $0.2, 0.2, 0.6$	$-\frac{27}{48}$ $\frac{25}{48}$
Quintic		$R = O(h^6)$	a b c d e f g	$\frac{1}{3}, \frac{1}{3}, \frac{1}{3}$ $\alpha_1, \beta_1, \beta_1$ $\beta_1, \alpha_1, \beta_1$ $\beta_1, \beta_1, \alpha_1$ $\alpha_2, \beta_2, \beta_2$ $\beta_2, \alpha_2, \beta_2$ $\beta_2, \beta_2, \alpha_2$	0.22500,00000 0.13239,41527 0.12593,91805

with

$\alpha_1 = 0.0597158717$
$\beta_1 = 0.4701420641$
$\alpha_2 = 0.7974269853$
$\beta_2 = 0.1012865073$

where A is the area of the triangle. The Gauss integration rule is written as

$$\int_0^{+1} \int_0^{1-L_1} f(L_1, L_2, L_3)\, dL_1\, dL_2 = \frac{1}{2} \sum_{i=1}^{n} H_i f(L_{1i}, L_{2i}, L_{3i}) \quad (5.4.14)$$

The adapted integration rules, Gauss point co-ordinates and weight coefficients are given in Table 5.4 (from Zienkiewicz, 1977). The corresponding information for tetrahedra is given in Table 5.5. The integration errors for triangles and tetrahedra are indicated in these tables, and for quadrilateral or parallelipipedic elements we have the following properties. The error due to the numerical integration is of the order $O(\Delta x^{2n-k+1})$ for an integration with n Gauss points of an integral containing derivatives up to the first order and interpolation functions of order k. In order for the numerical integration not to destroy the overall order of accuracy we should have

$$n \geqslant k \quad (5.4.15)$$

Hence for linear elements single-point integration will be sufficient, while for quadratic elements 2×2 (or $2 \times 2 \times 2$ in three dimensions) Gauss points will be required in order to maintain the overall accuracy. For quadratic triangles it is

Table 5.5. Gaussian integration formula from tetrahedral elements. (Reproduced by permission of McGraw-Hill Book Company from Zienkiewicz, 1977)

No.	Order	Fig.	Error	Points	Triangular co-ordinates	Weights
1	Linear		$R = O(h^2)$	a	$\frac{1}{4}, \frac{1}{4}, \frac{1}{4}, \frac{1}{4}$	1
2	Quadratic		$R = O(h^3)$	a b c d	$\alpha, \beta, \beta, \beta$ $\beta, \alpha, \beta, \beta$ $\beta, \beta, \alpha, \beta$ $\beta, \beta, \beta, \alpha$ $\alpha = 0.58541020$ $\beta = 0.13819660$	$\frac{1}{4}$ $\frac{1}{4}$ $\frac{1}{4}$ $\frac{1}{4}$
3	Cubic		$R = O(h^4)$	a b c d e	$\frac{1}{4}, \frac{1}{4}, \frac{1}{4}, \frac{1}{4}$ $\frac{1}{3}, \frac{1}{6}, \frac{1}{6}, \frac{1}{6}$ $\frac{1}{6}, \frac{1}{3}, \frac{1}{6}, \frac{1}{6}$ $\frac{1}{6}, \frac{1}{6}, \frac{1}{3}, \frac{1}{6}$ $\frac{1}{6}, \frac{1}{6}, \frac{1}{6}, \frac{1}{3}$	$-\frac{4}{5}$ $\frac{9}{20}$ $\frac{9}{20}$ $\frac{9}{20}$ $\frac{9}{20}$

seen from Table 5.4 that three Gauss points are need in two dimensions and four for the linear tetrahedra.

Note the following exact integration formulas for triangles and tetrahedra, which are very useful in practice:

$$\int L_1^m L_2^n L_3^p \, d\Omega = 2A \, \frac{m!n!p!}{(m+n+p+2)!} \tag{5.4.16}$$

valid in two dimensions, and for tetrahedra of volume V,

$$\int L_1^m L_2^n L_3^p L_4^q \, d\Omega = 6V \, \frac{m!n!p!q!}{(m+n+p+q+3)!} \tag{5.4.17}$$

References

Baker, A. J. (1983). *Finite Element Computational Fluid Mechanics*, New York: Hemisphere/McGraw-Hill.

Bristeau, M. D., Glowinski, R., Mantel, B., Periaux, J., Perrier, P., and Pironneau, O. (1980). In *Approximation Methods for Navier–Stokes Problems*, Lecture Notes in Mathematics, Vol. 771, New York: Springer Verlag.

Chung, T. J. (1978). *Finite Element Analysis in Fluid Dynamics*, New York: McGraw-Hill.

Ciarlet, P. G. (1978). *The Finite Element Method for Elliptic Problems*, Studies in Math. and Application, Vol. 4, Amsterdam: North-Holland.

Clough, R. W. (1960). 'The finite element method in plane stress analysis.' *Proc. 2nd ASCE Conf. on Electronic Computation, J. Struc. Div. ASCE*, 345–78.

Finlayson, B. A. (1972). *The Method of Weighted Residuals and Variation Principles*, New York: Academic Press.

Girault, V., and Raviart, P. A. (1979). *Finite Element Approximation of the Navier-Stokes Equation*, Lecture Notes in Mathematics, Vol. 749, New York: Springer Verlag.

Glowinski, R. (1983). *Numerical Methods for Non-linear Variational Problems*, 2nd edn, New York: Springer Verlag.

Hirsch, Ch., and Warzee, G. (1978). 'An orthogonal finite element method for transonic flow calculations.' *Proc. 6th Int. Conf. on Numerical Methods in Fluid Dynamics*, Lecture Notes in Physics, Vol. 90, New York: Springer Verlag.

Huebner, K. H. (1975). *The Finite Element Method for Engineers*, New York: John Wiley.

Marchuk, G. I. (1975). *Methods of Numerical Mathematics*, Berlin: Springer Verlag.

Oden, J. T. (1972). *Finite Elements of Non Linear Continua*, New York: McGraw-Hill.

Oden, J. T., and Reddy, J. N. (1976). *An Introduction to the Mathematical Theory of Finite Elements*, New York: John Wiley.

Strang, G., and Fix, G. (1973). *An Analysis of the Finite Element Method*, Englewood Clifts, NJ: Prentice-Hall.

Temam, R. (1977). *Navier–Stokes Equations*, Amsterdam: North-Holland.

Thomasset, F. (1981). *Implementation of Finite Element Methods for Navier–Stokes Equations*, Springer Series in Computational Physics, New York: Springer Verlag.

Turner, M. J., Clough, R. W., Martin, H. C., and Topp, L. P. (1956). 'Stiffness and deflection analysis of complex structures.' *J. Aeron. Soc.*, **23**, 805.

Zienkiewicz, O. C., and Cheung, Y. K. (1965). 'Finite elements in the solution of field problems.' *The Engineer*, 507–10.

Zienkiewicz, O. C. (1977).) *The Finite Element Method*, 3rd edn, New York: McGraw-Hill.

PROBLEMS

Problem 5.1

Derive the relations similar to equations (5.2.22) for the quadratic basis functions N_i and N_{i+1} defined by equation (5.2.17).

Problem 5.2

Obtain the approximations for the derivatives $(\partial u/\partial x)$ within the quadratic Lagrangian element $(i + 1, i, i - 1)$ following equation (5.2.23). Show that the following approximations are obtained:

$$(u_x)_{i+1} = \frac{u_{i-1} - 4u_i + 3u_{i+1}}{3\Delta x_{i+1} - \Delta x_i}$$

$$(u_x)_i = \frac{u_{i+1} - u_{i-1}}{\Delta x_{i+1} + \Delta x_i}$$

$$(u_x)_{i-1} = \frac{-3u_{i-1} + 4u_i - u_{i+1}}{3\Delta x_i - \Delta x_{i+1}}$$

Compare these with the finite difference formulas (4.2.19), (4.2.21) and (4.2.31) and show that the finite element formulas become of second-order accuracy on a uniform mesh $\Delta x_i = \Delta x_{i+1}$. Compare the truncation errors of the above approximation for $(u_x)_i$ with approximation (5.2.13) (see also Problem 4.11).

Problem 5.3

Apply the Galerkin method, with linear elements, to the first-order equation

$$a\frac{\partial u}{\partial x} = q$$

Show that on a uniform mesh, $\Delta x_i = \Delta x_{i+1} = \Delta x_i$, we obtain the same discretization as with central differences.

Problem 5.4

Apply the Galerkin method to equation (E5.3.1) for constant k with the second-order Lagrangian elements. Consider the two elements formed by the points $(i - 2, i - 1, i)$ and $(i, i + 1, i + 2)$. Show that for an interior point, such as $(i - 1)$, we obtain with constant k

$$\frac{k}{\Delta x^2}(u_{i-2} - 2u_{i-1} + u_i) = \tfrac{1}{10}(q_{i-2} + 8q_{i-1} + q_i)$$

and for a boundary point, such as point i,

$$\frac{k}{4\Delta x^2}(-u_{i-2} + 8u_{i-1} - 14u_i + 8u_{i+1} - u_{i+2})$$

$$= -\tfrac{1}{10}(-q_{i-2} + 2q_{i-1} + 8q_i + 2q_{i+1} - q_{i+2})$$

Show, by a Taylor expansion, that the left-hand side expression is a second-order approximation of a second derivative, and compare with the finite difference formula (4.2.47), which has fourth-order accuracy.

Problem 5.5

Apply the relations derived in Problem 5.1 in order to obtain the Galerkin equations for the convection equation for quadratic elements

$$a \frac{\partial u}{\partial x} = q$$

Show that we obtain the following finite element representation, referring to the elements defined in Problem 5.4, at an interior node $(i-1)$, for constant a:

$$\frac{a}{2\Delta x}(u_{i-2} - u_i) = \tfrac{1}{10}(q_{i-2} + 8q_{i-1} + q_i)$$

and at a boundary point i,

$$\frac{a}{6\Delta x}(u_{i-2} - 4u_{i-1} + 4u_{i+1} - u_{i+2}) = \tfrac{1}{10}(-q_{i-2} + 2q_{i-1} + 8q_i + 2q_{i+1} - q_{i+2})$$

Compare the left-hand side representation of a first derivative with the five-point finite difference formula (4.2.32) which has fourth-order accuracy:

$$\left(\frac{\partial u}{\partial x}\right)_i = \frac{1}{12\Delta x}(u_{i-2} - 8u_{i-1} + 8u_{i+1} - u_{i+2}) + O(\Delta x^4)$$

Show by a Taylor series expansion that the finite element formula, obtained from the Galerkin term,

$$\left(\frac{\partial u}{\partial x}, N_i\right) = \frac{1}{6\Delta x}(u_{i-2} - 4u_{i-1} + 4u_{i+1} - u_{i+2})$$

has only second-order accuracy.

Problem 5.6

Apply the Galerkin weak formulation with the cubic Hermitian interpolation functions to the first-order derivative operator u_x:

$$(u_x, N_J) = \sum_{I=1}^{4} u_I(\partial_x N_I, N_J)$$

With a first choice of $N_J = H_i^0$ show that we obtain, on a uniform mesh,

$$(u_x, H_i^0) = \frac{-[(u_x)_{i-1} - 3(u_x)_i + (u_x)_{i+1}]}{10} + \frac{(u_{i+1} - u_{i-1})}{2\Delta x}$$

Obtain also the second equation at node i for $N_J = H_i^1$:

$$(u_x, H_i^1) = \frac{(u_x)_{i-1} - (u_x)_{i+1}}{60} + \frac{(u_{i-1} - 2u_i + u_{i+1})}{10\Delta x}$$

Show from Taylor series expansions that the above equations are implicit formulas for the first-order derivatives with second-order accuracy. Compare with the third-order accurate implicit difference formula obtained from equation (4.3.22) with $\beta = -1$

$$\frac{(u_x)_{i-1} - (u_x)_{i+1}}{2} + \frac{u_{i+1} - 2u_i + u_{i-1}}{\Delta x} = -\frac{1}{12}\Delta x^3\left(\frac{\partial^4 u}{\partial x^4}\right)$$

Problem 5.7

Work out all the calculations of Example 5.3.2.

234

Problem 5.8

Apply the Galerkin method to the Laplace equation on a uniform Cartesian mesh, with bilinear quadrilateral elements. Show that we obtain the nine-point molecule of Figure 4.4.5.

Problem 5.9

Show, by an explicit calculation, that the average value of a quantity U over an element is approximated by

$$\frac{1}{\Omega} \int_\Omega U \, d\Omega = \frac{1}{3} \sum_{I=1}^{3} U_I \quad \text{for linear triangles}$$

$$= \frac{1}{4} \sum_{I=1}^{4} U_I \quad \text{for bilinear quadrilaterals}$$

Compare with the approximations obtained from Gauss point integration formulas for triangles and quadrilaterals using one Gauss point.

Hint: Take $U = \Sigma_I U_I N_I$ and perform exact integrations. With a single Gauss point the average of U is approximated by the value of U at the centre of the element instead of the average of the nodal values.

Problem 5.10

Calculate the mass matrix elements attached to node J of Figure 5.3.1 with linear triangles. Show that we obtain equation (E5.3.14) for the lumped mass approximation.

Hint: Obtain the following matrix, for a triangle of area A:

$$M_{IJ} = \frac{A}{12} \begin{vmatrix} 2 & 1 & 1 \\ 1 & 2 & 1 \\ 1 & 1 & 2 \end{vmatrix}$$

Problem 5.11

Referring to Figure 5.3.1 show that the average of $\partial f/\partial x$ and $\partial g/\partial y$ over the domain Ω_J, covered by the six linear triangles, can be defined as

$$\frac{1}{\Omega_J} \int_{\Omega_J} \frac{\partial f}{\partial x} \, d\Omega \equiv \overline{\left(\frac{\partial f}{\partial x}\right)}$$

$$= \frac{1}{2\Omega_J} \sum_I f_I(y_{I+1} - y_{I-1})$$

where the summation extends over all the nodes of the contour Γ_J, and

$$\overline{\left(\frac{\partial g}{\partial y}\right)} \equiv \frac{1}{\Omega_J} \int_{\Omega_J} \frac{\partial g}{\partial y} \, d\Omega = -\frac{1}{2\Omega_J} \sum_I g_I(x_{I+1} - x_{I-1})$$

Hint: Calculate $\int(\partial f/\partial x) \, d\Omega$ for each triangle $J12$ by taking $f = \Sigma f_I N_I$. Show that for each triangle we have

$$\int_{J12} \frac{\partial f}{\partial x} \, d\Omega = \tfrac{1}{2}[f_1(y_2 - y_J) + f_2(y_J - y_1) + f_J(y_1 - y_2)]$$

and sum these contributions over all the triangles.

Problem 5.12

Apply the results of Example 5.3.3 to a quadrilateral domain such as $J234$ in Figure 5.3.1(b), to obtain the following discretization of the flux integral on this quadrilateral:

$$\oint_{J234} \vec{F} \cdot d\vec{S} = \tfrac{1}{2}[(y_3 - y_J)(f_2 - f_4) + (f_3 - f_J)(y_4 - y_2)$$
$$- (x_3 - x_J)(g_2 - g_4) - (g_3 - g_J)(x_4 - x_2)]$$

Problem 5.13

Proof equation (5.2.16) for linear one-dimensional elements.

Problem 5.14

Apply the Galerkin method with linear elements to the conservation equation

$$\frac{\partial u}{\partial t} + \frac{\partial f}{\partial x} = 0$$

following Section 5.3.2. Show that we obtain the implicit formulation

$$\frac{1}{6}\left[\frac{du_{i-1}}{dt} + 4\frac{du_i}{dt} + \frac{du_{i+1}}{dt}\right] = \frac{1}{2\Delta x}(f_{i+1} - f_{i-1})$$

Problem 5.15

Apply the Galerkin formulation of a conservation law (equation (5.3.20)) to the Cartesian mesh of Figure 5.3.1(a), considered as bilinear elements, without source terms. Obtain the equation for node (i, j) and compare with the results of Example 5.3.3.

Perform the calculation, with and without mass lumping.

Problem 5.16

Consider a 2D cartesian mesh, such as Figure 5.3.1(a), and bilinear elements. Define the x-derivative in point $I(ij)$ by the following area average

$$\left(\frac{\partial u}{\partial x}\right)_{ij} = \frac{\displaystyle\int_{\Omega_I} \frac{\partial u}{\partial x} N_I \, d\Omega}{\displaystyle\int_{\Omega_I} N_I \, d\Omega}$$

where the integration domain covers the four quadrilateral elements having point $I(ij)$ in common. Show that this approximation corresponds to the central difference formula

$$\left(\frac{\partial u}{\partial x}\right)_{ij} = \frac{1}{12\Delta x}[(u_{i+1,j+1} - u_{i-1,j+1}) + 4(u_{i+1,j} - u_{i-1,j}) + (u_{i+1,j-1} - u_{i-1,j-1})]$$

Problem 5.17

Consider one-dimensional quadratic elements $(i - 2, i - 1, i)$ and $(i, i + 1, i + 2)$ and apply a Galerkin method to the discretization of the convection equation $u_t + au_x = 0$.

236

Show that the following schemes are obtained:

For end-node i

$$\frac{1}{10}\left[-\frac{du_{i-2}}{dt}+2\frac{du_{i-1}}{dt}+8\frac{du_i}{dt}+2\frac{du_{i+1}}{dt}-\frac{du_{i+2}}{dt}\right]$$

$$+\frac{a}{4\Delta x}(u_{i-2}-4u_{i-1}+4u_{i+1}-u_{i+2})=0$$

For midside-node i + 1

$$\frac{1}{10}\left[-\frac{du_{i+2}}{dt}+8\frac{du_{i+1}}{dt}+\frac{du_i}{dt}\right]+\frac{a}{2\Delta x}(u_{i+2}-u_i)=0$$

Chapter 6

Finite Volume Method and Conservative Discretizations

The finite volume method was apparently introduced into the field of numerical fluid dynamics independently by McDonald (1971) and Mac-Cormack and Paullay (1972) for the solution of two-dimensional, time-dependent Euler equations and extended by Rizzi and Inouye (1973) to three-dimensional flows. This is the name given to the technique by which the integral formulation of the conservation laws are discretized directly in the physical space. Although, according to one's point of view, it can be considered as a finite difference method applied to the differential, conservative form of the conservation laws, written in arbitrary co-ordinates, or as a variant of a weak formulation as described in the previous chapter, its importance and wide range of application justifies a separate presentation here.

The method takes full advantage of an arbitrary mesh, where a large number of options are open for the definition of the control volumes around which the conservation laws are expressed. Modifying the shape and location of the control volumes associated with a given mesh point, as well as varying the rules and accuracy for the evaluation of the fluxes through the control surfaces, gives considerable flexibility to the finite volume method. In addition, by the direct discretization of the integral form of the conservation laws we can ensure that the basic quantities mass, momentum and energy will also remain conserved at the discrete level. This is a most fundamental property for numerical schemes, and its precise meaning will be discussed prior to the introduction of the finite volume method.

6.1 THE CONSERVATIVE DISCRETIZATION

From the general presentation of Chapter 1 we know that the flow equations are the expression of a conservation law. Their general form for a scalar quantity U, with volume sources Q, is given by equation (1.1.1):

$$\frac{\partial}{\partial t} \int_\Omega U \, d\Omega + \oint_S \vec{F} \cdot d\vec{S} = \int_\Omega Q \, d\Omega \qquad (6.1.1)$$

The essential significance of this formulation lies in the presence of the surface

237

238

integral and the fact that the time variation of U inside the volume *only depends on the surface values of the fluxes.* Hence for an arbitrary subdivision of the volume Ω into, say, three subvolumes we can write the conservation law for each subvolume and recover the global conservation law by adding up the three subvolume conservation laws. Indeed, referring to Figure 6.1.1, the above equation for the subvolumes $\Omega_1, \Omega_2, \Omega_3$ becomes

$$\frac{\partial}{\partial t} \int_{\Omega_1} U \, d\Omega + \oint_{ABCA} \vec{F} \cdot d\vec{S} = \int_{\Omega_1} Q \, d\Omega$$

$$\frac{\partial}{\partial t} \int_{\Omega_2} U \, d\Omega + \oint_{DEBD} \vec{F} \cdot d\vec{S} = \int_{\Omega_2} Q \, d\Omega \qquad (6.1.2)$$

$$\frac{\partial}{\partial t} \int_{\Omega_3} U \, d\Omega + \oint_{AEDA} \vec{F} \cdot d\vec{S} = \int_{\Omega_3} Q \, d\Omega$$

When summing the surface integrals the contributions of the internal lines ADB and DE always appear twice but with opposite signs, and will cancel in the addition of the three subvolume conservation laws. Indeed, for colume Ω_2, for instance, we have a contribution of the fluxes

$$\int_{DE} \vec{F} \cdot d\vec{S}$$

while for Ω_3 we have a similar term:

$$\int_{ED} \vec{F} \cdot d\vec{S} = - \int_{DE} \vec{F} \cdot d\vec{S}$$

This essential property has to be satisfied by the numerical discretization of the flux contributions in order for a scheme to be *conservative.* When this is not the case, that is, when, after summation of the discretized equations over a certain number of adjacent mesh cells, the resulting equation still contains flux contributions from inside the total cell, the discretization is said to be *non-conservative,* and the internal flux contributions appear as *numerical internal volume sources.*

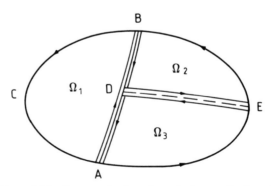

Figure 6.1.1 Conservation laws for subvolumes of volume Ω

Let us illustrate this on a one-dimensional form of the conservation law, written here as follows, where f is the x-component of the flux vector:

$$\frac{\partial u}{\partial t} + \frac{\partial f}{\partial x} = q \qquad (6.1.3)$$

With a central difference applied to the mesh of Figure 6.1.2. the following discretized equation is obtained at point i:

$$\frac{\partial u_i}{\partial t} + \frac{f_{i+1/2} - f_{i-1/2}}{\Delta x} = q_i \qquad (6.1.4)$$

The same discretization applied to point $(i + 1)$ will give

$$\frac{\partial u_{i+1}}{\partial t} + \frac{f_{i+3/2} - f_{i+1/2}}{\Delta x} = q_{i+1} \qquad (6.1.5)$$

and for $(i - 1)$

$$\frac{\partial u_{i-1}}{\partial t} + \frac{f_{i-1/2} - f_{i-3/2}}{\Delta x} = q_{i-1} \qquad (6.1.6)$$

The sum of these three equations is a consistent discretization of the conservation law for the cell AB $\equiv (i - 3/2, i + 3/2)$:

$$\frac{\partial}{\partial t} \frac{(u_i + u_{i+1} + u_{i-1})}{3} + \frac{f_{i+3/2} - f_{i-3/2}}{3\Delta x} = \frac{1}{3}(q_i + q_{i+1} + q_{i-1}) \qquad (6.1.7)$$

since the flux contributions at internal points have cancelled out. This is sometimes called the 'telescoping property' for the flux terms (Roache, 1972).

On the other hand, the *non-conservative* form (equation (1.1.7)) can be written as

$$\frac{\partial u}{\partial t} + a(u) \frac{\partial u}{\partial x} = q \qquad (6.1.8)$$

where the flux derivative has been expressed as

$$\frac{\partial f}{\partial x} = \left(\frac{\partial f}{\partial u}\right) \frac{\partial u}{\partial x}$$

and the function $a(u) = \partial f / \partial u$ is the derivative of the flux function with respect to the variable u. For instance, if $f = u^2/2$, $a(u) = u$.

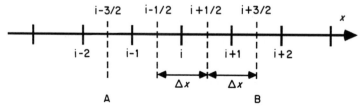

Figure 6.1.2 Subdivision of the one-dimensional space into mesh cells

Both formulations (6.1.3) and (6.1.8) are mathematically equivalent for arbitrary, non-linear fluxes, but their numerical implementation is not. Applying, for instance, a second-order central difference at mesh point i would give

$$\frac{\partial u_i}{\partial t} + a_i \frac{(u_{i+1/2} - u_{i-1/2})}{\Delta x} = q_i \qquad (6.1.9)$$

where a_i can be estimated as $a_i = (a_{i+1/2} + a_{i-1/2})/2$.

If similar equations are written for $(i+1)$ and $(i-1)$ and summed, a discretized equation for the cell AB in Figure 6.1.2 is obtained:

$$\frac{\partial}{\partial t}\left(\frac{u_i + u_{i+1} + u_{i-1}}{3}\right) + (a_{i+3/2} + a_{i-3/2})\frac{u_{i+3/2} - u_{i-3/2}}{6\Delta x} - \frac{q_i + q_{i+1} + q_{i-1}}{3} \qquad (6.1.10)$$

$$= -(a_{i+1/2} - a_{i-3/2})\frac{u_{i+3/2} - u_{i-1/2}}{6\Delta x} + (a_{i+3/2} - a_{i-1/2})\frac{u_{i+1/2} - u_{i-3/2}}{6\Delta x}$$

A direct discretization of equation (6.1.8) on the cell AB would have given the left-hand side of equation (6.1.10) with a vanishing right-hand side. It is therefore seen that the discretization of the non-conservative form of the equation gives rise to internal sources, equal in this case to the right-hand side of equation (6.1.10). These terms can be considered (by performing a Taylor expansion) as a discretization to second order of a term proportional to $\Delta x^2 [(a_x u_x)_x - (a_x u_{xx})]$ at mesh point i. For continuous flows, these numerical source terms are of the same order as the truncation error and hence could be neglected. However, numerical experiments and comparisons consistently show that non-conservative formulations are generally less accurate than conservative ones, particularly in the presence of strong gradients.

For discontinuous flows, such as transonic flows with shock waves, these numerical source terms can become important across the discontinuity and give rise to large errors. This is indeed the case, and the discretization of the non-conservative form will not lead to the correct shock intensities. Therefore in order to obtain, in the numerical computation, the correct discontinuities (such as the Rankine–Hugoniot relations for the Euler equations) it has been shown by Lax (1954) that it is necessary to discretize the conservative form of the flow equations.

Formal expression of a conservative discretization

The conservativity requirement on equation (6.1.3) will be satisfied if the scheme can be written as

$$\frac{\partial u_i}{\partial t} + \frac{(f^*_{i+1/2} - f^*_{i-1/2})}{\Delta x} = q_i \qquad (6.1.11)$$

where f^* is called the *numerical flux* and is a function of the values of u at

$(2k - 1)$ neighbouring points:

$$f^*_{i+1/2} = f^*(u_{i+k}, ..., u_{i-k+1})$$ (6.1.12)

In addition, the consistency of equation (6.1.11) with the original equation requires that, when all the u_{i+j} are equal, we should have

$$f^*(u, ..., u) = f(u)$$ (6.1.13)

The generalization to multi-dimensions is straightforward, and the above conditions must hold separately for all the components of the flux vector. The importance of this formalization of the conservativity condition is expressed by the following fundamental theorem of Lax and Wendroff (1960):

Theorem If the solution u_i of the discretized equation (6.1.11) converges boundedly almost everywhere to some function $u(x, t)$ when Δx, Δt tend to zero, then $u(x, t)$ is a *weak solution* of equation (6.1.3).

This theorem guarantees that when the numerical solution converges it will do so to a solution of the basic equations, with the correct satisfaction of the Rankine–Hugoniot relations in the presence of discontinuities. Indeed, by comparing the derivation of the Rankine–Hugoniot relations in Section 2.7.1 with the weak formulation of the basic flow equations, with $W = 1$ in equation (5.3.14), it is obvious that these relations are satisfied by the weak solutions, since the starting point of the derivation is the integral form of the conservation law.

6.2 THE FINITE VOLUME METHOD

The integral conservation laws are written for a discrete volume,

$$\frac{\partial}{\partial t} \int_\Omega U \, d\Omega + \oint_S \vec{F} \cdot d\vec{S} = \int_\Omega Q \, d\Omega$$ (6.2.1)

and applied to a control volume Ω_J, when the discretized equation associated with U_J is to be defined. Equation (6.2.1) is replaced by the discrete form:

$$\frac{\partial}{\partial t}(U_J \Omega_J) + \sum_{sides} (\vec{F} \cdot \vec{S}) = Q_J \Omega_J$$ (6.2.2)

where the sum of the flux terms refers to all the external sides of the control cell Ω_J. Referring to Figure 6.2.2(a) and to cell 1(i, j), we would identify U_J with $U_{i,j}$, Ω_J with the area of ABCD, and the flux terms are summed over the four sides AB, BC, CD, DA. On the mesh of Figure 6.2.2(d) Ω_J is the dotted area of the triangles having node J in common, and the flux summation extends over the six sides 12, 23, 34, 45, 56, 61. This is the general formulation of the finite volume method, and the user has to define, for a selected Ω_J, how to estimate the volume and cell face areas of the control volume Ω_J and how to approximate the fluxes at the faces. We will discuss some of the most current options, in two and three dimensions.

The following constraints on the choice of the Ω_J volumes for a conservative finite volume method have to be satisfied:

(1) Their sum should cover the whole domain Ω;
(2) Adjacent Ω_J may overlap if each internal surface Γ_I is common to two volumes;
(3) Fluxes along a cell surface have to be computed by formulas *independent* of the cell in which they are considered.

Requirement (3) ensures that the conservative property is satisfied, since the flux contributions of internal boundaries will cancel when the contributions of the associated finite volumes are added.

Referring to Figure 6.2.1, cells 1–4 have no common sides and their sum does not cover the whole volume. In addition, the sides are not common to two volumes. Cells 5–7 overlap, but have no common surfaces. Hence the conservative property will not be satisfied.

Equation (6.2.2) shows several interesting features which distinguish the interpretation of finite volume methods from the finite difference and finite element approaches:

(1) The co-ordinates of point J, that is, the precise location of the variable U inside the control volume Ω_J, do not appear explicitly. Consequently, U_J is not necessarily attached to a fixed point inside the control volume and can be considered as an *average value* of the flow variable U over the

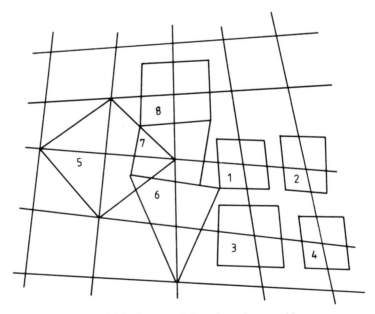

Figure 6.2.1 Incorrect finite volume decomposition

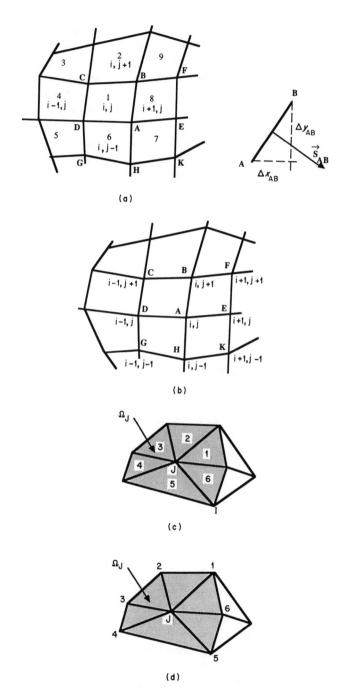

Figure 6.2.2 Two-dimensional finite volume mesh systems. (a) Cell centred structured finite volume mesh; (b) cell vertex structured finite volume mesh; (c) cell centred unstructured finite volume mesh; (d) cell vertex unstructured finite volume mesh

control cell. This is the interpretation taken in Figure 6.2.2(a). The first term of equation (6.2.2) therefore represents the time rate of change of the averaged flow variable over the selected finite volume.

(2) The mesh co-ordinates appear only in the determination of the cell volume and side areas. Hence, referring to Figure 6.2.2(a), and considering, for instance, the control cell ABCD around point 1, only the co-ordinates of A, B, C, D will be needed.

(3) In the absence of source terms, the finite volume formulation expresses that the variation of the average value U over a time interval Δt is equal to the sum of the fluxes exchanged between neighbouring cells. For stationary flows the numerical solution is obtained as a result of the balance of all the fluxes entering the control volume. That is,

$$\sum_{\text{sides}} (\vec{F} \cdot \vec{S}) = 0 \qquad (6.2.3)$$

When adjacent cells are considered, for instance cells ABCD and AEFB in Figure 6.2.2(a), the flux through face AB contributes to the two cells but with opposite signs. It is therefore convenient to program the method by sweeping through the cell faces and, when calculating the flux through side AB, to add this contribution to the flux balance of cell 1 and substract it from the flux balance of cell 8. This automatically guarantees global conservation.

(4) The finite volume method also allows a natural introduction of boundary conditions, for instance at solid walls where certain normal components are zero. For the mass conservation equation, $\vec{F} = \rho \vec{v}$ and at a solid boundary $\vec{F} \cdot \mathrm{d}\vec{S} = 0$. Hence the corresponding contribution to equations (6.2.2) or (6.2.3) would vanish.

Mesh and control volume definitions

Due to its generality, the finite volume method can handle any type of mesh and has therefore the same flexibility, in this respect, as the finite element method, restricted to elements with rectilinear sides. Two types of meshes can be considered:

(1) A 'finite difference' type mesh, where all mesh points lie on the intersection of two (or three) families of lines, considered as defining curvilinear co-ordinate lines. They are currently designated as *structured* meshes, and examples are shown in Figures 6.2.2(a) and 6.2.2(b).

(2) A 'finite element' type mesh formed by combinations of triangular and quadrilateral cells (or tetrahedra and pyramids in three dimensions), where the mesh points cannot be identified with co-ordinate lines. Therefore they cannot be represented by a set of integers, such as i, j (or i, j, k in three dimensions), but have to be numbered individually in a certain order. This type of mesh is designated as *unstructured,* and

examples are shown in Figures 6.2.2(c) and 6.2.2(d). Although requiring a more complicated bookkeeping, unstructured meshes can offer greater flexibility for complicated geometrical configurations.

An interesting example is provided by the internal flow in a circular duct where a structured mesh, formed by circles and radial lines as shown in Figure 6.2.3(a), cannot avoid the 'singular' point at the centre, which requires special treatment. This difficulty is avoided with the unstructured mesh of Figure 6.2.3(b).

Once the mesh is selected, we have to decide where to define the variables:

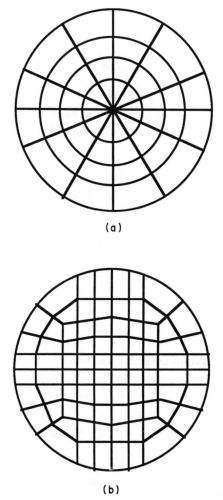

(a)

(b)

Figure 6.2.3 Mesh configurations for flow computation in a circular duct. (a) Structured and (b) unstructured mesh

246

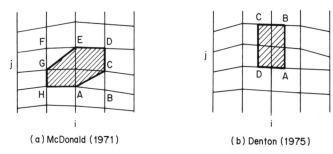

(a) McDonald (1971)   (b) Denton (1975)

Figure 6.2.4 Examples of two-dimensional control surfaces with cell-vertex finite volume method

(1) When the variables are associated with a cell, as in Figures 6.2.2(a) and 6.2.2(c), a *cell-centred* finite volume method is defined. The flow variables are averaged values over the cell and can be considered as representative of some point inside the cell (for instance, the central point of the cell).

(2) When the variables are attached to the mesh points, that is, to the cell vertices, we speak of a *cell-vertex* finite volume method, as shown in Figures 6.2.2(b) and 6.2.2(d).

With the first choice the mesh cells coincide with the control volume. With the second, a larger flexibility exists for the definition of the control volumes. Referring to Figure 6.2.2(b), an obvious choice would be to consider the four cells having mesh point (i, j) in common as the control volume GHKEFBCDG, associated with point (i, j). Many other choices are, however, possible and two of them are shown in Figure 6.2.4. Figure 6.2.4(a) is from McDonald (1971), who selected an hexagonal control volume, while Denton (1975) used a trapezoidal control surface covering two half-mesh cells (Figure 6.2.4(b)).

6.2.1 Two-dimensional finite volume method

Equation (6.2.1), considered for control cell ABCD of Figure 6.2.2, can be written as

$$\frac{\partial}{\partial t} \int_{\Omega_{ij}} U \, d\Omega + \oint_{ABCD} (f \, dy - g \, dx) = \int_{\Omega_{ij}} Q \, d\Omega \qquad (6.2.4)$$

where f and g are the Cartesian components of the flux vector $\vec{F}$. Equation (6.2.4) is the most appropriate for a direct discretization. The surface vector for a side AB can be defined as

$$\vec{S}_{AB} = \Delta y_{AB} \, \vec{1}_x - \Delta x_{AB} \, \vec{1}_y = (y_B - y_A) \, \vec{1}_x - (x_B - x_A) \, \vec{1}_y \qquad (6.2.5)$$

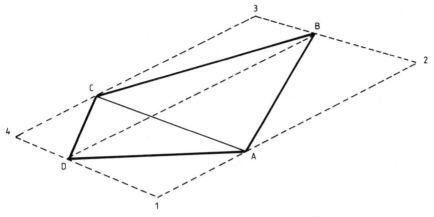

Figure 6.2.5 Area of an arbitrary plane quadrilateral

and we obtain the finite volume equation for cell Ω_{ij}:

$$\frac{\partial}{\partial t}(U\Omega)_{ij} + \sum_{ABCD}[f_{AB}(y_B - y_A) - g_{AB}(x_B - x_A)] = (Q\Omega)_{ij} \qquad (6.2.6)$$

The sum Σ_{ABCD} extends over the four sides of the quadrilateral ABCD.

For a general quadrilateral ABCD the area Ω can be evaluated from the vector products of the diagonals. As seen from Figure 6.2.5, the parallelogram 1234 built on the diagonals is twice the area of the quadrilateral ABCD. Hence with $\vec{x}_{AB} = \vec{x}_B - \vec{x}_A$, where $\vec{x}_A$ is the position vector of point A,

$$\begin{aligned}\Omega_{ABCD} &= \tfrac{1}{2}|\vec{x}_{AC} \times \vec{x}_{BD}| \\ &= \tfrac{1}{2}[(x_C - x_A)(y_D - y_B) - (y_C - y_A)(x_D - x_B)] \qquad (6.2.7) \\ &\equiv \tfrac{1}{2}(\Delta x_{AC}\,\Delta y_{BD} - \Delta x_{BD}\,\Delta y_{AC})\end{aligned}$$

The right-hand side of equation (6.2.7) should be positive for a cell ABCD, where A, B, C, D are located counterclockwise.

Evaluation of fluxes through cell faces

The evaluation of flux components along the sides, such as f_{AB}, g_{AB}, depends on the selected scheme as well as on the location of the flow variables with respect to the mesh. As will be seen more in detail in the following chapters and in Volume 2 for the systems of Euler and Navier–Stokes equations, we can distinguish essentially between central and upwind discretization schemes. Central schemes are based on local flux estimations, while upwind schemes determine the cell face fluxes according to the propagation direction of the wave components.

For *central schemes* and *cell-centred* finite volume methods the following alternatives can be considered:

(1)   Average of fluxes:

$$f_{AB} = \tfrac{1}{2}(f_{ij} + f_{i+1,j}) \qquad (6.2.8)$$

$$f_{ij} = f(U_{ij}) \qquad (6.2.9)$$

(2)   Since the flux components are generally non-linear functions of U, the following choice is not identical to equation (6.2.8):

$$f_{AB} = f\left(\frac{U_{ij} + U_{i+1,j}}{2}\right) \qquad (6.2.10)$$

(3)   Take f as the average of the fluxes in A and B:

$$f_{AB} = \tfrac{1}{2}\,(f_A + f_B) \qquad (6.2.11)$$

where either the variables are evaluated in A and B

$$U_A = \tfrac{1}{4}(U_{ij} + U_{i+1,j} + U_{i+1,j-1} + U_{i,j-1}) \qquad (6.2.12)$$

and

$$f_A = f(U_A) \qquad (6.2.13)$$

or the fluxes are averaged, as

$$f_A = \tfrac{1}{4}\,(f_{ij} + f_{i+1,j} + f_{i+1,j-1} + f_{i,j-1}) \qquad (6.2.14)$$

Observe that equations (6.2.10) and (6.2.13) will generally lead to schemes requiring a lower number of flux evaluations compared with the application of equations (6.2.8) and (6.2.14).

For *central schemes* and *cell-vertex* finite volume methods equations (6.2.10) or (6.2.11) are straightforward approximations to the flux f_{AB}. The choice (6.2.11) corresponds to the application of a trapezium formula for the integral $\int_{AB} f \, dy = (f_A + f_B)(y_B - y_A)/2$.

By summing the contributions of these integrals over the four sides of cell ABCD of Figure 6.2.2(b) we obtain the various discretizations already derived in Example 5.3.3; for instance, the flux terms in equation (E5.3.23), which become here:

$$\oint_{ABCD} \vec{F} \cdot d\vec{S} = \tfrac{1}{2}[(f_A - f_C)\Delta y_{DB} + (f_B - f_D)\Delta y_{AC}$$
$$- (g_A - g_C)\Delta x_{DB} - (g_B - g_D)\Delta x_{AC}] = 0 \qquad (6.2.15)$$

This also shows the equivalence of the flux evaluations in the cell-vertex, finite volume approach, with the finite element Galerkin method on linear triangles or bilinear quadrilaterals.

Example 6.2.1   Central scheme on a Cartesian mesh

Over a Cartesian, uniform mesh the above finite volume formulation is identical to a finite difference formula. Indeed, with

$$\Delta y_{AB} = y_{i+1/2,j+1/2} - y_{i+1/2,j-1/2} = \Delta y$$
$$\Delta x_{AB} = 0 \qquad\qquad \Delta x_{BC} = -\Delta x \qquad\qquad \text{(E6.2.1)}$$
$$\Omega_{ij} = \Delta x \cdot \Delta y \qquad \Delta y_{CB} = 0$$

we obtain, writing $f_{AB} = f_{i+1/2,j}$ and similary for the other components,

$$\frac{\partial}{\partial t} U_{ij}\Delta x\,\Delta y + (f_{i+1/2,j} - f_{i-1/2,j})\Delta y + (g_{i,j+1/2} - g_{i,j-1/2})\Delta x = Q_{ij}\,\Delta x\,\Delta y$$

$$\text{(E6.2.2)}$$

After division by $\Delta x\,\Delta y$ this reduces to the central difference form:

$$\frac{\partial U_{ij}}{\partial t} + \frac{(f_{i+1/2,j} - f_{i-1/2,j})}{\Delta x} + \frac{g_{i,j+1/2} - g_{i,j-1/2}}{\Delta y} = Q_{ij} \qquad \text{(E6.2.3)}$$

We have still to define how to calculate the flux components at the side centres $f_{i\pm1/2,j}$, $g_{i,j\pm1/2}$. With the choice (6.2.8) applied to Figure 6.2.2(a), equation (E6.2.3) becomes

$$\frac{\partial U_{ij}}{\partial t} + \frac{f_{i+1,j} - f_{i-1,j}}{2\Delta x} + \frac{g_{i,j+1} - g_{i,j-1}}{2\Delta y} = Q_{ij} \qquad \text{(E6.2.4)}$$

while equation (6.2.11) with equation (6.2.14) leads to

$$\frac{\partial U_{ij}}{\partial t} + \frac{1}{4}\left[2\,\frac{f_{i+1,j} - f_{i-1,j}}{2\Delta x} + \frac{f_{i+1,j+1} - f_{i-1,j+1}}{2\Delta x} + \frac{f_{i+1,j-1} - f_{i-1,j-1}}{2\Delta x}\right]$$

$$\text{(E6.2.5)}$$

$$+ \frac{1}{4}\left[2\,\frac{g_{i,j+1} - g_{i,j-1}}{2\Delta y} + \frac{g_{i+1,j+1} - g_{i+1,j-1}}{2\Delta y} + \frac{g_{i-1,j+1} - g_{i-1,j-1}}{2\Delta y}\right] = Q_{ij}$$

The central finite volume method therefore leads to second-order accurate space discretizations on Cartesian meshes.

Observe that f_{ij}, g_{ij} do not appear in equation (E6.2.4), and if $(i + j)$ is even, this equation contains only nodes with $(i + j)$ odd. Hence even- and odd-numbered nodes are separated, and this could lead to oscillations in the solution. This separation is not present with equation (E6.2.5). For applications to cell-vertex meshes, the reader is referred to Problems 6.1–6.4.

For *upwind schemes* and *cell-centred* finite volume methods a convective flux is evaluated as a function of the propagation direction of the associated convection speed. The latter is determined by the flux Jacobian

$$\vec{A}(U) = \frac{\partial \vec{F}}{\partial U} = a\,\vec{1}_x + b\,\vec{1}_y \qquad \text{(6.2.16)}$$

with $a(U) = \partial f/\partial U$ and $b(U) = \partial g/\partial U$.

The simplest upwind scheme takes the cell side flux equal to the flux generated in the upstream cell. This expresses that the cell side flux is fully determined by contributions transported in the direction of the convection velocity.

Considering Figure 6.2.2(a), we could define

$$(\vec{F}\cdot\vec{S})_{AB}=(\vec{F}\cdot\vec{S})_{ij} \qquad \text{if} \qquad (\vec{A}\cdot\vec{S})_{AB}>0$$
$$(\vec{F}\cdot\vec{S})_{AB}=(\vec{F}\cdot\vec{S})_{i+1,j} \qquad \text{if} \qquad (\vec{A}\cdot\vec{S})_{AB}<0$$

(6.2.17)

For *upwind schemes* and *cell-vertex* finite volume methods (Figure 6.2.2(b)) we could define

$$(\vec{F}\cdot\vec{S})_{AB}=(\vec{F}\cdot\vec{S})_{CD} \qquad \text{if} \qquad (\vec{A}\cdot\vec{S})_{AB}>0$$
$$(\vec{F}\cdot\vec{S})_{AB}=(\vec{F}\cdot\vec{S})_{EF} \qquad \text{if} \qquad (\vec{A}\cdot\vec{S})_{AB}<0$$

(6.2.18)

When applied to the control volume GHKEFBCD of Figure 6.2.2(b), we obtain contributions from points such as $(i-2,j)$ and $(i,j-2)$ for positive convection speeds. This leads to schemes with an unnecessary large support for the same accuracy. Therefore this option is not applied in practice (see Problem 6.14).

Example 6.2.2   Upwind scheme on a Cartesian mesh

We consider the discretization of the two-dimensional linear convection equation

$$\frac{\partial U}{\partial t}+a\frac{\partial U}{\partial x}+b\frac{\partial U}{\partial y}=0 \text{ with } a>0 \text{ and } b>0 \qquad (E6.2.6)$$

by a finite volume formulation on the cell ABCD of Figure 6.2.2(a), defined as a Cartesian cell following Example 6.2.1.

The fluxes are defined by $f=aU$ and $g=bU$ and with the choice of equation (6.2.17) we have, for AB and CD taken as vertical sides,

$$(\vec{F}\cdot\vec{S})_{AB}=f_{ij}\,\Delta y=aU_{ij}\,\Delta y$$
$$(\vec{F}\cdot\vec{S})_{CD}=-f_{i-1,}\,\Delta y=-aU_{i-1,j}\,\Delta y$$

(E6.2.7)

and similarly for the two horizontal sides BC and DA:

$$(\vec{F}\cdot\vec{S})_{BC}=g_{ij}\,\Delta x=bU_{ij}\,\Delta x$$
$$(\vec{F}\cdot\vec{S})_{DA}=-g_{i,j-1}\,\Delta x=-bU_{i-1,j}\,\Delta x$$

(E6.2.8)

The resulting scheme, obtained after division by the cell area $\Delta x\,\Delta y$, is only first-order accurate, and is a straightforward generalization of the first-order upwind scheme, to be introduced in Chapter 7 (equation (7.2.7)):

$$\frac{\partial U_{ij}}{\partial t}+\frac{1}{\Delta x}(f_{ij}-f_{i-1,j})+\frac{1}{\Delta y}(g_{ij}-g_{i,j-1})=0 \qquad (E6.2.9)$$

or

$$\frac{\partial U_{ij}}{\partial t} + \frac{a}{\Delta x}(U_{ij} - U_{i-1,j}) + \frac{b}{\Delta y}(U_{ij} - U_{i,j-1}) = 0 \qquad \text{(E6.2.10)}$$

Non-uniform mesh Although the finite volume formulation applies to arbitrary grids, the above equations for the determination of the fluxes nevertheless imply some regularity of the mesh. Referring, for instance, to equations (6.2.8) or (6.2.10) as applied to cell-centred finite volume methods, and interpreting the cell-averaged values U_{ij} in Figure 6.2.2(a) as mid-cell values, it is seen that these equations perform an arithmetic average of the fluxes (or the variables) on both sides of the cell face AB. This leads to a second-order approximation on a Cartesian mesh (see Example 6.2.1) if AB is at mid-distance from the cell centres 1 and 8. However, this will seldom be the case on non-uniform meshes, as shown in Figure 6.2.6(a), and a loss of accuracy will result from the application of these equations. Similar considerations apply to equations (6.2.12) and (6.2.14), based on the assumption that point A is in the centre of cell 1678. An analysis of the truncation errors for certain finite volume discretizations on non-uniform meshes can be found in Arts (1984), and more general analysis can be found in Turkel (1985), Turkel *et al.* (1985) and Roe (1987).

A straightforward generalization of equations (6.2.8) and (6.2.10) can be defined through a linear interpolation of f_{AB} (or U_{AB}) between the cell values f_{ij} and $f_{i+1,j}$ (or U_{ij} and $U_{i+1,j}$). This is uniquely defined on a one-dimensional basis, which is of application for an orthogonal mesh such as shown in Figure 6.2.6(b). Hence we can define

$$f_{AB} = \frac{b}{a+b}f_{ij} + \frac{a}{a+b}f_{i+1,j} \qquad (6.2.19)$$

or

$$f_{AB} = f\left(\frac{b}{a+b}U_{ij} + \frac{a}{a+b}U_{i+1,j}\right) \qquad (6.2.20)$$

For more general meshes, the distances a and b could be defined as shown in Figure 6.2.6(a), where M is the mid-point of AB, but in any case the second-order accuracy can only be maintained for sufficiently smooth varying mesh sizes (see Turkel, 1985, and Turkel *et al.*, 1985, for a more detailed discussion).

Generalizations of equations (6.2.12) and (6.2.14) to non-uniform meshes can be considered via an area weighted average instead of an arithmetic one. For instance, we could define

$$U_A = \sum_I \Omega_I U_I / \Omega_T \qquad (6.2.21)$$

where the summation ranges over the four points 1, 6, 7, 8, with Ω_I being the area of cell I and Ω_T is the total area of the four cells 1, 6, 7, 8. However, this formula, although natural, has the drawback of giving the lowest weight to the

252

smaller cell. Applying equation (6.2.21) to Figure 6.2.6(a) will give to point A a stronger dependence on point $7(i + 1, j - 1)$ than on the much closer point $1(i, j)$. This should be avoided, for instance through the application of finite element interpolations. We could consider the quadrilateral 1678 as a bilinear element and define, instead of equation (6.2.14),

$$U_A = \sum_I U_I N_I(x_A, y_A) \tag{6.2.22}$$

where the shape functions N_I are the bilinear polynomials of the Q_1 elements (see Table 5.1) and where the summation ranges over the four points $1, 6, 7, 8$.

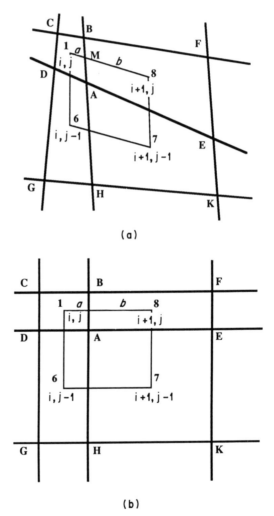

(a)

(b)

Figure 6.2.6 Non-uniform finite volume meshes. (a) Non-uniform finite volume mesh; (b) orthogonal non-uniform finite volume mesh

By definition of the finite element interpolation functions, if point A is close to one of the corner points of the cell, say point 1, than $U_A \approx U_1$, since $N_I(x_J, y_J) \approx \delta_{IJ}$ if points I and J are close enough. Alternatively, we could define a similar relation for the fluxes as a generalization of equation (6.2.12). For cell-vertex methods equation (6.2.11) can remain unchanged for arbitrary cell configurations.

In all cases, even with the above equations, a loss of accuracy will be unavoidable on strongly distorted meshes. However, cell-vertex schemes will generally maintain their accuracy for broader families of non-uniform meshes. As a general guideline one should avoid, if possible, discontinuous mesh size variations, for instance by generating grids analytically. With regard to the flux integrals, the application of finite element interpolations and integration rules can be considered as a valid guideline for the obtention of equations on strong non-uniform meshes.

6.2.2 General integration formulas for finite volumes

It is often necessary in finite volume discretizations to define numerically cell averages of derivatives of mesh variables. Particularly with the Navier–Stokes equations, the viscous flux components are functions of the velocity gradients, and we have to estimate appropriate values of these gradients on the cell faces. A general procedure, valid for an arbitrary control volume in two and three dimensions, can be derived by application of the divergence theorem.

This theorem can be considered as defining the average of the gradient of a scalar U as a function of its values at the boundaries of the volume under consideration. Since for an arbitrary volume Ω.

$$\int_\Omega \vec{\nabla} U \, d\Omega = \oint_s U \, d\vec{S} \tag{6.2.23}$$

where S is the closed boundary surface, we can define the averaged gradients as

$$\left(\overline{\frac{\partial U}{\partial x}}\right)_\Omega \equiv \frac{1}{\Omega} \int_\Omega \frac{\partial U}{\partial x} \, d\Omega = \frac{1}{\Omega} \oint_s U \, \vec{1}_x \cdot d\vec{S} \tag{6.2.24a}$$

and

$$\left(\overline{\frac{\partial U}{\partial y}}\right)_\Omega \equiv \frac{1}{\Omega} \int_\Omega \frac{\partial U}{\partial y} \, d\Omega = \frac{1}{\Omega} \oint_s U \, \vec{1}_y \cdot d\vec{S} \tag{6.2.24b}$$

For two-dimensional control cells Ω we obtain

$$\left(\overline{\frac{\partial U}{\partial x}}\right)_\Omega = \frac{1}{\Omega} \oint_s U \, dy = -\frac{1}{\Omega} \oint_s y \, dU \tag{6.2.25a}$$

after partial integration. Similarly, the averaged y-derivatives are obtained from

$$\left(\overline{\frac{\partial U}{\partial y}}\right)_\Omega \equiv -\frac{1}{\Omega} \oint_s U \, dx = \frac{1}{\Omega} \oint_s x \, dU \tag{6.2.25b}$$

Considering the control cell of Figure 6.2.2(d) and applying trapezoidal integration formulas along each side, the following equations are obtained, in agreement with the relations of the previous section:

$$\left(\frac{\overline{\partial U}}{\partial x}\right)_\Omega \equiv \frac{1}{\Omega} \int_\Omega \frac{\partial U}{\partial x} d\Omega = \frac{1}{2\Omega} \sum_l (U_l + U_{l+1})(y_{l+1} - y_l)$$

$$= \frac{-1}{2\Omega} \sum_l (y_l + y_{l+1})(U_{l+1} - U_l)$$

$$= \frac{1}{2\Omega} \sum_l U_l(y_{l+1} - y_{l-1}) \qquad (6.2.26a)$$

$$= \frac{-1}{2\Omega} \sum_l y_l(U_{l+1} - U_{l-1})$$

where the summation extends over all the vertices, from 1 to 6 with $U_0 = U_6$ and $U_7 = U_1$. The two last relations are obtained by rearranging the sums, as has been shown in Example 5.3.3.

The corresponding relations for the y-derivatives are derived after replacing x by y and changing the signs of the various expressions:

$$\left(\frac{\overline{\partial U}}{\partial y}\right)_\Omega \equiv \frac{1}{\Omega} \int_\Omega \frac{\partial U}{\partial y} d\Omega = \frac{-1}{2\Omega} \sum_l (U_l + U_{l+1})(x_{l+1} - x_l)$$

$$= \frac{1}{2\Omega} \sum_l (x_l + x_{l+1})(U_{l+1} - U_l)$$

$$= \frac{-1}{2\Omega} \sum_l U_l(x_{l+1} - x_{l-1}) \qquad (6.2.26b)$$

$$= \frac{1}{2\Omega} \sum_l x_l(U_{l+1} - U_{l-1})$$

The area of the cells can be obtained by equations similar to the above by noting that for $U = x$ the left-hand side of equation (6.2.26b) is equal to 1. Hence the following expressions can be used for the estimation of the area of an arbitrary cell:

$$\Omega = \frac{1}{2} \sum_l (x_l + x_{l+1})(y_{l+1} - y_l)$$

$$= \frac{-1}{2} \sum_l (y_l + y_{l+1})(x_{l+1} - x_l)$$

$$= \frac{1}{2} \sum_l x_l (y_{l+1} - y_{l-1}) \qquad (6.2.27)$$

$$= \frac{-1}{2} \sum_l y_l(x_{l+1} - x_{l-1})$$

For an arbitrary quadrilateral ABCD, as shown in Figure 6.2.5, an interesting formula is obtained by applying the third of the above relations, noting that the differences Δy can be grouped for opposite nodes, leading to

$$\oint_{ABCD} U \, dy = \frac{1}{2}[(U_A - U_C)(y_B - y_D) - (U_B - U_D)(y_A - y_C)] \quad (6.2.28)$$

and

$$\left(\frac{\overline{\partial U}}{\partial x}\right)_{ABCD} = \frac{(U_A - U_C)(y_B - y_D) - (U_B - U_D)(y_A - y_C)}{(x_A - x_C)(y_B - y_D) - (x_B - x_D)(y_A - y_C)} \quad (6.2.29a)$$

with a similar relation for the y-derivative:

$$\left(\frac{\overline{\partial U}}{\partial y}\right)_{ABCD} = \frac{(x_A - x_C)(U_B - U_D) - (x_B - x_D)(U_A - U_C)}{(x_A - x_C)(y_B - y_D) - (x_B - x_D)(y_A - y_C)} \quad (6.2.29b)$$

Example 6.2.3   Two-dimensional diffusion equation

We consider the two-dimensional diffusion equation

$$\frac{\partial U}{\partial t} + \frac{\partial}{\partial x}\left(k \frac{\partial U}{\partial x}\right) + \frac{\partial}{\partial y}\left(k \frac{\partial U}{\partial y}\right) = 0 \quad (E6.2.11)$$

with diffusive flux components $f = k\partial U/\partial x$ and $g = k\partial U/\partial y$, where k is a constant. We would like to construct a finite volume discretization on the mesh of Figure 6.2.2(a), considered as Cartesian, by expressing the balance of fluxes around the cell ABCD with the choice

$$f_{AB} = \frac{1}{2}(f_A + f_B) \quad (E6.2.12)$$

and an evaluation of the derivatives $\partial U/\partial x$ and $\partial U/\partial y$ in the cell corners A, B. Equation (6.2.6) for cell (i, j) is written here as

$$\left(\frac{\partial U}{\partial t}\right)_{ij} \Delta x \, \Delta y + (f_{AB} - f_{CD}) \, \Delta y + (g_{BC} - g_{DA}) \, \Delta x = 0 \quad (E6.2.13)$$

For point A the derivatives of U are taken as the average value over the cell 1678 and with equation (6.2.29):

$$f_A = k\left(\frac{\partial U}{\partial x}\right)_A = \frac{k}{2\Delta x}(U_{i+1,j} + U_{i+1,j-1} - U_{i,j} - U_{i,j-1}) \quad (E6.2.14)$$

A similar relation is obtained for point B:

$$f_B = k\left(\frac{\partial U}{\partial x}\right)_B = \frac{k}{2\Delta x}(U_{i+1,j} + U_{i+1,j+1} - U_{i,j} - U_{i,j+1}) \quad (E6.2.15)$$

and the flux contribution through the side AB is given by the sum of the two equations (E6.2.14) and (E6.2.15) multiplied by Δy.

The contributions of the other sides are obtained in a similar way. For instance, the flux through BC is given by the sum

$$g_{BC} \, \Delta x = \frac{1}{2}(g_B + g_C) \, \Delta x \quad (E6.2.16)$$

with

$$g_B = k\left(\frac{\partial U}{\partial y}\right)_B = \frac{k}{2\Delta y}\left(U_{i+1,j+1} + U_{i,j+1} - U_{i,j} - U_{i+1,j}\right) \quad \text{(E6.2.17)}$$

A similar relation is obtained for point C:

$$g_C = k\left(\frac{\partial U}{\partial y}\right)_C = \frac{k}{2\Delta y}\left(U_{i,j+1} + U_{i-1,j+1} - U_{i,j} - U_{i-1,j}\right) \quad \text{(E6.2.18)}$$

Finally, equation (E6.2.13) becomes, with $\Delta x = \Delta y$,

$$\frac{\partial U_{ij}}{\partial t} + k\,\frac{U_{i+1,j+1} + U_{i+1,j-1} + U_{i-1,j+1} + U_{i-1,j-1} - 4U_{ij}}{4\Delta x^2} = 0$$

$$\text{(E6.2.19)}$$

This scheme corresponds to the discretization of Figure 4.4.3 for the Laplace operator.

Note that the alternative, simpler choice,

$$f_{AB} = k\left(\frac{\partial U}{\partial x}\right)_{AB} = \frac{k}{\Delta x}\left(U_{i+1,j} - U_{i,j}\right) \quad \text{(E6.2.20)}$$

leads to the standard finite difference discretization of the diffusion equation, corresponding to Figure 4.4.2:

$$\frac{\partial U_{ij}}{\partial t} + \frac{U_{i+1,j} + U_{i,j-1} + U_{i-1,j} + U_{i,j+1} - 4U_{ij}}{\Delta x^2} = 0 \quad \text{(E6.2.21)}$$

The vector version of the divergence relation, written for an arbitrary vector $\vec{a}$, is also of interest:

$$\int_\Omega \vec{\nabla}\cdot\vec{a}\,\,d\Omega = \oint_S \vec{a}\cdot d\vec{S} \quad \text{(6.2.30)}$$

since it can be applied, particularly for the derivation of equations for cell face areas and volumes. For a two-dimensional cell, taking $\vec{a} = \vec{x}$ with $\vec{\nabla}\cdot\vec{x} = 2$, leads to

$$2\Omega = \oint_S \vec{x}\cdot d\vec{S} = \oint_S (x\,dy - y\,dx) \quad \text{(6.2.31)}$$

which reproduces the above relations when a trapezium formula is applied. Applications to three-dimensional volumes are discussed in the next section.

6.2.3 Three-dimensional finite volume method

In three dimensions the geometrical space is mostly divided into six-sided hexahedral control volumes (Figure 6.2.7), where the four points forming a cell face are not necessarily coplanar, or into tetrahedral volumes. Equation (6.2.2) remains unchanged, but some care has to be exercised in the evaluation

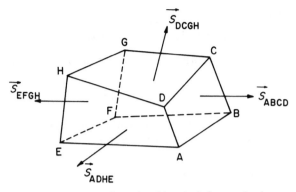

Figure 6.2.7 Three-dimensional hexahedral control volume

of volumes and cell surface areas in order to ensure that the sum of the computed volumes of adjacent cells is indeed equal to the total volume of the combined cells.

Evaluation of cell face areas

An important property of the area vector $\vec{S}$ attached to a cell face is derived from the divergence theorem. Equation (6.2.23), with $U = 1$, becomes

$$\oint_S d\vec{S} = 0 \qquad (6.2.32)$$

showing that the outward surface vector of a given face contained in the closed surface S:

$$\vec{S}_{\text{face}} = \int_{\text{face}} d\vec{S} \qquad (6.2.33)$$

is only dependent on the boundaries of the face. Hence for face ABCD of Figure 6.2.7 we could apply equation (6.2.7) or other alternatives:

$$\vec{S}_{\text{ABCD}} = \tfrac{1}{2} (\vec{x}_{\text{AC}} \times \vec{x}_{\text{BD}}) \qquad (6.2.34)$$

or

$$\vec{S}_{\text{ABCD}} = \tfrac{1}{2} [(\vec{x}_{\text{AB}} \times \vec{x}_{\text{BC}}) + (\vec{x}_{\text{CD}} \times \vec{x}_{\text{DA}})] \qquad (6.2.35)$$

The last equation expresses the surface vector $\vec{S}_{\text{ABCD}}$ as the average of the surface vectors of the two parallelograms constructed on the adjacent sides (AB, BC) and (CD, DA). In the general case the two normals will not be in the same direction, since (ABC) and (CDA) are not in the same plane. Hence equation (6.2.35) takes the vector $\vec{S}_{\text{ABCD}}$ as the average vector of these two normals, while equation (6.2.18) expresses $\vec{S}_{\text{ABCD}}$ as the vector product of the two diagonals. Note, however, that the two equations lead to identical results, even for non-coplanar cell faces (see Problem 6.8).

Similarly to equation (6.2.35), we have

$$\vec{S}_{ABCD} = \tfrac{1}{2}\,[\,(\vec{x}_{BC} \times \vec{x}_{CD}) + (\vec{x}_{DA} \times \vec{x}_{AB})\,] \qquad (6.2.36)$$

and also, combining opposite instead of adjacent sides, by averaging equations (6.2.35) and (6.2.36):

$$\vec{S}_{ABCD} = \tfrac{1}{4}\,[\,(\vec{x}_{AB} + \vec{x}_{DC}) \times (\vec{x}_{BC} + \vec{x}_{AD})\,] \qquad (6.2.37)$$

All these equations are applied in practical computations, the first one (equation (6.2.34)) being less expensive in number of arithmetic operations.

Evaluation of control cell volumes

Different equations can be applied to obtain the volume of the hexahedral cell, the most current approach consisting of a subdivision into tetrahedra or pyramids (Figure 6.2.8). The volume of the tetrahedron Ω_{PABC} is obtained by applying equation (6.2.30) for a vector $\vec{a}$ equal to the position vector $\vec{x}$. We obtain, since $\vec{\nabla} \cdot \vec{x} = 3$,

$$\Omega_{PABC} = \frac{1}{3} \oint_{PABC} \vec{x} \cdot d\vec{S} = \frac{1}{3} \sum_{faces} \vec{x} \cdot \vec{S}_{faces} \qquad (6.2.38)$$

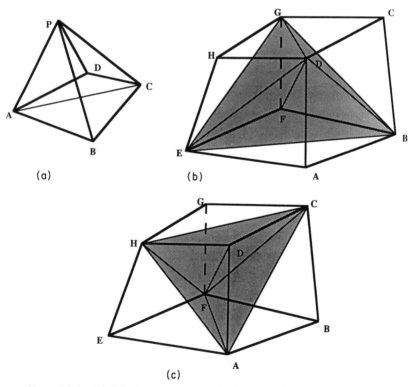

Figure 6.2.8 Subdivision of an hexahedral volume into tetrahedra or pyramids

or

$$\Omega_{PABC} = \frac{1}{3} \vec{x}_{(P)} \cdot \vec{S}_{ABC} \qquad (6.2.39)$$

if $\vec{x}_{(P)}$ represents a vector originating in P. This relation results from the fact that when $\vec{x}_{(P)}$ lies in the faces containing P, it is orthogonal to the associated $\vec{S}$ vector. The only remaining contribution comes from the face ABC opposite P. Hence with $\vec{x}_{(P)} = \vec{x}_{PA}$

$$\Omega_{PABC} = \frac{1}{6} \vec{x}_{PA} \cdot (\vec{x}_{AB} \times \vec{x}_{BC}) = \frac{1}{6} \vec{x}_{PA} \cdot (\vec{x}_{BC} \times \vec{x}_{CA}) \qquad (6.2.40)$$

Equation (6.2.40) can also be expressed as a determinant:

$$\Omega_{PABC} = \frac{1}{6} \begin{vmatrix} x_P & y_P & z_P & 1 \\ x_A & y_A & z_A & 1 \\ x_B & y_B & z_B & 1 \\ x_C & y_C & z_C & 1 \end{vmatrix} \qquad (6.2.41)$$

In a similar way, for a pyramid PABCD, we have

$$\Omega_{PABCD} = \frac{1}{3} \oint_{PABCD} \vec{x} \cdot d\vec{S}$$

$$= \frac{1}{3} \vec{x}_{(P)} \cdot \vec{S}_{ABCD} \qquad (6.2.42)$$

Since ABCD is not necessarily coplanar, $\vec{x}_{(P)}$ has to be estimated by an appropriate approximation. For instance,

$$\vec{x}_{(P)} = \tfrac{1}{4} (\vec{x}_{PA} + \vec{x}_{PB} + \vec{x}_{PC} + \vec{x}_{PD}) \qquad (6.2.43)$$

and with expression (6.2.34) for $\vec{S}_{ABCD}$ we obtain

$$\Omega_{PABCD} = \tfrac{1}{24} (\vec{x}_{PA} + \vec{x}_{PB} + \vec{x}_{PC} + \vec{x}_{PD}) \cdot (\vec{x}_{AC} \times \vec{x}_{BD})$$

$$= \tfrac{1}{12} (\vec{x}_{PA} + \vec{x}_{PB}) \cdot (\vec{x}_{AC} \times \vec{x}_{BD}) \qquad (6.2.44)$$

If the face ABCD is coplanar, then equation (6.2.44) reduces to

$$\Omega_{PABCD} = \frac{1}{6} \vec{x}_{PA} \cdot (\vec{x}_{AC} \times \vec{x}_{BD}) \qquad (6.2.45)$$

The volume equations for the pyramids are actually expressed as the sum of the two tetrahedra.

Referring to Figure 6.2.8, the hexahedron can be divided into three pyramids, for instance with point D as summit:

$$\Omega_{HEX} = \Omega_{DABFE} + \Omega_{DBCGF} + \Omega_{DEFGH} \qquad (6.2.46)$$

Dividing each pyramid into two tetrahedra leads to a decomposition of the hexahedron into six tetrahedra, originating, for instance, in D, as

$$\Omega_{HEX} = \Omega_{DABE} + \Omega_{DBFE} + \Omega_{DBCG} + \Omega_{DBGF} + \Omega_{DEFG} + \Omega_{DEGH} \qquad (6.2.47)$$

Extreme care has to be exercised in the evaluation of the tetrahedra volumes, since the sign of the volumes Ω_{PABC} in equations (6.2.39)–(6.2.41) depends on the orientation of the triangular decompositions. In addition, when the cell surfaces are not coplanar the same diagonal has to be used in the evaluations of the tetrahedra in the two cells which share this surface, otherwise gaps or overlaps would occur in the summation of volumes. A useful guideline, in order to avoid sign errors, consists of applying a right-hand rotation (screwdriver) rule from the base towards the summit of each tetrahedron.

Another alternative is to decompose the volume of the hexahedron into five tetrahedra originating in D, for instance, referring to Figure 6.2.8(b) as

$$\Omega_{HEX} = \Omega_{DABE} + \Omega_{DBCG} + \Omega_{DEGH} + \Omega_{DBGE} + \Omega_{FBEG} \qquad (6.2.48)$$

In this decomposition four tetrahedra have D as a summit and one tetrahedron originates in point F, opposite to D. Considering the same two points D and F as references, there is a unique second decomposition into five tetrahedra, shown in Figure 6.2.8(c):

$$\Omega_{HEX} = \Omega_{FACB} + \Omega_{FAEH} + \Omega_{FCHG} + \Omega_{FAHC} + \Omega_{DACH} \qquad (6.2.49)$$

For a general hexahedral volume, where points of a same cell face are not coplanar the two equations (6.2.48) and (6.2.49) will not give identical volume values. It is therefore recommended to take an average of both.

In this context it is interesting to observe that volumes of hexahedral cells can also be evaluated from a finite element isoparametric trilinear transformation, applying a $2 \times 2 \times 2$ Gauss point integration rule, as described in Section 5.4. Although very tedious to prove analytically, numerical experiments consistently show that this finite element procedure leads to volume values equal to the average of the equations (6.2.48) and (6.2.49).

An investigation of more elaborate decompositions of hexahedral volumes in pyramids can be found in Davies and Salmond (1985), while some of the above-mentioned decompositions are also discussed in Rizzi and Ericksson (1981) and Kordulla and Vinokur (1983).

References

Arts, T. (1984). 'On the consistency of four different control surfaces used for finite area blade-to-blade calculations.' *Intern. Journal for Numerical Methods in Fluids,* **4**, 1083–96.

Davies, D. E., and Salmond, D. J. (1985). 'Calculation of the volume of a general hexahedron for flow predictions.' *AIAA Journal,* **23**, 954–6.

Denton, J. (1975). 'A time marching method for two and three dimensional blade to blade flows.' *R & M 3775,* Aeronautical Research Council.

Kordulla, W., and Vinokur, M. (1983). 'Efficient computation of volumes in flow predictions.' *AIAA Journal,* **21**, 917–18.

Lax, P. D., and Wendroff, B. (1960). 'Systems of conservation laws.' *Comm. Pure and Applied Mathematics,* **13**, 217–37.

Lax, P. D. (1954). 'Weak solutions of non-linear hyperbolic equations and their numerical computation.' *Comm. Pure and Applied Mathematics,* **7**, 159–93.

MacCormack, R. W., and Paullay, A. J. (1972). 'Computational efficiency achieved by time splitting of finite difference operators.' *AIAA Paper 72-154*, San Diego.

McDonald, P. W. (1971). 'The computation of transonic flow through two-dimensional gas turbine cascades.' *ASME Paper 71-GT-89*.

Roache P. J. (1972). *Computational Fluid Dynamics*, Albuquerque, New Mexico: Hermosa.

Rizzi A. W., and Inouye M. (1973). 'Time split finite volume method for three-dimensional blunt-body flows.' *AIAA Journal*, 11, 1478–85.

Rizzi, A. W., and Eriksson, L. E. (1981). 'Transfinite mesh generation and damped Euler equation algorithm for transonic flow around wing–body configurations.' *Proc. AIAA Fifth Computational Fluid Dynamics Conference, AIAA Paper 81-0999*, pp. 43–68.

Roe P.L. (1987). 'Error estimates for cell-vertex solutions of the compressible Euler equations.' *ICASE Report No. 87-6*, NASA Langley Research Center.

Turkel, E. (1985). 'Accuracy of schemes with non-uniform meshes for compressible fluid flows.' *ICASE Report No. 85-43*, NASA Langley Research Center.

Turkel, E., Yaniv, S., and Landau, U. (1985). 'Accuracy of schemes for the Euler equations with non-uniform meshes,' *ICASE Report No. 85-59*, NASA Langley Research Center; also in *SIAM Journal Scientific and Statistical Computing*.

PROBLEMS

Problem 6.1

Apply the finite volume formula (6.2.2) to the contour ACDEGH of Figure 6.2.4(a) and derive the discretization for node (i, j) Compare with equation (E6.2.5) when the variables are defined at the nodes of the control volume, with the side fluxes defined by the average of the corner points value; that is,

$$f_{AC} = \tfrac{1}{2}(f_A + f_C)$$

Problem 6.2

Apply the finite volume method to the contour of Figure 6.2.4(b) and compare the different assumptions for the evaluation of fluxes at the mid-side. Derive four schemes by combining the two options for the vertical sides with the two options for the horizontal sides:

$$f_{AB} = (f_{i+1,j} + f_{ij})/2$$

or

$$f_{AB} = \tfrac{1}{6}(f_{i+1,j} + f_{i+1,j+1} + f_{i+1,j-1} + f_{ij} + f_{i,j+1} + f_{i,j-1})$$

with

$$f_{DA} = f_{i,j-1}$$

or

$$f_{DA} = \tfrac{1}{4}(2f_{i,j-1} + f_{i+1,j-1} + f_{i-1,j-1})$$

with similar expressions for g and for the two other sides.

Problem 6.3

Determine the different formulas obtained in Problem 6.2 when the mesh is Cartesian and compare with the results of Problem 6.1.

Problem 6.4

Develop a finite volume discretization for mesh point $A(i, j)$, with the control volume BCDGHKEF of Figure 6.2.2(b). Compare the results from the evaluation of the side fluxes by the following three options, written, for instance, for side K(E)F:

$$f_{KF}\,\Delta y_{KF} = \tfrac{1}{2}\,(f_{i+1,j+1} + f_{i+1,j-1})(y_{i+1,j+1} - y_{i+1,j-1}) \tag{a}$$

or

$$f_{KF}\,\Delta y_{KF} = f_{KE}\,\Delta y_{KE} + f_{EF}\,\Delta y_{EF}$$
$$= \tfrac{1}{2}\,(f_{i+1,j} + f_{i+1,j-1})(y_{i+1,j} - y_{i+1,j-1})$$
$$+ \tfrac{1}{2}\,(f_{i+1,j+1} + f_{i+1,j})(y_{i+1,j+1} - y_{i+1,j}) \tag{b}$$

or

$$f_{KF}\,\Delta y_{KF} = f_E\,\Delta y_{KF} = f_{i+1,j}(y_{i+1,j+1} - y_{i+1,j-1}) \tag{c}$$

Compare the three results for a Cartesian mesh and refer also to Example 6.2.1.

Problem 6.5

Apply the results (6.2.26) in order to derive average values of the first derivatives $\partial f/\partial x$ and $\partial f/\partial y$ over the triangle J_{12} of Figure 6.2.2(d). Compare with the expressions obtained in equations (E5.3.21) and (E5.3.22) and with the results of Problem 5.11. Note that the results are identical and comment on the reason behind the validity of the derivation by the finite element method with linear triangles.

Problem 6.6

Consider the two-dimensional diffusion equation treated in Example 6.2.3 with diffusive flux components $f = k\partial u/\partial x$ and $g = k\partial u/\partial y$, where k is a function of the co-ordinates. Construct the discrete equation by the finite volume approach on the mesh of Figure 6.2.2(a), considered as Cartesian, by generalizing the development of Example 6.2.3. Consider the quadrilateral control surface ABCD for the mesh point $1(i, j)$ and consider the values of k defined at the corners of the cell, that is in A, B, C, D. If necessary, define

$$k_{AB} = \tfrac{1}{2}\,(k_A + k_B)$$

Problem 6.7

Consider the diffusion equation of the previous problem and apply it to the cell BCDGHKEF of Figure 6.2.2(b), considered as Cartesian, with constant k. Define the derivatives on the cell sides by one-sided formulas from inside the control cell. Apply successively the three options of Problem 6.4 and compare with the results of Example 6.2.3. Show in particular that options a, b and c reproduce the schemes derived in this example.

Hint: For a point F, define the derivatives as

$$\left(\frac{\partial u}{\partial x}\right)_F = (u_{i+1,j+1} - u_{i,j+1})/\Delta x$$

$$\left(\frac{\partial u}{\partial y}\right)_F = (u_{i+1,j+1} - u_{i+1,j})/\Delta y$$

and similar relations for the other points.

Problem 6.8

Show that equations (6.2.35)–(6.2.37) lead to results identical to the simplest expression (6.2.34).
Hint: Apply vector relations such as $\vec{x}_{AC} = \vec{x}_{AB} + \vec{x}_{BC}$ and the properties of the vector products.

Problem 6.9

Show that the relation (6.2.37) for the area of a quadrilateral element ABCD can be obtained by applying a 2×2 Gauss-point integration rule with bilinear interpolation functions to the area integral (6.2.33).
Hint: Apply the integration techniques of Section 5.4 together with relation (5.4.8). In this relation $d\Omega$ is the magnitude of the vector $d\vec{S}$ of equation (6.2.33). Calculate the Jacobian matrix with the isoparameteric transformation.

Problem 6.10

Apply equation (6.2.31) to the quadrilateral ABCD of Figure 6.2.5 and take point A as the origin of the position vector $\vec{x}$. Show that the contour integral reduces to the contributions along BC and CD with $\vec{x} = \vec{x}_{AC}$ and that

$$2\Omega = \vec{x}_{AC} \cdot \int_{BCD} d\vec{S}$$

By working out the integral, obtain relation (6.2.7):

$$\Omega_{ABCD} = \tfrac{1}{2}(\Delta x_{AC}\, \Delta y_{BD} - \Delta x_{BD}\, \Delta y_{AC})$$

Hint: Observe that, with A as origin, the position vector is aligned with sides AB and AD and hence normal to the vector $d\vec{S}$. Therefore there is no contributions from these two sides.

Problem 6.11

Repeat Problem 6.10 for triangle ABC of Figure 6.2.5 and show that we can write

$$2\Omega_{ABC} = \vec{x}_{AB} \cdot \int_{BC} d\vec{S}$$

obtaining

$$\Omega_{ABC} = \tfrac{1}{2}(\Delta x_{AB}\, \Delta y_{BC} - \Delta x_{BC}\, \Delta y_{AB})$$

Problem 6.12

Consider the quadrilateral BDHE in Figure 6.2.6(b) and apply relations (6.2.29) in order to define the average value of the x-derivative of a function U. Consider $y_E = y_D$ and obtain

$$\left(\frac{\overline{\partial U}}{\partial x}\right)_{BDHE} = \frac{(U_E - U_D)}{(x_E - x_D)}$$

Comment on the accuracy of this formula when applied to point A.

Problem 6.13

Repeat Problem 6.12 for the contour BCDGHKEF of Figure 6.2.6(b) by applying equations (6.2.26). Obtain the following approximation:

$$\left(\frac{\partial U}{\partial x}\right) = \frac{(U_E - U_D)}{(x_E - x_D)} + \frac{\Delta y_{BA}}{2\Delta y_{BH}}\frac{(U_F - U_C)}{(x_F - x_C)} + \frac{\Delta y_{AH}}{2\Delta y_{BH}}\frac{(U_K - U_G)}{(x_K - x_G)}$$

Derive also the corresponding expression for a Cartesian mesh.

Problem 6.14

Apply the upwind flux evaluation (equation (6.2.18)) to derive a finite volume scheme for the cell GHKEFBCD of Figure 6.2.2(b), considered as Cartesian. Compare the obtained discretization with the results of Example 6.2.2.

Problem 6.15

Repeat the calculations of Example 6.2.3 for the arbitrary finite volume mesh of Figure 6.2.2(a).

Problem 6.16

Show that the integral conservation law over the one-dimensional domain $a \leqslant x \leqslant b$, applied to

$$\frac{\partial u}{\partial t} + \frac{\partial f}{\partial x} = 0$$

with the condition $f(b) = f(a)$, reduces to the condition:

$$\int_a^b u \, dx$$

is constant in time.

Apply this condition to a discretized x-space, with an arbitrary mesh point distribution, and show that this condition reduces to

$$\tfrac{1}{2}\sum_i \Delta u_i (x_{i+1} - x_{i-1}) = 0$$

where

$$\Delta u_i = u_i^{n+1} - u_i^n = \left(\frac{\partial u_i}{\partial t}\right)\Delta t$$

Hint: Apply a trapezoidal rule to evaluate the integral

$$\frac{\partial}{\partial t}\int_a^b u \, dx = 0$$

and rearrange the sum to isolate the u_i-terms.

PART III: THE ANALYSIS OF NUMERICAL SCHEMES

The reader who has followed the steps of the previous chapters in the development of a numerical simulation of a flow problem is, at this stage, faced with a set of discretized equations. This has now to be analysed for *consistency, accuracy, stability* and *convergence*.

Chapter 7 will present basic definitions and simplified model equations to serve as a basis for the illustration of various methods of analysis. Chapters 8–10 will be devoted to the problem of the analysis of stability and accuracy of the numerical scheme. This is a most fundamental step in the development of an algorithm, and constitutes an essential aspect of the numerical simulation. As soon as a scheme has been defined the first task is to analyse its stability and accuracy. Various methods are available, the most popular and useful being the Von Neumann method based on a Fourier analysis in space (Chapter 8). Next to the assessment of stability, this approach also allows a detailed investigation of the accuracy and the error structure of the scheme.

A second method (Chapter 9) is based on the *equivalent differential equation* and the obtained *truncation error*. The method is attributed to Hirt (1968) and Yanenko and Shokin (1969). It leads mostly to sufficient conditions for stability and to the establishment of the order of accuracy of the scheme. However, it delivers information on error and accuracy which is partly complementary to that obtained by the Von Neumann method. This approach does not allow us to define the influence of boundary conditions on the stability of the scheme but has the advantage of allowing the non-linear contributions to the error generation in non-linear equations to be distinguished. Also, as will be seen in Chapter 9, conditions can be derived for the establishment of whole families of schemes having a predetermined support and order of accuracy.

A third method, to be discussed in Chapter 10, is the *matrix* method. This is a most general approach, whereby the scheme is written as a system of first-order differential equations in time with, in the right-hand side, the matrix operator representing the discretized space differential operators, acting on the vector of mesh point variables u_i. This matrix also contains the numerical boundary treatment, and the stability is investigated through its eigenvalue spectrum. This approach is therefore general, including the influence of the boundary conditions, but its practical application is often made difficult by the impossibility of obtaining analytically the eigenvalues of the matrix. The precise definition of stability is also shown to have a profound impact on the stability conditions imposed on the parameters of the schemes and the predominant role of the Von Neumann method will be stressed.

If the Von Neumann method is based on a decomposition of the numerical solution in Fourier modes, implying periodic boundary conditions, the matrix method is based on a decomposition in eigenmodes of the space-discretization operator, including the boundary conditions, and on the properties of its eigenvalues. This generalizes the Von Neumann method and it is shown that both methods are identical for periodic boundary conditions.

A third decomposition of the numerical solution is defined by the normal mode representation. This decomposition is of a more general form than the one based on the eigenmodes of the space discretization and allows a detailed investigation of the influence of boundary conditions on stability. This representation leads, in addition, to the analysis of the spatial error propagation in stationary schemes or of resolution algorithms based on selected sweep directions through the mesh. These various topics are also illustrated with the important model of the one-dimensional convection–diffusion equation.

Chapter 7

The Concepts of Consistency, Stability and Convergence

In order to analyse the properties of numerical schemes and sustain the general procedures to be presented in the following chapters we will illustrate them on typical and representative examples of the various forms of equations which can result from the large variety of mathematical flow models. Since all fluid flow equations can be classified into elliptic, parabolic or hyperbolic, typical examples of each of them will cover the whole range of possible systems. The non-linearity of most of the flow models adds an additional level of difficulty, and its treatment and influence will have to be dealt with separately.

7.1 MODEL EQUATIONS

7.1.1 One-dimensional simplified models

Within the one-dimensional approximation the system of flow equations can be written as follows:

$$\frac{\partial \rho}{\partial t} + \frac{\partial (\rho u)}{\partial x} = 0 \qquad (7.1.1)$$

$$\frac{\partial (\rho u)}{\partial t} + \frac{\partial}{\partial x} (\rho u^2 + p) = \frac{\partial}{\partial x} \left(\mu \frac{\partial u}{\partial x} \right) \qquad (7.1.2)$$

$$\rho \left| \frac{\partial e}{\partial t} + u \frac{\partial e}{\partial x} \right| = -p \frac{\partial u}{\partial x} + \tau \frac{\partial u}{\partial x} + \frac{\partial}{\partial x} \left(k \frac{\partial T}{\partial x} \right) + q_H \qquad (7.1.3)$$

where equation (1.5.15) for the internal energy is applied.

(1)   If the velocity $u(x, t)$ is considered as constant, equation (7.1.1) takes the typical form of a *linear convection* equation:

$$\frac{\partial \rho}{\partial t} + u \frac{\partial \rho}{\partial x} = 0 \qquad (7.1.4)$$

describing the transport of mass (ρ) by a flow of velocity u. This is a typical first-order *hyperbolic* equation in (x, t), and can also be viewed as describing a wave of amplitude ρ propagating with a wave speed equal to u.

(2) If the energy equation is considered, either in a fluid at rest or in a very slowly moving fluid, we can neglect the terms containing the velocity in equation (7.1.3). We then obtain the *one-dimensional heat conduction equation*, with $e = c_v T$, where c_v is the specific heat at constant volume, and for constant conductivity k,

$$\frac{\partial T}{\partial t} = \alpha \frac{\partial^2 T}{\partial x^2} + q \tag{7.1.5}$$

with

$$\alpha = \frac{k}{\rho c_v} \tag{7.1.6}$$

This equation is a typical *parabolic equation in time*, in the (x, t) space, for sources which do not depend on the derivatives of the temperature. It describes a *diffusion* of heat in the medium.

(3) If the velocity is not small but the flow is incompressible then from equation (7.1.1) u is constant. The equation for the temperature then becomes

$$\frac{\partial T}{\partial t} + u \frac{\partial T}{\partial x} = \alpha \frac{\partial^2 T}{\partial x^2} + q \tag{7.1.7}$$

This equation is still *parabolic* in time in the (x, t) space but is called the *convection–diffusion* equation, because of the simultaneous presence of convection by the velocity u and diffusion through the diffusivity coefficient α. The dimensionless ratio UL/α, where L is a representative length, is called the *Peclet number*, and plays the same role as the Reynolds number UL/ν for the momentum equation, as a measure of the ratio of the convective flux to the diffusive flux. Actually, we have

$$Pe = \frac{UL}{\alpha} = \left(\frac{\nu}{\alpha}\right)\left(\frac{UL}{\nu}\right) = \left(\frac{\nu}{\alpha}\right) \cdot Re \tag{7.1.8}$$

where (ν/α) is proportional to the Prandtl number (see Section 1.5).

The influence of this number is considerable, since at very high values of the Peclet, or Reynolds, numbers the equation will be mostly of a convective nature and therefore close to hyperbolic, while in the opposite situation, for very low values of Pe, equation (7.1.7) is close to the purely diffusive equation (7.1.5). At intermediate values of Pe the presence of both contributions poses severe problems of accuracy to the numerical simulation, and this will be discussed in Chapter 10.

Note also that the general form of a transport equation for a scalar quantity u, as given by equation (1.1.7), reduces to the same form as equation (7.1.7) in the one-dimensional case. That is,

$$\frac{\partial u}{\partial t} + a \frac{\partial u}{\partial x} = \alpha \frac{\partial^2 u}{\partial x^2} + q \tag{7.1.9}$$

where a is the convection velocity. Hence equations (7.1.7) or (7.1.9) represent the most general form of the one-dimensional convection–diffusion transport equation.

(4) For time-independent conditions the transport equation becomes

$$a \frac{\partial u}{\partial x} = \alpha \frac{\partial^2 u}{\partial x^2} + q \qquad (7.1.10)$$

The solution of both equations (7.1.10) and (7.1.9) should be identical for a stationary field u and time-independent boundary conditions.

(5) If the effect of the pressure gradient on the momentum equation can be neglected or considered as an external force equation (7.1.2) becomes

$$\frac{\partial u}{\partial t} + u \frac{\partial u}{\partial x} = \nu \frac{\partial^2 u}{\partial x^2} \qquad (7.1.11)$$

This equation has the structure of a convection–diffusion equation but is non-linear. This is known as *Burger's equation*, and contains the full non-linearity of the one-dimensional flow equations. This equation, as well as the following one, are very often used as test cases for numerical schemes since a large number of exact solutions are known (Whitham, 1974).

(6) If viscosity can be neglected we obtain the non-linear, hyperbolic equation:

$$\frac{\partial u}{\partial t} + u \frac{\partial u}{\partial x} = 0 \qquad (7.1.12)$$

known as the *'inviscid' Burger's equation*.

7.1.2 Two-dimensional simplified models

A variety of simplified linear models for flow and temperature evolution can be obtained through some of the following assumptions.

Incompressible, potential flows

In this case it is seen from equation (2.9.5) that the potential ϕ satisfies the Laplace equation:

$$\frac{\partial^2 \phi}{\partial x^2} + \frac{\partial^2 \phi}{\partial y^2} = 0 \qquad (7.1.13)$$

This is a typical elliptic equation, describing an isotropic diffusion in the (x, y) space.

(7) Note that a similar equation is obtained for the temperature field in a medium at rest and for stationary conditions. In the presence of source terms we obtain the Poisson equation for constant conductivity coefficients k:

$$\frac{\partial^2 T}{\partial x^2} + \frac{\partial^2 T}{\partial y^2} = -\frac{1}{k} q_H \qquad (7.1.14)$$

This same equation is obtained, under similar assumptions, from the general transport equation (1.1.7) for the quantity u. The streamfunction equation for incompressible flows has also the same structure as equation (7.1.14).

(8) If the problem is time-dependent the heat conduction equation becomes, for inviscid flows and for a homogeneous conductivity,

$$\frac{\partial T}{\partial t} + u \frac{\partial T}{\partial x} + v \frac{\partial T}{\partial y} = \alpha \left(\frac{\partial^2 T}{\partial x^2} + \frac{\partial^2 T}{\partial y^2} \right) + q \qquad (7.1.15)$$

This is a form of the two-dimensional convection–diffusion equation in a flow of velocity $\vec{v}(u, v)$.

This equation is parabolic in time in the (x, y, t) space. For stationary coefficients and boundary conditions the numerical solution of equation (7.1.15) in a medium at rest ($u = v = 0$) should approach, at large time values, the solution of the stationary equation (7.1.14).

Supersonic, steady, two-dimensional potential flow

If we consider a stationary potential flow predominantly in the x-direction, with a nearly constant supersonic velocity, equation (2.9.25) reduces in two dimensions to an equation of the form

$$\frac{\partial^2 \phi}{\partial y^2} - K^2 \frac{\partial^2 \phi}{\partial x^2} = 0 \qquad (7.1.16)$$

with $K^2 = (M_\infty^2 - 1)$, M_∞ being the upstream Mach number of the flow. This equation is hyperbolic in the (x, y) space and is known as the two-dimensional *wave equation*. Observe that the same equation is elliptic at subsonic speeds, since K^2 is then negative. The fact that this transition can occur for non-linear potential flows makes the physical problem much more difficult. (See Volume 2 for a detailed presentation of various methods developed in order to solve the transonic potential flow problem.)

The above types of equations are representative of most of the approximation models occurring in practice describing flow and temperature behaviour.

7.2 BASIC DEFINITIONS: CONSISTENCY, STABILITY, CONVERGENCE

Let us consider one of the representative model equations, for instance the convective, hyperbolic equation (7.1.4), written here as follows:

$$\frac{\partial u}{\partial t} + a \frac{\partial u}{\partial x} = 0 \qquad (7.2.1)$$

where u is the unknown function of (x, t) and a the convection speed, or the wave speed according to the interpretation given to equation (7.2.1). In the following, when no danger of ambiguity can arise, we will also use a shorthand notation, where the derivatives are indicated as subscripts. Here we will write for equation (7.2.1)

$$u_t + au_x = 0 \qquad (7.2.2)$$

Considering an initial boundary value problem this equation has to be substantiated by the following initial and boundary conditions for $a > 0$:

$$
\begin{aligned}
t = 0 \qquad & u(x, 0) = f(x) \qquad 0 \leqslant x \leqslant L \\
x = 0 \qquad & u(0, t) = g(t) \qquad t \geqslant 0
\end{aligned}
\qquad (7.2.3)
$$

In order to apply a finite difference method to this equation we could select, for instance, a central, second-order difference formula for the discretization of the space derivative u_x at mesh point i after subdivision of the space domain into cells of length Δx. This leads to the semi-discrete scheme (also called *method of lines*):

$$(u_t)_i = -\frac{a}{2\Delta x}(u_{i+1} - u_{i-1}) \qquad (7.2.4)$$

The left-hand side represents the time derivatives evaluated at point i, and the next step is to define a discretization in time. This implies the replacement of $(u_t)_i$ by a discrete form but also a decision as to the time level at which the right-hand side will be evaluated.

Selecting a forward difference formula for $(u_t)_i$, the simplest scheme would be obtained with an evaluation of the right-hand side of equation (7.2.4) at time step n; such a method is known as the Euler method for the time integration of ordinary differential equations defining $u_i^n = u(x_i, n\Delta t)$:

$$\frac{u_i^{n+1} - u_i^n}{\Delta t} = -\frac{a}{2\Delta x}(u_{i+1}^n - u_{i-1}^n) \qquad (7.2.5)$$

This is an *explicit* scheme, since each discretized equation contains only one unknown at level $(n + 1)$.

Evaluating the right-hand side at level $(n + 1)$ leads to the *implicit* scheme, known as the *backward* or implicit Euler method:

$$\frac{u_i^{n+1} - u_i^n}{\Delta t} = -\frac{a}{2\Delta x}(u_{i+1}^{n+1} - u_{i-1}^{n+1}) \qquad (7.2.6)$$

where three unknowns appear simultaneously at time level $(n + 1)$.

Note that this equation could also be obtained from equation (7.2.4) by applying a backward difference in time for the discretization of u_t. Equation (7.2.6) leads to a system of equations with a *tridiagonal* matrix, and we will present in Chapter 12 an algorithm leading to an efficient solution of tridiagonal systems.

From the definitions of the order of accuracy of the finite difference formulas we expect schemes (7.2.5) and (7.2.6) to be first order in time and second order in space at points i and time level n. Another alternative, with a first-order approximation for the space derivative, would be obtained with a backward difference in space, leading to the semi-discrete form:

$$(u_t)_i = -\frac{a}{\Delta x}(u_i - u_{i-1}) \tag{7.2.7}$$

With a forward difference in time we obtain the following explicit scheme, known as the *first-order upwind scheme*:

$$\frac{u_i^{n+1} - u_i^n}{\Delta t} = -\frac{a}{\Delta x}(u_i^n - u_{i-1}^n) \tag{7.2.8}$$

The corresponding implicit version, evaluating the right-hand side at $(n + 1)$, would be

$$\frac{u_i^{n+1} - u_i^n}{\Delta t} = -\frac{a}{\Delta x}(u_i^{n+1} - u_{i-1}^{n+1}) \tag{7.2.9}$$

Let us investigate the behaviour of schemes (7.2.5) and (7.2.8) on a simple example. We consider an initial solution of triangular form illustrated in Figure 7.2.1:

$$
\begin{aligned}
f(x) &= 0 & x &\leqslant -0.2 \\
&= 1 + \frac{x}{2} & -0.2 &\leqslant x < 0 \\
&= 1 - \frac{x}{2} & 0 &\leqslant x \leqslant 0.2 \\
&= 0 & x &\geqslant 0.2
\end{aligned}
\tag{7.2.10}
$$

and calculate the numerical solution of the convection equation for $a = 1$ and $\Delta x = 0.1$. Observe that the exact solution $\bar{u}$ of $u_t + u_x = 0$ is a pure translation, and hence

$$\bar{u}(x, t) = f(x - t) \tag{7.2.11}$$

Writing scheme (7.2.5) under the form

$$u_i^{n+1} = u_i^n - \frac{\sigma}{2}(u_{i+1}^n - u_{i-1}^n) \qquad \sigma = \frac{a\,\Delta t}{\Delta x} \tag{7.2.12}$$

the scheme is seen to depend only on the parameter σ, called the *Courant number*. We obtain the results shown in Figure 7.2.1, after two, three and four time steps for $\sigma = 1$ and an initial solution centred around $x = 1$.

As can be seen, the simple scheme (7.2.12) leads to an increasing error when compared with the exact solution. A similar situation would be obtained for any other value of σ (see also Problem 7.1), and the scheme is therefore *unstable*. The same test can be performed with scheme (7.2.8), which can be written as

$$u_i^{n+1} = u_i^n - \sigma(u_i^n - u_{i-1}^n) \tag{7.2.13}$$

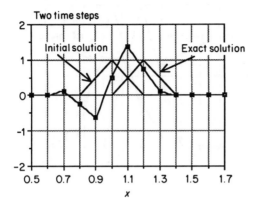

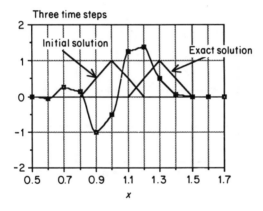

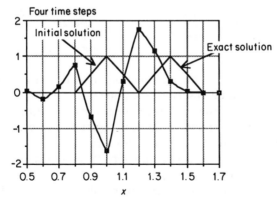

Figure 7.2.1 Numerical solution of the convection problem
with scheme (7.2.5), after two, three and four time steps, for an
initial triangular shape and $\sigma = 1$

274

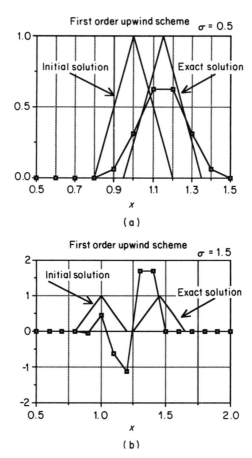

Figure 7.2.2 Numerical solution of the convection problem with scheme (7.2.8), after three time steps, for an initial triangular shape. (a) Solution for a Courant number $\sigma = 1/2$; (b) solution for a Courant number $\sigma = 3/2$

For $\sigma = 1/2$ we obtain the numerical solution, after three time steps, shown in Figure 7.2.2(a). As can be observed, the numerical solution follows the exact solution, although with poor accuracy. Repeating the same calculation with $\sigma = 3/2$ leads to an unstable numerical solution, as can be seen from Figure 7.2.2(b), also after three time steps (see also Problems 7.2 and 7.3). This is a typical example of a *conditionally stable* scheme since its stability depends on the parameter σ.

These elementary examples, which can be obtained by hand calculations, show nevertheless the extreme complexity of numerical schemes and raise some basic questions with regard to the analysis of discretized equations:

(1) What are the conditions we have to impose on a numerical scheme in order to obtain an acceptable approximation to the differential problem?

(2) Why do these two simple schemes have completely different behaviours, and how can we predict their stability limits?
(3) For a stable calculation, such as shown in Figure 7.2.2(a), how can we obtain quantitative information on the *accuracy* of the numerical simulation?

In order to provide answers to these questions it is necessary to define more precisely the requirements to be applied to a numerical scheme. These requirements are defined as *consistency, stability and convergence*. These three conditions cover different aspects of the relations between the discretized equations, the numerical solution and the exact, analytical solution of the differential equation. They are summarized in Figure 7.2.3, which expresses, in short, that the consistency condition defines a relation between the differential equation and its discrete formulation; that the stability condition establishes a relation between the computed solution and the exact solution of the discretized equations; while the convergence condition connects the computed solution to the exact solution of the differential equation.

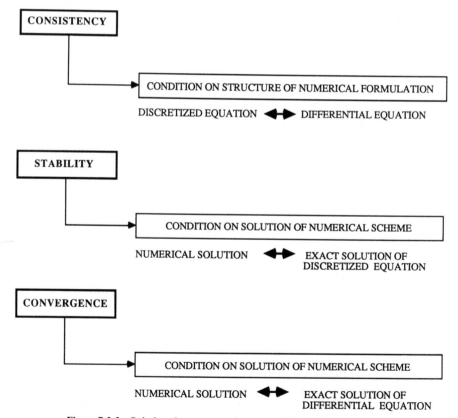

Figure 7.2.3 Relations between consistency, stability and convergence

7.2.1 Consistency

Consistency expresses that the discretized equations should tend to the differential equations to which they are related when Δt and Δx tend to zero. In order to check for consistency the various function values u_j^m occurring in the discretized equations are developed in a Taylor series around the value u_i^n and the high-order terms are maintained in substituting these developments back in the numerical equations.

Let us illustrate this by the example of scheme (7.2.5). If the function $u(x, t)$ is sufficiently smooth we can write the following Taylor expansions:

$$u_i^{n+1} = u_i^n + \Delta t (u_t)_i^n + \frac{\Delta t^2}{2} (u_{tt})_i^n + \cdots \qquad (7.2.14a)$$

$$u_{i+1}^n = u_i^n + \Delta x (u_x)_i^n + \frac{\Delta x^2}{2} (u_{xx})_i^n + \frac{\Delta x^3}{6} (u_{xxx})_i^n + \cdots \qquad (7.2.14b)$$

$$u_{i-1}^n = u_i^n - \Delta x (u_x)_i^n + \frac{\Delta x^2}{2} (u_{xx})_i^n - \frac{\Delta x^3}{6} (u_{xxx})_i^n + \cdots \qquad (7.2.14c)$$

where the x and t subscripts indicate partial derivatives. Substituting these developments in equation (7.2.5) we obtain

$$\frac{u_i^{n+1} - u_i^n}{\Delta t} + a \frac{u_{i+1}^n - u_{i-1}^n}{2\Delta x} - (u_t + au_x)_i^n$$

$$= +\frac{\Delta t}{2} (u_{tt})_i^n + \frac{\Delta x^2}{6} a(u_{xxx})_i + 0(\Delta t^2, \Delta x^4) \qquad (7.2.15)$$

It is clearly seen from the above equation that the right-hand side vanishes when Δt and Δx tend to zero, and therefore scheme (7.2.5) is consistent. As expected, the *accuracy* of the scheme is first order in time and second order in space, since the right-hand side goes to zero as the first power of Δt and the second power of Δx. Note that if a relation is established between Δt and Δx, when they both tend to zero, then the *overall* accuracy of the scheme might be different. If $\Delta t / \Delta x$ is kept constant, then the scheme has first-order accuracy, while it would be second order if $\Delta t / \Delta x^2$ were kept constant.

The consistency equation (7.2.15) can be interpreted in two equivalent ways:

(1) The values u_i^n are considered as *exact solutions of the discretized equation*. In this case, denoting by $\bar{u}_i^n$ the exact solution of the numerical scheme, equation (7.2.15) reduces to,

$$(\bar{u}_t + a\bar{u}_x)_i^n = -\frac{\Delta t}{2} (u_{tt})_i^n - a \frac{\Delta x^2}{6} (u_{xxx})_i^n + 0(\Delta t^2, \Delta x^4) \,(7.2.16)$$

showing that the exact solution of the difference equation does not satisfy exactly the differential equation at finite values of Δt and Δx (which is always the case in practical computations). However, the solution of the difference equation satisfies an equivalent differential equation (also called a *modified* differential equation), which differs from the original (differential) equation by a truncation error represented by the terms on the right-hand side.

In the present example the truncation error ε_T is equal to

$$\varepsilon_T = - \frac{\Delta t}{2} (u_{tt})_i^n - a \frac{\Delta x^2}{6} (u_{xxx})_i^n + 0(\Delta t^2, \Delta x^4) \qquad (7.2.17)$$

and can be written in an equivalent form, up to higher-order correction terms, by applying the equivalent differential equation (7.2.16) to eliminate the time derivatives; for instance,

$$(u_t)_i^n = - a(u_x)_i^n + 0(\Delta t, \Delta x^2) \qquad (7.2.18)$$

and

$$
\begin{aligned}
(u_{tt})_i^n &= - a(u_{xt})_i^n + 0(\Delta t, \Delta x^2) \\
&= + a^2 (u_{xx})_i^n + 0(\Delta t, \Delta x^2)
\end{aligned}
\qquad (7.2.19)
$$

Hence the truncation error can be written as

$$\varepsilon_T = - \frac{\Delta t}{2} a^2 (u_{xx})_i^n - a \frac{\Delta x^2}{6} (u_{xxx})_i^n + 0(\Delta t^2, \Delta x^2) \qquad (7.2.20)$$

Up to the lowest order the equivalent differential equation becomes

$$u_t + a u_x = - \frac{\Delta t}{2} a^2 \cdot u_{xx} + 0(\Delta t^2, \Delta x^2) \qquad (7.2.21)$$

and this shows why the corresponding scheme is unstable. Indeed, the right-hand side represents a viscosity term, with a negative viscosity coefficient equal to $[- (\Delta t/2) a^2]$. A positive viscosity is known to damp oscillations and strong gradients; a negative viscosity, on the other hand, will amplify any disturbance and, since the numerical solution satisfies the above equation, its behaviour is unstable (see also Problem 7.4). Therefore the determination of the equivalent differential equation and, in particular, the truncation error provides essential information as to the behaviour of the numerical solution, and will generally lead to necessary conditions for stability (see Chapter 9).

The condition for consistency can be restated as follows: a *scheme is consistent if the truncation error tends to zero for* Δt, Δx *tending to zero.*

The order of accuracy of a scheme is also defined by the truncation error. If the truncation error is of the form

$$\varepsilon_T = 0(\Delta t^q, \Delta x^p) \qquad (7.2.22)$$

where p and q are the lowest values occurring in the development of the truncation error, the scheme is said to be of order q in time and p in space. The consistency condition can also be stated as the requirement that the order of accuracy in time and (or) in space, should be positive for any combinations of Δt and Δx when they both tend to zero.

The second interpretation of the consistency condition is obtained as follows

(2) The values u_i^n in equation (7.2.15) are *exact solutions of the differential equation*, that is, $u_i^n = \tilde{u}(i \, \Delta x, n \, \Delta t)$, where $\tilde{u}(x, t)$ is the analytical solution. Equation (7.2.15) becomes

$$\frac{\tilde{u}_i^{n+1} - \tilde{u}_i^n}{\Delta t} + a \frac{\tilde{u}_{i+1}^n - \tilde{u}_{i-1}^n}{2\Delta x} = - \varepsilon_{\mathrm{T}} \qquad (7.2.23)$$

which shows that the exact values $\tilde{u}_i^n$ do not satisfy the difference equations exactly but are solutions of a modified discrete equation with the truncation error in the right-hand side.

We can also view equation (7.2.23) as a definition of the truncation error: the truncation error is equal to the *residual* of the discretized equation for values of $\tilde{u}_i^n$ equal to the exact, analytical solution.

7.2.2 Stability

The difference scheme should not allow errors to grow indefinitely, that is, to be amplified without bound, as we progress from one time step to another. A condition for stability can be formulated by the requirement that any error ε_i^n between u and $\bar{u}$ *should remain uniformly bounded* for $n \to \infty$ at fixed Δt. Actually, early definitions of stability (O'Brien *et al.*, 1950) were defined on the basis of non-amplification of the round-off errors.

If we define the error ε as the difference between the computed solution and the exact solution of the discretized equation

$$\varepsilon_i^n = u_i^n - \bar{u}_i^n \qquad (7.2.24)$$

the stability condition can be written as

$$\lim_{n \to \infty} | \varepsilon_i^n | \leqslant K \qquad \text{at fixed } \Delta t \qquad (7.2.25)$$

with K independent of n. This stability condition is a requirement solely on the numerical scheme and contains no condition on the differential equation. Actually, the stability condition (7.2.25) has to be valid for any kind of error. However, this condition does not ensure that the error will not become unacceptably large at fixed intermediate times $t_n = n\Delta t$. Practical examples showing that condition (7.2.25) is not always acceptable will be discussed in Chapter 10.

Therefore a more general definition of stability, introduced by Lax and Richtmyer (1956) and developed in Richtmyer and Morton (1967), is based on the time behaviour of the solution itself instead of the error's behaviour. This stability criterion states that *any component of the initial solution should not be amplified without bound*. In order to express the mathematical statement of this condition the schemes have to be written in matrix or operator form as follows.

All the unknowns, at every mesh point, at a given time level, are grouped into a vector, say U, defined as follows at time $n\Delta t$:

$$U^n \equiv \begin{vmatrix} u_1^n \\ \vdots \\ u_{i-1}^n \\ u_i^n \\ u_{i+1}^n \\ \vdots \end{vmatrix}$$

(7.2.26)

As a function of U^n the discretized scheme can be written into the operator or matrix form:

$$U^{n+1} = C \cdot U^n$$

(7.2.27)

where the operator C is a function of the time step Δt and mesh size Δx.

Example of scheme (7.2.12): $u_i^{n+1} = u_i^n - \sigma(u_{i+1}^n - u_{i-1}^n)/2$

The matrix form of this scheme is easily seen to be defined by the operator:

$$CU^n = \begin{vmatrix} \cdot & \cdot & \cdot & & & \\ \sigma/2 & 1 & -\sigma/2 & & & \\ & \sigma/2 & 1 & -\sigma/2 & & \\ & & \sigma/2 & 1 & -\sigma/2 & \\ 0 & & & \cdot & \cdot & \cdot \\ & & & & \cdot & \cdot \end{vmatrix} \begin{vmatrix} u_{i-1}^n \\ u_i^n \\ u_{i+1}^n \\ \cdot \end{vmatrix}$$

(7.2.28)

Example of scheme (7.2.13): $u_i^{n+1} = u_i^n - \sigma(u_i^n - u_{i-1}^n)$

In this case the operator C takes the following form:

$$CU^n = \begin{vmatrix} \cdot & \cdot & & & 0 \\ \sigma & 1-\sigma & & & \\ & \sigma & 1-\sigma & & \\ & & \sigma & 1-\sigma & \\ 0 & & & \cdot & \cdot \end{vmatrix} \begin{vmatrix} \vdots \\ u_{i-1}^n \\ u_i^n \\ u_{i+1}^n \\ \vdots \end{vmatrix}$$

(7.2.29)

Example of implicit scheme (7.2.6): $u_i^{n+1} = u_i^n - \sigma/2(u_{i+1}^{n+1} - u_{i-1}^{n+1})$

For implicit schemes it is somewhat more difficult to obtain the matrix form of the operator C explicitly, but it can be defined formally in a straightforward way by writing first the scheme as an operator on U^{n+1}. For scheme (7.2.6) this leads to an expression of the form

$$B \cdot U^{n+1} = U^n$$

(7.2.30)

where

$$B \cdot U^{n+1} = \begin{vmatrix} \ddots & & & \\ & -\sigma/2 & 1 & +\sigma/2 & \\ & & -\sigma/2 & 1 & +\sigma/2 \\ & & & -\sigma/2 & 1 & +\sigma/2 \\ & & & & & \ddots \end{vmatrix} \begin{vmatrix} \vdots \\ u_{i-1}^{n+1} \\ u_i^{n+1} \\ u_{i+1}^{n+1} \\ \vdots \end{vmatrix} \qquad (7.2.31)$$

The operator C is defined by the inverse of B, that is,

$$C = B^{-1} \qquad (7.2.32)$$

Example of implicit scheme (7.2.9): $u_i^{n+1} = u_i^n - \sigma(u_i^{n+1} - u_{i-1}^{n+1})$

In this case the B matrix has the following structure:

$$B \cdot U^{n+1} = \begin{vmatrix} \ddots & & & \\ & -\sigma & 1+\sigma & \\ & & -\sigma & 1+\sigma \\ & & & -\sigma & 1+\sigma \end{vmatrix} \begin{vmatrix} \vdots \\ u_{i-1}^{n+1} \\ u_i^{n+1} \\ u_{i+1}^{n+1} \\ \vdots \end{vmatrix} \qquad (7.2.33)$$

and $C = B^{-1}$. The stability condition can now be stated as follows. If U^0 represents the initial solution at time $t = 0$ then, by repeated action of the operator C, we have

$$\begin{aligned} U^1 &= C \cdot U^0 \\ U^2 &= C \cdot U^1 = CCU^0 = (C)^2 \cdot U^0 \\ &\vdots \\ U^n &= (C)^n \cdot U^0 \end{aligned} \qquad (7.2.34)$$

where $(C)^n$ denotes the operator C to the power n.

In order for all U^n to remain bounded and the scheme, defined by the operator C, to remain stable the infinite set of operators $(C)^n$ has to be uniformly bounded. That is, a constant K exists, such that

$$\| (C)^n \| < K \qquad \text{for} \qquad \begin{matrix} 0 < \Delta t < \tau \\ 0 \leqslant n\,\Delta t \leqslant T \end{matrix} \qquad (7.2.35)$$

for fixed values of τ and T and for all n. This condition implies the definition of some norm in the considered functional space; for instance, a Hilbert norm in the space of square integrable functions.

7.2.3 Convergence

The numerical solution u_i^n should approach the exact solution $\tilde{u}(x, t)$ of the differential equation at any point $x_i = i\Delta x$ and time $t_n = n\Delta t$ *when Δx and Δt tend to zero*, that is, when the mesh is refined, x_i and t_n being fixed. This condition implies that i and n tend to infinity while Δx and Δt tend to zero, such that the products $(i\Delta x)$ and $(n\Delta t)$ remain constant. This condition for convergence of the numerical solution to the exact solution of the differential equation expresses that the error

$$\tilde{\varepsilon}_i^n = u_i^n - \tilde{u}(i\Delta x, n\Delta t) \tag{7.2.36}$$

satisfies the following convergence condition:

$$\lim_{\substack{\Delta t \to 0 \\ \Delta x \to 0}} |\tilde{\varepsilon}_i^n| = 0 \text{ at fixed values of } x_i = i\Delta x \text{ and } t_n = n\Delta t \tag{7.2.37}$$

Following Richtmyer and Morton (1967) the convergence condition can also be restated as a condition on C as follows:

$$\lim_{\substack{\Delta t \to 0 \\ n \to \infty}} \|[C(\Delta t)]^n U^0 - \tilde{U}(t)\| = 0 \tag{7.2.38}$$

with $n\Delta t = t$ fixed. The notation $\|\cdot\|$ indicates the selected norm.

Clearly, the conditions of consistency, stability and convergence are related to each other, and the precise relation is contained in the fundamental *Equivalence Theorem of Lax*, a proof of which can be found in the now-classical book of Richtmyer and Morton (1967).

For a well-posed initial value problem and a consistent discretization scheme, stability is the necessary and sufficient condition for convergence.

This fundamental theorem shows that in order to analyse a time-dependent or initial value problem two tasks have to be performed:

(1) Analyse the consistency condition; this leads to the determination of the order of accuracy of the scheme and its truncation error.
(2) Analyse the stability properties; this leads, in addition, to detailed information on the frequency distribution of the error, that is, its behaviour as a function of the frequency content of the computed solution.

From these two steps convergence can be defined without additional analysis.

References

Lax, P. D., and Richtmyer, R. D. (1956). 'Survey of the stability of linear finite difference equations.' *Comm. Pure and Applied Mathematics*, **17**, 267–93.
O'Brien, G. G., Hyman, M. A., and Kaplan, S. (1950).) 'A study of the numerical solution of partial differential equations.' *Journal Mathematics and Physics*, **29**, 223–51.
Richtmyer, R. D., and Morton, K. W. (1967). *Difference Methods for Initial Value Problems*, 2nd edn, London: Wiley–Interscience.

282

PROBLEMS

Problem 7.1

Obtain the results of Figure 7.2.1 by applying successively scheme (7.2.5) under the form (7.2.12) for $\sigma = 1$. Extend the calculation up to five time steps. Repeat the same calculations for $\sigma = 1/2$ and $\sigma = 3/2$ and observe that in all cases the scheme is unstable.

Problem 7.2

Obtain the results of Figure 7.2.2 for the same initial solution as in the previous problem for scheme (7.2.13) with $\sigma = 1/2$ and $\sigma = 3/2$. Repeat the calculation with $\sigma = 1$ and observe that we obtain numerically the exact solution.

Problem 7.3

Apply scheme (7.2.13) with $\sigma = 1/2$ to the same test case as in Problem 7.2 with $\Delta x = 0.05$ and $\Delta x = 0.025$. Compare the numerical solutions after four, eight and sixteen time steps.

Problem 7.4

Determine the equivalent differential equation and the truncation error for scheme (7.2.13). Show that it is stable for $\sigma \leqslant 1$.

Problem 7.5

Apply a generalized trapezium formula in time to the central discretized space derivative of the linear convection equation $u_t + a u_x = 0$

$$\frac{u_i^{n+1} - u_i^n}{\Delta t} + \theta a \frac{u_{i+1}^{n+1} - u_{i-1}^{n+1}}{2\Delta x} + (1-\theta)a \frac{u_{i+1}^n - u_{i-1}^n}{2\Delta x} = 0$$

with θ as free parameter.

Apply a Taylor series expansion to obtain the truncation error and show that the scheme is unstable for $\theta < \frac{1}{2}$.

Chapter 8

The Von Neumann Method for Stability Analysis

Various methods have been developed for the analysis of stability, nearly all of them limited to linear problems. However, even within this restriction the complete investigation of stability for initial, boundary value problems can be extremely complicated, particularly in the presence of boundary conditions and their numerical representation.

The problem of stability for a linear problem with constant coefficients is now well understood when the influence of boundaries can be neglected or removed. This is the case either for an infinite domain or for periodic conditions on a finite domain. In the latter case we consider that the computational domain on the x-axis of length L is repeated periodically, and therefore all quantities, the solution, as well as the errors, can be developed in a finite Fourier series over the domain $2L$. This development in the frequency domain (in space) forms the basis of the *Von Neumann method* for stability analysis (Sections 8.1 and 8.2). This method was developed in Los Alamos during World War II by Von Neumann and was considered classified until its brief description in Cranck and Nicholson (1947) and in a publication in 1950 by Charney *et al.* (1950). At present this is the most widely applied technique for stability analysis, and furthermore allows an extensive investigation of the behaviour of the error as a function of the frequency content of the initial data and of the solution, as will be seen in Section 8.3. The generalization of the Von Neumann method to multidimensional problems is presented in Section 8.4.

It the problem of stability analysis can be treated generally for linear equations with constant coefficients and with periodic boundary conditions, as soon as we have to deal with non-constant coefficients and (or) non-linear terms in the basic equations the information on stability becomes very limited. Hence we have to resort to a local stability analysis, with frozen values of the non-linear and non-constant coefficients, to make the formulation linear. In any case, linear stability is a necessary condition for non-linear problems but it is certainly not sufficient. We will touch on this difficult problem in Section 8.5.

Finally, Section 8.6 presents certain general techniques in order to obtain the stability conditions from the Von Neumann analysis.

8.1 FOURIER DECOMPOSITION OF THE ERROR

If $\bar{u}_i^n$ is the exact solution of the difference equation and u_i^n the actual computed solution the difference might be due to round-off errors and to errors in the initial data. Hence,

$$u_i^n = \bar{u}_i^n + \varepsilon_i^n \tag{8.1.1}$$

where ε_i^n indicates the error at time level n in mesh point i. Clearly, any linear numerical scheme for u_i^n is satisfied exactly by $\bar{u}_i^n$, and therefore the errors ε_i^n are also solutions of the same discretized equation.

In order to present the essentials of the method we will first refer to the previous examples. Considering scheme (7.2.5) and inserting equation (8.1.1) leads to

$$\frac{\bar{u}_i^{n+1} - \bar{u}_i^n}{\Delta t} + \frac{\varepsilon_i^{n+1} - \varepsilon_i^n}{\Delta t} = -\frac{a}{2\Delta x}(\bar{u}_{i+1}^n - \bar{u}_{i-1}^n) - \frac{a}{2\Delta x}(\varepsilon_{i+1}^n - \varepsilon_i^n) \tag{8.1.2}$$

Since $\bar{u}_i^n$ satisfies exactly equation (7.2.5) we obtain the equation for the errors ε_i^n:

$$\frac{\varepsilon_i^{n+1} - \varepsilon_i^n}{\Delta t} = -\frac{a}{2\Delta x}(\varepsilon_{i+1}^n - \varepsilon_{i-1}^n) \tag{8.1.3}$$

which is identical to the basic scheme. Hence the errors ε_i^n do evolve over time in the same way as the numerical solution u_i^n.

The general demonstration of this property is obvious when the operator form (equation (7.2.27)) is applied, considering the operator C to be linear. If e^n designates the column vector of the errors at time level n:

$$e^n = \begin{vmatrix} \cdot \\ \varepsilon_{i-1}^n \\ \varepsilon_i^n \\ \varepsilon_{i+1}^n \\ \cdot \end{vmatrix} \tag{8.1.4}$$

relation (8.1.1) can be written, with $\bar{U}^n$ indicating the exact solution,

$$U^n = \bar{U}^n + e^n \tag{8.1.5}$$

Inserting this equation into the basic scheme leads to

$$\bar{U}^{n+1} + e^{n+1} = C\bar{U}^n + Ce^n \tag{8.1.6}$$

or

$$e^{n+1} = Ce^n \tag{8.1.7}$$

by definition of $\bar{U}^n$ as a solution of

$$\bar{U}^{n+1} = C\bar{U}^n \tag{8.1.8}$$

Hence time evolution of the error is determined by the same operator C as the solution of the numerical problem.

If the boundary conditions are considered as periodic the error ε_i^n can be decomposed into a Fourier series in space at each time level n. Since the space domain is of a finite length we will have a discrete Fourier representation summed over a finite number of harmonics.

In a one-dimensional domain of length L the complex Fourier representation reflects the region $(0, L)$ onto the negative part $(-L, 0)$, and the fundamental frequency corresponds to the maximum wavelength of $\lambda_{max} = 2L$. The associated wavenumber $k = 2\pi/\lambda$ attains its minimum value $k_{min} = \pi/L$. On the other hand, the maximum value of the wavenumber k_{max} of the finite spectrum on the interval $(-L, L)$ is associated with the shortest resolvable wavelength on a mesh with spacing Δx. This shortest wavelength is clearly equal to $\lambda_{min} = 2\Delta x$ (see Figure 8.1.1), and consequently, $k_{max} = \pi/\Delta x$.

Therefore with the mesh index i, ranging from 0 to N, with $x_i = i \cdot \Delta x$ and

$$\Delta x = L/N \tag{8.1.9}$$

all the harmonics represented on a finite mesh are given by

$$k_j = jk_{min} = j\,\frac{\pi}{L} = j\,\frac{\pi}{N\,\Delta x} \qquad j = 0, 1, 2, ..., N \tag{8.1.10}$$

with the maximum value of j being associated with the maximum frequency. Hence with $k_{max} = \pi/\Delta x$ the highest value of j is equal to the number of mesh

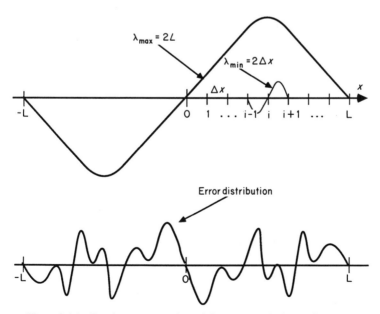

Figure 8.1.1 Fourier representation of the error on the interval $(-L, L)$

intervals N. Any finite mesh function, such as ε_i^n or the full solution u_i^n, will be decomposed into a Fourier series as

$$\varepsilon_i^n = \sum_{j=-N}^{N} E_j^n \, e^{Ik_j \cdot i\Delta x} = \sum_{j=-N}^{N} E_j^n \, e^{Iij \, \pi/N} \qquad (8.1.11)$$

where $I = \sqrt{-1}$ and E_j^n is the amplitude of the jth harmonic.

The harmonic associated with $j = 0$ represents a constant function in space. The produce $k_j \Delta x$ is often represented as a phase angle:

$$\phi \equiv k_j \cdot \Delta x = \frac{j\pi}{N} \qquad (8.1.12)$$

and covers the domain $(-\pi, \pi)$ in steps of π/N. The region around $\phi = 0$ corresponds to the low frequencies while the region close to $\phi = \pi$ is associated with the high-frequency range of the spectrum. In particular, the value $\phi = \pi$ corresponds to the highest frequency resolvable on the mesh, namely the frequency of the wavelength $2\Delta x$. Since we deal with linear schemes the discretized equation (8.1.7), which is satisfied by the error ε_i^n, must also be satisfied by each individual harmonic.

8.1.1 Amplification factor

Considering a single harmonic $E_j^n \, e^{Ii\phi}$, its time evolution is determined by the same numerical scheme as the full solution u_i^n. Hence inserting a representation of this form into equation (8.1.3) for the example considered we obtain, dropping the subscript j,

$$\frac{(E^{n+1} - E^n)}{\Delta t} \, e^{Ii\phi} + \frac{a}{2\Delta x} \, (E^n \, e^{I(i+1)\phi} - E^n \, e^{I(i-1)\phi})] = 0$$

or, dividing by $e^{Ii\phi}$,

$$(E^{n+1} - E^n) + \frac{\sigma}{2} E^n (e^{I\phi} - e^{-I\phi}) = 0 \qquad (8.1.13)$$

where the parameter

$$\sigma = \frac{a\Delta t}{\Delta x} \qquad (8.1.14)$$

has been introduced.

The stability condition (7.2.25) will be satisfied if the amplitude of any error harmonic E^n does not grow in time, that is, if the ratio

$$|G| \equiv \left| \frac{E^{n+1}}{E^n} \right| \leqslant 1 \qquad \text{for all } \phi \qquad (8.1.15)$$

The quantity G, defined by,

$$G = \frac{E^{n+1}}{E^n} \qquad (8.1.16)$$

is the *amplification factor*, and is a function of time step Δt, frequency and mesh size Δx. In the present case from equation (8.1.13) we have

$$G - 1 + \frac{\sigma}{2} \cdot 2I \sin \phi = 0$$

or

$$G = 1 - I\sigma \sin \phi \qquad (8.1.17)$$

The stability condition (8.1.15) requires the modulus of G to be lower or equal to one. For the present example,

$$|G|^2 = 1 + \sigma^2 \sin^2\phi \qquad (8.1.18)$$

and is clearly never satisfied. Hence the centred scheme (7.2.5), for the convection equation with forward difference in time is *unconditionally unstable*.

Example of scheme (7.2.8): conditional stability

Inserting the single harmonic $E^n \, e^{Ii\phi}$ into scheme (7.2.8) written for the error we obtain

$$(E^{n+1} - E^n)e^{Ii\phi} + \sigma E^n(e^{Ii\phi} - e^{I(i-1)\phi}) = 0$$

or after division by $E^n \, e^{Ii\phi}$,

$$\begin{aligned} G &= 1 - \sigma + \sigma e^{-I\phi} \\ &= 1 - 2\sigma \sin^2\phi/2 - I \sigma \sin \phi \end{aligned} \qquad (8.1.19)$$

In order to analyse the stability of scheme (7.2.8), that is, the regions where the modulus of the amplification factor G is lower than one, a representation of G in the complex plane is a convenient approach. Writing ξ and η, respectively, for the real and imaginary parts of G we have

$$\begin{aligned} \xi &= 1 - 2\sigma \sin^2\phi/2 = (1 - \sigma) + \sigma \cos \phi \\ \eta &= -\sigma \sin \phi \end{aligned} \qquad (8.1.20)$$

which can be considered as parametric equations for G with ϕ as a parameter. We recognize the parametric equations of a circle centred on the real axis ξ at $(1 - \sigma)$ with radius σ.

In the complex plane of G the stability condition (8.1.15) states that the curve representing G for all values of $\phi = k\Delta x$ should remain within the unit circle (see Figure 8.1.2). It is clearly seen from Figure 8.1.2 that the scheme is stable for

$$0 < \sigma \leqslant 1 \qquad (8.1.21)$$

Hence scheme (7.2.8) is conditionally stable and condition (8.1.21) is known as the *Courant–Friedrichs–Lewy* or *CFL* condition. The parameter σ is called

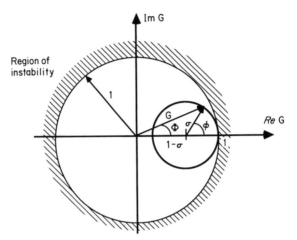

Figure 8.1.2 Complex *G* plane representation of upwind scheme (7.2.8), with unit circle defining the stability region

the *Courant number*. This condition for stability was introduced for the first time in 1928 in a paper by Courant *et al.* (1928), which can be considered as laying the foundations of the concepts of convergence and stability for finite difference schemes, although the authors were using finite difference concepts as a mathematical tool for proving existence theorems of continuous problems. Observe that the upwind scheme (7.2.8) is unstable for $a < 0$ (see also Problem 8.1).

8.1.2 Comment on the CFL condition

This fundamental stability condition of most explicit schemes for wave and convection equations expresses that the distance covered during the time

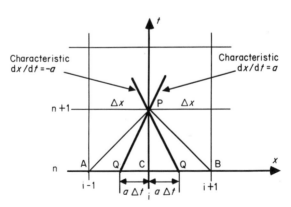

Figure 8.1.3 Geometrical interpretation of the CFL condition, $\sigma \leqslant 1$

interval Δt, by the disturbances propagating with speed a, should be lower than the minimum distance between two mesh points. Referring to Figure 8.1.3, the line PQ is the characteristic $dx/dt = a$, through P, and defines the domain of dependence of the differential equation in P. On the other hand, the difference equation defines a numerical domain of dependence of P which is the domain between PAC.

The CFL stability condition $\sigma \leqslant 1$ expresses that the mesh ratio $\Delta t/\Delta x$ has to be chosen in such a way that *the domain of dependence of the differential equation should be contained in the domain of dependence of the discretized equations.* In other words, the numerical scheme defining the approximation u_i^{n+1} in (mesh point i) must be able to include all the physical information which influences the behaviour of the system in this point.

Example of scheme (7.2.6): unconditional stability

The implicit, backward Euler scheme with central space differencing of the convection equation offers a third situation with respect to stability properties. Performing the same stability analysis with scheme (7.2.6), the error amplitude E^{n+1} becomes, after introduction of an harmonic of the form $E^n\, e^{Ii\phi}$,

$$e^{Ii\phi}(E^{n+1} - E^n) + \frac{\sigma}{2} E^{n+1}(e^{I\phi} - e^{-I\phi})e^{Ii\phi} = 0$$

or

$$G - 1 + \frac{\sigma}{2} G(e^{I\phi} - e^{-I\phi}) = 0$$

leading to

$$G = \frac{1}{1 + I\sigma \sin \phi} \tag{8.1.22}$$

The modulus of G is always lower than one, for all values of σ, since

$$|G|^2 = G \cdot G^* = \frac{1}{1 + \sigma^2 \sin^2 \phi} \tag{8.1.23}$$

and therefore the implicit scheme (7.2.6) is unconditionally stable. Hence it is seen that schemes can have either conditional stability, unconditional stability or unconditional instability.

The Von Neumann method offers an easy and simple way of assessing the stability properties of linear schemes with constant coefficients when the boundary conditions are assumed periodic.

8.2 GENERAL FORMULATION OF VON NEUMANN'S METHOD: SYSTEM OF EQUATIONS

Referring to the second definition of stability (equation (7.2.35)), the Von Neumann method can be restated on the basis of the development of the solution u_i^n in a Fourier series, that is, writing

$$u_i^n = \sum_{m=-N}^{N} v_m^n \, e^{Iik_m \Delta x} = \sum_{m=-N}^{N} v_m^n \, e^{Ii\phi} \tag{8.2.1}$$

where v_m^n is the amplitude of the mth harmonic of u_i^n. An arbitrary harmonic can be singled out and, when introduced into the scheme, stability requires that no harmonic should be allowed to increase in time without bound. Since u_i^n and the error ε_i^n satisfy the same numerical equation, the results obtained from equation (8.2.1) are identical to those obtained above. The amplification factor G is defined here as the ratio of the amplitudes v_m^n, that is, omitting the m subscript,

$$G = \frac{v^{n+1}}{v^n} = G(\phi, \Delta t, \Delta x) \tag{8.2.2}$$

and definition (7.2.35) leads to the stability condition (8.1.15).

In order to formulate the general Von Neumann stability condition it is necessary to write the discretized equations in operator and matrix forms.

8.2.1 Matrix and operator formulation

We consider that the numerical scheme is obtained in two steps: a space discretization, followed by a time integration.

(1) When a space discretization is applied (for instance, a finite difference method) the differential space operator is aproximated by a discretized space operator S, leading to the method of line formulation for the discrete values $u_i^n = u(\vec{x}_i, n\Delta t)$, where $\vec{x}_i$ is the co-ordinate of mesh point i:

$$\frac{du_i}{dt} = Su_i + q_i \tag{8.2.3}$$

The q_i term contains eventual sources and the contributions from boundary conditions. The matrix representation of the above system of ordinary differential equations in time is written with the vector U^n, defined by equation (7.2.26) as

$$\frac{dU}{dt} = SU + Q \tag{8.2.4}$$

where we use the same notation for the discretized space operator and its matrix representation.

(2) When a time-integration scheme is applied to the above space-discretized equations, corresponding to a two-level scheme connecting time levels $(n+1)$ and n, the numerical scheme associated with the differential problem generalizes equation (7.2.27):

$$u_i^{n+1} = C \cdot u_i^n + \bar{q}_i \tag{8.2.5}$$

or, in matrix form,

$$U^{n+1} = CU^n + \bar{Q} \tag{8.2.6}$$

where C can be considered as a discretization operator of the scheme.

For a two-level implicit scheme, of the form $B_1 U^{n+1} = B_0 U^n$ the difference operator C is defined by $C = B_1^{-1} B_0$. Note that for the Euler method we have $C = 1 + \Delta t S$. Some examples of the matrix representation of C have been given in Chapter 7 and we illustrate these various representations with a few additional examples.

The linear diffusion equation

$$\frac{\partial u}{\partial t} = a \frac{\partial^2 u}{\partial x^2} \tag{8.2.7}$$

The one-dimensional linearized shallow-water equations

These equations have been treated in Example 3.4.1 and we write here v for the x-component of the velocity, keeping the notation U for the column of the two dependent variables. The equations are linearized by setting v and h equal to v_0 and h_0 in the non-linear terms:

$$\left|\begin{array}{l} \dfrac{\partial h}{\partial t} + v_0 \dfrac{\partial h}{\partial x} + h_0 \dfrac{\partial v}{\partial x} = 0 \\[3mm] \dfrac{\partial v}{\partial t} + v_0 \dfrac{\partial v}{\partial x} + g \dfrac{\partial h}{\partial x} = 0 \end{array}\right. \tag{8.2.8}$$

Here the vector u is defined by

$$U = \left|\begin{array}{c} h \\ v \end{array}\right| \tag{8.2.9}$$

and the system is written as

$$\frac{\partial U}{\partial t} + A \frac{\partial U}{\partial x} = 0 \tag{8.2.10a}$$

where

$$A = \left|\begin{array}{cc} v_0 & h_0 \\ g & v_0 \end{array}\right| \tag{8.2.10b}$$

It is seen that, under this form, equation (8.2.10) generalizes the single convection equation.

Wave equation

$$\frac{\partial^2 w}{\partial t^2} - a^2 \frac{\partial^2 w}{\partial x^2} = 0 \tag{8.2.11a}$$

Table 8.1.

Differential equation	Space discretization operator S	Matrix representation of S (excluding boundary conditions)

Heat diffusion

$$\frac{\partial u}{\partial t} = \alpha \frac{\partial^2 u}{\partial x^2}$$

$$L = \alpha \frac{\partial^2}{\partial x^2}$$

Second order central difference

$$\frac{du_i}{dt} = \frac{\alpha}{\Delta x^2}(u_{i+1} - 2u_i + u_{i-1})$$

$$= \frac{\alpha}{\Delta x^2}(E - 2 + E^{-1})u_i$$

$$S = \frac{\alpha}{\Delta x^2}(E - 2 + E^{-1})$$

$$S = \begin{vmatrix} +1 & -2 & 1 & \\ & 1 & -2 & 1 \\ & & 1 & -2 & 1 \\ & & & & \cdot \end{vmatrix} \frac{\alpha}{\Delta x^2}$$

Shallow water equation

$$\frac{\partial U}{\partial t} + A \frac{\partial U}{\partial x} = 0$$

$$U = \begin{vmatrix} h \\ v \end{vmatrix}$$

$$A = \begin{vmatrix} v_0 & h_0 \\ g & v_0 \end{vmatrix}$$

$$L = -A \frac{\partial}{\partial x}$$

Central scheme

$$\frac{du_i}{dt} = -A \frac{u_{i+1} - u_{i-1}}{2\,\Delta x}$$

$$S = \frac{-A}{2\,\Delta x}(E - E^{-1})$$

$$U = \begin{vmatrix} \cdot \\ u_{i-1} \\ u_i \\ u_{i+1} \\ \cdot \end{vmatrix} = \begin{vmatrix} \begin{vmatrix} v \\ h \end{vmatrix}_{i-1} \\ \begin{vmatrix} v \\ h \end{vmatrix}_{i} \\ \begin{vmatrix} v \\ h \end{vmatrix}_{i+1} \\ \cdot \end{vmatrix}$$

$$S = \begin{vmatrix} \cdot \\ \frac{+A}{2\,\Delta x} & 0 & \frac{-A}{2\,\Delta x} \\ & \frac{+A}{2\,\Delta x} & 0 & \frac{-A}{2\,\Delta x} \\ & & & \cdot \end{vmatrix}$$

each element is a (2×2) matrix

Table 8.1. (*continued*)

Discretization operator C of time integrated scheme	Matrix representation of C	Amplification matrix

Euler method

$$\beta = \frac{\alpha \, \Delta t}{\Delta x^2}$$

$$u_i^{n+1} = u_i^n + \frac{\alpha \, \Delta t}{\Delta x^2} (u_{i+1}^n - 2u_i^n + u_{i-1}^n)$$

$$= [1 + \beta(E - 2 + E^{-1})]u_i^n$$

$$C \equiv 1 + \beta(E - 2 + E^{-1})$$

$$C = \begin{vmatrix} \cdot & \cdot & & & 0 \\ \beta & (1 - 2\beta) & \beta & & \\ & \beta & (1 - 2\beta) & \beta & \\ & & \beta & (1 - 2\beta) & \beta \\ 0 & & & \cdot & \cdot \end{vmatrix}$$

$$G = 1 + \beta(e^{I\phi} - 2 + e^{-I\phi})$$
$$G = 1 - 2\beta(1 - \cos \phi)$$
$$G = 1 - 4\beta \sin^2 \phi/2$$

Trapezoidal (Cranck–Nicholson) method

$$u_i^{n+1} = u_i^n + \frac{\Delta t}{2} Su_i^n + \frac{\Delta t}{2} Su_i^{n+1}$$

$$\left(1 - \frac{\Delta t}{2} S\right)u_i^{n+1} = \left(1 + \frac{\Delta t}{2} S\right)u_i^{n+1}$$

$$B = 1 - \frac{\Delta t}{2} S, \quad A = 1 + \frac{\Delta t}{2} S$$

$$C = B^{-1} \cdot A$$

$$A = \begin{vmatrix} \cdot & & & \\ \beta/2 & (1 - \beta) & \beta/2 & \\ & \beta/2 & (1 - \beta) & \beta/2 \\ & & \cdot & \end{vmatrix}$$

$$B = \begin{vmatrix} \cdot & & & \\ -\beta/2 & (1 + \beta) & -\beta/2 & \\ & -\beta/2 & (1 + \beta) & -\beta/2 \\ & & \cdot & \end{vmatrix}$$

$$C = B^{-1}A$$

$$G = \frac{1 + \dfrac{\beta}{2}(e^{I\phi} - 2 + e^{-I\phi})}{1 - \dfrac{\beta}{2}(e^{I\phi} - 2 + e^{-I\phi})}$$

Euler method

$$u_i^{n+1} = u_i^n - \frac{A \, \Delta t}{2 \, \Delta x} (u_{i+1}^n - u_{i-1}^n)$$

$$C = 1 - \frac{A \, \Delta t}{2 \, \Delta x} (E - E^{-1})$$

$$C =$$

$$\begin{vmatrix} 1 - \dfrac{v_0 \, \Delta t}{2 \, \Delta x}(E - E^{-1}) & -\dfrac{h_0 \, \Delta t}{2 \, \Delta x}(E - E^{-1}) \\ -\dfrac{g \, \Delta t}{\Delta x} \dfrac{1}{2}(E - E^{-1}) & 1 - \dfrac{v_0 \, \Delta t}{2 \, \Delta x}(E - E^{-1}) \end{vmatrix}$$

$$C = \begin{vmatrix} \cdot & & & & \\ \dfrac{A \, \Delta t}{2 \, \Delta x} & 1 & -\dfrac{A \, \Delta t}{2 \, \Delta x} & & \\ & \dfrac{A \, \Delta t}{2 \, \Delta x} & 1 & -\dfrac{A \, \Delta t}{2 \, \Delta x} & \\ & & \dfrac{A \, \Delta t}{2 \, \Delta x} & 1 & -\dfrac{A \, \Delta t}{2 \, \Delta x} \\ & & & \cdot & \end{vmatrix}$$

$$G = 1 - \frac{A \, \Delta t}{2 \, \Delta x}(e^{I\phi} - e^{-I\phi})$$

$$G =$$

$$\begin{vmatrix} 1 - I \dfrac{v_0 \, \Delta t}{\Delta x} \sin \phi & -I \dfrac{h_0 \, \Delta t}{\Delta x} \sin \phi \\ -I \dfrac{g \, \Delta t}{\Delta x} \sin \phi & 1 - I \dfrac{v_0 \, \Delta t}{\Delta x} \sin \phi \end{vmatrix}$$

Lax-Friedrichs scheme

$$u_i^{n+1} = (u_{i+1}^n + u_{i-1}^n)/2$$
$$- \frac{A \, \Delta t}{2 \, \Delta x}(u_{i+1}^n - u_{i-1}^n)$$

$$C =$$
$$\frac{1}{2}\left(1 - \frac{A \, \Delta t}{\Delta x}\right)E + \frac{1}{2}\left(1 + \frac{A \, \Delta t}{\Delta x}\right)E^{-1}$$

$$C = \begin{vmatrix} \cdot & & & \\ \dfrac{1}{2}\left(1 + \dfrac{A \, \Delta t}{\Delta x}\right) & 0 & \dfrac{1}{2}\left(1 - \dfrac{A \, \Delta t}{\Delta x}\right) & \\ & \dfrac{1}{2}\left(1 + \dfrac{A \, \Delta t}{\Delta x}\right) & 0 & \dfrac{1}{2}\left(1 - \dfrac{A \, \Delta t}{\Delta x}\right) \\ & & \cdot & \end{vmatrix}$$

$$G = \frac{1}{2}\left(1 - \frac{A \, \Delta t}{\Delta x}\right)e^{I\phi} + \frac{1}{2}\left(1 + \frac{A \, \Delta t}{\Delta x}\right)e^{-I\phi}$$

$$G = \cos \phi - I \frac{\Delta t}{\Delta x} A \sin \phi$$

$$G =$$

$$\begin{vmatrix} \cos \phi - \dfrac{v_0 \, \Delta t}{\Delta x} I \sin \phi & -I \dfrac{\Delta t}{\Delta x} h_0 \sin \phi \\ -I \dfrac{\Delta t}{\Delta x} g \sin \phi & \cos \phi - v_0 \dfrac{\Delta t}{\Delta x} I \sin \phi \end{vmatrix}$$

(*continued*)

Table 8.1. (*continued*)

Differential equation	Space discretization operator S	Matrix representation of S (excluding boundary conditions)

Wave equation

$$\frac{\partial^2 w}{\partial t^2} - a^2 \frac{\partial^2 w}{\partial x^2} = 0$$

or

$$\frac{\partial U}{\partial t} = A \frac{\partial U}{\partial x}$$

$$U = \begin{vmatrix} v \\ w \end{vmatrix}$$

$$A = \begin{vmatrix} 0 & a \\ a & 0 \end{vmatrix}$$

$$L = A \frac{\partial}{\partial x}$$

Central scheme

$$\frac{du_i}{dt} = A \frac{u_{i+1} - u_{i-1}}{2 \Delta x}$$

$$S = \frac{A}{2 \Delta x} (E - E^{-1})$$

$$S = \begin{vmatrix} & & 0 & & \\ -\dfrac{A}{2\Delta x} & 0 & \dfrac{A}{2\Delta x} & \\ & -\dfrac{A}{2\Delta x} & 0 & \dfrac{A}{2\Delta x} \\ & & 0 & & \end{vmatrix}$$

Forward/backward scheme
The two components v, w are discretized separately.

$$\frac{dv_i}{dt} = \frac{a}{\Delta x}(w_{i+1} - w_i)$$

$$\frac{dw_i}{dt} = \frac{a}{\Delta x}(v_i - v_{i-1})$$

$$S = \begin{vmatrix} 0 & a(E-1) \\ a(1 - E^{-1}) & 0 \end{vmatrix}$$

$$S = E^{-1} \begin{vmatrix} 0 & 0 \\ -a & 0 \end{vmatrix} + \begin{vmatrix} 0 & -a \\ a & 0 \end{vmatrix} + \begin{vmatrix} 0 & a \\ 0 & 0 \end{vmatrix} E$$

$$\equiv A_- E^{-1} + A_0 + A_+ E$$

$$S = \begin{vmatrix} A_- & A_0 & A_+ \\ A_- & A_0 & A_+ \\ A_- & A_0 & A_+ \end{vmatrix}$$

---

with the initial boundary conditions, for $t = 0$,

$$w(x, 0) = f(x)$$

$$\frac{\partial w}{\partial t}(x, 0) = g(x) \qquad (8.2.11b)$$

This wave equation is written as a system of first-order equations; for instance,

$$\frac{\partial v}{\partial t} = a \frac{\partial w}{\partial x}$$

$$\frac{\partial w}{\partial t} = a \frac{\partial v}{\partial x} \qquad (8.2.12)$$

Defining

$$U = \begin{vmatrix} v \\ w \end{vmatrix} \qquad (8.2.13)$$

we can write the system as

$$\frac{\partial U}{\partial t} = A \frac{\partial U}{\partial x} \qquad (8.2.14a)$$

Table 8.1. *(continued)*

Discretization operator C of time integrated scheme	Matrix representation of C	Amplification matrix
Backward Euler $U_i^{n+1} = U_i^n + A \dfrac{U_{i+1}^{n+1} - U_{i-1}^{n+1}}{2\Delta x}$ $C^{-1} = 1 - \dfrac{A\Delta t}{2\Delta x}(E - E^{-1})$	$C^{-1} = \begin{vmatrix} \ddots & & & \\ \frac{A\Delta t}{2\Delta x} & 1 & \frac{-A\Delta t}{2\Delta x} & \\ & \frac{A\Delta t}{2\Delta x} & 1 & \frac{-A\Delta t}{2\Delta x} \\ & & & \ddots \end{vmatrix}$	$G^{-1} = 1 - \dfrac{A\Delta t}{2\Delta x}(e^{I\phi} - e^{-I\phi})$ $G^{-1} = 1 - I\dfrac{A\Delta t}{\Delta x}\sin\phi$
Forward Euler scheme—two step/ semi-implicit $v_i^{n+1} - v_i^n = \dfrac{a\Delta t}{\Delta x}(w_{i+1}^n - w_i^n)$ $w_i^{n+1} - w_i^n = \dfrac{a\Delta t}{\Delta x}(v_i^{n+1} - v_{i-1}^{n+1})$ or, with $\sigma = a\Delta t/\Delta x$ $C = \begin{vmatrix} 1 & \sigma(E-1) \\ (1 - E^{-1})\sigma & 1 + \sigma^2(1 - E^{-1})(E-1) \end{vmatrix}$ $C = \begin{vmatrix} 1 & -\sigma \\ \sigma & 1-2\sigma^2 \end{vmatrix} + \begin{vmatrix} 0 & 0 \\ -\sigma & \sigma^2 \end{vmatrix}E^{-1}$ $\qquad + \begin{vmatrix} 0 & \sigma \\ 0 & \sigma^2 \end{vmatrix}E$	$C = \begin{vmatrix} \ddots & & & \\ C_- & C_0 & C_+ & \\ & C_- & C_0 & C_+ \\ & & C_- & C_0 & C_+ \\ & & & \ddots \end{vmatrix}$ $C_0 = \begin{vmatrix} 1 & -\sigma \\ \sigma & 1-2\sigma^2 \end{vmatrix}$ $C_- = \begin{vmatrix} 0 & 0 \\ -\sigma & \sigma^2 \end{vmatrix}$ $C_+ = \begin{vmatrix} 0 & \sigma \\ 0 & \sigma^2 \end{vmatrix}$	$G = \begin{vmatrix} 1 & I\gamma e^{I\phi/2} \\ I\gamma e^{-I\phi/2} & 1-\gamma^2 \end{vmatrix}$ $\gamma = 2\sigma \sin\phi/2$

with

$$A = \begin{vmatrix} 0 & a \\ a & 0 \end{vmatrix} \qquad (8.2.14b)$$

These operators are summarized in Table 8.1 for some representative schemes and the operators S and C are expressed as a function of the shift operator E defined in Chapter 4. Note that the matrix representation of the operators S and C of Table 8.1 do not contain the boundary points. This will be dealt with in Chapter 10.

8.2.2 The general Von Neumann stability condition

When a single harmonic is applied to scheme (8.2.5) the operator C will act on the space index i, since C can be considered as a polynomial in the displacement operator E, as can be seen from Table 8.1. Hence we obtain, inserting

$$u_i^n = v^n e^{Ii\phi} \qquad (8.2.15)$$

into the homogeneous part of scheme (8.2.5),

$$e^{Ii\phi} \cdot v^{n+1} = C(E)\, e^{Ii\phi} \cdot v^n \equiv G(\phi) \cdot v^n \cdot e^{Ii\phi}$$

and after division by $e^{li\phi}$,

$$v^{n+1} = G(\phi) \cdot v^n = [G(\phi)]^n v^1 \qquad (8.2.16)$$

with

$$G(\phi) = C(e^{I\phi}) \qquad (8.2.17)$$

The matrix $G(\phi)$ is called the *amplification* matrix, and reduces to the previously defined amplification factor when there is only one equation to be discretized. Observe that $G(\phi)$ or $G(k)$ can be considered as the discrete Fourier symbol of the discretization operator C, and is obtained from C by replacing E^j by $e^{Ij\phi}$ (see Table 8.1 for several examples).

The stability condition (7.2.35) requires that the matrix $[G(\phi)]^n$ remains uniformly bounded for all values of ϕ. The bound of a matrix G is defined by the maximum value of the ratio of the two vector magnitudes

$$\| G \| = \underset{u \neq 0}{\text{Max}} \frac{|G \cdot u|}{|u|} \qquad (8.2.18)$$

where $|u|$ is any vector norm. For instance, the L_2 norm is defined by the square root of the sum of the components squared $|u|_{L_2} = (|u_1|^2 + \cdots + |u_p|^2)^{1/2}$ if u is a vector with p components.

Since G is a $(p \times p)$ matrix with p eigenvalues $\lambda_1, \ldots, \lambda_j, \ldots, \lambda_p$ obtained as solutions of the polynomial

$$\det | G - \Lambda I | = 0 \qquad (8.2.19)$$

its spectral radius is defined by the modulus of the largest eigenvalue:

$$\rho(G) = \underset{j = 1, p}{\text{Max}} | \lambda_j | \qquad (8.2.20)$$

We have the following properties (see, for instance, Varga, 1962):

$$\| G \| \geqslant \underset{j}{\text{Max}} \frac{|G g_j|}{|g_j|} = \underset{j}{\text{Max}} | \lambda_j | = \rho(G) \qquad (8.2.21)$$

where g_i are the eigenvectors of G, and

$$\| G \|^n \geqslant \| G^n \| \geqslant \rho^n(G) \qquad (8.2.22)$$

The Von Neumann necessary condition for stability can be stated as the condition that the spectral radius of the amplification matrix satisfies (Richtmyer and Morton, 1967)

$$\rho(G) \leqslant 1 + 0(\Delta t) \qquad (8.2.23)$$

for finite Δt and for *all* values of ϕ, in the range $(-\pi, \pi)$. This condition is less severe than the previous one (equation (8.1.15)), which corresponds to a condition

$$\rho(G) \leqslant 1 \qquad (8.2.24)$$

The possibility for the spectral radius to be slightly higher than one for stability allows the treatment of problems where the exact solution grows exponentially (for instance, equation (7.1.5), with a source term q proportional to the temperature, $q = bT$, $b > 0$). However, in other cases condition (8.2.23) allows numerical modes to grow exponentially in time for finite values of Δt. Therefore the practical, or strict, stability condition (8.2.24) is recommended in order to prevent numerical modes growing faster than physical modes solution of the differential equation. (We will return to this important aspect in Chapter 10.) In this connection, when some eigenvalues are equal to one they would generate a growth of the form $\Delta t^{(m-1)}$, where m is the multiplicity. Hence eigenvalues $\lambda_j = 1$ should be simple.

Conditions (8.2.23) or (8.2.24) are also sufficient for stability if G is a normal matrix, that is, if G commutes with its Hermitian conjugate. In this case, equation (8.2.22) is valid with an equality sign in the L_2-norm, that is, $\| G \|_{L_2} = \rho(G)$ and $\| G^2 \|_{L_2} = \rho^2(G)$. In particular, for a single equation this is satisfied, and therefore condition (8.2.24) is sufficient and necessary for the stability of two-level schemes of linear equations with constant coefficients. Other cases for which the above condition is also sufficient for stability can be found in Richtmyer and Morton (1967).

Properties

(1) If G can be expressed as a polynomial of a matrix A, $G = P(A)$, then the spectral mapping theorem (Varga, 1962) states that

$$\lambda(G) = P(\lambda(A)) \qquad (8.2.25)$$

where $\lambda(A)$ are the eigenvalues of A. For example, if G is of the form

$$G = 1 - I\alpha A + \beta A^2$$

then

$$\lambda(G) = 1 - I\alpha\lambda(A) + \beta\lambda^2(A)$$

(2) If G can be expressed as a function of several *commuting* matrices the above property remains valid. That is, if

$$G = P(A, B) \qquad \text{with } AB = BA \qquad (8.2.26)$$

the two matrices have the same set of eigenvectors, and

$$\lambda(G) = P(\lambda(A), \lambda(B)) \qquad (8.2.27)$$

This property ceases to be valid when the matrices do not commute. Unfortunately this is the case for the system of flow equations in two and three dimensions. Therefore additional conjectures have to be introduced in order to derive stability conditions for schemes applied to the linearized flow equations in multi-dimensions. More details will be found in Volume 2 when dealing with the discretization of Euler equations.

Note that this condition of strict stability is called *zero* stability by Lambert (1973) when applied to the discretization of initial value problems in systems of ordinary differential equations (see also Chapter 11).

Example 8.2.1   Shallow-water equations

Referring to Table 8.1 we deduce readily the amplification matrix for the two schemes considered. The steps can easily be followed and we leave it to the reader to reproduce this table as an exercise.

Euler method: For the Euler method in time the amplification factor is

$$G = \begin{vmatrix} 1 - I \dfrac{v_0 \Delta t}{\Delta x} \sin \phi & - h_0 \dfrac{\Delta t}{\Delta x} I \sin \phi \\[2ex] - g \dfrac{\Delta t}{\Delta x} I \sin \phi & 1 - I \dfrac{v_0 \Delta t}{\Delta x} \sin \phi \end{vmatrix} \qquad \text{(E8.2.1)}$$

The stability condition (8.2.24) requires a knowledge of the eigenvalues of G, and these are obtained from

$$[\lambda - (1 - I\sigma_0 \sin \phi)]^2 + \sigma^2 \sin^2 \phi = 0 \qquad \text{(E8.2.2)}$$

where

$$\sigma_0 = \frac{v_0 \Delta t}{\Delta x} \qquad \text{(E8.2.3)}$$

$$\sigma = (gh_0)^{1/2} \frac{\Delta t}{\Delta x} \qquad \text{(E8.2.4)}$$

Hence the two eigenvalues are

$$\lambda_\pm = 1 - I(\sigma_0 \pm \sigma) \sin \phi \qquad \text{(E8.2.5)}$$

and the spectral radius is given by

$$\rho(G) = |\lambda_+| = 1 + \left(\frac{\Delta t}{\Delta x}\right)^2 (v_0 + \sqrt{(gh_0)})^2 \sin^2 \phi \geqslant 1 \qquad \text{(E8.2.6)}$$

The scheme is therefore *unstable*, as might be expected from the previous analysis of the central, Euler scheme for the convection equation.

Lax–Friedrichs scheme: This scheme was introduced by Lax (1954) as a way of stabilizing the unstable, forward in time, central scheme of the previous example. It consists of replacing u_i^n in the right-hand side by the average value $(u_{i+1}^n + u_{i-1}^n)/2$, maintaining the scheme as first order in time and space. It is

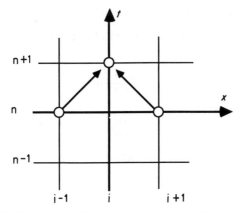

Figure 8.2.1 Lax–Friedrichs scheme for convection equations

schematically represented in Figure 8.2.1:

$$U_i^{n+1} = \frac{1}{2}(U_{i+1}^n + U_{i-1}^n) - \frac{\Delta t}{2\Delta x} A(U_{i+1}^n - U_{i-1}^n) \qquad (E8.2.7)$$

The reader can deduce the amplification matrix following the steps of Table 8.1, obtaining

$$G = \begin{vmatrix} \cos\phi - \sigma_0 I \sin\phi & -I\dfrac{\Delta t}{\Delta x} h_0 \sin\phi \\[2mm] -I\dfrac{\Delta t}{\Delta x} g \sin\phi & \cos\phi - \sigma_0 I \sin\phi \end{vmatrix} \qquad (E8.2.8)$$

The eigenvalues λ of G are given by

$$(\lambda - \cos\phi + \sigma_0 I \sin\phi)^2 + \sigma^2 \sin^2\phi = 0$$

or

$$\lambda_\pm = \cos\phi - I(\sigma_0 \pm \sigma)\sin\phi \qquad (E8.2.9)$$

The spectral radius is given by

$$\rho(G) = |\lambda_+| = [\cos^2\phi + (\sigma_0 + \sigma)^2 \sin^2\phi]^{1/2} \qquad (E8.2.10)$$

The stability condition $\rho(G) \leqslant 1$ will be satisfied if (for $v_0 > 0$)

$$(\sigma_0 + \sigma) \leqslant 1$$

or

$$(v_0 + \sqrt{(gh_0)})\frac{\Delta t}{\Delta x} \leqslant 1 \qquad (E8.2.11)$$

This is the CFL condition for the wave speed $(v_0 + \sqrt{(gh_0)})$, which is the largest eigenvalue of A.

Example 8.2.2   Second-order wave equation $(\partial^2 w/\partial t^2) - a^2(\partial^2 w/\partial x^2) = 0$

The forward–backward scheme, with semi-implicit time integration, of Table 8.1:

$$v_i^{n+1} - v_i^n = \frac{a\Delta t}{\Delta x}(w_{i+1}^n - w_i^n)$$

$$w_i^{n+1} - w_i^n = \frac{a\Delta t}{\Delta x}(v_i^{n+1} - v_{i-1}^{n+1})$$

(E8.2.12)

is equivalent to the three-level, centred scheme for the second-order wave equation, that is, to the scheme

$$w_i^{n+1} - 2w_i^n + w_i^{n-1} = \sigma^2(w_{i+1}^n - 2w_i^n + w_{i-1}^n) \qquad \text{(E8.2.13)}$$

where $\sigma = a\Delta t/\Delta x$ (see also Problems 8.3 and 8.4). The amplification matrix is obtained from Table 8.1 as

$$G = \begin{vmatrix} 1 & I\gamma\, e^{I\phi/2} \\ I\gamma\, e^{-I\phi/2} & 1 - \gamma^2 \end{vmatrix} \qquad \text{(E8.2.14)}$$

where

$$\gamma = 2\sigma \sin \phi/2 \qquad \text{(E8.2.15)}$$

The eigenvalues of G are obtained from

$$(1 - \lambda)(1 - \gamma^2 - \lambda) + \gamma^2 = 0$$

leading to the two solutions

$$\lambda_\pm = \tfrac{1}{2}[(2 - \gamma^2) \pm I\gamma\sqrt{(4 - \gamma^2)}] \qquad \text{(E8.2.16)}$$

For $\gamma^2 > 4$, that is, for $|\sigma \sin \phi/2| > 1$ or $|\sigma| > 1$, the spectral radius

$$\rho(G) = |\lambda_+| > 1$$

and the scheme is unstable. On the other hand, when $\gamma^2 \leqslant 4$, that is, for

$$|\sigma| \leqslant 1 \qquad \text{(E8.2.17)}$$
$$\rho(G) = |\lambda_+| = 1$$

the scheme is stable, although only marginally, since the norm of G is equal to one.

For negative values of a^2, that is, for negative values of σ^2, the wave equation becomes elliptic:

$$\frac{\partial^2 w}{\partial t^2} + |a^2|\frac{\partial^2 w}{\partial x^2} = 0 \qquad \text{(E8.2.18)}$$

and the scheme

$$(w_{i+1}^{n+1} - 2w_i^n + w_{i-1}^{n-1}) + |\sigma^2|(w_{i+1}^n - 2w_i^n + w_{i+1}^n) = 0 \qquad \text{(E8.2.19)}$$

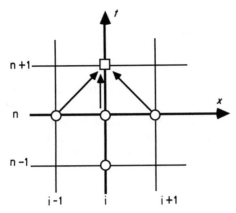

Figure 8.2.2 Unstable resolution scheme for Laplace
equation

is *unstable*. Indeed, when a^2 is negative, the positive eigenvalue λ_+ becomes

$$\lambda_+ = 1 + 2|\sigma|^2\sin^2\phi/2 + 2|\sigma|\sin\phi/2\sqrt{(1 + |\sigma|^2\sin^2\phi/2)} \geqslant 1 \quad \text{(E8.2.20)}$$

This shows that an elliptic problem cannot be treated numerically as an initial value problem. This is not surprising, since it is known that the Cauchy or initial value problem is not well posed for an elliptic equation (see, for instance, Courant and Hilbert, 1962, Volume II).

Observe that the above scheme, with $|a^2| = 1$, is the five-point difference operator for the Laplace equation, in the space (x, t). This scheme, as it stands, can be solved in a stable way for the associated boundary value problem, say on a rectangle $0 \leqslant x \leqslant L, 0 \leqslant t \leqslant T$, with any of the methods to be described in Chapter 12.

What the above results show is that the numerical solution of the elliptic problem cannot be obtained by a propagation from the points indicated by a circle in Figure 8.2.2 towards the point $(i, n + 1)$. Such an algorithm is basically unstable. A resolution method for elliptic equations based on this marching scheme has nevertheless been developed by Roache (1971) and is called the error vector propagation method (EVP). This is based on a computation of the error generated in the marching procedure from $t = 0$ to $t = T$ and a comparison with the imposed boundary condition on $t = T$. However, this method cannot be stabilized when the number of grid points in the marching direction increases (McAveney and Leslie, 1972). The reader will find a recent account of this approach in the monograph edited by Book (1981), chapter 7 by Madela and McDonald.

8.3 THE SPECTRAL ANALYSIS OF NUMERICAL ERRORS

The amplification matrix G allows, next to an assessment of stability, an evaluation of the frequency distribution of the discretization errors generated by the numerical scheme. Definition (8.2.16) of the amplification matrix defines the numerical representation of the time evolution of the solution, and the amplitude v^n of the harmonic corresponding to the wavenumber k can be written as

$$v^n = \hat{v} \, e^{-I\omega t_n} = \hat{v} \, e^{-I\omega \cdot n\Delta t} \tag{8.3.1}$$

where $\omega = \omega(k)$ is a complex function of the real number k, representing the numerical dispersion relation. The function $\hat{v}(k)$ is obtained from the Fourier decomposition of the initial solution, since for $u(x,0) = f(x)$ at $t = 0$ we have, assuming that the initial solution is represented exactly in the numerical scheme, with the exception of round-off errors:

$$\hat{v}(k) = \frac{1}{2L} \int_{-L}^{L} f(x) \, e^{-Ikx} \, dx \tag{8.3.2}$$

Actually, this defines the harmonic k of the solution u_i^n following equation (8.2.15) as

$$(u_i^n)_k = \hat{v}(k) \, e^{-I\omega(n\Delta t)} \, e^{Ik(i\Delta x)} \tag{8.3.3}$$

and is a discrete formulation of the single-wave representation applied in equation (3.4.13). In this latter form the exact solution is represented as

$$\tilde{u}_i^n = \hat{v} \, e^{-I\tilde{\omega}\Delta t} \, e^{Ik(i\Delta x)} \tag{8.3.4}$$

As seen in Chapter 3, the exact dispersion relation $\tilde{\omega} = \tilde{\omega}(k)$ can be obtained from the differential system as a solution of the eigenvalue equation (3.4.20), while the approximate relation between ω and k, obtained from the amplification matrix G, is the numerical dispersion relation of the scheme.

From equation (8.2.16) we have

$$v^n = G^n \cdot v^0 = G^n \cdot \hat{v} = e^{-I\omega n\Delta t} \cdot \hat{v} \tag{8.3.5}$$

and G can be written as

$$G = e^{-I\omega \Delta t} \tag{8.3.6}$$

A comparison with the exact amplification function

$$\tilde{G} = e^{-I\tilde{\omega}\Delta t} \tag{8.3.7}$$

will allow us to investigate the nature and frequency spectrum of the numerical errors. Since ω is a complex function the amplification matrix can be separated into an amplitude $|G|$ and a phase Φ. With

$$\omega = \xi + I\eta \tag{8.3.8}$$

we have

$$G = e^{+\eta \Delta t} \cdot e^{-I\xi \Delta t}$$
$$= |G| \, e^{-I\Phi} \tag{8.3.9a}$$

where

$$|G| = e^{\eta \Delta t}$$
$$\Phi = \xi \Delta t \tag{8.3.9b}$$

A similar decomposition, performed for the exact solution

$$\tilde{\omega} = \tilde{\xi} + I\tilde{\eta} \qquad (8.3.10)$$

following equation (3.4.24), leads to

$$|\tilde{G}| = e^{\tilde{\eta}\Delta t} \quad \text{and} \quad \tilde{\Phi} = \tilde{\xi}\Delta t \qquad (8.3.11)$$

The error in amplitude, called the *diffusion* or *dissipation* error, is defined by the ratio of the computed amplitude to the exact amplitude:

$$\varepsilon_D = \frac{|G|}{e^{\tilde{\eta}\Delta t}} \qquad (8.3.12)$$

The error on the phase of the solution, the *dispersion* error, can be defined as the difference

$$\varepsilon_\phi = \Phi - \tilde{\Phi} \qquad (8.3.13)$$

suitable for pure parabolic problems, where $\tilde{\Phi} = 0$, in the absence of convective terms. For convection-dominated problems the definition

$$\varepsilon_\phi = \Phi/\tilde{\Phi} \qquad (8.3.14)$$

is better adapted. In particular, for hyperbolic problems such as the scalar convection equation (7.2.1) the exact solution is a single wave propagating with the velocity a. Hence

$$\tilde{\phi} = ka\Delta t \qquad (8.3.15)$$

8.3.1 Error analysis for parabolic problems

Let us consider as an example the error analysis for the explicit central discretization of the heat diffusion equation (8.2.7). Consider the explicit scheme, with space-centred differences

$$u_i^{n+1} = u_i^n + \frac{\alpha\Delta t}{\Delta x^2}(u_{i+1}^n - 2u_i^n + u_{i-1}^n) \qquad (8.3.16)$$

the amplification factor is obtained from Table 8.1 as

$$G = 1 - 4\beta \sin^2\phi/2 \qquad (8.3.17)$$

with

$$\beta = \frac{\alpha\Delta t}{\Delta x^2} \qquad (8.3.18)$$

The stability condition is

$$|1 - 4\beta \sin^2\phi/2| \leqslant 1$$

which is satisfied for

$$1 \geqslant (1 - 4\beta \sin^2\phi/2) \geqslant -1$$

that is,

$$0 \leqslant \beta \leqslant 1/2 \tag{8.3.19}$$

Hence the above scheme is stable for

$$\alpha \geqslant 0 \quad \text{and} \quad \beta = \frac{\alpha \Delta t}{\Delta x^2} \leqslant \frac{1}{2} \tag{8.3.20}$$

The first condition expresses the stability of the physical problem, since for $\alpha < 0$ the analytical solution is exponentially increasing with time.

The exact solution corresponding to a wavenumber k is obtained by searching a solution of the type

$$\tilde{u} = \hat{v}\, e^{-I\tilde{\omega}t}\, e^{Ikx} \tag{8.3.21}$$

Inserting into equation (8.2.7) we have

$$\tilde{\omega}(k) = -I\alpha k^2 = -I\beta \cdot \phi^2/\Delta t \tag{8.3.22}$$

The exact solution of this parabolic problem is associated with a purely imaginary eigenvalue $\tilde{\omega}$, that is, with an exponential decay in time of the initial amplitude if $\alpha > 0$:

$$\tilde{u} = \hat{v}\, e^{Ikx}\, e^{-\alpha k^2 t} \tag{8.3.23}$$

Hence the error in the amplitude is measured by the ratio

$$\varepsilon_D = \frac{1 - 4\beta \sin^2 \phi/2}{e^{-\beta \cdot \phi^2}} \tag{8.3.24}$$

Expanding in powers of ϕ we obtain

$$\varepsilon_D = \frac{1 - \beta\phi^2 + \beta\phi^4/12 + \cdots}{1 - \beta\phi^2 + (\beta^2\phi^4/2) + \cdots} \approx 1 - \frac{\beta^2\phi^4}{2} + \frac{\beta\phi^4}{12} + \cdots$$

$$\approx 1 - \frac{\alpha^2 k^4 \Delta t^2}{2} + \frac{\alpha k^4}{12}\, \Delta t \Delta x^2 \tag{8.3.25}$$

For the low frequencies ($\phi \approx 0$) the error in amplitude remains small; while at high frequencies ($\phi \approx \pi$) the error could become unacceptably high, particularly for the larger values of $\beta \leqslant 1/2$. However, for $\beta = 1/6$ the two first terms of the expansion cancel, and the error is minimized, becoming of higher order, namely of the order $0(\Delta t^2, \Delta x^4)$ for constant values of $\beta = \alpha \Delta t/\Delta x^2$ and proportional to k^6.

Since G is real there is no error in phase, that is, there is no dispersive error for this scheme. It is seen that the error is proportional to the fourth and sixth power of the wavenumber, indicating that the high frequencies are computed with large errors. However, the amplitudes of these high frequencies are strongly damped since they are equal to $e^{-\alpha k^2 t}$. Therefore this will generally not greatly affect the overall accuracy, with the exception of situations where the initial solution $u(x, 0)$ contains a large number of high-frequency components (see also Problem 8.5).

8.3.2 Error analysis for hyperbolic problems

A hyperbolic problem such as the convection equation $u_t + au_x = 0$ represents a wave travelling at constant speed without damping, that is, with constant amplitude. The exact solution for a wave of the form $u = e^{-I\tilde{\omega}t} e^{Ikx}$ is given by

$$\tilde{u} = e^{-Ikat} e^{Ikx} \qquad (8.3.26)$$

Hence the exact amplification function is defined by the real value of $\tilde{\omega}$:

$$\tilde{\omega} = ka = \tilde{\xi}$$
$$\tilde{\eta} = 0 \qquad (8.3.27)$$

The error in amplitude will be given by the modulus of the amplification factor

$$\varepsilon_D = |G| \qquad (8.3.28)$$

and the error in phase (the dispersive error) is defined by

$$\varepsilon_\phi = \frac{\Phi}{ka\,\Delta t} = \frac{\Phi}{\sigma\phi} \qquad (8.3.29)$$

An initial sinusoidal wave will be damped in the numerical simulation by a factor $|G|$ per time step and its propagation speed will be modified by the dispertion error ε_ϕ. When this ratio is larger than one ($\varepsilon_\phi > 1$) the phase error is a *leading* error and the numerical computed wave speed, $\bar{a}$, is larger than the exact speed, since

$$\bar{a} = \Phi/(k\Delta t) = a\Phi/(\sigma\phi) \qquad (8.3.30)$$

and

$$\varepsilon_\phi = \bar{a}/a \qquad (8.3.31)$$

This means that the computed waves appear to travel faster than the physical waves. On the other hand, when $\varepsilon_\phi < 1$ the phase error is said to be a *lagging* error, and the computed waves travel at a lower velocity than the physical ones.

Example 8.3.1 Lax–Friedrichs scheme for the convection equation

Applying the Lax–Friedrichs scheme to the single convection equation (see Table 8.1) leads to

$$u_i^{n+1} = \frac{1}{2}(u_{i+1}^n + u_{i-1}^n) - \frac{\sigma}{2}(u_{i+1}^n - u_{i-1}^n) \qquad (E8.3.1)$$

The amplification factor is obtained by inserting a single harmonic $v^n e^{Iki\Delta x}$:

$$G = \cos\phi - I\sigma\sin\phi \qquad (E8.3.2)$$

leading to the CFL stability condition $|\sigma| \leqslant 1$.

The accuracy of the scheme is obtained from the modulus and phase of the amplification factor:

$$| G | = | \cos^2\phi + \sigma^2\sin^2\phi |^{1/2}$$
$$\Phi = \tan^{-1}(\sigma \tan \phi)$$
(E8.3.3)

This defines the dissipation error

$$\varepsilon_D = | G | = | \cos^2\phi + \sigma^2\sin^2\phi |^{1/2}$$
(E8.3.4)

and the dispersion error

$$\varepsilon_\phi = \frac{\Phi}{\sigma\phi} = \frac{\tan^{-1}(\sigma \tan \phi)}{\sigma\phi}$$
(E8.3.5)

As can be seen, the choice $\sigma = 1$ gives the exact solution, but lower values of σ will generate amplitude and phase errors.

Two equivalent graphical representations for the amplification factor are applied in practice. Cartesian representation of $| G |$ and ε_ϕ as a function of the parameter $\phi = k\Delta x$, ranging from 0 to π or a polar representation for $| G |$ and

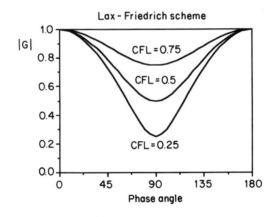

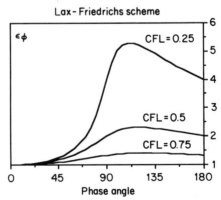

Figure 8.3.1 Amplitude and phase errors for Lax–Friedrichs scheme applied to the convection equation

ε_ϕ, where ϕ is represented as the polar angle. Figure 8.3.1 shows the Cartesian representation of $|G|$ and ε_ϕ for the Lax–Friedrichs scheme. For small values of σ the waves are strongly damped, indicating that this scheme is generating a strong numerical dissipation. The phase error is everywhere larger or equal to one, showing a leading phase error, particularly for $\phi = \pi$, $\varepsilon_\phi = 1/\sigma$ (see also Problem 8.6).

Example 8.3.2   Explicit upwind scheme (7.2.8)

The amplification factor for this scheme is defined by equation (8.1.19). Its modulus is given by

$$|G| = [(1 - \sigma + \sigma \cos \phi)^2 + \sigma^2\sin^2\phi]^{1/2} = [1 - 4\sigma(1 - \sigma)\sin^2\phi/2]^{1/2}$$

$$(E8.3.6)$$

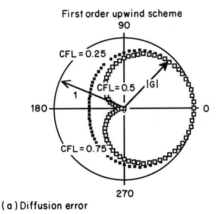

(a) Diffusion error

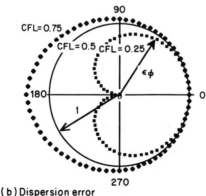

(b) Dispersion error

Figure 8.3.2 Polar representation of amplitude and phase errors for the upwind scheme applied to the convection equation

and the phase error is

$$\varepsilon_\phi = \frac{\tan^{-1}[(\sigma \sin \phi)/(1 - \sigma + \sigma \cos \phi)]}{\sigma\phi} \qquad \text{(E8.3.7)}$$

A polar representation is shown in Figure 8.3.2.

For $\sigma = 0.5$ the phase error $\varepsilon_\phi = 1$, but for $\sigma < 0.5$, $\varepsilon_\phi < 1$, indicating a lagging error, while the numerical speed of propagation becomes larger than the physical speed, $\varepsilon_\phi > 1$ for Courant numbers $\sigma > 0.5$ (see also Problem 8.7).

Example 8.3.3   The Lax–Wendroff scheme for the convection equation

The schemes of the two previous examples are of first-order accuracy, which is generally insufficient for practical purposes. The first second-order scheme for the convection equation with two time levels is due to Lax and Wendroff (1960). The original derivation of Lax and Wendroff was based on a Taylor expansion in time up to the third order such to achieve second-order accuracy. In the development

$$u_i^{n+1} = u_i^n + \Delta t(u_t)_i + \frac{\Delta t^2}{2}(u_{tt})_i + 0(\Delta t^3) \qquad \text{(E8.3.8)}$$

the second derivative is replaced by

$$u_{tt} = a^2 u_{xx} \qquad \text{(E8.3.9)}$$

leading to

$$u_i^{n+1} = u_i^n - a\,\Delta t(u_x)_i + \frac{a^2\,\Delta t^2}{2}(u_{xx})_i + 0(\Delta t^3) \qquad \text{(E8.3.10)}$$

When this is discretized centrally in mesh point i we obtain

$$u_i^{n+1} = u_i^n - \frac{\sigma}{2}(u_{i+1}^n - u_{i-1}^n) + \frac{\sigma^2}{2}(u_{i+1}^n - 2u_i^n + u_{i-1}^n) \qquad \text{(E.8.3.11)}$$

As can be seen, the third term, which stabilizes the instability generated by the first two terms, is the discretization of an additional dissipative term of the form $(a^2\,\Delta t/2)u_{xx}$.

The amplification matrix from the Von Neumann method is

$$G = 1 - I\sigma \sin \phi - \sigma^2(1 - \cos \phi) \qquad \text{(E.8.3.12)}$$

In the complex G-plane this represents an ellipse centred on the real axis at the abscissa $(1 - \sigma^2)$ and having a semi-axis length of σ^2 along the real axis and σ along the vertical axis. Hence this ellipse will always be contained in the unit circle if the CFL condition is satisfied (Figure 8.3.3). For $\sigma = 1$ the ellipse becomes identical to the unit circle. The stability condition is therefore

$$|\sigma| \leq 1 \qquad \text{(E8.3.13)}$$

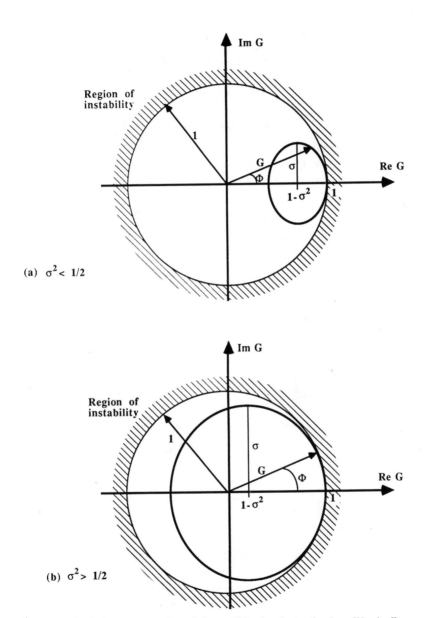

Figure 8.3.3 Polar representation of the amplification factor for Lax–Wendroff scheme. (a) $\sigma < 1$ and $\sigma^2 < 1/2$; (b) $\sigma < 1$ and $\sigma^2 > 1/2$

The dissipation error is given by

$$|G|^2 = 1 - 4\sigma^2(1 - \sigma^2)\sin^4\phi/2 \qquad \text{(E8.3.14)}$$

and the phase error by

$$\varepsilon_\phi = \frac{\tan^{-1}[(\sigma \sin \phi)/(1 - 2\sigma^2\sin^2\phi/2)]}{\sigma\phi} \qquad \text{(E8.3.15)}$$

To the lowest order we have

$$\varepsilon_\phi \approx 1 - \tfrac{1}{6}(1 - \sigma^2)\phi^2 + 0(\phi^4) \qquad \text{(E8.3.16)}$$

This relative phase error is mostly lower than one, indicating a dominating lagging phase error. On the high-frequency end the phase angle Φ goes to zero if $\sigma^2 < 1/2$ and tends to π if $\sigma^2 > 1/2$. These diffusion and dispersion errors are represented in Figure 8.3.4.

The phase error is the largest at the high frequencies, hence this will tend to accumulate high-frequency errors (for instance, those generated at a moving

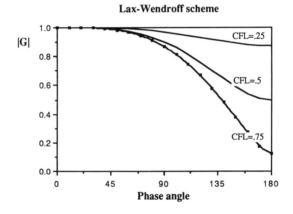

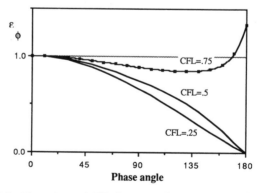

Figure 8.3.4 Dispersion and diffusion errors for Lax–Wendroff scheme

discontinuity). When the linear equation $u_t + au_x = 0$ is solved for a propagating discontinuity, oscillations will appear being the shock as can be seen from Figure 8.3.6 (to be discussed in Section 8.3.4, which compares the results computed with four different schemes).

8.3.3 Extension to three-level schemes

The properties of the amplification factor in the previous sections were based on two-level schemes, allowing a straightforward definition of G. However, many schemes can be defined which involve more than two time levels, particularly when the time derivatives are discretized with central difference formulas. A general form, generalizing equations (8.2.6), would be

$$U^{n+1} + b_0 U^n + b_1 U^{n-1} = CU^n + \tilde{Q} \qquad (8.3.32)$$

For instance, for the convection equation $u_t + au_x = 0$ and a central difference in space we can define a scheme

$$\frac{u_i^{n+1} - u_i^{n-1}}{2\Delta t} = -\frac{a}{2\Delta x}(u_{i+1}^n - u_{i-1}^n) \qquad (8.3.33)$$

which is second-order accurate in space and time. This scheme is known as the *leapfrog* scheme, because of the particular structure of its computational molecule (Figure 8.3.5) where the nodal value u_i^n does *not* contribute to the computation of u_i^{n+1}.

This scheme treats three levels simultaneously and, in order to start the calculation, two time levels $n = 0$ and $n = 1$ have to be known. In practical computations this can be obtained by applying another, two-level, scheme for the first time step. The method applied for the determination of the amplification matrix, consists of replacing the multi-level scheme by a two-step

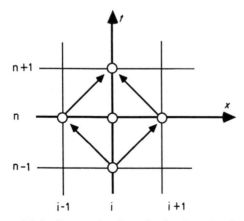

Figure 8.3.5 Computational molecule for the leapfrog scheme

system through the introduction of a new variable Z:

$$Z = U^{n-1} \tag{8.3.34}$$

Equation (7.3.32) then becomes

$$U^{n+1} = -b_1 Z^n + (C - b_0) U^n + \tilde{Q}$$
$$Z^{n+1} = U^n \tag{8.3.35}$$

and by defining a new variable

$$W = \left| \begin{matrix} U \\ Z \end{matrix} \right| \tag{8.3.36}$$

the system is rewritten as

$$W^{n+1} = \bar{C} W^n + \bar{Q} \tag{8.3.37}$$

and analysed as in the previous cases.

Alternatively, the method of introducing an additional variable is fully equivalent to a more direct approach, whereby we write for the amplitudes v^n of a single harmonic

$$v^{n-1} = G^{-1} \cdot v^n \tag{8.3.38}$$

and

$$v^{n+1} = G \cdot v^n \tag{8.3.39}$$

When this is introduced into the three-level scheme a quadratic equation for G is obtained.

Example 8.3.4   The leapfrog scheme for the convection equation

Scheme (8.3.33) will be written with the new variable Z as follows:

$$\left| \begin{matrix} u_i^{n+1} = z_i^n - \sigma(u_{i+1}^n - u_{i-1}^n) \\ z_i^{n+1} = u_i^n \end{matrix} \right. \tag{E8.3.17}$$

and as a function of the vector W we obtain the system

$$w_i^{n+1} = \bar{C} w_i^n \tag{E8.3.18}$$

With the introduction of the shift operator E the operator $\bar{C}$ becomes

$$\bar{C} = \left| \begin{matrix} -\sigma(E - E^{-1}) & 1 \\ 1 & 0 \end{matrix} \right| \tag{E8.3.19}$$

The amplification matrix becomes

$$G = \left| \begin{matrix} -\sigma(e^{I\phi} - e^{-I\phi}) & 1 \\ 1 & 0 \end{matrix} \right| \tag{E8.3.20}$$

The eigenvalues of G are readily obtained as

$$\lambda_\pm = -I\sigma \sin \phi \pm \sqrt{(1 - \sigma^2 \sin^2 \phi)} \tag{E8.3.21}$$

and are to be considered as the *amplification factors of the three-level scheme*. Indeed, applying the second approach (8.3.38) and (8.3.39) to equation (8.3.33), for a harmonic $\phi = k \, \Delta x$, leads to

$$(G - 1/G) = -\sigma(e^{I\phi} - e^{-I\phi}) \tag{E8.3.22}$$

with the two solutions

$$G = -I\sigma \sin \phi \pm \sqrt{(1 - \sigma^2 \sin^2 \phi)} \tag{E8.3.23}$$

If $\sigma > 1$ the scheme is unstable, since the term under the square root can become negative, and for these values G is purely imaginary and in magnitude larger than one. This is best seen for the particular value $\phi = \pi/2$.

For $|\sigma| \leqslant 1$ the scheme is *neutrally stable*, since

$$|G| = 1 \qquad \text{for} \qquad |\sigma| \leqslant 1 \tag{E8.3.24}$$

The phase error is given by

$$\varepsilon_\phi = \frac{\pm \tan^{-1}[\sigma \sin \phi / \sqrt{(1 - \sigma^2 \sin^2 \phi)}]}{\sigma \phi} = \pm \frac{\sin^{-1}(\sigma \sin \phi)}{\sigma \phi} \tag{E8.3.25}$$

Hence the leapfrog scheme should give accurate results when the function u has a smooth variation, since the amplitudes are correctly modelled, so much that for low frequencies the phase error is close to one since $\varepsilon_\phi = \pm 1$ for $\phi \to 0$. However, high-frequency errors tend to remain stationary since $\varepsilon_\phi \to 0$ for $\phi \to \pi$ and, since they are undamped, they can accumulate and destroy the accuracy of the numerical solution. This is clearly seen in Figure 8.3.6.

Example 8.3.5 Du Fort and Frankel scheme for the heat-conduction equation

The scheme of Du Fort and Frankel (1953) is obtained from the unstable 'leapfrog' explicit scheme applied to the diffusion equation (8.2.7) (see Problem 8.9):

$$u_i^{n+1} - u_i^{n-1} = 2\left(\frac{\alpha \, \Delta t}{\Delta x^2}\right)(u_{i+1}^n - 2u_i^n + u_{i-1}^n) \tag{E8.3.26}$$

by averaging out the term u_i^n in time as $(u_i^{n+1} + u_i^{n-1})/2$. This leads to the scheme $\beta = \alpha \, \Delta t / \Delta x^2$:

$$u_i^{n+1} - u_i^{n-1} = 2 \frac{\alpha \, \Delta t}{\Delta x^2} (u_{i+1}^n - u_i^{n+1} - u_i^{n-1} + u_{i-1}^n) \tag{E8.3.27a}$$

or

$$u_i^{n+1}(1 + 2\beta) = u_i^{n-1}(1 - 2\beta) + 2\beta(u_{i+1}^n + u_{i-1}^n) \tag{E8.3.27b}$$

The amplification matrix is obtained from the system, with $Z^n = U^{n-1}$,

$$w_i^{n+1} = \bar{C} \cdot w_i^n \tag{E8.3.28}$$

where

$$\bar{C} = \begin{vmatrix} \dfrac{2\beta(E + E^{-1})}{1 + 2\beta} & \dfrac{1 - 2\beta}{1 + 2\beta} \\ 1 & 0 \end{vmatrix} \tag{E8.3.29}$$

Hence

$$G = \begin{vmatrix} \dfrac{4\beta \cos \phi}{1 + 2\beta} & \dfrac{1 - 2\beta}{1 + 2\beta} \\ 1 & 0 \end{vmatrix} \tag{E8.3.30}$$

and the two eigenvalues, representing the amplification factors of the scheme, are given by

$$\lambda_\pm = \frac{2\beta \cos \phi \pm \sqrt{(1 - 4\beta^2 \sin^2 \phi)}}{1 + 2\beta} \tag{E8.3.31}$$

A plot of the eigenvalues $\lambda_\pm$ for different values of β as a function of ϕ, or a direct calculation of the condition $|\lambda_\pm| < 1$, shows that the scheme of Du Fort and Frankel is unconditionally stable for $\beta > 0$. This is very unusual for an explicit scheme. However, as will be seen in Chapter 10, this scheme is not always consistent.

Note that, for three-level schemes, there are two amplification factors, although the exact solution has a single value of the amplification. For the leapfrog scheme applied to the wave equation it can be observed that one of the two solutions has a negative phase error, that is, it propagates in the wrong direction. Hence the solution with the + sign corresponds to the physical solution, while the other is a *spurious* solution generated by the scheme. More insight into this aspect will appear from the stability analysis of Chapter 10 dealing with the matrix method.

8.3.4 A comparison of different schemes for the linear convection equation

It is instructive to compare the results obtained with the four schemes described in Examples 8.3.1–8.3.4 when applied to the linear convection equation. The effects of the diffusion and dispersion errors can be demonstrated, as a function of frequency, with the following two test cases, a propagating discontinuity and a sinusoidal wave packet.

The former is typical of a signal with a high-frequency content, since the Fourier decomposition of a discontinuity contains essentially high-order harmonics. On the other hand, the sinusoidal wave packet can be chosen to correspond to a selected value of the wavenumber and hence to a fixed value of the phase angle ϕ for a given mesh size Δx.

Figure 8.3.6 compares the computed results for the propagating discontinuity at a Courant number of 0.8 after 50 time steps on a mesh size

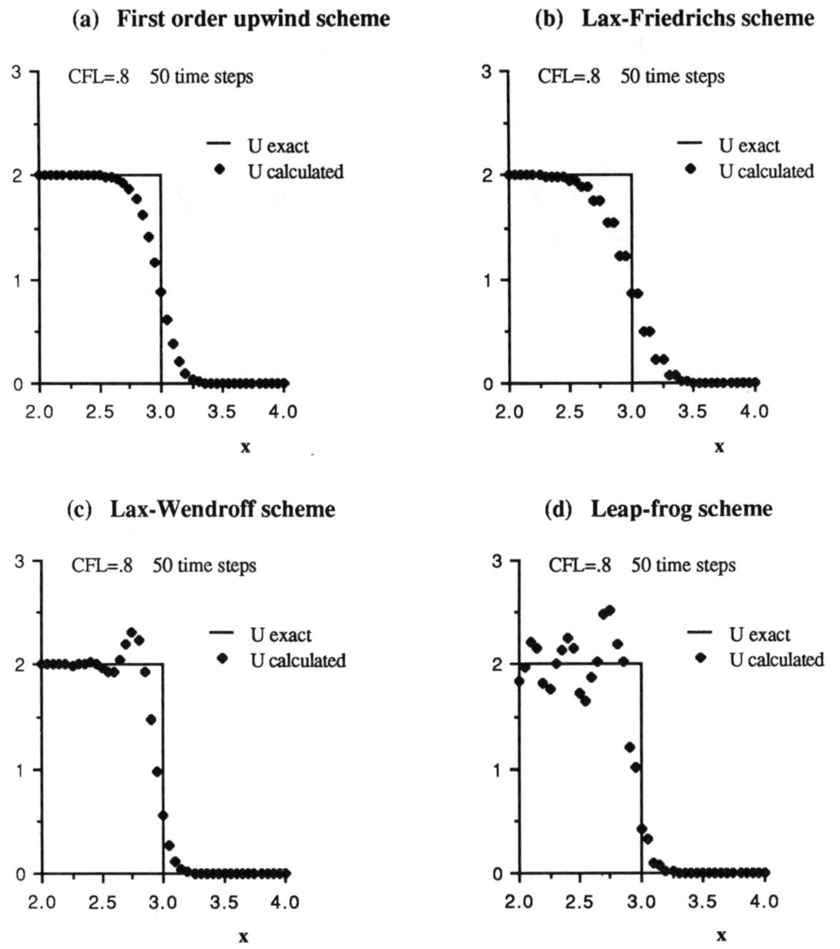

(a) First order upwind scheme

(b) Lax-Friedrichs scheme

CFL=.8   50 time steps

— U exact
• U calculated

CFL=.8   50 time steps

— U exact
• U calculated

(c) Lax-Wendroff scheme

(d) Leap-frog scheme

CFL=.8   50 time steps

— U exact
• U calculated

CFL=.8   50 time steps

— U exact
• U calculated

Figure 8.3.6 Comparison of four schemes on the linear convection equation for a propagating discontinuity

$\Delta x = 0.05$. The strong dissipation of the first-order upwind and Lax–Friedrichs schemes is clearly seen from the way the discontinuity is smoothed out. Observe also the 'double' solution obtained with the Lax–Friedrichs scheme, illustrating the odd–even decoupling discussed in Section 4.4 (Figure 4.4.4). Looking at Figure 8.2.1 it can be seen that u_i^{n+1} does not depend on u_i^n but on the neighbouring points u_{i-1}^n and u_{i+1}^n. These points also influence the solutions u_{i+2}^{n+1}, u_{i+4}^{n+1}, ..., while u_i^n will influence independently the points $u_{i+1}^{n+1}, u_{i+3}^{n+1}, ...$ The solutions obtained at the even- and odd-numbered points can therefore differ by a small constant without preventing convergence and such a difference appears on the solution shown in Figure 8.3.6(b).

The second-order Lax–Wendroff and leapfrog schemes generate oscillations due to the dominating high-frequency dispersion errors, which are mostly

316

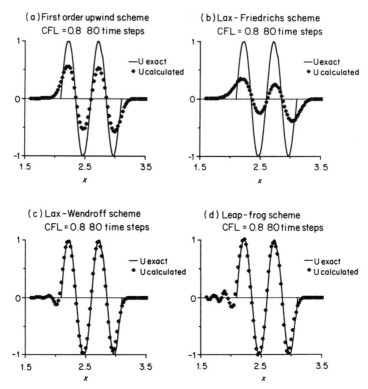

Figure 8.3.7 Comparison of four schemes on the linear convection equation for a propagating wave packet for $\phi = \pi/10$

lagging. The leapfrog scheme, which has no damping, generates stronger high-frequency oscillations compared with the Lax–Wendroff scheme, whose amplification factor is lower than one at the phase angle $\phi = \pi$, where $G(\pi) = 1 - 2\sigma^2$.

The test cases of the moving wave packet allow us to experiment with the frequency dependence of the schemes at the low end of the spectrum. Figure 8.3.7 compares the four schemes for a phase angle ϕ equal to $\pi/10$ at a Courant number of 0.8 after 80 time steps on a mesh $\Delta x = 0.025$. The strong diffusion error of the first-order schemes is clearly seen, showing that they are useless for time-dependent propagation problems of this kind. The second-order schemes give accurate results at these low frequencies, the oscillations at the beginning of the wave packet being created by the high-frequency errors generated by the slope discontinuity of the solution at this point. Hence a behaviour similar to the propagating discontinuity of the previous figure appears.

The same computations performed at a higher frequency corresponding to a phase angle of $\phi = \pi/5$, are shown in Figure 8.3.8. The first-order schemes are

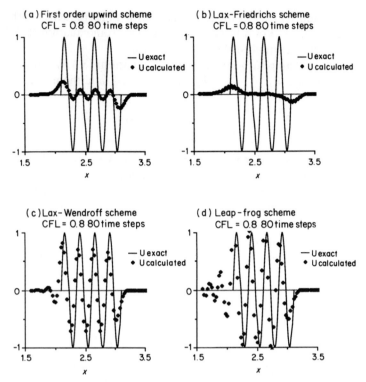

Figure 8.3.8 Comparison of four schemes on the linear convection equation for a propagating wave packet for $\phi = \pi/5$

more severely damped while the increasing, lagging dispersion errors of the two- second-order schemes can be seen by the phase shift of the computed solutions. The Lax–Wendroff scheme has a diffusion error which increases with frequency, as can be seen in Figure 8.3.4, and an amplitude error develops. The leapfrog scheme has a better behaviour with regard to the amplitude of the wave, as can be seen from the amplitudes of the second and third periods, although the first period of the wave is spoiled by the high-frequency oscillations generated at the initial slope discontinuity.

8.3.5 The numerical group velocity

The group velocity of a wave packet, containing more than one frequency, has been defined in Chapter 3 (equation (3.4.35)) and is also the velocity at which the energy of the wave is travelling. For a one-dimensional wave we have

$$\bar{v}_G(k) = \frac{d\bar{\omega}}{dk} \tag{8.3.40}$$

318

defining the group velocity as the derivative of the time frequency with respect to the wavenumber k. For a linear wave it is seen from equation (8.3.27) that the group velocity is equal to the phase speed a.

By writing the amplification factor as equation (8.3.9) the numerical dispersion relation $\omega = \omega(k) = \xi + I\eta$ can be defined, and the numerical group velocity

$$v_G(k) = \frac{d\xi}{dk} = \mathrm{Re}\left(\frac{d\omega}{dk}\right) \qquad (8.3.41)$$

will represent the travelling speed of wave packets centred around the wavenumber k. Since the errors generated by a numerical scheme generally contain a variety of frequencies it is more likely that they will travel at the numerical group velocity instead of the numerical phase speed $\bar{a}$, defined by equation (8.3.30).

For the leapfrog scheme (equation (8.3.33)) the introduction of equation (8.3.6) into equation (E8.3.23) leads to the numerical dispersion relation:

$$\sin \omega \Delta t = \sigma \sin \phi \qquad (8.3.42)$$

from which we derive

$$v_G = a \frac{\cos \phi}{\cos \omega \Delta t} = a \frac{\cos \phi}{(1 - \sigma^2 \sin^2 \phi)^{1/2}} \qquad (8.3.43)$$

For low frequencies the group velocity is close to the phase speed a, but for the high frequencies ($\phi \approx \pi$) the group velocity is close to $-a$, indicating that the high wavenumber packets will travel in the opposite direction to the wave phase speed a. This can be observed in Figure 8.5.2, where it is seen that the high-frequency errors, generated upstream of the stationary shock, travel in the upstream direction.

An instructive example is provided by the exponential wave packet

$$u(x, t = 0) = \exp(-\alpha x^2)\sin 2\pi k_w x \qquad (8.3.44)$$

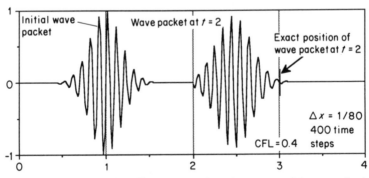

Figure 8.3.9 Solution of the linear propagation of an exponential wave packet by the leapfrog scheme, after 400 time steps, for $\phi = \pi/4$

shown in Figure 8.3.9 for a phase angle $\phi = k\Delta x = \pi/4$, corresponding to a wavelength of $\lambda = 8\Delta x$. The solution of the linear wave equation $u_t + au_x = 0$ with the leapfrog scheme is shown in the same figure after 400 time steps for a Courant number of 0.4 and $\Delta x = 1/80$ for $a = 1$. If the initial solution is centred at $x = 1$ the exact solution should be located around $x = 3$ at time $t = 400\Delta t = 2$. However, the numerical solution is seen to have travelled only to the point $x \approx 2.475$, which indicates a propagating speed of 0.7375 instead of the phase speed $a = 1$. This corresponds exactly to the computed group velocity from equation (8.3.43), which gives a value of $v_G = 0.7372$ at $\phi = \pi/4$.

These properties of the group velocity should be kept in mind when analysing numerical data associated with high-frequency solutions. More details on the applications of the concept of group velocity to the analysis of numerical schemes can be found in Vichnevetsky and Bowles (1982), Trefethen (1982) and Cathers and O'Connor (1985). The last reference presents a detailed, comparative analysis of the group velocity properties of various finite element and finite difference schemes applied to the one-dimensional, linear convection equation. Trefethen (1983, 1984) has derived some important relations between group velocity and the stability of numerical boundary conditions of hyperbolic problems. His results can be expressed by the condition that the numerical boundary treatment should not allow group velocities at these boundaries to transport energy into the computational domain. We refer the reader to the original references for more details and derivations.

8.4 MULTI-DIMENSIONAL VON NEUMANN ANALYSIS

For problems in more than one space dimension the Fourier decomposition at the basis of the Von Neumann stability analysis can be performed separately in each space direction through the introduction of a wavenumber vector $\vec{\varkappa}$. For instance, the solution $u(\vec{x}, t)$ will be represented as a superposition of harmonics of the form

$$u(\vec{x}, t) \sim \hat{v}\, e^{-I\omega t}\, e^{I\vec{\varkappa} \cdot \vec{x}} \tag{8.4.1}$$

where the scalar product $\vec{\varkappa} \cdot \vec{x}$ is defined as

$$\vec{\varkappa} \cdot \vec{x} = \varkappa_x x + \varkappa_y y + \varkappa_x z \tag{8.4.2}$$

In discretized form, with mesh point indexes i, j, k, we have

$$(\vec{\varkappa} \cdot \vec{x})_{i,j,k} = i(\varkappa_x \Delta x) + j(\varkappa_y \Delta y) + k(\varkappa_z \Delta z) \equiv i \cdot \phi_x + j \cdot \phi_y + k \cdot \phi_z \tag{8.4.3}$$

and the three parameters ϕ_x, ϕ_y, ϕ_z range from $-\pi$ to π is each of the three space directions. The further determination of the amplification matrix remains unchanged from the one-dimensional case.

8.4.1 Parabolic equations

Let us consider as an example the two-dimensional heat diffusion equation
(7.1.15). The obvious generalization of the one-dimensional explicit central
scheme (8.3.16) for the parabolic equation, written as

$$\frac{\partial u}{\partial t} = \alpha \left(\frac{\partial^2 u}{\partial x^2} + \frac{\partial^2 u}{\partial y^2} \right) \tag{8.4.4}$$

is

$$u_{ij}^{n+1} - u_{ij}^n = \alpha \, \Delta t \left[\frac{u_{i+1,j}^n - 2u_{ij}^n + u_{i-1,j}^n}{\Delta x^2} + \frac{u_{i,j+1}^n - 2u_{ij}^n + u_{i,j-1}^n}{\Delta y^2} \right] \tag{8.4.5}$$

A discrete Fourier decomposition is defined by

$$u_{ij}^n = \sum_{\varkappa_x, \varkappa_y} v^n \, e^{I\varkappa_x i \, \Delta x} \, e^{I\varkappa_y j \, \Delta y} \tag{8.4.6}$$

where the range of $\varkappa_x$ and $\varkappa_y$ is defined separately for each direction, as in the
one-dimensional case. Inserting a single component into the discretized
scheme, the amplification matrix G is still defined, as in the one-dimensional
case, as

$$v^{n+1} = Gv^n \tag{8.4.7}$$

We obtain, from equation (8.4.5), after division by $v^n \, e^{iI\phi_x} \, e^{jI\phi_y}$,

$$G - 1 = \beta [(e^{I\phi_x} + e^{-I\phi_x} - 2) + \left(\frac{\Delta x}{\Delta y} \right)^2 (e^{I\phi_y} + e^{-I\phi_y} - 2)]$$

$$\tag{8.4.8}$$

$$G - 1 = -4\beta \left(\sin^2\phi_x/2 + \left(\frac{\Delta x}{\Delta y} \right)^2 \sin^2\phi_y/2 \right)$$

The strict stability condition becomes

$$\left| 1 - 4\beta \left(\sin^2\phi_x/2 + \left(\frac{\Delta x}{\Delta y} \right)^2 \sin^2\phi_y/2 \right) \right| \leqslant 1 \tag{8.4.9}$$

which leads to

$$\alpha > 0 \tag{8.4.10}$$

and

$$\beta \left(1 + \left(\frac{\Delta x}{\Delta y} \right)^2 \right) \leqslant \frac{1}{2}$$

or

$$\alpha \left(\frac{1}{\Delta x^2} + \frac{1}{\Delta y^2} \right) \Delta t \leqslant \frac{1}{2} \tag{8.4.11}$$

This stability condition is necessary and sufficient and is analogous to
condition (8.3.20) but puts a more severe requirement on the time step. For

instance, if $\Delta x = \Delta y$ the time step is reduced by a factor of two, compared with the one-dimensional case:

$$\Delta t \leqslant \frac{\Delta x^2}{4\alpha} \qquad (8.4.12)$$

8.4.2 The two-dimensional convection equation

Consider the system of p equations

$$\frac{\partial U}{\partial t} + A \frac{\partial U}{\partial x} + B \frac{\partial U}{\partial y} = 0 \qquad (8.4.13)$$

where A and B are $(p \times p)$ constant matrices, with the property $AB = BA$. Applying a Lax–Friedrichs scheme to this system leads to

$$U_{ij}^{n+1} = \frac{1}{4} (U_{i,j+1}^n + U_{i+1,j}^n + U_{i-1,j}^n + U_{i,j-1}^n) - \frac{\Delta t}{2\Delta x} A(U_{i+1,j}^n - U_{i-1,j}^n)$$

$$- \frac{\Delta t}{2\Delta y} B(U_{i,j+1}^n - U_{i,j-1}^n) \qquad (8.4.14)$$

With the decomposition (8.4.6) for a single harmonic the amplification matrix becomes

$$G = \frac{1}{2} (\cos \phi_x + \cos \phi_y) - \frac{\Delta t}{\Delta x} A I \sin \phi_x - \frac{\Delta t}{\Delta y} B I \sin \phi_y \qquad (8.4.15)$$

The spectral radius ρ can be obtained from equation (8.2.22) and the fact that G is a normal matrix. Hence with

$$\| G \|_{L_2}^2 = \rho(G^*G) = \rho^2(G)$$

$$\rho(G^*G) = \tfrac{1}{4}(\cos \phi_x + \cos \phi_y)^2 + (\sigma_x \sin \phi_x + \sigma_y \sin \phi_y)^2 \qquad (8.4.16)$$

where

$$\sigma_x = \frac{\Delta t}{\Delta x} \rho(A) \qquad \sigma_y = \frac{\Delta t}{\Delta y} \rho(B) \qquad (8.4.17)$$

A necessary condition is obtained by looking at the most unfavourable situation, namely ϕ_x and ϕ_y independent but small. Expanding the sine and cosine functions up to higher order.

$$\| G \|_{L_2}^2 = 1 - [(\tfrac{1}{2} - \sigma_x^2)\phi_x^2 + (\tfrac{1}{2} - \sigma_y^2)\phi_y^2 - 2\sigma_x\sigma_y\phi_x\phi_y] + 0(\phi_x^4, \phi_y^4) \qquad (8.4.18)$$

The quadratic form in ϕ_x, ϕ_y between parentheses has to be positive for stability. Thus if the discriminant is negative the quadratic form never goes through zero avoiding a change of sign. This will occur if

$$(\sigma_x^2 + \sigma_y^2) \leqslant \tfrac{1}{2} \qquad (8.4.19)$$

representing the interior of a circle in the (σ_x, σ_y) plane of radius $\sqrt{2}/2$, centred at the origin. This condition is also shown to be sufficient in Section 8.6. Here again, this condition is far more severe than the corresponding one-dimensional case.

As can be seen from these examples, it is much more difficult to obtain the stability limits for multi-dimensional problems, even for linear equations, and several non-sufficient stability conditions can be found in the literature. Actually, even for one-dimensional problems, controversial results from Von Neumann analysis have appeared in the literature (see Chapter 10 for a discussion of a famous example concerning the convection–diffusion equation).

8.5 STABILITY CONDITIONS FOR NON-LINEAR PROBLEMS

Most of the mathematical models describing the various approximations to a flow problem contain non-linear terms, or eventually non-constant coefficients. In these cases the Von Neumann method for stability analysis based on the Fourier expansion cannot strictly be applied since we can no longer isolate single harmonics. Nevertheless, if we introduced a complete Fourier series into the discretized scheme with non-constant coefficients the amplification matrix would become a function of all wavenumbers, instead of a linear superposition of amplification matrices for single harmonics. In addition, for non-linear problems the amplification matrix would also become a function of the amplitude of the solutions and not only of their frequency as in the constant-coefficient, linear case. Hence these contributions could generate instabilities, even with schemes which are basically linearly stable.

8.5.1 Non-constant coefficients

Consider a linear problem with non-constant coefficients, for instance, the one-dimensional, parabolic problem

$$\frac{\partial u}{\partial t} = \frac{\partial}{\partial x}\left(\alpha(x)\,\frac{\partial u}{\partial x}\right) \tag{8.5.1}$$

or the hyperbolic problem

$$\frac{\partial u}{\partial t} + a(x)\,\frac{\partial u}{\partial x} = 0 \tag{8.5.2}$$

A two-step numerical scheme applied to these equations will be written as

$$u_i^{n+1} = C(x, E)u_i^n \tag{8.5.3}$$

For instance, for an explicit, central scheme the parabolic equation (8.5.1) becomes

$$u_i^{n+1} = u_i^n + \frac{\Delta t}{\Delta x^2}\,[\alpha_{i+1/2}(u_{i+1}^n - u_i^n) - \alpha_{i-1/2}(u_i^n - u_{i-1}^n)] \tag{8.5.4}$$

where

$$\alpha_{i+1/2} = \alpha(x_{i+1/2}) \qquad (8.5.5)$$

Hence

$$C(x, E) = 1 + \frac{\Delta t}{\Delta x^2} \left[\alpha(x_{i+1/2})(E - 1) - \alpha(x_{i-1/2})(1 - E^{-1}) \right] \qquad (8.5.6)$$

The hyperbolic equation (8.5.2) with an explicit, upwind scheme for $a > 0$ will be written as

$$u_i^{n+1} = u_i^n - \frac{\Delta t}{\Delta x} a(x_{i-1/2})(u_i^n - u_{i-1}^n) \qquad (8.5.7)$$

or

$$C(x, E) = 1 - \frac{\Delta t}{\Delta x} a(x_{i+1/2})(1 - E^{-1}) \qquad (8.5.8)$$

The amplification matrix is now a function of x and not only of the wavenumber k. Indeed, introducing a single harmonic $(u_i^n)_k = v^n e^{lik\,\Delta x}$, a *local* amplification matrix can be defined by

$$G(x, k) = C(x, e^{l\phi}) \qquad (8.5.9)$$

where the variable coefficients are formally retained as functions of x.

In the two examples above we have

$$G(x, \phi) = 1 + \frac{\Delta t}{\Delta x^2} \left[\alpha(x)(e^{l\phi} - 1) - \alpha(x)(1 - e^{-l\phi}) \right] \qquad (8.5.10)$$

and for the hyperbolic example

$$G(x, \phi) = 1 - \frac{\Delta t}{\Delta x} a(x)(1 - e^{-l\phi}) \qquad (8.5.11)$$

Under general conditions (see Richtmyer and Morton, 1967) it can be proved that for linear, non-constant coefficient problems a *local* Von Neumann analysis will provide a necessary condition for stability. That is, freezing the coefficients at their value at a certain point and applying the Von Neumann method provides a local stability condition. However, in order to obtain also sufficient conditions for stability, additional restrictions on the amplification matrix G have to be introduced. These conditions are connected with the generation of high-frequency harmonics due to the non-linear behaviour and to the necessity of damping these frequencies in order to maintain stability. This is particularly urgent for non-linear hyperbolic problems, since they describe essentially propagating waves without physical damping. Even with parabolic problems, where such a physical damping exists an additional condition on the amplification matrix is required. This is provided by the concept of *dissipative schemes*.

8.5.2 Dissipative schemes (Kreiss, 1964)

A scheme is called dissipative (in the sense of Kreiss) of order $2r$, where r is a positive integer, if there exists a constant $\delta > 0$ such that for wavenumbers $\vec{x}$ with $\phi_j = (x_j \Delta x_j) \leqslant \pi$ for each space component j ($j = 1, 2, 3$ in a three-dimensional space) the eigenvalues λ of the amplification matrix satisfy the condition

$$| \lambda(\vec{x}, \Delta t, \vec{x}) | \leqslant 1 - \delta | \vec{x} \cdot \Delta \vec{x} |^{2r} \qquad (8.5.12)$$

for all $\vec{x}$ and for $0 < \Delta t < \tau$. This condition ensures that for $\phi = \pi$, that is, for the high frequencies associated with the $(2\Delta x_j)$ waves (the shortest waves to be resolved on the mesh), enough dissipation is provided by the discretization to avoid their negative impact on the stability.

For parabolic problems we can show, under fairly general conditions (Richtmyer and Morton, 1967), that if a $\delta > 0$ exists such that

$$| G(x, \phi) | \leqslant 1 - \delta\phi^2 \qquad \text{for } -\pi \leqslant \phi \leqslant \pi \qquad (8.5.13)$$

the corresponding schemes are stable. In particular, a scheme with an amplification matrix, such that the spectral radius $\rho(G) = 1$ for $\phi = \pi$, is not dissipative in the sense of Kreiss.

For hyperbolic problems we have the following theorem of Kreiss (1964): If the matrix A is Hermitian, uniformly bounded and Lipshitz continuous in x, then if the scheme is dissipative of order $2r$ and accurate of order $(2r - 1)$, it is stable.

This is a sufficient condition for stability, but many schemes applied in practice do not satisfy this condition.

Lax–Friedrichs scheme

From the amplification factor of the Lax scheme (equation (E8.3.2))

$$G(\phi) = \cos \phi - I\sigma \sin \phi$$

we deduce that $G(\pi) = 1$ for all σ. Hence the Lax scheme is not dissipative in Kreiss's sense. However, since this scheme damps strongly all frequencies, as seen earlier, it remains generally stable even for non-linear problems such as the Euler equations (Di Perna, 1983).

Upwind scheme

According to equation (E8.3.6) the modulus of the amplification factor becomes, for small values of ϕ,

$$| G | \approx 1 - \sigma(1 - \sigma)\phi^2 + \cdots \qquad (8.5.14)$$

and is dissipative of order 2 for $0 < \sigma < 1$. Since

$$| G(\pi) | = | 1 - 2\sigma | \qquad (8.5.15)$$

the upwind scheme is dissipative in the sense of Kreiss. The order of the scheme being one, the conditions of Kreiss's theorem are satisfied and the upwind scheme will be stable for functions $a(x)$ such that

$$0 < \frac{a(x)\Delta t}{\Delta x} < 1$$

for all values of x in the computational domain.

Lax–Wendroff scheme

The dissipation of the scheme is of fourth order, since for small ϕ, from equation (E8.3.13)

$$|G| \approx 1 - \frac{\sigma^2}{8}(1 - \sigma^2)\phi^4 + 0(\phi^6) \qquad (8.5.16)$$

showing that the Lax–Wendroff scheme is dissipative to the fourth order. Since $G(\pi) = 1 - 2\sigma^2$ the Lax–Wendroff scheme is dissipative in the sense of Kreiss for non-zero values of σ.

8.5.3 Non-linear problems

Very little information is available on the stability of general non-linear discretized schemes. Within the framework of the Von Neumann method it can be said that the stability of the linearized equations, with frozen coefficients, is necessary for the stability of the non-linear form but that it is certainly not sufficient. Products of the form $u(\partial u/\partial x)$ will generate high-frequency waves which, through a combination of the Fourier modes on a finite mesh, will reappear as low-frequency waves and could deteriorate the solutions. Indeed, a discretization of the form

$$\left(u\frac{\partial u}{\partial x}\right)_i = u_i\left(\frac{u_{i+1} - u_{i-1}}{2\Delta x}\right) \qquad (8.5.17)$$

becomes, when the Fourier expansion (8.2.1) is introduced,

$$\left(u\frac{\partial u}{\partial x}\right)_i = \sum_{k_1}\left(\sum_{k_2} v(k_2)e^{Ik_2 i\,\Delta x}\right)v(k_1)\frac{e^{Ik_1 i\,\Delta x}}{2\Delta x}(e^{Ik_1\,\Delta x} - e^{-Ik_1\,\Delta x})$$

$$\qquad (8.5.18)$$

$$= \frac{I}{\Delta x}\sum_{k_1}\sum_{k_2} v(k_1)v(k_2)\, e^{I(k_1 + k_2)\Delta x}\sin k_1\Delta x$$

The sum $(k_1 + k_2)\Delta x$ can become larger than the maximum value π associated with the $(2\Delta x)$ wavelength. In this case the corresponding harmonic will behave as a frequency $[2\pi - (k_1 + k_2)\Delta x]$ and will therefore appear as a low-frequency contribution. This non-linear phenomenon is called *aliasing*, and is to be avoided by providing enough disipation in the scheme to damp the high frequencies.

For non-linear problems we also observe that the coefficient of a single harmonic k_1 is a function of the amplitude of the signal through the factor $v(k_2)$ in the above development of the non-linear term uu_x. Hence for small amplitudes the non-linear version of a linearly stable scheme could remain stable, while an unstable behaviour could appear for larger amplitudes of the solution. In this case the scheme could be generally stabilized by adding additional dissipation to the scheme without affecting the order of accuracy.

A typical example is the leapfrog scheme, which is neutrally stable, $|G| = 1$ for all $|\sigma| < 1$. Hence this scheme is not dissipative in the sense of Kreiss, and when applied to the inviscid Burger's equation $u_t + uu_x = 0$, the computations become unstable in certain circumstances, as can be seen from Figure 8.5.1. This figure shows the computed solutions of Burger's equation for a stationary shock, after 10, 20 and 30 time steps at a Courant number of 0.8 and a mesh size of $\Delta x = 0.05$. The open squares indicate the exact solution. The amplitude of the errors increases continuously, and the solution is completely destroyed after 50 time steps. The instability is entirely due to the non-linearity of the equation, since the same scheme applied to the linear convection equation does not diverge, although strong oscillations are generated, as shown in Figure 8.3.6(d).

In the present case the high-frequency errors are generated by the fact that the shock is located on a mesh point. This point has zero velocity and, with an initial solution passing through this point, a computed shock structure is enforced with this internal point fixed, creating high-frequency errors at the two adjacent points. This is clearly seen in Figure 8.5.1, looking at the evolution of the computed solutions, and also by comparing it with Figure 8.5.2, which displays the results of an identical computation for a stationary shock located between two mesh points. This computation does not become unstable, since the shock structure is not constrained by an internal point. Observe also the propagation of the generated high-frequency errors away

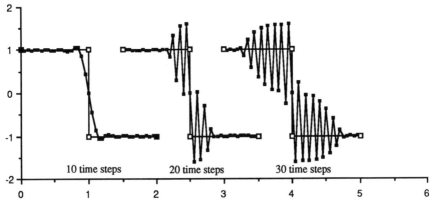

Figure 8.5.1   Solutions of Burger's equation with the leapfrog scheme, after 10, 20 and 30 time steps, for a stationary shock located on a mesh point

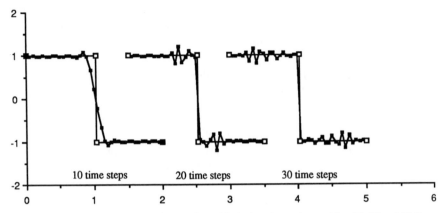

Figure 8.5.2 Solutions of Burger's equation with the leapfrog scheme, after 10, 20 and 30 time steps, for a stationary shock located between mesh points

from the shock position. They propagate at a velocity equal to the numerical group velocity of the scheme, associated with the errors with the shortest $2\Delta x$ wavelength.

Other sources of non-linear instability have been identified for the leapfrog scheme applied to Burger's equation and are described as a 'focusing' mechansim by Briggs *et al.* (1983). The structure of this mechanism has been further investigated by Sloan and Mitchell (1986).

This mechanism is not the classical, finite amplitude instability generated by terms of the form of equation (8.5.18). This instability can be analysed by considering group of modes which are closed under aliasing, that is, modes $k_1, k_2, k_3, \ldots$, such that

$$2\pi(k_1 + k_2)\Delta x = k_3 \, \Delta x \qquad (8.5.19)$$

For instance, referring to definition (8.1.10) of the discrete wavenumber k_j it is seen that the modes $k_1 \, \Delta x = 2\pi/3$, $k_2 \, \Delta x = \pi$ and $k_3 \, \Delta x = \pi/3$ satisfy equation (8.5.19) for all permutations of the three modes.

By investigating solutions which contain a finite number of closed modes, the non-linear contributions from terms of the form (8.5.18) can lead to exponentially growing amplitudes, for Courant numbers below one, when the amplitudes reach a certain critical threshold function of σ. This is the mechanism which generates the instability of Figure 8.5.1.

The 'focusing' mechanism, described by Briggs *et al.*, is of a different nature. It corresponds to an amplification and a concentration of the initial errors at isolated points in the grid. This generates sharp local peaks as a result of the non-linear interaction between the original stable modes and their immediate neighbours in wavenumber space, even for initial amplitudes below the critical threshold for finite amplitude instabilities. Once the critical amplitude is reached locally it starts growing exponentially. The particular

328

character of this focusing property lies in its local aspect, while other non-linear instabilities are global in that the breakdown, for a continuous solution, occurs uniformly throughout the grid.

It has to be added that this focusing process can take a long time, several thousand time steps, depending on the initial error level. Figure 8.5.3, from Briggs *et al.* (1983), illustrates this process for an initial solution composed of three modes ($\pi/3$, $2\pi/3$, and π) with amplitudes below critical such that the computed solution should remain stable. The dashed line indicates the critical level above which finite amplitude instability develops. The computed results are shown for a Courant number of 0.9 and $\Delta x = 1/300$ after 400, 1000, 2000, 2200, 2400 and 2680 time steps. Until 1000 time steps the solution still retains its periodic structure; by 2000 time steps the envelope of the initial profile begins to oscillate, and local amplitudes start to concentrate until the critical threshold is reached at a single point after 2680 time steps. From this stage onwards the classical mechanism takes over and the solution diverges rapidly.

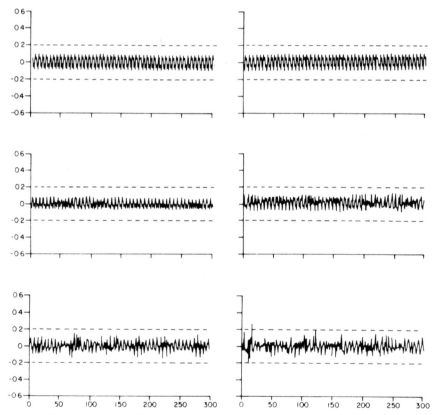

Figure 8.5.3 Solutions of Burger's equation with the leapfrog scheme for a wave solution with three modes, after 400, 1000, 2000, 2200, 2400 and 2680 time steps. (From Briggs *et al.*, 1983)

This mechanism has a strong resemblance to the chaotic behaviour of solutions of non-linear equations and their multiple bifurcations, which are also the basis of descriptions of the generation of turbulence. The reader might refer in this relation to a recent study of McDonough and Bywater (1986) on the chaotic solutions to Burger's equation.

The above examples indicate the degree of complexity involved in the analysis of the stability of non-linear equations and the need for methods which would prevent the development of instabilities for long-term computations. A frequently applied method consists of adding higher-order terms which provide additional dissipation in order to damp the non-linear instabilities without affecting the accuracy. Examples of this approach will be presented in Volume 2, when dealing with the discretization of the Euler equations.

8.6 SOME GENERAL METHODS FOR THE DETERMINATION OF VON NEUMANN STABILITY CONDITIONS

Although simple in principle and in its derivation, the Von Neumann amplification matrix is often very tedious and complicated to analyse in order to obtain the practical stability limits on the parameters of the scheme. If it is straightforward to obtain necessary conditions, it is much more difficult to derive the sufficient conditions for stability. The variety of imprecise conditions found in the literature for relatively simple problems, such as the one-dimensional convection–diffusion equation, testify to these difficulties. The situation is still worse for multi-dimensional problems. For instance, the correct, necessary and sufficient stability limits for the convection–diffusion equation in any number of dimensions had been obtained only recently (Hindmarsh *et al.*, 1984).

Due to the importance of the Von Neumann analysis, we will present a few methods which allow the derivation of precise, necessary and sufficient stability conditions for some limited, but still frequently occurring, discretizations. The first case treats the general two-level, three-point central schemes in one dimension, while the second will present the stability criteria for multi-dimensional, centrally discretized convection–diffusion equations, and, because of its importance, we will reproduce the derivation of Hindmarsh *et al.* (1984). The last case is of a more general nature, and applies to any amplification matrix obtained for an arbitrary system of discretized equation, allowing the reduction of the polynomial (8.2.19) to simpler forms.

8.6.1 One-dimensional, two-level, three-point schemes

We consider here the general scheme

$$b_3 u_{i+1}^{n+1} + b_2 u_i^{n+1} + b_1 u_{i-1}^{n+1} = a_3 u_{i+1}^n + a_2 u_i^n + a_1 u_{i-1}^n$$

$$(8.6.1)$$

where, for consistency, we should have (u_i = constant should be a solution)

$$b_3 + b_2 + b_1 = a_3 + a_2 + a_1 = 1 \qquad (8.6.2)$$

with an arbitrary normalization at one. After elimination of b_2 and a_2, the amplification factor is

$$G = \frac{a_3(e^{I\phi} - 1) + a_1(e^{-I\phi} - 1) + 1}{b_3(e^{I\phi} - 1) + b_1(e^{-I\phi} - 1) + 1}$$

$$= \frac{1 - (a_3 + a_1)(1 - \cos \phi) + I(a_3 - a_1)\sin \phi}{1 - (b_3 + b_1)(1 - \cos \phi) + I(b_3 - b_1)\sin \phi} \qquad (8.6.3)$$

Hence

$$|G|^2 = |GG^*| = \frac{A_1\beta^2 + A_2\beta + 1}{B_1\beta^2 + B_2\beta + 1} \qquad (8.6.4)$$

where

$$\beta = \sin^2\phi/2$$

$$A_1 = 16a_3a_1 \qquad\qquad B_1 = 16b_3b_1 \qquad (8.6.5)$$

$$A_2 = 4[(a_3 - a_1)^2 - (a_3 + a_1)] \qquad B_2 = 4[(b_3 - b_1)^2 - (b_3 + b_1)]$$

Note that the denominator $(B_1\beta^2 + B_2\beta + 1) \geqslant 0$ in the range $0 \leqslant \beta \leqslant 1$, since $(1 + B_1 + B_2) = (1 - 2b_2)^2$ is always non-negative. Hence the condition $|G|^2 \leqslant 1$ leads to

$$(A_1 - B_1)\beta^2 + (A_2 - B_2)\beta \leqslant 0 \qquad (8.6.6)$$

and for all values of $0 \leqslant \beta \leqslant 1$ the necessary and sufficient Von Neumann stability conditions are

$$(A_2 - B_2) \leqslant 0$$
$$(A_1 - B_1 + A_2 - B_2) \leqslant 0 \qquad (8.6.7)$$

Example 8.6.1 Diffusion equation

Considering scheme (8.3.16) we have $b_3 = b_1 = 0$, $a_3 = a_1 = \beta$. Hence

$$B_1 = 0 \qquad\qquad B_2 = 0$$
$$A_1 = 16\beta^2 \qquad A_2 = -8\beta \qquad (E8.6.1)$$

and we obtain the necessary and sufficient conditions

$$B > 0$$
$$8\beta(2\beta - 1) \leqslant 0 \qquad (E8.6.2)$$

leading to the earlier obtained relation

$$0 < \beta \leqslant 1/2$$

8.6.2 Multi-dimensional space-centred, convection–diffusion equation

We consider here a general scalar equation in a space of dimension M of the form

$$\frac{\partial u}{\partial t} + (\vec{a} \cdot \vec{\nabla})u = \vec{\nabla} \cdot (\bar{\bar{\alpha}} \, \vec{\nabla} u) \qquad (8.6.8)$$

where $\bar{\bar{\alpha}}$ is a *diagonal* diffusivity tensor and a central discretization of second-order accuracy:

$$\frac{1}{\Delta t}(u_J^{n+1} - u_J) + \sum_{n=1}^{M} a_m \frac{\bar{\delta}_m u_J}{\Delta x_m} = \sum_{m=1}^{M} \alpha_m \frac{\delta_m^2 u_J}{\Delta x_m^2} \qquad (8.6.9)$$

where J represents a mesh point index (for instance, in two dimensions $J(i, j)$ and $J(i, j, k)$ in a three-dimensional Cartesian mesh). The operator $\bar{\delta}_m$ is the central difference acting on the variable x_m, that is

$$\frac{\bar{\delta}_m u_J}{\Delta x_m} = \frac{1}{2\Delta x_j}(u_{i,j+1,k} - u_{i,j-1,k}) \qquad \text{if } m = j \qquad (8.6.10)$$

and the second derivative operator δ_m^2 is similarly defined:

$$\frac{\delta_m^2 u_J}{\Delta x_m^2} = \frac{1}{\Delta x_j^2}(u_{i,j+1,k} - 2u_{ijk} + u_{i,j-1,k}) \qquad \text{if } m = j \qquad (8.6.11)$$

Defining

$$\begin{aligned} \sigma_m &= a_m \, \Delta t / \Delta x_m \\ \beta_m &= \alpha_m \, \Delta t / \Delta x_m^2 \end{aligned} \qquad (8.6.12)$$

the above scheme becomes

$$u_J^{n+1} = u_J^n - \sum_{m=1}^{M} (\sigma_m \, \bar{\delta}_m u_J - \beta_m \delta_m^2 u_J) \qquad (8.6.13)$$

This discretized equation represents the scheme to be analysed independently of the original equation (8.6.8), used as a starting point. Hence the following results can be applied to a wider range of problems; for instance, the two-dimensional convection equation, discretized with the Lax–Friedrichs scheme (8.4.14), can clearly be written in the above form. As will be seen in Volume 2, many numerical schemes for the inviscid system of Euler equations can also be written in this way.

With representation (8.4.1), and $\phi_m = \varkappa_m \, \Delta x_m$, the amplification factor becomes

$$G = 1 - I \sum_{m=1}^{M} \sigma_m \sin \phi_m - 4 \sum_{m=1}^{M} \beta_m \sin^2 \phi_m / 2 \qquad (8.6.14)$$

The modulus squared is given by

$$|G|^2 = \left[1 - 4 \sum_{m=1}^{M} \beta_m \sin^2 \phi_m / 2 \right]^2 + \left[\sum_{m=1}^{M} \sigma_m \sin \phi_m \right]^2 \qquad (8.6.15)$$

The extreme conditions are obtained when all $\phi_m = \pi$, on the one hand, and when all ϕ_m go to zero on the other. In the first case we obtain the condition

$$\left(1 - 4 \sum_{m=1}^{M} \beta_m\right)^2 \leqslant 1 \tag{8.6.16}$$

Hence this leads to

$$0 \leqslant \sum_{m=1}^{M} \beta_m \leqslant \frac{1}{2} \tag{8.6.17}$$

In the second case, performing a Taylor expansion around $\phi_m = 0$, and neglecting higher-order terms, we obtain

$$|G|^2 = \left[1 - \sum_{m=1}^{M} \beta_m \phi_m^2\right]^2 + \left[\sum_{m=1}^{M} \sigma_m \phi_m\right]^2 + 0(\phi_m^4)$$
$$= 1 - 2 \sum_{m=1}^{M} \beta_m \phi_m^2 + \left(\sum_{m=1}^{M} \sigma_m \phi_m\right)^2 + 0(\phi_m^4) \tag{8.6.18}$$

The right-hand side is a quadratic expression in the ϕ_m. Following Hindmarsh *et al.* (1984), it can be written as follows, introducing the vectors $\vec{\phi} = (\phi_1, ..., \phi_M)^T$, $\vec{\sigma} = (\sigma_1, ..., \sigma_M)^T$ and the diagonal matrix $\beta = \text{diag}(\beta_1, ..., \beta_M)$, neglecting higher-order terms:

$$|G|^2 = 1 - \vec{\phi}^T(2\bar{\bar{\beta}} - \vec{\sigma} \otimes \vec{\sigma}^T)\vec{\phi} \tag{8.6.19}$$

For $|G|^2$ to be lower than one, the symmetric matrix $(2\bar{\bar{\beta}} - \vec{\sigma}x\vec{\sigma}^T)$ must be non-negative definite. In particular, the diagonal elements $(2\beta_m - \sigma_m^2)$ must be non-negative, implying $\sigma_m^2 < 2\beta_m$. If one of the β_m is zero, then σ_m (or a_m) is also zero and the corresponding mth dimension can be dropped from the problem. Therefore we can assume all $\beta_m > 0$ and the equality sign on the lower bound of equation (8.6.17) has to be removed.

Defining the diagonal matrix $\bar{\bar{\gamma}}$ by $\bar{\bar{\gamma}} = \text{diag}((2\beta_1)^{1/2}, ..., (2\beta_M)^{1/2})$, we have

$$2\bar{\bar{\beta}} - \vec{\sigma} \otimes \vec{\sigma}^T = \bar{\bar{\gamma}}(\bar{\bar{I}} - \bar{\bar{\gamma}}^{-1}\vec{\sigma} \otimes \vec{\sigma}^T\bar{\bar{\gamma}}^{-1})\bar{\bar{\gamma}} \tag{8.6.20}$$

and the matrix

$$\bar{\bar{A}} \equiv \bar{\bar{I}} - (\bar{\bar{\gamma}}^{-1}\vec{\sigma}) \otimes (\bar{\bar{\gamma}}^{-1}\vec{\sigma})^T \equiv \bar{\bar{I}} - \vec{d} \otimes \vec{d}^T \tag{8.6.21}$$

should also be non-negative definite. Considering the associated quadratic form, for any M-dimensional vector $\vec{x}$,

$$\vec{x}^T\bar{\bar{A}}\,\vec{x} = \vec{x}^T \cdot \vec{x} - (\vec{d}^T \cdot \vec{x})^2 \tag{8.6.22}$$

the matrix $\bar{\bar{A}}$ is non-negative definite if and only if

$$\vec{d}^T \cdot \vec{d} = \sum_{m=1}^{M} \frac{\sigma_m^2}{2\beta_m} \leqslant 1 \tag{8.6.23}$$

The Von Neumann stability conditions are therefore

$$0 < \sum_{m=1}^{M} \beta_m \leqslant \frac{1}{2} \tag{8.6.24a}$$

and

$$\sum_{m=1}^{M} \frac{\sigma_m^2}{\beta_m} \leqslant 2 \tag{8.6.24b}$$

Assuming all α_m positive, we can easily prove that these conditions are sufficient from the Schwartz inequality applied to the sum

$$\left[\sum_{m=1}^{M} \sigma_m \sin \phi_m \right]^2 \leqslant \left[\sum_{m=1}^{M} \left(\frac{|\sigma_m|}{\sqrt{\beta_m}} \right) (\sqrt{(\beta_m)} \, |\sin \phi_m|) \right]^2$$

$$\leqslant \sum_{m=1}^{M} \frac{\sigma_m^2}{\beta_m} \cdot \sum_{m=1}^{M} \beta_m \sin^2 \phi_m$$

$$< \sum_{m=1}^{M} 2\beta_m \sin^2 \phi_m \tag{8.6.25}$$

where the second condition (8.6.24b) has been applied.

If any $\beta_m = 0$ the above condition implies that $\sigma_m = 0$ and the sum (8.6.25) is obtained by summing first only over those m for which $\beta_m > 0$. Inserting this relation into the expression of $|G|^2$, we obtain

$$|G|^2 \leqslant \left[1 - 4 \sum_{m=1}^{M} \beta_m \sin^2 \phi_m/2 \right]^2 + 8 \sum_{m=1}^{M} \beta_m \sin^2 \phi_m/2 \cdot \cos^2 \phi_m/2$$

$$\tag{8.6.26}$$

$$= 1 - 8 \sum_{m=1}^{M} \beta_m \sin^4 \phi_m/2 + \left[4 \sum_{m=1}^{M} \beta_m \sin^2 \phi_m/2 \right]^2$$

Applying the first stability condition (8.6.24a) with the Schwartz inequality on the last term we obtain

$$|G|^2 \leqslant 1 - 8 \sum_{m=1}^{M} \beta_m \sin^4 \phi_m/2 + 16 \sum_{m=1}^{M} \beta_m \sin^4 \phi_m/2 \cdot \sum_{m=1}^{M} \beta_m$$

$$< 1 - 8 \sum_{\mu=1}^{M} \beta_m \sin^4 \phi_m/2 + 8 \sum_{m=1}^{M} \beta_m \sin^4 \phi_m/2 \tag{8.6.27}$$

$$< 1$$

This completes the proof that conditions (8.6.24) are necessary and sufficient for the strict Von Neumann stability of scheme (8.6.13).

Example 8.6.2   The two-dimensional Lax–Friedrichs scheme (8.4.14)

Writing (8.4.14) in the above form we have

$$\sigma_1 = \frac{\Delta t}{\Delta x} \lambda_{\max}^{(A)} \qquad \sigma_2 = \frac{\Delta t}{\Delta y} \lambda_{\max}^{(B)} \tag{E8.6.3}$$

and

$$\beta_1 = \beta_2 = \tfrac{1}{4} \qquad \text{(E8.6.4)}$$

leading to condition (8.4.19).

Example 8.6.3   Two-dimensional convection–diffusion equation

Consider the energy equation

$$\frac{\partial T}{\partial t} + u \frac{\partial T}{\partial x} + v \frac{\partial T}{\partial y} = \alpha \left(\frac{\partial^2 T}{\partial x^2} + \frac{\partial^2 T}{\partial y^2} \right) \qquad \text{(E8.6.5)}$$

discretized with central differences and an Euler explicit time integration,

$$T_{ij}^{n+1} - T_{ij}^n = -\frac{\sigma_1}{2}\,(T_{i+1,j}^n - T_{i-1,j}^n) - \frac{\sigma_2}{2}\,(T_{i,j+1}^n - T_{i,j-1}^n)$$

$$+ \beta_1 (T_{i+1,j} - 2T_{ij} + T_{i-1,j}) + \beta_2 (T_{i,j+1} - 2T_{ij} + T_{i,j-1})$$

$$\text{(E8.6.6)}$$

where

$$\sigma_1 = \frac{u\,\Delta t}{\Delta x} \qquad \sigma_2 = \frac{v\,\Delta t}{\Delta y}$$

$$\text{(E8.6.7)}$$

$$\beta_1 = \frac{\alpha\,\Delta t}{\Delta x^2} \qquad \beta_2 = \frac{\alpha\,\Delta t}{\Delta y^2}$$

The necessary and sufficient stability conditions are as follows:

$$(\beta_1 + \beta_2) = \alpha\,\Delta t \left(\frac{1}{\Delta x^2} + \frac{1}{\Delta y^2} \right) \leqslant \frac{1}{2} \qquad \text{(E8.6.8a)}$$

$$\frac{\sigma_1^2}{\beta_1} + \frac{\sigma_2^2}{\beta_2} = \frac{\Delta t}{\alpha}\,(u^2 + v^2) \leqslant 2 \qquad \text{(E8.6.8b)}$$

Hence the maximum allowable time step is given by

$$\Delta t \leqslant \text{Min}\left(\frac{1}{2\alpha}\,\frac{\Delta x^2\,\Delta y^2}{\Delta x^2 + \Delta y^2}, \frac{2\alpha}{q^2} \right) \qquad \text{(E8.6.9)}$$

where $q^2 = u^2 + v^2$ is the square of the velocity $\vec{v}(u, v)$. Observe that the second condition (E8.6.8b) is independent of the mesh sizes Δx, Δy.

Additional remarks: If all β_m are equal, we have the necessary and sufficient condition

$$\sum_{m=1}^{M} \sigma_m^2 \leqslant 2\beta \leqslant \frac{1}{M} \qquad (\beta_m = \beta) \qquad \text{(8.6.28)}$$

Otherwise for $\beta = \text{Max}_m \beta_m$ the above condition is sufficient.

Introducing the mesh Reynolds, or Peclet, numbers $R_m = (a_m\,\Delta x_m/\alpha_m)$ the stability condition (8.6.24b) can be written as

$$0 < \sum_{m=1}^{M} \sigma_m R_m \leqslant 2 \qquad (8.6.29)$$

or as

$$\sum_{m=1}^{M} \beta_m R_m^2 \leqslant 2 \qquad (8.6.30)$$

When all R_m are lower than two ($R_m < 2$) it is seen that condition (8.6.24a) is more restrictive. The second condition (8.6.24b) will be the more restrictive one when all the R_m are larger than two. Otherwise both conditions have to be satisfied, that is,

$$\Delta t \leqslant \text{Min}\left(\frac{1}{2\Sigma_m (\alpha_m/\Delta x_m^2)}, \frac{2}{\Sigma_m (a_m^2/\alpha_m)}\right) \qquad (8.6.31)$$

8.6.3 General multi-level, multi-dimensional schemes

In the general case (discussed in Section 8.2) the strict Von Neumann stability condition is expressed by requirements on the eigenvalues of the matrix G, obtained as a solution of $\det | G - \lambda I | = 0$. These eigenvalues are the zeros of the polynomial of degree p, when G is a $p \times p$ matrix,

$$P(\lambda) = \det | G - \lambda I | = 0 \qquad (8.6.32)$$

The stability condition (8.2.24) requires that all the eigenvalues should be lower than or equal to one, and the eigenvalues $\lambda_j = 1$ should be single. Hence this condition has to be satisfied by the zeros of the polynomial $P(\lambda)$. A polynomial satisfying this condition is called a *Von Neumann polynomial*.

The following remarkable theorem, based on the Schur theory of the zeros of a polynomial, can be found in Miller (1971). Let $\bar{P}(\lambda)$ be the associated polynomial of

$$P(\lambda) = \sum_{k=0}^{p} a_k \lambda^k \qquad (8.6.33)$$

$$\bar{P}(\lambda) = \sum_{k=0}^{p} a_{p-k}^* \lambda^k \qquad (8.6.34)$$

where a_k^* is the complex conjugate of a_k, and define a reduced polynomial of degree not higher than $(p - 1)$:

$$P_1(\lambda) = \frac{1}{\lambda} [\bar{P}(0)P(\lambda) - P(0)\bar{P}(\lambda)] \qquad (8.6.35)$$

Then the zeros of $P(\lambda)$ satisfy the stability conditions ($P(\lambda)$ is a Von Neumann polynomial) if and only if

(1) $| \bar{P}(0) | > | P(0) |$ and $P_1(\lambda)$ is a Von Neumann polynomial; or
(2) $P_1(\lambda) \equiv 0$ and the zeros of $dP/d\lambda = 0$ are such that $| \lambda | \leqslant 1$.

Hence applying this theorem reduces the analysis to the investigation of the properties of a polynomial of a lower degree (at least $p - 1$). Repeating the application of this theorem to P_1 or to $dP/d\lambda$, the degree of the resulting polynomials is further reduced until a polynomial of degree one, which can be more easily analysed, is obtained.

Examples of applications of this technique to various schemes for the convection–diffusion equation can be found in Chan (1984), where some stability conditions for higher-order schemes are obtained for the first time.

Example 8.6.4   Leapfrog scheme applied to the convection–diffusion equation

The equation

$$\frac{\partial u}{\partial t} + a\frac{\partial u}{\partial x} = \alpha\frac{\partial^2 u}{\partial x^2} \qquad (E8.6.10)$$

is discretized with central differences in space and time, leading to a leapfrog scheme with the diffusion terms discretized at level $(n - 1)$:

$$u_i^{n+1} - u_i^{n-1} = -\sigma(u_{i+1}^n - u_{i-1}^n) + 2\beta(u_{i+1}^{n-1} - 2u_i^{n-1} + u_{i-1}^{n-1}) \qquad (E8.6.11)$$

The amplification factors or eigenvalues are solutions of the second-order polynomial (applying the method of Section 8.3.3)

$$P(\lambda) \equiv \lambda^2 + 2\lambda\sigma I \sin\phi - 1 - 4\beta(\cos\phi - 1) = 0 \qquad (E8.6.12)$$

We obtain

$$\bar{P}(\lambda) = 1 - 2\lambda\sigma I \sin\phi - [1 + 4\beta(\cos\phi - 1)]\lambda^2 \qquad (E8.6.13)$$

The condition $|\bar{P}(0)| > |P(0)|$ leads to

$$|1 - 4\beta(1 - \cos\phi)| < 1$$

or

$$4\beta < 1 \qquad (E8.6.14)$$

which is a necessary condition for stability.
Constructing $P_1(\lambda)$, we obtain, with $\gamma = 1 - 4\beta(1 - \cos\phi)$,

$$P_1(\lambda) = \lambda(1 - \gamma^2) + 2\sigma I(1 - \gamma)\sin\phi \qquad (E8.6.15)$$

and the stability condition becomes $|\bar{P}_1(0)| > |P_1(0)|$, or

$$|\sigma \sin\phi| \leqslant 1 - 2\beta(1 - \cos\phi) \qquad (E8.6.16)$$

Following Chan (1984), this leads to the necessary and sufficient condition, for all ϕ,

$$\sigma^2 + 4\beta \leqslant 1 \qquad (E8.6.17)$$

Summary

The Von Neumann stability method, based on a Fourier analysis in the space domain, has been developed for linear, one- and multi-dimensional problems. This method is the most widely applied technique and the amplification factor is easily obtained. Although the stability conditions cannot always be derived analytically, we could, if necessary, analyse the properties of the amplification matrix numerically. These properties also contain information on the dispersion and diffusion errors of a numerical scheme, allowing the selection of a scheme as a function of the desired properties. For non-linear problems it has been shown that a local, linearized stability analysis will lead to necessary conditions.

References

Book, D. L. (Ed.) (1981). *Finite Difference Techniques for Vectorized Fluid Dynamics Calculations*, New York: Springer Verlag.

Briggs, W. L., Newell, A. C., and Sarie, T. (1983). 'Focusing: a mechanism for instability of nonlinear finite difference equations.' *Journal of Computational Physics*, **51**, 83–106.

Cathers, B., and O'Connor, B. A. (1985). 'The group velocity of some numerical schemes.' *Int. Journal for Numerical Methods in Fluids*, **5**, 201–24

Chan, T. F. (1984). 'Stability analysis of finite difference schemes for the advection–diffusion equation.' *SIAM Journal of Numerical Analysis*, **21**, 272–83.

Charney, J. G., Fjortoft, R., and Von Neumann, J. (1950). 'Numerical integration of the barotropic vorticity equation.' *Tellus*, **2**, 237–54.

Courant, R., Friedrichs, K. O., and Lewy, H. (1928). 'Uber die partiellen differenzgleichungen der mathematischen Physik.' *Mathematische Annalen*, **100**, 32–74. English translation in *IBM Journal* (1967), 215–34.

Courant, R., and Hilbert, D. (1962). *Methods of Mathematical Physics*, Vols I and II, New York: John Wiley Interscience.

Cranck, J., and Nicholson, P. (1947). 'A practical method for numerical evaluation of solutions of partial differential equations of the heat conduction type.' *Proceedings of the Cambridge Philosophical Society*, **43**, 50–67.

Du Fort, E. C., and Frankel, S. P. (1953). 'Stability conditions in the numerical treatment of parabolic equations.' *Math. Tables and Other Aids to Computation*, **7**, 135–52.

Hindmarsh, A. C., Gresho, P. M., and Griffiths, D. F. (1984). 'The stability of explicit Euler time integration for certain finite difference approximations of the multidimensional advection-diffusion equation.' *Int. Journal for Numerical Methods in Fluids*, **4**, 853–97.

Hirt, C. W. (1968). 'Heuristic stability theory for finite difference equations.' *Journal of Computational Physics*, **2**, 339–55.

Kreiss, H. O. (1964). 'On difference approximations of the dissipative type for hyperbolic differential equations.' *Comm. Pure and Applied Mathematics*, **17**, 335–53.

Lambert, J. D. (1973). *Computational Methods in Ordinary Differential Equations*, Chichester: John Wiley.

Lax, P. D. (1954). 'Weak solutions of nonlinear hyperbolic equations and their numerical computation.' *Comm. Pure and Applied Mathematics*, **7**, 159–93.

Lax, P. D., and Wendroff, B. (1960). 'Systems of conservation laws.' *Comm. Pure and Applied Mathematics*, **13**, 217–37.

McAveney, B. J., and Leslie, L. M. (1972). 'Comments on a direct solution of Poisson's equation by generalized sweep-out method.' *Journal Meteor, Society of Japan*, **50**, 136.

McDonough, J. M., and Bywater, R. J. (1986). 'Large-scale effects on local small-scale chaotic solutions to Burger's equation.' *AIAA Journal*, **24**, 1924–30.

Miller, J. J. H. (1971). 'On the location of zeros of certain classes of polynomials with applications to numerical analysis.' *Journal Inst. Math. Applic.*, **8**, 397–406.

Peyret, R., and Taylor, T. D. (1983). *Computational Methods for Fluid Flow*, New York: Springer Verlag.

Richtmyer, R. D., and Morton, K. W. (1967). *Difference Methods for Initial Value Problems*, 2nd edn, Chichester: John Wiley/Interscience.

Roache, P. J. (1971). 'A new direct method for the discretized Poisson equation.' *Second Int. Conf. on Num. Methods in Fluid Dynamics*, New York: Springer Verlag.

Sloan, D. M., and Mitchell, A. R. (1986). 'On nonlinear instabilities in leap-frog finite difference schemes.' *Journal of Computational Physics*, **67**, 372–95.

Trefethen, L. N. (1982). 'Group velocity in finite difference schemes.' *SIAM Review*, **24**, 113–36

Trefethen, L. N. (1983). 'Group velocity interpretation of the stability theory of Gustafsson, Kreiss and Sundstrom.' *Journal of Computational Physics*, **49**, 199–217.

Trefethen, L. N. (1984). 'Instability of difference models for hyperbolic initial boundary value problems.' *Comm. Pure and Applied Mathematics*, **37**, 329–67.

Varga, R. S. (1962). *Matrix Iterative Analysis*, Englewood Cliffs, NJ: Prentice-Hall.

Vichnevetsky, R., and Bowles, J. B (1982). *Fourier Analysis of Numerical Approximations of Hyberbolic Equations*, Philadelphia: SIAM Publications.

PROBLEMS

Problem 8.1

Derive the succession of operators for the various examples of Table 8.1.

Problem 8.2

Apply a forward space differencing with a forward time difference (Euler method) to the convective equation $u_t + au_x = 0$. Analyse the stability with the Von Neumann method and show that the scheme is unconditionally unstable for $a > 0$ and conditionally stable for $a < 0$. Derive also the equivalent differential equation and show why this scheme is unstable when $a > 0$.

Problem 8.3

Show that the forward/backward scheme for the second-order wave equation (E8.2.12) is

$$w_i^{n+1} - 2w_i^n + w_i^{n-1} = \left(\frac{a\,\Delta t}{\Delta x}\right)^2 (w_{i+1}^n - 2w_i^n + w_{i-1}^n)$$

referring to Table 8.1. Obtain the explicit form of the operators and matrices S, C and G.

Problem 8.4

Consider the same space discretization of the second-order wave equation as in the previous problem but apply a full Euler scheme (forward in time):

$$v_i^{n+1} - v_i^n = \sigma(w_{i+1}^n - w_i^n)$$
$$w_i^{n+1} - w_i^n = \sigma(v_i^n - v_{i-1}^n)$$

Calculate for this scheme the operators and corresponding matrices C and G.

Problem 8.5

Solve the one-dimensional heat conduction equation $u_t = \alpha u_{xx}$ for the following conditions, with k an integer:

$$u(x,0) = \sin k\pi x \qquad 0 \leq x \leq 1$$
$$u(0,t) = 0$$
$$u(1,t) = 0$$

applying the explicit central scheme (8.3.16). Compare with the exact solution for different values of β, in particular $\beta = 1/3$ and $\beta = 1/6$ (which is the optimal value). Consider initial functions with different wavenumbers k, namely $k = 1, 5, 10$.
   The exact solution is $u = e^{-\alpha k^2 \pi^2 t} \sin k\pi x$. Compute with $x_i = i\Delta x$ and i ranging from 0 to 30. Make plots of the computed solution as a function of x and of the exact solution. Perform the calculations for five and ten time steps and control the error by comparing with equation (8.3.25) for ε_D in the case of $\beta = 1/3$. Calculate the higher-order terms in ε_D for $\beta = 1/6$ by taking more terms in the expansion.

Problem 8.6

Calculate the amplitude and phase errors for the Lax–Friedrichs scheme (E8.3.1) after ten time steps for an initial wave of the form

$$u(x,0) = \sin k\pi x \qquad 0 \leq x \leq 1$$

for $k = 1, 10$. Consider $\Delta x = 0.02$ and a velocity $a = 1$. Perform the calculations for $\sigma = 0.25$ and $\sigma = 0.75$. Plot the computed and exact solutions for these various cases and compare and comment on the results.
Hint: The exact solution is $\tilde{u} = \sin \pi k(x - t)$. The exact numerical solution is $\bar{u}_i^n = |G|^n \sin \pi k(x_i - \bar{a}n \Delta t)$ where $\bar{a}$ is the numerical speed of propagation and is equal to a ε_ϕ (equation (E8.3.5)). Show that we can write $\bar{u}_i^n = |G|^n \sin[\pi k(x_i - n\Delta t) + n(\bar{\phi} - \phi)]$.

Problem 8.7

Repeat Problem 8.6 with the upwind scheme (7.2.8).

Problem 8.8

Repeat Problem 8.6 with the leapfrog scheme (8.3.33).

Problem 8.9

Apply the central difference in time (leapfrog scheme) to the heat-conduction equation with the space differences of second-order accuracy:

$$u_i^{n+1} - u_i^{n-1} = 2\frac{\alpha\Delta t}{\Delta x^2}(u_{i+1}^n - 2u_i^n + u_{i-1}^n)$$

Calculate the amplification matrix and show that the scheme is unconditionally unstable.

Problem 8.10

Analyse the leapfrog scheme with the upwind space discretization of the convection equation $u_t + au_x = 0$. This is the scheme

$$u_i^{n+1} - u_i^{n-1} = -2\sigma(u_i^n - u_{i-1}^n)$$

Calculate the amplification matrix and show that the scheme is unstable.

Problem 8.11

Consider the implicit upwind scheme (7.2.9) and analyse its stability. Show that the scheme is unconditionally stable for $a > 0$ and unstable for $a < 0$.

Problem 8.12

Write a program to solve the linear convection equation and obtain Figure 8.3.6.

Problem 8.13

Write a program to solve the linear convection equation and obtain Figures 8.3.7 and 8.3.8 for the wave packet problem. Compare with similar calculations at $CFL = 0.2$

Problem 8.14

Apply an upwind (backward in space) discretization to the two-dimensional convection equation

$$u_t + au_x + bu_y = 0$$

with an explicit Euler time integration. Perform a Von Neumann stability analysis and show that we obtain the CFL conditions in both directions.

Problem 8.15

Apply the Dufort–Frankel scheme to the leapfrog convection-diffusion equation

$$u_i^{n+1} - u_i^{n-1} = -\sigma(u_{i+1}^n - u_{i-1}^n) + 2\beta(u_{i+1}^n - u_i^{n+1} - u_i^{n-1} + u_{i-1}^n)$$

and show, by application of Section 8.6.3, that the stability condition is $|\sigma| < 1$, independently of the diffusion related coefficient β.

Hint: Write the scheme as

$$u_i^{n+1} - u_i^{n-1} = -\sigma(u_{i+1}^n - u_{i-1}^n) + 2\beta(u_{i+1}^n - 2u_i^n + u_{i-1}^n)$$
$$- 2\beta(u_i^{n+1} - 2u_i^n + u_i^{n-1})$$

Problem 8.16

Write a program to solve Burger's equation for the stationary shock and obtain the results of Figures 8.5.1 and 8.5.2.

Problem 8.17

Obtain the results of Example 8.3.3 for the Lax-Wendroff scheme.

Problem 8.18

Obtain the results of Example 8.3.4 for the leapfrog scheme and derive the expansion of the dispersion error in powers of the phase angle.

Problem 8.19

Obtain the relations of Example 8.3.5.

Problem 8.20

Apply a central time integration (leapfrog) method to the finite element scheme of Problem 5.14, considering the linear equation $f = au$. The scheme is

$$(u_{i-1}^{n+1} - u_{i-1}^{n-1}) + 4(u_i^{n+1} - u_i^{n-1}) + (u_{i+1}^{n+1} - u_{i+1}^{n-1}) + 6\sigma(u_{i+1}^n - u_{i-1}^n) = 0$$

Determine the amplification factor and obtain the stability condition $\sigma \leqslant 1/\sqrt{3}$. Determine the dispersion and diffusion errors and obtain the numerical group velocity. Compare with the leapfrog scheme (8.3.33).
Hint: The amplification factor is

$$G = -Ib \pm \sqrt{1 - b^2} \qquad b = \frac{3\sigma \sin \phi}{2 + \cos \phi}$$

Problem 8.21

Apply a generalized trapezium formula in time as defined by Problem 7.5, to the finite element discretization of Problem 5.14. Obtain the scheme

$$(u_{i-1}^{n+1} - u_{i-1}^n) + 4(u_i^{n+1} - u_i^n) + (u_{i+1}^{n+1} - u_{i+1}^n) + 3\theta\sigma(u_{i+1}^{n+1} - u_{i-1}^{n+1})$$
$$+ 3(1 - \theta)\sigma(u_{i+1}^n - u_{i-1}^n) = 0$$

and derive the amplification factor

$$G = \frac{2 + \cos \phi - 3(1 - \theta)I\sigma \sin \theta}{2 + \cos \phi + 3\theta I\sigma \sin \phi}$$

Show that the scheme is unconditionally stable for $\theta \geqslant 1/2$.

Show that for $\theta = 1/2$ (the trapezium formula), there is no dissipation error, and obtain the numerical dispersion relation as $\tan(\omega\Delta t/2) = 3\sigma \sin \phi/[2(2 + \cos \phi)]$.

Problem 8.22

Find the numerical group velocities for the upwind, Lax-Friedrichs and Lax-Wendroff schemes for the linear convection equation, applying the relation (8.3.41) to the real part of the numerical dispersion relation. Plot the ratios v_G/a in function of ϕ and observe the deviations from the exact value of 1.
Hint: Obtain

First order upwind scheme $\quad v_G = a[(1 - \sigma)\cos \phi + \sigma]/[(1 + \sigma(\cos \phi - 1))^2 + \sigma^2 \sin^2 \phi]$

Lax-Friedrichs scheme $\quad v_G = a/(\cos^2 \phi + \sigma^2 \sin^2 \phi)$

Lax-Wendroff scheme $\quad v_G = a[(1 - 2\sigma^2 \sin^2 \varphi/2)\cos \varphi + \sigma^2 \sin^2 \varphi]/$
$$[(1 - 2\sigma^2 \sin^2 \varphi/2)^2 + \sigma^2 \sin^2 \varphi]$$

Chapter 9

The Method of the Equivalent Differential Equation for the Analysis of Stability

The concept of the equivalent or modified, differential equation of a numerical scheme has been introduced in Chapter 7 in relation with the definition of consistency. The equivalent differential equation is obtained from the discretized equations by introducing Taylor series developments for all function values u_j^m around the local value at node i and time level n, u_i^n. Following the discussion of Chapter 7, the quantities u_j^m are considered as *exact solutions* of the discretized equations. If the differential problem is written as

$$\frac{\partial u}{\partial t} = Lu \qquad (9.1.1)$$

the equivalent differential equation of the selected scheme, which is satisfied by the numerical solution, is

$$\frac{\partial u}{\partial t} = Lu + \varepsilon_T \qquad (9.1.2)$$

where ε_T is the truncation error.

The stability of the scheme can be *partly* analysed by an investigation of the properties of the truncation error, as shown initially by Hirt (1968). An example has been shown already with equation (7.2.21), where the truncation error of the explicit central scheme for the convection equation was shown to correspond to a negative viscosity coefficient and hence could only be unstable. More detailed investigations relating the structure of the truncation error to the stability of the scheme have been developed by Warming and Hyett (1974) and in a very systematic way by Yanenko and Shokin (1969). Extensive applications of the method of the equivalent differential equation developed by the Russian authors, called the method of *differential approximation*, can be found in Shokin (1983).

Generally, this method will allow us to define necessary conditions for stability, although sufficient conditions can, in some cases, also be derived, in particular for hyperbolic equations. A most important application of this method is the analysis of the nature and properties of the truncation error. In particular, the errors generated from non-linear terms can be investigated by this approach and schemes can be defined in order to minimize the non-linear error sources (Lerat and Peyret, 1974, 1975; Lerat, 1979; Shokin, 1983).

The order of accuracy of the scheme, will be defined according to equation (7.2.22) by the lowest-order terms of the equivalent differential equation.

9.1 STABILITY ANALYSIS FOR PARABOLIC PROBLEMS

Considering the one-dimensional heat diffusion equation and an explicit second-order discretization (equation (8.3.16)) with the developments

$$u_i^{n+1} = u_i^n + \Delta t (u_t)_i + \frac{\Delta t^2}{2} (u_{tt})_i + \frac{\Delta t^3}{6} (u_{ttt})_i + \cdots \qquad (9.1.3)$$

$$u_{i\pm1}^n = u_i^n \pm \Delta t (u_x)_i + \frac{\Delta x^2}{2} (u_{xx})_i \pm \frac{\Delta x^3}{2} (u_{xxx})_i + \cdots \qquad (9.1.4)$$

we have the following form for the equivalent differential equation, removing the index i:

$$u_t - \alpha u_{xx} = -\frac{\Delta t}{2} u_{tt} + \frac{\alpha \Delta x^2}{12} \left(\frac{\partial^4 u}{\partial x^4}\right) + \frac{\alpha \Delta x^4}{360} \left(\frac{\partial^6 u}{\partial x^6}\right) - \frac{\Delta t^2}{6} u_{ttt} + \cdots$$

$$(9.1.5)$$

Equation (9.1.5) is used to eliminate the time derivatives in the truncation error terms of the right-hand side. Taking the time derivative of equation (9.1.5) and inserting into the right-hand side leads to

$$u_t - \alpha u_{xx} = -\frac{1}{2}\alpha \Delta x^2 \left(\beta - \frac{1}{6}\right)\left(\frac{\partial^4 u}{\partial x^4}\right) + \frac{\alpha \Delta x^4}{3}\left(\beta^2 - \frac{\beta}{4} + \frac{1}{120}\right)\left(\frac{\partial^6 u}{\partial x^6}\right) + \cdots$$

$$(9.1.6)$$

with

$$\beta = \frac{\alpha \Delta t}{\Delta x^2} \qquad (9.1.7)$$

Note that, due to the central discretization, there are only even-order derivatives in the expansion of the truncation error.

The interpretation to be given to this equation is that the exact solution of the numerical scheme is a solution of the modified differential equation. Hence the stability and accuracy of the numerical scheme can be analysed by the amplification function of a single harmonic of this differential equation. Actually this corresponds to an investigation of the well-posedness of this equivalent differential equation.

Consider the differential equation

$$\frac{\partial u}{\partial t} = \alpha \left(\frac{\partial^2 u}{\partial x^2}\right) + \gamma_1 \left(\frac{\partial^4 u}{\partial x^4}\right) + \gamma_2 \left(\frac{\partial^6 u}{\partial x^6}\right) \qquad (9.1.8)$$

The amplification factor of this equation is the function

$$G = e^{-I\omega t} \tag{9.1.9}$$

such that the single harmonic k

$$u_k(t) = e^{-I\omega t}\, e^{Ikx} \tag{9.1.10}$$

is a solution of the considered equation.

Inserting into the above equation, we obtain

$$-I\omega = -\alpha k^2 + \gamma_1 k^4 - \gamma_2 k^6 \tag{9.1.11}$$

and the corresponding solution is

$$u_k(t) = e^{-(\alpha k^2 - \gamma_1 k^4 + \gamma_2 k^6)}\, e^{Ikx} \tag{9.1.12}$$

This solution remains bounded for all k if

$$\alpha - \gamma_1 k^2 + \gamma_2 k^4 \geq 0 \tag{9.1.13}$$

This will be satisfied if

$$\alpha \geq 0, \qquad \gamma_1 \leq 0, \qquad \gamma_2 \geq 0 \tag{9.1.14}$$

showing that condition (9.1.13) is satisfied if the coefficients of the even-order derivatives are of alternate sign.

In particular, the fourth-order derivative in the truncation error, which is the lowest-order term, must have a negative coefficient for stability. This leads to

$$0 \leq \beta \leq \tfrac{1}{6} \tag{9.1.15}$$

which is a sufficient condition. (Remember that the Von Neumann condition led to the restriction $0 \leq \beta \leq 1/2$.)

It is also important to observe, in the present context, that there are no odd-order derivatives in the expansion of the space terms, showing that the central explicit scheme (8.3.16) has no dispersion errors. Indeed, if an odd-order derivative would have occurred in the right-hand side of equation (9.1.8) it would generate an imaginary contribution to $I\omega$ in equation (9.1.11) and represent a dispersion error. One can conclude, on the basis of this example, *that even-order derivatives in the truncation error expansion represent dissipative errors, while odd-order derivatives are associated with dispersion or phase errors.*

Example 9.1.1   DuFort–Frankel scheme for the diffusion equation

This scheme has been introduced in Example 8.3.5:

$$u_i^{n+1} - u_i^{n-1} = 2\beta(u_{i+1}^n - u_i^{n+1} - u_i^{n-1} + u_{i-1}^n) \tag{E9.1.1}$$

Applying equations (9.1.3) and (9.1.4), the following equivalent differential equation, dropping the index i, is obtained:

$$u_t - \alpha u_{xx} = -\frac{\Delta t^2}{6} u_{ttt} + \alpha \frac{\Delta x^2}{12} \left(\frac{\partial^4 u}{\partial x^4}\right) + \alpha \frac{\Delta x^4}{360} \left(\frac{\partial^6 u}{\partial x^6}\right)$$

$$- \alpha \frac{\Delta t^2}{\Delta x^2} u_{tt} - \frac{\alpha \Delta t^4}{12 \Delta x^2} \left(\frac{\partial^4 u}{\partial t^4}\right) + \cdots \qquad \text{(E9.1.2)}$$

If the ratio $\Delta t / \Delta x$ is kept constant when Δt and Δx tend towards zero, then the equivalent differential equation (E9.1.2) reduces to

$$\frac{\partial u}{\partial t} + \alpha \left(\frac{\Delta t}{\Delta x}\right)^2 \frac{\partial^2 u}{\partial t^2} = \alpha \frac{\partial^2 u}{\partial x^2} \qquad \text{(E9.1.3)}$$

Under these conditions this equation is *not* consistent with the original diffusion equation $u_t = \alpha u_{xx}$. However, the method can be applied for stationary solutions with an algorithm achieving convergence when $u^{n+1} \rightarrow u^n$ or when $u_t \rightarrow 0$.

On the other hand, if β is held fixed, that is, $\Delta t / \Delta x^2$ is kept constant when Δt and Δx tend to zero, then the DuFort–Frankel scheme will be consistent. If equation (E9.1.2) is again applied to eliminate the time derivatives, we obtain successively

$$u_{tt} = \alpha^2 \frac{\partial^4 u}{\partial x^4} + \alpha^2 \frac{\Delta x^2}{12} \left(\frac{\partial^6 u}{\partial x^6}\right) - \alpha \frac{\Delta t^2}{\Delta x^2} u_{ttt} + \cdots \qquad \text{(E9.1.4)}$$

$$u_{ttt} = \alpha^3 \left(\frac{\partial^6 u}{\partial x^6}\right) - \alpha \frac{\Delta t^2}{\Delta x^2} \left(\frac{\partial^4 u}{\partial t^4}\right) + \cdots \qquad \text{(E9.1.5)}$$

introduced into equation (E9.1.2) the modified equation for the DuFort–Frankel scheme, for constant ratio $\Delta t / \Delta x^2$, becomes

$$u_t - \alpha u_{xx} = -\alpha \Delta x^2 \left(\beta^2 - \frac{1}{12}\right) \left(\frac{\partial^4 u}{\partial x^4}\right) + \alpha \Delta x^4 \left(\frac{1}{360} - \frac{\beta^2}{4} + \beta^4\right) \left(\frac{\partial^6 u}{\partial x^6}\right) + \cdots$$
$$\text{(E9.1.6)}$$

As shown in the previous chapter, this scheme is unconditionally stable.

The practical applications of the method of the equivalent differential equation are more important for hyperbolic equations, since the generation of errors is not to be associated with the presence of physical damping, as is the case with the parabolic equations. Therefore we will, in the rest of this chapter concentrate on hyperbolic systems.

9.2 STABILITY AND ACCURACY ANALYSIS FOR HYPERBOLIC PROBLEMS

If we consider a hyperbolic linear, scalar equation $u_t + a u_x = 0$, with constant a, the general form of the equivalent differential equation can be written, after elimination of the time-derivative terms following the procedures of Chapter 7 and absorbing Δt in the variable σ, as

$$u_t + au_x = \sum_{l=1}^{\infty} \left[a_{2l}\left(\frac{\partial^{2l}u}{\partial x^{2l}}\right) \Delta x^{2l-1} + a_{2l+1}\left(\frac{\partial^{2l+1}u}{\partial x^{2l+1}}\right) \Delta x^{2l} \right] \quad (9.2.1)$$

If the scheme is of order p in space, the first non-zero coefficient is proportional to Δx^p, according to definition (7.2.22). Therefore

$$a_{2l} = a_{2l+1} = 0 \quad \text{for} \quad 2l, 2l - 1 < p \quad (9.2.2)$$

The analytical amplification factor of the above equation for an harmonic k is obtained by inserting a solution of the form $\mathrm{e}^{-I\omega t}\,\mathrm{e}^{Ikx}$. Observe that the analytical amplification factor of the equivalent equation represents the Von Neumann amplification factor of the numerical scheme. Hence

$$G = \mathrm{e}^{-I\omega \Delta t} \quad (9.2.3)$$

where

$$I\omega = Ika - \frac{1}{\Delta x} \sum_l \left[a_{2l}(-)^l \phi^{2l} + I(-)^l a_{2l+1}\phi^{2l+1} \right] \quad (9.2.4)$$

Here again, the even derivatives lead to a real exponent, that is, a damping or dissipation if the associated coefficients are negative, and the uneven coefficients contribute to the dispersion error. Hence, to the lowest order, a necessary condition for stability is

$$(-)^r a_{2r} < 0 \quad \text{with} \quad 2r = (2l)_{\min} \quad (9.2.5)$$

if $2r$ is the lowest even derivative of the expansion. For a first-order scheme, $p = 1$, the lowest value of $2l$ is 2, hence $r = 1$; for second- or third-order schemes, $p = 2$ or 3, the lowest value of $2l$ is equal to 4 and $r = 2$.

From the definition of the dispersion error (equation (8.3.29)) the amplification over a time step Δt is written as

$$G = |G|\mathrm{e}^{-I\Phi} \quad (9.2.6)$$

and

$$\varepsilon_\phi = \frac{\Phi}{ka\Delta t} = 1 - \sum_l (-)^l \frac{\phi^{2l}}{a} a_{2l+1} \quad (9.2.7)$$

The diffusion error, on the other hand, is defined by equation (8.3.28):

$$\varepsilon_D = |G| = \exp\left[\sum_l a_{2l}(-)^l \phi^{2l} \frac{\Delta t}{\Delta x} \right] \quad (9.2.8)$$

Observe that the dispersion error contains only odd-order coefficients, while the diffusion error is totally defined by the even-order ones.

The order of accuracy of the scheme, as determined by equation (7.2.22), is given here by the lowest-order term of expansion (9.2.1). In particular, for a first-order scheme the coefficient a_2 will be different from zero, and correspond to a term $a_2 u_{xx}$. This term is interpreted as a *numerical viscosity* generated by the discretization error of the scheme, and has to be positive for stability.

For second-order schemes, $a_2 = 0$, the first non-zero term in the expansion of the truncation error is a third-order derivative $a_3 u_{xxx}$, associated with a dispersion error. Since the phase error can be of either sign no stability condition can be deduced from the dispersion term. However, the fourth-order derivative term, $a_4(\partial^4 u/\partial x^4)$, contributes a dissipation error

$$\varepsilon_D = e^{a_4 \phi^4 (\Delta t/\Delta x)} \qquad (9.2.9)$$

and the coefficient a_4 has to be negative, according to equation (9.2.5).

Example 9.2.1   The leapfrog scheme for the scalar convection equation

The leapfrog scheme

$$u_i^{n+1} = u_i^{n-1} - \sigma(u_{i+1}^n - u_{i-1}^n) \qquad (E9.2.1)$$

has the following equivalent differential equation:

$$u_t + au_x = \frac{a\,\Delta x^2}{6}(\sigma^2 - 1)\,u_{xxx} - \frac{a\,\Delta x^4}{120}(9\sigma^2 - 1)(\sigma - 1)\frac{\partial^5 u}{\partial x^5} + 0(\Delta x^6)\,(E9.2.2)$$

The first coefficients of expansion (9.2.1) are $a_2 = 0$, $a_3 = a(\sigma^2 - 1)/6$, $a_5 = (a/120)(9\sigma^2 - 1)(\sigma - 1)$. It can be observed that this scheme is second-order accurate and has the particularity of having no even-order derivatives in the truncation error terms. This is a consequence of the fact that the modulus of the amplification factor, $|G| = 1$, as seen earlier, and equation (9.2.8) implies that all even-order coefficients have to be zero.

9.2.1   General Formulation of the equivalent differential equation for linear hyperbolic problems

The coefficients of the Taylor series development of the truncation error can be obtained in a general form for linear hyperbolic equations with constant coefficients. A general two-level explicit scheme for the equation $u_t + au_x = 0$ can be written as

$$u_i^{n+1} = \sum_j b_j u_{i+j}^n \qquad (9.2.10)$$

where the sum over j involves the mesh points defining the numerical scheme. The range of j on the x-axis is called 'the support of the scheme'. For instance, for the upwind scheme (equation (7.2.8)) j takes the values $-1, 0, +1$:

$$b_{-1} = \sigma, \qquad b_0 = (1 - \sigma), \qquad b_1 = 0 \qquad (9.2.11a)$$

and the scheme is written as

$$u_i^{n+1} = \sigma u_{i-1}^n + (1 - \sigma)u_i^n \qquad (9.2.11b)$$

For the Lax–Friedrichs scheme we have

$$b_{-1} = \tfrac{1}{2}(1 + \sigma), \qquad b_0 = 0, \qquad b_1 = \tfrac{1}{2}(1 - \sigma) \qquad (9.2.12a)$$

and the scheme is written as

$$u_i^{n+1} = \tfrac{1}{2}(1 + \sigma)u_{i-1}^n + \tfrac{1}{2}(1 - \sigma)u_{i+1}^n \qquad (9.2.12b)$$

For the Lax–Wendroff scheme we have

$$b_{-1} = \frac{\sigma}{2}(1 + \sigma), \qquad b_0 = 1 - \sigma^2, \qquad b_1 = -\frac{\sigma}{2}(1 - \sigma) \qquad (9.2.13a)$$

and the scheme can be written as

$$u_i^{n+1} = \frac{\sigma}{2}(1 + \sigma)u_{i-1}^n + (1 - \sigma^2)u_i^n - \frac{\sigma}{2}(1 - \sigma)u_{i+1}^n \qquad (9.2.13b)$$

The leapfrog scheme cannot be put into this form since it is a three-level scheme, and the general development to follow will not be valid in this case. For an implicit scheme we could also write equation (9.2.10), but the sum over j would cover all the mesh points.

Equation (9.2.10) represents the way the new function values at level $n + 1$ is obtained from the known function values at time level n. The coefficient b_j is the weight of the contribution of point $(i + j)$ to the new value at point i and time level $n + 1$. This is illustrated in Figure 9.2.1 in general and for the upwind, Lax–Friedrichs and Lax–Wendroff schemes in Figure 9.2.2.

The b_j coefficients are obviously not arbitrary, and have to satisfy a certain number of consistency conditions, depending on the order of accuracy of the scheme. If p is the order of the scheme there are clearly $(p + 1)$ relations to be satisfied. A first condition is obtained from the requirement that a constant should be a solution of the numerical scheme. This leads to the first consistency condition:

$$\sum_j b_j = 1 \qquad (9.2.14)$$

In order to obtain the equivalent differential equation for scheme (9.2.10) the Taylor series expansions have to be introduced:

$$u_{i+j}^n = u_i^n + \sum_{m=1}^{\infty} \frac{(j \cdot \Delta x)^m}{m!} \left(\frac{\partial^m u}{\partial x^m}\right) \qquad (9.2.15)$$

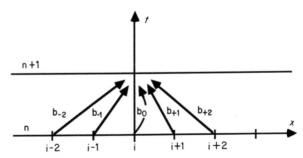

Figure 9.2.1 Weight coefficients of contributions of function values at level n to solution at level $n + 1$

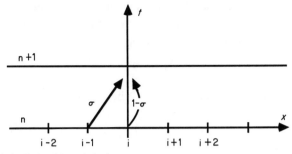

(a) First-order upwind scheme

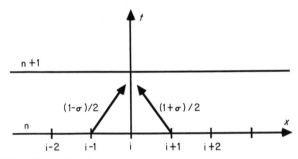

(b) Lax–Friedrichs scheme

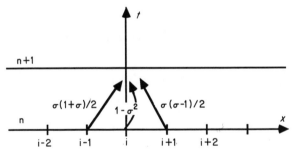

(c) Lax-Wendroff scheme

Figure 9.2.2 Weight coefficients of contributions of function values
at level n to solution at level $n + 1$ for (a) first-order upwind scheme,
(b) Lax–Friedrichs scheme and (c) Lax–Wendroff scheme

Also for the time development we have

$$u_i^{n+1} = u_i^n + \sum_{m=1}^{\infty} \frac{\Delta t^m}{m!} \left(\frac{\partial^m u}{\partial t^m}\right) \tag{9.2.16}$$

Inserting these developments into equation (9.2.10) leads to

$$\Delta t \cdot \frac{\partial u}{\partial t} + \sum_{m=2}^{\infty} \frac{\Delta t^m}{m!} \left(\frac{\partial^m u}{\partial t^m}\right) = \sum_j (b_j j \cdot \Delta x) \frac{\partial u}{\partial x} + \sum_{m=2}^{\infty} b_j \frac{(j \cdot \Delta x)^m}{m!} \left(\frac{\partial^m u}{\partial x^m}\right) \tag{9.2.17}$$

A second consistency condition is obtained from the above equation, since the coefficient of the first-order space derivative of u has to be equal to $(-a \, \Delta t)$. Hence we obtain

$$\sum_j j b_j = -\sigma \tag{9.2.18}$$

With this condition development (9.2.17) becomes

$$u_t + a u_x = \frac{1}{\Delta t} \sum_{m=2}^{\infty} b_j \frac{(j \cdot \Delta x)^m}{m!} \left(\frac{\partial^m u}{\partial x^m}\right) - \sum_{m=2}^{\infty} \frac{\Delta t^{m-1}}{m!} \left(\frac{\partial^m u}{\partial t^m}\right) \tag{9.2.19}$$

The full equivalent differential equation is obtained by replacing time derivatives in equation (9.2.19) by space derivatives derived from the *equivalent differential equation*. Remember that the numerical solution u_i^n is an exact solution of this equation, which we write, according to equation (9.2.1), as

$$u_t + a u_x = \sum_{l=p}^{\infty} \Delta x^l \cdot a_{l+1} \left(\frac{\partial^{l+1} u}{\partial x^{l+1}}\right) \tag{9.2.20}$$

for a scheme of order p.

The time derivatives in equation (9.2.19) are therefore calculated from equation (9.2.20) as follows:

$$\frac{\partial^m u}{\partial t^m} = \left[-a \frac{\partial}{\partial x} + \sum_{l=p}^{\infty} \Delta x^l \cdot a_{l+1} \frac{\partial^{l+1}}{\partial x^{l+1}}\right]^m u$$

$$= (-a)^m \frac{\partial^m u}{\partial x^m} + m(-a)^{m-1} \sum_{l=p}^{\infty} \Delta x^l a_{l+1} \left(\frac{\partial^{l+m} u}{\partial x^{l+m}}\right) \tag{9.2.21}$$

$$+ \frac{m(m-1)}{2} (-a)^{m-2} \sum_{l,k=p}^{\infty} (\Delta x)^{l+k} a_{l+1} a_{k+1} \frac{\partial^{l+k+m} u}{\partial x^{l+k+m}} + 0(\Delta x^{3p})$$

Inserting the development (9.2.21), limited here to terms of the order of 3p, into equation (9.2.19) leads to the equivalent equation:

$$u_t + a u_x = \left(\frac{\Delta x}{\Delta t}\right) \sum_{m=2}^{\infty} \left[\sum_J (b_j j^m) - (-\sigma)^m\right] \frac{\Delta x^{m-1}}{m!} \left(\frac{\partial^m u}{\partial x^m}\right)$$

$$- \sum_{m=2}^{\infty} \frac{(-\sigma)^{m-1}}{(m-1)!} \sum_{l=p}^{\infty} \Delta x^{l+m-1} a_{l+1} \left(\frac{\partial^{l+m} u}{\partial x^{l+m}}\right)$$

$$- \frac{\Delta t}{2\Delta x} \sum_{m=2}^{\infty} \frac{(-\sigma)^{m-2}}{(m-2)!} \sum_{l,k=p}^{\infty} \Delta x^{l+k+m-1} a_{l+1} a_{k+1} \left(\frac{\partial^{l+k+m} u}{\partial x^{l+k+m}}\right) + 0(\Delta x^{3p+1})$$

$$\tag{9.2.22}$$

The second summation contributes terms of the order of Δx^{p+1} and higher, while the third summation starts at Δx^{2p+1}.

By comparing with development (9.2.20) additional consistency conditions are derived, since all terms with order lower than p should vanish. Therefore the coefficients b_j of the numerical scheme of order p have to satisfy the following $(p+1)$ relations:

$$\sum_j j^m b_j = (-\sigma)^m \qquad \text{for} \qquad m = 0, 1, \ldots, p \qquad (9.2.23)$$

The relations for $m = 0$ and 1 reproduce conditions (9.2.14) and (9.2.18).

The remaining term in the first summation of equation (9.2.22) therefore starts at $m = p + 1$, for a consistent scheme, and by comparing with equation (9.2.20), we obtain the coefficient of the first non-zero term of the truncation error:

$$a_{p+1} = \frac{\Delta x}{\Delta t} \left[\sum_j b_j j^{p+1} - (-\sigma)^{p+1} \right] \frac{1}{(p+1)!} \qquad (9.2.24)$$

The second highest term (Δx^{p+1}) is obtained as follows, introducing the quantity α_{p+1} defined by

$$\alpha_{p+1} = \left[\sum_j b_j j^{p+1} - (-\sigma)^{p+1} \right] \frac{1}{(p+1)!} = \frac{\Delta t}{\Delta x} a_{p+1} \qquad (9.2.25)$$

$$a_{p+2} = \frac{\Delta x}{\Delta t} \cdot \alpha_{p+2} + \sigma a_{p+1} \qquad (9.2.26)$$

The next term is of order $(p+2)$ and the third summation will contribute to the coefficient a_{p+3} if the scheme is of first-order accuracy, since for $p = 1$, $p + 2 = 2p + 1$.

Hence for a first-order scheme $(p = 1)$, the coefficient of $\Delta x^3 (\partial^4 u / \partial x^4)$ is given by

$$a_{p+3} = \frac{\Delta x}{\Delta t} \alpha_{p+3} - \frac{\sigma^2}{2} a_{p+1} + \sigma a_{p+2} - \frac{\Delta t}{2 \Delta x} a_{p+1}^2 \qquad (9.2.27)$$

For a second-order scheme $(p = 2)$ the last term does not contribute, hence

$$a_{p+3} = \frac{a}{\sigma} \left(\alpha_{p+3} + \sigma \alpha_{p+2} + \frac{\sigma^2}{2} \alpha_{p+1} \right) \qquad (9.2.28)$$

Higher-order terms can be obtained by inspection, if necessary.

Example 9.2.2   Upwind scheme for the convection equation $u_t + au_x = 0$

The upwind scheme, for $a > 0$,

$$u_i^{n+1} = u_i^n - \sigma(u_i^n - u_{i-1}^n) \qquad (E9.2.3)$$

or

$$u_i^{n+1} = u_i^n - \frac{\sigma}{2} (u_{i+1}^n - u_{i-1}^n) + \frac{\sigma}{2} (u_{i+1}^n - 2u_i^n + u_{i-1}^n) \qquad (E9.2.4)$$

has been shown to be stable, by a Von Neumann analysis, for $0 < \sigma \leqslant 1$. The equivalent differential equation is obtained from the above relations.

With $b_{-1} = \sigma$, $b_0 = (1 - \sigma)$, $b_1 = 0$ we verify for this first-order scheme,

$$\sum_j jb_j = -\sigma \qquad (p = 1)$$

For $l = 2$ and $l = 3$ the above relations become

$$a_2 = \frac{a}{2}(1 - \sigma) \equiv \frac{\Delta x}{\Delta t}\alpha_2 \tag{E9.2.5}$$

$$a_3 = \frac{a}{6}(\sigma^2 - 1) + \frac{a}{2}\sigma(1 - \sigma) = +\frac{a}{6}(\sigma - 1)(1 - 2\sigma) \tag{E9.2.6}$$

corresponding to the values obtained earlier.

The next term, a_4 is given by equation (9.2.27):

$$a_4 = \frac{a}{24}(1 - \sigma)(1 + 6\sigma^2 - 6\sigma) \tag{E9.2.7}$$

Up to the third-order terms we have

$$u_t + au_x = \frac{a\,\Delta x}{2}(1 - \sigma)u_{xx} + \frac{a\,\Delta x^2}{6}(2\sigma - 1)(1 - \sigma)u_{xxx}$$

$$+ \frac{a\,\Delta x^3}{24}(1 - \sigma)(1 + 6\sigma^2 - 6\sigma)\frac{\partial^4 u}{\partial x^4} + 0(\Delta x^4) \tag{E9.2.8}$$

The scheme is first-order accurate for constant ratios $\Delta x/\Delta t$. The first term of the truncation error indicates a *numerical viscosity* equal to

$$\frac{a\,\Delta x}{2}(1 - \sigma) = \frac{1}{2}\frac{\Delta x^2}{\Delta t}\sigma(1 - \sigma)$$

which has to be positive for stability. Hence we obtain the stability conditions $0 < \sigma \leqslant 1$; identical to the Von Neumann conditions. Observe that for $\sigma = 1$ the truncation error is zero, and the scheme has the exact solution $u_i^{n+1} = u_{i-1}^n$.

Example 9.2.3 Lax–Friedrichs scheme for the convection equation
$$u_t + au_x = 0$$

The scheme

$$u_i^{n+1} = \frac{1}{2}(u_{i+1}^n + u_{i-1}^n) - \frac{\sigma}{2}(u_{i+1}^n - u_{i-1}^n) \tag{E9.2.9}$$

which can also be written as a correction to the unstable central scheme

$$u_i^{n+1} = u_i^n - \frac{\sigma}{2}(u_{i+1}^n - u_{i-1}^n) + \frac{1}{2}(u_{i+1}^n - 2u_i^n + u_{i-1}^n) \tag{E9.2.10}$$

has the following equivalent differential equation:

$$u_t + au_x = \frac{\Delta x^2}{2\Delta t}(1 - \sigma^2)u_{xx} + \frac{a\,\Delta x^2}{3}(1 - \sigma^2)u_{xxx} + \cdots \tag{E9.2.11}$$

The coefficients of expansion (9.2.1) are $a_2 = a(1 - \sigma^2)/2\sigma$, $a_3 = a(1 - \sigma^2)/3$. This scheme is first-order accurate for constant ratios $\Delta x/\Delta t$.

The numerical viscosity of the scheme is

$$v_{num} = \frac{\Delta x^2}{2\Delta t}(1 - \sigma^2) = \frac{a}{2\sigma}\Delta x(1 - \sigma^2) \tag{E9.2.12}$$

and is higher than the viscosity generated by the first-order upwind scheme by a factor equal to $(1 + \sigma)/\sigma$. For this coefficient to be non-negative the CFL condition, $|\sigma| \leqslant 1$, has to be satisfied for stability.

For low values of ϕ the leading term of the dispersion error is always positive, since

$$\varepsilon_\phi \simeq 1 + \frac{\phi^2}{3}(1 - \sigma^2) + 0(\Delta x^4) \tag{E9.2.13}$$

indicating a leading phase error. This again confirms the results obtained from the complete expression of the phase error (equation (E8.3.5)), which leads to the above result when developed in a series of powers of ϕ.

9.2.2 Error estimations for two-level explicit schemes

From equations (9.2.7) and (9.2.8) we can define the discretization errors as a function of the coefficients b_j. The exact form of the amplification matrix of the scheme is obtained from the Von Neumann method by analysing the behaviour of Fourier modes $u_i^n = v^n e^{Ii\phi}$ ($\phi = k\,\Delta x$). Hence with $G(\phi) = v^{n+1}/v^n$:

$$G(\phi) = \sum_j b_j\, e^{Ij\phi} = \sum_j b_j \cos j\phi + I \sum_j b_j \sin j\phi \tag{9.2.29}$$

The expansion as a function of ϕ gives, from equation (9.2.8), the amplitude of the diffusion error for small values of ϕ:

$$\varepsilon_D = 1 + \sum_{l=1} a_{21}(-)^l \phi^{21} \cdot \frac{\Delta t}{\Delta x} + \cdots \tag{9.2.30}$$

and the dispersion error (9.2.7) is rewritten here for convenience:

$$\varepsilon_\phi = 1 - \sum_{l=1} \frac{(-)^l}{a} a_{21+1}\phi^{2l} \tag{9.2.31}$$

If the order of accuracy p is odd the first term in the above development starts at $2l = p + 1$, and

$$\varepsilon_D = 1 + (-)^{(p+1)/2}\alpha_{p+1}\phi^{p+1} + 0(\phi^{p+3}) \tag{9.2.32}$$

The associated dispersion error is of order $(p + 1)$ and is given by

$$\varepsilon_\phi = 1 - (-)^{(p+1)/2}a_{p+2}\,\phi^{p+1}/a + 0(\phi^{p+3})$$

$$= 1 - (-)^{(p+1)/2}\left(\frac{\alpha_{p+2}}{\sigma} + \alpha_{p+1}\right)\phi^{p+1} + 0(\phi^{p+3}) \tag{9.2.33}$$

A necessary condition for stability is therefore

$$(-)^{(p+1)/2}\alpha_{p+1} < 0 \tag{9.2.34}$$

For even orders of accuracy the first term in the development of the amplification factor starts at $p = 2l$ and this is an odd-derivative term contributing to the dispersion error. The amplitude error is therefore obtained, to the lowest order in ϕ, from the term a_{p+2}:

$$\varepsilon_D = 1 - (-)^{p/2}(\alpha_{p+2} + \sigma\alpha_{p+1})\phi^{p+2} + 0(\phi^{p+4}) \tag{9.2.35}$$

The dispersion error is defined by the term a_{p+1}:

$$\varepsilon_\phi = 1 - (-)^{p/2}\frac{1}{\sigma}\alpha_{p+1}\phi^p + 0(\phi^{p+2}) \tag{9.2.36}$$

The necessary condition for stability is, in this case, for p even, applying equation (9.2.5):

$$\frac{\Delta t}{\Delta x}(-)^{p/2}a_{p+2} = (-)^{p/2}(\alpha_{p+2} + \sigma\alpha_{p+1}) > 0 \tag{9.2.37}$$

An important criterion for stability, when the coefficients are not constant or when the problem is not linear, is related to the dissipation property in the sense of Kreiss, as defined in Section 8.5. In particular, this property requires that the amplification factor be different from one for the high-frequency waves associated with the $2\Delta x$ waves, or $\phi \approx \pi$. From equation (9.2.29) we have

$$G(\pi) = \sum_j b_j \cos j\pi = \sum_{j \text{ even}} b_j - \sum_{j \text{ odd}} b_j \tag{9.2.38}$$

The condition

$$|G(\pi)| < 1$$

is satisfied if

$$-1 < \left(\sum_{j \text{ even}} b_j - \sum_{j \text{ odd}} b_j\right) < 1 \tag{9.2.39}$$

or, taking into account the consistency condition, $\sum_j b_j = 1$, we obtain (Roe, 1981)

$$0 < \sum_{j \text{ even}} b_j < 1 \tag{9.2.40}$$

The equality signs on these limits are valid for stability but have to be excluded for the Kreiss dissipative property. For Lax–Friedrichs scheme $b_0 = 0$, and condition (9.2.40) is not satisfied, indicating that this scheme is not dissipative in the sense of Kreiss. For the upwind scheme $b_0 = 1 - \sigma$, and the scheme is dissipative for $0 < \sigma < 1$. For Lax-Wendroff's scheme $b_0 = 1 - \sigma^2$, and the scheme is dissipative for $0 < \sigma < 1$.

9.2.3   Stability analysis for two-level explicit schemes

Warming and Hyett (1974) have shown that for a wide range of schemes necessary as well as sufficient stability conditions can be derived from the coefficients of the equivalent differential equation. In these cases the method of the equivalent differential equation can be considered on an equal footing with the Von Neumann method with regard to the stability analysis.

From equation (9.2.29) the modulus squared of the amplification factor can be written as

$$| G(\phi)|^2 = |\sum_j b_j e^{Ij\phi}|^2 = \left[\sum_j b_j \cos j\phi\right]^2 + \left[\sum_j b_j \sin j\phi\right]^2$$

$$= \sum_j \sum_k b_j b_k \cos(j - k)\phi \tag{9.2.41}$$

With the trigonometric relations

$$\cos n\phi = 1 + \sum_{l=1}^{n} (-1)^l 2^{2l-1} \sin^{2l}\phi/2 \tag{9.2.42}$$

equation (9.2.41) can be written as a polynomial in the variable

$$z = \sin^2\phi/2 \tag{9.2.43}$$

under the form

$$| G(\phi)|^2 = 1 + \sum_{l=1}^{m} \beta_l z^l \tag{9.2.44}$$

The polynomial in z is, at most, of degree m, where m is the total number of points included in scheme (9.2.10), located at the left and at the right of mesh point i. That is, m is equal to the total number of points involved at level n, excluding point i. The coefficients β_l are the sum of products of (two) b_j coefficients.

It is seen from equation (9.2.44) that a term z^r can always be factored out from the sum over m, leading to an expression of the form

$$| G(\phi)|^2 = 1 - z^r S(z) \tag{9.2.45}$$

where $S(z)$ is a polynomial of, at most, order $s = m - r$.

By comparing equation (9.2.45) for $z \to 0$, that is, for $\phi^2 \to 0$, with equation (9.2.8) it is seen that $2r$ is equal to the order of the lowest even derivative appearing in the truncation error and hence is equal to the value defined in equation (9.2.5). For the same reasons the scheme is also dissipative of order $2r$, according to equation (8.5.12).

As an example, for all schemes based on three points including point i, $m = 2$ and therefore $r = 1$ or 2 and $s = 1$ or 0. For first-order schemes $r = 1$ and for second-order schemes $r = 2$. In particular, Lax-Wendroff's scheme corresponds to $s = 0$ and the polynomial $S(z)$ reduces to a constant $S(0) = -16b_1 b_{-1} = 4\sigma^2(1 - \sigma^2)$.

The polynomial $S(z)$ determines the Von Neumann stability of the scheme, since the stability condition $|G(\phi)| \leqslant 1$ will be satisfied for the following necessary and sufficient conditions:

$$S(0) > 0 \text{ and } S(z) \geqslant 0 \text{ for } 0 < z \leqslant 1 \qquad (9.2.46)$$

By expanding equation (9.2.45) in powers of ϕ^2 and comparing with development (9.2.8) we can express the polynomial $S(z)$ as a function of the coefficients a_{2l} of the truncation error and obtain necessary and sufficient stability conditions on the a_{2l} coefficients in this way.

The first condition (9.2.46) leads to equation (9.2.5), since the lowest term of the expansion of equation (9.2.45) is uniquely defined by $S(0)$, through

$$|G(\phi)|^2 \approx 1 - S(0) \sin^{2r}\phi/2 \approx 1 - S(0)(\phi/2)^{2r} + 0(\phi^{2r+2}) \qquad (9.2.47)$$

By comparison with equation (9.2.8) we obtain

$$S(0) = a_{2r}(-1)^{r-1}2^{2r+1} \, \Delta t / \Delta x \qquad (9.2.48)$$

and the condition $S(0) > 0$ gives equation (9.2.5).

The second condition (9.2.46) might not be easy to achieve for high-order schemes when $S(z)$ is a high-order polynomial, but can readily be worked out if $S(z)$ is, at most, of degree one. In this case the conditions $S(0) > 0$ and $S(1) \geqslant 0$ are both necessary and sufficient for stability. For instance, for $r = 1$ we obtain (see Problem 9.6)

$$S(1) = 2 \frac{\Delta t}{\Delta x} \left[\frac{1}{3} a_2 - \frac{\Delta t}{\Delta x} a_2^2 - a_4 \right] \geqslant 0 \qquad (9.2.49)$$

for stability. An extension of this analysis for two-level implicit schemes can be found in Warming and Hyett (1974).

9.3 THE GENERATION OF NEW ALGORITHMS WITH A PRESCRIBED ORDER OF ACCURACY

One of the consequences of the consistency conditions derived in the previous section is the possibility of generating new families of algorithms having a given support on the x-axis and a given order of accuracy. The consistency conditions (9.2.23), for a scheme of order of accuracy p, define $(p + 1)$ relations for the b_j coefficients. If the support of the scheme covers M points there are M values b_j, and $(M - p - 1)$ coefficients b_j can be chosen arbitrarily. If $M = p + 1$, there is a unique solution and therefore there is always a unique scheme on a support of M points having the maximum possible order of accuracy of $p = M - 1$. This interesting analysis is due to Roe (1981).

When $M > p + 1$ families of schemes can be generated by adding an arbitrary multiple of the homogeneous part of the system of equations (9.2.23) to a particular solution of this system, generating a family of schemes with $(M - p - 1)$ parameters. For instance, on the three-point support $j = -1, 0, 1$

$(M = 3)$ the first-order accurate schemes satisfy the conditions

$$b_{-1} + b_0 + b_1 = 1$$
$$b_{-1} - b_1 = \sigma \tag{9.3.1}$$

Hence we have a one-parameter family of first-order schemes with three-point support. If, following Roe (1981), the schemes are identified by the set $S(b_{-1}, b_0, b_1)$ the Lax–Friedrichs scheme corresponds to $S_{LF}((1 + \sigma)/2, 0, (1 - \sigma)/2)$ and the first order upwind scheme to $S_U(\sigma, (1 - \sigma), 0)$. The unstable, central scheme (7.2.5) can be represented by S_c $(\sigma/2, 1, -\sigma/2)$.

Since the homogeneous solution of system (9.3.1) is given by the set $H(1, -2, 1)$, all schemes of the form

$$S(b_{-1}, b_0, b_1) = S_c(\sigma/2, 1, -\sigma/2) + \gamma H(1, -2, 1) \tag{9.3.2}$$

are valid first-order schemes. For instance, with $\gamma = \sigma/2$ we recover the upwind scheme and with $\gamma = 1/2$ the Lax–Friedrichs scheme. Hence the possible schemes on the three-point support $(i - 1, i, i + 1)$ with first-order accuracy are defined by the parameter γ and the b_j values $b_{-1} = \sigma/2 + \gamma$, $b_0 = 1 - 2\gamma$, $b_1 = \gamma - \sigma/2$. They can be written as follows:

$$u_i^{n+1} = u_i^n - \frac{\sigma}{2} (u_{i+1}^n - u_{i-1}^n) + \gamma(u_{i+1}^n - 2u_i^n + u_{i-1}^n) \tag{9.3.3}$$

where γ appears as an *artificial viscosity* coefficient. The stability condition, from equations (9.2.24) and (9.2.34), is

$$(2\gamma - \sigma^2) \geqslant 0 \tag{9.3.4}$$

Various properties can be defined with regard to these schemes, in particular the important concept of monoticity of a scheme introduced by Godunov (1959). We will discuss this more in detail with reference to the Euler equations in Volume 2.

Two explicit schemes of second-order accuracy for the convection equation $u_t + au_x = 0$ have been discussed up to now, the Lax–Wendroff and the leapfrog schemes. The latter was shown to be neutrally stable and to have severe limitations due to its three-level nature as well as its lack of dissipation for strongly varying functions $u(x, t)$ (for instance, with discontinuous variations). The two other schemes, the upwind and the Lax–Friedrichs schemes, are only first-order accurate, which is often insufficient for practical purposes. From the above considerations there is only one explicit scheme on the domain $j(-1, 0, 1)$, with second-order accuracy and two time levels, that is *centrally defined with respect to the mesh point i*. This is the Lax–Wendroff scheme obtained from equation (9.3.1) by adding the third consistency condition (9.3.23), which reads here for the support $j(-1, 0, 1)$:

$$b_{-1} + b_1 = \sigma^2 \tag{9.3.5}$$

leading to $\gamma = \sigma^2/2$. The Lax–Wendroff scheme can be written as

$$u_i^{n+1} = \frac{\sigma}{2}(1 + \sigma) u_{i-1}^n + (1 - \sigma^2) u_i^n - \frac{\sigma}{2}(1 - \sigma) u_{i+1}^n \qquad (9.3.6)$$

or, as a correction to the unstable central scheme,

$$u_i^{n+1} = u_i^n - \frac{\sigma}{2}(u_{i+1}^n - u_{i-1}^n) + \frac{\sigma^2}{2}(u_{i+1}^n - 2u_i^n + u_{i-1}^n) \qquad (9.3.7)$$

As can be seen, the third term, which stabilizes the instability generated by the first two terms, is the discretization of an additional dissipative term of the form $(a^2 \, \Delta t^2/2) \, u_{xx}$.

With the above relations the equivalent equation is

$$u_t + au_x = -\frac{a}{6} \Delta x^2 (1 - \sigma^2)u_{xxx} - a \frac{\Delta x^3}{8} \sigma(1 - \sigma^2)u_{xxxx} + 0(\Delta x^4) \qquad (9.3.8)$$

showing the second-order accuracy of the scheme.

Second-order upwind scheme

Other second-order schemes can be generated, for instance on the support $j(-2, -1, 0)$, generalizing the one-sided scheme (7.2.8). We attempt a scheme of the form, for $a > 0$,

$$u_i^{n+1} = b_{-2}u_{i-2} + b_{-1}u_{i-1} + b_0 u_i \qquad (9.3.9)$$

with the conditions

$$
\begin{aligned}
b_0 + b_{-1} + b_{-2} &= 1 \\
-2b_{-2} - b_{-1} &= -\sigma \\
4b_{-2} + b_{-1} &= \sigma^2
\end{aligned}
\qquad (9.3.10)
$$

There is only one solution, namely,

$$b_{-2} = \frac{\sigma}{2}(\sigma - 1), \qquad b_{-1} = \sigma(2 - \sigma), \qquad b_0 = \tfrac{1}{2}(1 - \sigma)(2 - \sigma) \qquad (9.3.11)$$

leading to the *second-order accurate upwind scheme* of Warming and Beam (1976):

$$u_i^{n+1} = \frac{\sigma}{2}(\sigma - 1)u_{i-2}^n + \sigma(2 - \sigma)u_{i-1}^n + \left(1 - \frac{3\sigma}{2} + \frac{\sigma^2}{2}\right)u_i^n \qquad (9.3.12)$$

This scheme can be rewritten as a correction to the first-order upwind scheme as

$$u_i^{n+1} = u_i^n - \sigma(u_i^n - u_{i-1}^n) + \tfrac{1}{2}\sigma(\sigma - 1)(u_i^n - 2u_{i-1}^n + u_{i-2}^n) \qquad (9.3.13)$$

The last term appears as a second-order derivative correction taken at the point $(i - 1)$, and compensates for the numerical viscosity generated by the first-order scheme (7.2.8), which is equal to $(a/2)\,\Delta x(1-\sigma)u_{xx}$. The truncation error is obtained from equations (9.2.24)–(9.2.26), leading to the equivalent equation:

$$u_t + au_x = \frac{a}{6}\,\Delta x^2(1-\sigma)(2-\sigma)u_{xxx} - \frac{a}{8}\,\Delta x^3(1-\sigma)^2(2-\sigma)\left(\frac{\partial^4 u}{\partial x^4}\right) + \cdots$$
(9.3.14)

The stability condition (9.2.5) reduces to the condition

$$0 \leqslant \sigma \leqslant 2 \tag{9.3.15}$$

This is confirmed by a Von Neumann analysis and the amplification factor

$$G = 1 - 2\sigma[1 - (1-\sigma)\cos\phi]\sin^2\phi/2 - I\sigma\sin\phi[1 + 2(1-\sigma)\sin^2\phi/2] \tag{9.3.16}$$

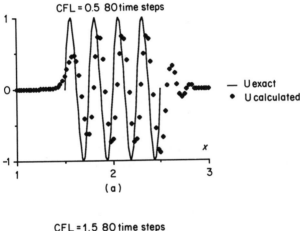

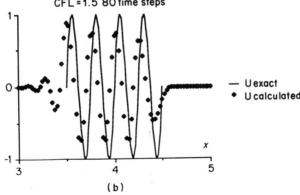

Figure 9.3.1 Solution of the linear propagation of a wave packet for the second-order Beam–Warming upwind scheme. (a) $\sigma = 0.5$; (b) $\sigma = 1.5$

with

$$|G|^2 = 1 - 4\sigma(1 - \sigma)^2(2 - \sigma)\sin^4\phi/2 \qquad (9.3.17)$$

At the high-frequency end of the spectrum:

$$G(\pi) = 1 - 2\sigma(2 - \sigma) \qquad (9.3.18)$$

and the scheme is dissipative in the sense of Kreiss for $0 < \sigma < 2$. The leading term of the phase error is given by equation (9.2.36) as

$$\varepsilon_\phi \approx 1 + \tfrac{1}{6}(1 - \sigma)(2 - \sigma)\phi^2 + 0(\phi^4) \qquad (9.3.19)$$

and shows a predominantly leading phase error ($\varepsilon_\phi > 1$), for $0 < \sigma < 1$ and a lagging phase error ($\varepsilon_\phi < 1$) for $1 < \sigma < 2$.

An example of a computation with this scheme for a wave packet is shown in Figure 9.3.1 for two values of the Courant number, $\sigma = 0.5$ and $\sigma = 1.5$. The first value corresponds to a leading phase error while the second generates a solution with a lagging phase error (see also Problem 9.8).

Both the Lax–Wendroff and the Warming–Beam schemes are the unique schemes of second-order accuracy on the supports $(i - 1, i, i + 1)$ and $(i - 2, i - 1, i)$, respectively.

A larger number of schemes can be generated by allowing a support of the schemes with a number of points larger than $(p + 1)$. For instance, schemes with support on $(i - 2, i - 1, i, i + 1)$ have been generated with second-order accuracy, namely that of Fromm (1968), which combines the Lax–Wendroff and Warming–Beam upwind scheme to reduce the dispersion errors, since these two schemes have phase errors of opposite signs in the range $0 < \sigma < 1$ (see Problem 9.10). On the four-point support $(i - 2, i - 1, i, i + 1)$ we can derive a unique third-order accurate scheme. This is left as an exercise to the reader (see Problem 9.12).

9.4 THE EQUIVALENT DIFFERENTIAL EQUATION FOR NON-LINEAR HYPERBOLIC PROBLEMS

The Taylor expansion which leads to the equivalent differential equation can still be applied to non-linear problems, allowing an investigation of the influence of the non-linear terms on the dispersion and dissipation errors of the scheme. The present approach is one of the sole methods allowing such a systematic investigation, and has been applied by Lerat and Peyret (1974, 1975) and by Lerat (1979) in order to optimize schemes for the Euler equations with respect to non-linear error generation. Of particular concern is the fact that the non-linearities generate additional terms in the truncation error, which can influence strongly the behaviour of the scheme.

In order to investigate these influences we consider a general non-linear hyperbolic equation, with a flux function $f = f(u)$, of the form

$$u_t + f_x = 0 \qquad (9.4.1)$$

In the linear case $f = au$ and for Burger's equation $f = u^2/2$.

The generalization of the general scheme (9.2.10) to non-linear problems is written under the conservative form with a numerical flux function as defined in Section 6.1:

$$u_i^{n+1} = u_i^n - \frac{\Delta t}{\Delta x} [g(u_i^n, u_{i+1}^n) - g(u_{i-1}^n, u_i^n)] \qquad (9.4.2)$$

for a three-point scheme with support $(i - 1, i, i + 1)$.

The numerical flux, written here as g instead of f^* in equation (6.1.11), has to satisfy the consistency relation

$$g(u, u) = f(u) \qquad (9.4.3)$$

The Jacobian $A = A(u)$:

$$A(u) = \frac{\partial f}{\partial u} \qquad (9.4.4)$$

plays an important role, since equation (9.4.1) can also be written as

$$u_t + A(u)u_x = 0 \qquad (9.4.5)$$

In the linear case $A = a$ is constant and for Burger's equation $A = u$.

In order to find the truncation error of scheme (9.4.2) and define its equivalent differential equation we have to develop the numerical flux terms in a Taylor series as a function of the variable u and its derivatives at mesh point i. Developing the numerical flux function g in a Taylor series gives

$$g(u_i, u_{i+1}) = g(u_i, u_i) + \left(\frac{\partial g}{\partial u_{i+1}}\right)_{u_i} \Delta u + \frac{1}{2}\left(\frac{\partial^2 g}{\partial u_{i+1}^2}\right)_{u_i} \Delta u^2$$
$$+ \frac{1}{6}\left(\frac{\partial^3 g}{\partial^2 u_{i+1}}\right)_{u_i} \Delta u^3 + \cdots \qquad (9.4.6)$$

where

$$\Delta u = (u_{i+1} - u_i) \qquad (9.4.7)$$

We will define

$$\left(\frac{\partial g}{\partial u_{i+1}}\right)_{u_i} = g_2(u_i, u_i); \qquad \left(\frac{\partial g}{\partial u_i}\right)_{u_i} = g_1(u_i, u_i) \qquad (9.4.8)$$

as the derivatives of g with respect to the second and first arguments, respectively, taken at the common value u_i. Similarly,

$$g_{11} = \left(\frac{\partial^2 g}{\partial u_i \partial u_i}\right)_{u_i}; \qquad g_{12} = \left(\frac{\partial^2 g}{\partial u_i \partial u_{i+1}}\right)_{u_i}; \qquad g_{22} = \left(\frac{\partial^2 g}{\partial u_{i+1}^2}\right)_{u_i} \qquad (9.4.9)$$

Hence introducing the Taylor expansion for u_{i+1} leads to

$$g(u_i, u_{i+1}) = g(u_i, u_i) + \left(u_x \Delta x + u_{xx}\frac{\Delta x^2}{2} + \frac{\Delta x^3}{6} u_{xxx} + \cdots\right)_i \cdot g_2$$
$$+ \frac{1}{2}(u_x^2 \Delta x^2 + u_x u_{xx} \Delta x^3 + \cdots)_i \cdot g_{22}$$
$$+ \frac{1}{6}(u_x^3 \Delta x^3 + \ldots)_i \cdot g_{222} \qquad (9.4.10)$$

or with the consistency condition (9.4.3):

$$g(u_i, u_{i+1}) = f_i + \Delta x g_2 u_x + \frac{\Delta x^2}{2}(g_2 u_{xx} + g_{22} u_x^2)$$

$$+ \frac{\Delta x^3}{6}(g_2 u_{xxx} + 3g_{22} u_x u_{xx} + g_{222} u_x^3) + 0(\Delta x^4) \quad (9.4.11)$$

Subtracting a similar development for $g(u_{i-1}, u_i)$ we obtain

$$g(u_i, u_{i+1}) - g(u_{i-1}, u_i) = \Delta x(g_1 + g_2)u_x + \frac{\Delta x^2}{2}[u_{xx}(g_2 - g_1) + (g_{22} - g_{11})u_x^2]$$

$$+ \frac{\Delta x^3}{6}[(g_1 + g_2)u_{xxx} + 3(g_{22} + g_{11})u_x u_{xx} + (g_{222} + g_{111})u_x^3] + 0(\Delta x^4)$$

$$(9.4.12)$$

The term in $\Delta x^2/2$ can be rewritten as $(\partial/\partial x)[(g_2 - g_1)u_x]$ and that in $\Delta x^3/6$ as $(\partial/\partial x)(f_{xx} - 3g_{12}u_x^2)$ by application of the derivative chain rule and of equation (9.4.3), leading to

$$g(u_i, u_{i+1}) - g(u_{i-1}, u_i) = \Delta x(g_1 + g_2)u_x + \frac{\Delta x^2}{2}\frac{\partial}{\partial x}[(g_2 - g_1)u_x]$$

$$+ \frac{\Delta x^3}{6}\frac{\partial}{\partial x}[f_{xx} - 3g_{12}u_x^2] + 0(\Delta x^4) \quad (9.4.13)$$

Inserting this relation into scheme (9.4.2), together with the time series (9.2.16), leads to

$$\Delta t\, u_t + \frac{\Delta t^2}{2}u_{tt} + \frac{\Delta t^3}{6}u_{ttt} + 0(\Delta t^4) = -\frac{\Delta t}{\Delta x}[g(u_i, u_{i+1}) - g(u_{i-1}, u_i)]$$

$$(9.4.14)$$

where the right-hand side of equation (9.4.14) is replaced by equation (9.4.13). By comparison, it is seen that a second consistency condition appears, namely

$$g_1 + g_2 = A \quad (9.4.15)$$

or

$$\left[\frac{\partial g(u_i, u_{i+1})}{\partial u_i} + \frac{\partial g(u_i, u_{i+1})}{\partial u_{i+1}}\right]_{u_i} = A(u_i) \quad (9.4.16)$$

and we obtain as a first form of the truncation error expansion

$$u_t + f_x = -\frac{\Delta x}{2}\frac{\partial}{\partial x}[(g_2 - g_1)u_x] - \frac{\Delta x^2}{6}\frac{\partial}{\partial x}(f_{xx} - 3g_{12}u_x^2)$$

$$- \frac{\Delta t}{2}u_{tt} - \frac{\Delta t^2}{6}u_{ttt} + 0(\Delta x^3) \quad (9.4.17)$$

If the truncation error is written to the lowest order p under the form

$$u_t + f_x = \Delta x^p \cdot Q(u) \qquad (9.4.18)$$

where $Q(u)$ is a differential operator on u, reducing to $\alpha(\partial^{p+1}u/\partial x^{p+1})$ for a linear equation, as seen from equation (9.2.20), we can eliminate the time derivatives as a function of the space derivatives in the above equation. We have, successively, with $A_u = (\partial A/\partial u)$

$$u_{tt} = -f_{xt} + \Delta x^p \frac{\partial}{\partial t} Q(u) + 0(\Delta x^{p+1})$$

$$= -(Au_t)_x + \Delta x^p \frac{\partial}{\partial t} Q(u) + 0(\Delta x^{p+1})$$

$$= (A^2 u_x)_x + \Delta x^p \frac{\partial}{\partial t} Q(u) + 0(\Delta x^{p+1}) \qquad (9.4.19)$$

If we limit the non-linear investigations to the first term in the expansion we also have

$$u_{ttt} = (A^2 u_x)_{xt} + 0(\Delta x^p)$$
$$u_{ttt} = -(2A^2 A_u u_x^2 + A^2 f_{xx})_x + 0(\Delta x^p) \qquad (9.4.20)$$
$$u_{ttt} = -3A(A A_{uu} + 2A_u^2)u_x^3 - 9A^2 A_u u_x u_{xx} - A^3 u_{xxx} + 0(\Delta x^p)$$

The last term of this sum is the only remaining contribution in a linear case.

Inserting into expansion (9.4.17) we obtain, finally, with $\tau = (\Delta t/\Delta x)$, the equivalent differential equation for scheme (9.4.2):

$$u_t + f_x = -\frac{\Delta x}{2} \frac{\partial}{\partial x} [(g_2 - g_1 + \tau A^2)u_x]$$

$$-\frac{\Delta x^2}{6} \frac{\partial}{\partial x} [(1 - \tau^2 A^2)f_{xx} - (3g_{12} + 2\tau^2 A^2 A_u)u_x^2] + 0(\Delta x^3) \quad (9.4.21)$$

If the scheme is second-order accurate, the following consistency condition has to be satisfied, since the expansion must start with the Δx^2 term:

$$g_1 - g_2 = \tau A^2 \qquad (9.4.22)$$

Observe that equations (9.4.3), (9.4.15) and (9.4.22) are the non-linear generalizations of the consistency conditions (9.2.23) up to $p = 2$.

When the scheme is first-order accurate the first term acts as a numerical viscosity of the form $(\partial/\partial x)(\bar{\nu}u_x)$, with the numerical viscosity coefficient equal to

$$\bar{\nu} = (g_1 - g_2 - \tau A^2) \qquad (9.4.23)$$

and has to be positive for stability. For second-order schemes the term in Δx^2 contains non-linear contributions proportional to u_{xx}, arising from the derivatives of the Jacobian, A_u.

The Δx^2 term can also be worked as

$$-\frac{\Delta x^2}{6} \{[A(1 - \tau^2 A^2)u_{xxx} - [(9\tau^2 A^2 - 3)A_u - 6g_{12}]u_x u_{xx}$$

$$+ [(1 - 3\tau^2 A^2)A_{uu} - 6\tau^2 A_u^2 - 3(g_{112} + g_{122})]u_x^3\} \qquad (9.4.24)$$

where g_{122} is a third derivative of the numerical flux g, with respect to the first argument and twice with respect to the second, taken at the common value u, and similarly for g_{112}. In the linear case the first term is the only remaining one and is *independent of the scheme*. Otherwise a non-linear *numerical viscosity* term appears proportional to Δx^2, equal to

$$\bar{\nu}_{NL} u_{xx} = \frac{\Delta x^2}{2} [(3\tau^2 A^2 - 1)A_u - 2g_{12}]u_x \cdot u_{xx} \qquad (9.4.25)$$

When this coefficient is negative it produces an anti-dissipation effect which tends to destabilize the scheme if this term becomes dominant. For instance, when $\tau A = 1$, which corresponds to a Courant number of one, the first dispersive error term $A(1 - \tau^2 A^2)$ vanishes, and non-linear oscillations can appear.

An explicit calculation has been performed by Lerat and Peyret (1975) for the inviscid Burger's equation for different schemes, particularly for Lax–Wendroff's scheme. Applied to the general non-linear equation $u_t + f_x = 0$, Lax–Wendroff's scheme can be written, by applying central differences in space to equation (E8.3.8) with the time derivative replaced by flux derivatives, as

$$u_i^{n+1} = u_i^n - \Delta t \, (f_x)_i + \frac{\Delta t^2}{2} \frac{\partial}{\partial x} (Af_x) \qquad (9.4.26)$$

or

$$u_i^{n+1} = u_i^n - \frac{\Delta t}{2\Delta x} (f_{i+1} - f_{i-1})$$

$$+ \frac{\Delta t^2}{2\Delta x^2} [A_{i+1/2}(f_{i+1} - f_i) - A_{i-1/2}(f_i - f_{i-1})] \qquad (9.4.27)$$

The numerical flux $g(u_i, u_{i-1})$ for this non-linear Lax–Wendroff scheme is equal to

$$g(u_i, u_{i+1}) = \frac{f_i + f_{i+1}}{2} - \frac{\tau}{2} A_{i+1/2}(f_{i+1} - f_i) \qquad (9.4.28)$$

with

$$A_{i+1/2} = A\left(\frac{u_i + u_{i+1}}{2}\right) \qquad (9.4.29)$$

A direct calculation for Burger's equation $u_t + (u^2/2)_x = 0$ shows that $g_{12} = 0$

when $u_i = u_{i+1} = u$ and the truncation error becomes, with $A = u$, $A_u = 1$, $\sigma = \tau u$,

$$u_t + \left(\frac{u^2}{2}\right)_x = -\frac{\Delta x^2}{6}(1 - \sigma^2)u_{xxx} + \frac{\Delta x^2}{2}(3\sigma^2 - 1)u_x \cdot u_{xx} + \Delta x^2\sigma^2 u_x^3$$

$$(9.4.30)$$

The non-linear dissipation term is negative for $|\sigma| < 1/\sqrt{3}$ and is proprtional to u_x. Hence around a discontinuity where u would jump from 3 to 1, say, the derivative u_x can become very high and negative, generating an anti-dissipation of non-linear origin for $|\sigma| > 1/\sqrt{3}$. This will reduce the dissipation present in the linear terms and can generate non-linear oscillations. These, however, are generally not sufficient to destabilize completely the scheme when linear dissipation is present, as is the case with the Lax–Wendroff scheme.

Figure 9.4.1 illustrates these properties by comparing the behaviour of the Lax–Wendroff scheme for a propagating discontinuity solution of the linear convection and Burger's equations at two values of the Courant number $\sigma = 0.2$ and $\sigma = 0.8$, with $\Delta x = 1/30$. For the linear equation stronger oscillations appear at low values of σ since the scheme has less dissipation at high frequencies for the lower values of the Courant number, as can be seen from Figure 8.3.4. Comparing with results from Burger's equation, the anti-dissipation generated by the non-linear terms is clearly seen at $\sigma = 0.8$, where a

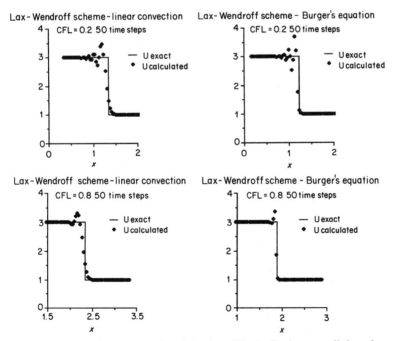

Figure 9.4.1 Non-linear properties of the Lax–Wendroff scheme applied to the inviscid Burger's equation and a propagating discontinuity

sharper profile is obtained, while at $\sigma = 0.2$ this effect is not clearly present, although the amplitude of the oscillations is increased. Schemes which do not have linear dissipation properties (such as the leapfrog scheme) are more sensitive to a destabilization by the non-linear effects, as has been shown by the example of Figure 8.5.1.

Summary

The equivalent differential equation allows the determination of the truncation error and of the consistency of a given scheme but becomes of significant interest essentially for hyperbolic problems. In the latter a detailed investigation of the structure of the truncation error delivers consistency and stability conditions in the most general case. A family of two-level, explicit schemes, with preset support and order of accuracy can be derived on a fairly general basis and their accuracy and stability properties can be expressed as a function of the coefficients of the scheme. A unique property of the equivalent differential equation approach is the possibility of defining the influence of non-linearities on the stability of schemes. Non-linear contributions to the truncation error can be derived from the numerical fluxes of a conservative scheme in a very general way and sources of instabilities can be detected.

References

Fromm, J. E. (1968). 'A method for reducing dispersion in convective difference schemes.' *Journal Computational Physics,* **3**, 176–89.

Godunov, S. K. (1959). 'A difference scheme for numerical computation of discontinuous solutions of hydrodynamic equations.' *Math. Sibornik,* **47**, 271–306 (in Russian). Translated US Joint Publ. Res. Service, *JPRS 7226* (1960).

Hirt, C. W. (1968). 'Heuristic stability theory for finite difference equations.' *Journal Computational Physics,* **2**, 339–55.

Lerat, A. (1979). 'Numerical shock structure and non linear corrections for difference schemes in conservation form.' *Lecture Notes in Physics,* **20**, 345–51, New York: Springer Verlag.

Lerat, A., and Peyret, R. (1974). 'Non centered schemes and shock propagation problems.' *Computers and Fluids,* **2**, 35–52.

Lerat, A., and Peyret, R. (1975). 'The problem of spurious oscillations in the numerical solution of the equations of gas dynamics.' *Lecture Notes in Physics,* **35**, 251–6, New York: Springer Verlag.

Richtmyer, R. D., and Morton, K. W. (1967). *Difference Methods for Initial Value Problems,* New York: John Wiley.

Roe, P. L. (1981). 'Numerical algorithms for the linear wave equation.' *TR 81047,* Royal Aircraft Establishment, April.

Shokin, Yu. I. (1983). *The Method of Differential Approximation,* New York: Springer Verlag.

Warming, R. F., and Hyett, B. J. (1974) 'The modified equation approach to the stability and accuracy of finite difference methods.' *Journal of Computational Physics,* **14**, 159–79.

Warming, R. F., and Beam, R. W. (1976). 'Upwind second order difference schemes and applications in aerodynamic flows.' *AIAA Journal,* **24**, 1241–9.

Warming, R. F., Kutler, P., and Lomax, H. (1973), 'Second and third order non centered difference schemes for non linear hyperbolic equations.' *AIAA Journal*, **11**, 189–95.

Yanenko, N. N., and Shokin, Yu. I. (1969). 'First differential approximation method and approximate viscosity of difference schemes.' *Physics of Fluids*, **12**, Suppl. II, 28–33.

PROBLEMS

Problem 9.1

Derive by a Taylor expansion the equivalent differential equation for the leapfrog scheme and obtain equation (E9.2.2). Compare with the series development of the compact expression (E8.3.25) for the phase error ε_ϕ.

Problem 9.2

Derive the equivalent differential equation for the upwind and Lax–Friedrichs schemes following Examples 9.2.2 and 9.2.3. Compare with the power expansions of $|G|$ and ε_ϕ. Obtain the next term a_4 for the Lax–Friedrichs scheme.

Problem 9.3

Develop the general relations of Section 9.2 for a three-level scheme of the form

$$u_i^{n+1} = u_i^{n-1} + \sum_j b_j u_{i+j}^n$$

by applying equation (9.2.15) at $t = (n + 1)\Delta t$ and $t = (n - 1)\Delta t$ to the convective linear equation $u_t + au_x = 0$. Obtain the consistency relations on the b_j coefficients. Apply to the leapfrog scheme.

Problem 9.4

Derive the general development of Section 9.2 for explicit schemes of the diffusion equation $u_t = \alpha u_{xx}$ of the form

$$u_i^{n+1} = \sum_j b_j u_{i+j}^n$$

Derive the consistency conditions.

Problem 9.5

Apply the form (9.2.45) of the modulus squared of the Von Neumann amplification factor to the first-order upwind, Lax–Friedrichs and Lax–Wendroff schemes for the linear convection equation. Determine for each case the polynomial $S(z)$.

Problem 9.6

Obtain equation (9.2.49), valid for $r = 1$, and show also that for $r \geqslant 2$:

$$S(1) = (-)^{r-1} 2^{2r-1} \frac{\Delta t}{\Delta x} [(3 + r)a_{2r}/12 - a_{2r+2}]$$

Problem 9.7

Obtain the amplification factor (9.3.16) for the second-order upwind scheme of Warming and Beam. Generate a polar plot of the amplitude and phase errors.

Problem 9.8

Solve the convective equation $u_t + au_x = 0$ for a sinus wave packet with four cycles on a mesh with $\Delta x = 1/60$ and $a = 1$. The wave packet is defined by

$$u(t = 0) = \sin 2\pi kx \qquad 0 \leqslant x \leqslant 1 \qquad \text{and} \qquad k = 4$$
$$= 0 \qquad x < 0 \qquad \text{and} \qquad x > 1$$

Compute the numerical transported wave after 50, 100 and 200 time steps and plot the exact solutions with the numerical solution. Take $\sigma = 0.1, 0.5, 0.9$ and apply the following schemes:

(1) Upwind first order;
(2) Lax–Friedrichs;
(3) Lax–Wendroff and
(4) Second-order upwind.

Hint: The exact solution after n time steps is

$$u^n = \sin 2\pi k(x - n\Delta t) \qquad n\,\Delta t < x \leqslant (n\,\Delta t + 1)$$
$$= 0 \qquad x \leqslant n\,\Delta t \qquad \text{and} \qquad x \geqslant n\,\Delta t + 1$$

Problem 9.9

Repeat the same problem for a moving discontinuity:

$$u(t = 0) = 1 \qquad x < 0$$
$$= 0 \qquad x > 0$$

after 10, 50 and 100 time steps for $\sigma = 0.1, 0.5$ and 0.9.

Problem 9.10.

Define a family of schemes on the support $(i - 2, i - 1, i, i + 1)$ with second-order accuracy. Follow the developments of Section 9.3 and write all the schemes as a particular solution to system (9.2.23) plus a parameter times the homogeneous solution. Take the Lax–Wendroff scheme S_{LW} as a particular solution $b_{-2} = 0$, $b_{-1} = \sigma(1 + \sigma)/2$, $b_0 = (1 - \sigma^2)$, $b_1 = -\sigma(1 - \sigma)/2$ and write a scheme as

$$S(b_{-2}, b_{-1}, b_0, b_1) = S_{LW} + \gamma H$$

where H is the homogeneous solution. Determine the value of γ for the upwind scheme of Warming and Beam.
    Derive Fromm's (1968) scheme by selecting $\gamma = \frac{1}{4}\sigma(1 - \sigma)$ and show that it can be obtained as the arithmetic average of the schemes of Lax–Wendroff, S_{LW}, and of the Warming–Beam second-order upwind scheme, S_{WB}. Analyse this scheme by calculating and plotting the amplitude and phase errors obtained from a Von Neumann analysis and obtain the stability condition. Apply also the relations of Section 9.2 to obtain the lowest-order terms of the dissipation and dispersion errors and compare with the two other schemes S_{LW}, S_{WB}. Obtain the equivalent differential equation.
Hint: Show that $H = (-1, 3, -3, 1)$. Fromm's scheme corresponds to

$$S_F(\tfrac{1}{4}\sigma(\sigma - 1), \tfrac{1}{4}\sigma(5 - \sigma), \tfrac{1}{4}(1 - \sigma)(4 + \sigma), \tfrac{1}{4}\sigma(\sigma - 1))$$

and is stable for $0 \leqslant \sigma \leqslant 1$.

Problem 9.11

Solve Problem 9.8 with Fromm's scheme and compare the results. Observe that the phase error is considerably reduced compared with S_{LW} and S_{WB}.

Problem 9.12

Refer to Problem 9.10 and obtain the unique, third-order accurate scheme on the support $(i-2, i-1, i, i+1)$(Warming *et al.*, 1973). Apply a Von Neumann analysis and plot the phase and amplitude errors; obtain the stability limits. Determine the lowest-order terms of these errors and the equivalent differential equation by applying the relations of Section 9.3.

Hint: The scheme is defined by

$$S_3\left[\frac{-\sigma}{6}(1-\sigma^2), \frac{\sigma(2-\sigma)(\sigma+1)}{2}, \frac{(2-\sigma)(1-\sigma^2)}{2}, \frac{-\sigma}{6}(1-\sigma)(2-\sigma)\right]$$

and is stable for $0 \leqslant \sigma \leqslant 1$.

Problem 9.13

Šolve Problem 9.8 with the third-order accurate scheme of Problem 9.12.

Problem 9.14

Solve Problem 9.9 with Fromm's scheme (Problem 9.10).

Problem 9.15

Solve Problem 9.9 with the third-order accurate scheme of Problem 9.12.

Problem 9.16

Derive the relations of Example 9.1.1, in particular equation (E9.1.6).

Problem 9.17

Obtain the truncation error (9.3.8) of the Lax–Wendroff scheme.

Problem 9.18

Obtain the truncation error of the Warming- and Beam scheme as given by equation (9.3.14).

Problem 9.19

Obtain equations (9.4.20), (9.4.21) and (9.4.24).

Problem 9.20

Obtain by a direct computation the truncation error for Burger's equation discretized by the Lax–Wendroff scheme, equation (9.4.30).

Chapter 10

The Matrix Method for Stability Analysis

The methods for stability analysis, described in Chapters 8 and 9, do not take into account the influence of the numerical representation of the boundary conditions on the overall stability of the scheme. The Von Neumann method is based on the assumptions of the existence of a Fourier decomposition of the solution over the finite computational domain in space. This implies the presence of periodic boundary conditions or, from another point of view, that we investigate the stability of the scheme applied at the interior points far enough from the boundaries. A similar position can be taken with respect to the equivalent differential equation approach. Obviously, this method does not provide any information on the influence of boundary conditions.

The matrix method, to be discussed in this chapter, provides, on the other hand, a means of determining this influence, although in practical situations it is generally not easy to derive analytically the corresponding stability conditions.

The denomination of the method comes from its starting point: the matrix representation of a scheme (equation (8.2.4)) considered as a system of ordinary differential equations in time. In addition, the analysis behind the matrix method provides an answer to the fundamental question on the criteria to be satisfied by a time-integration method, applied to a given space discretization, in order to lead to a stable scheme.

10.1 PRINCIPLE OF THE MATRIX METHOD—ANALYSIS OF THE SPACE DISCRETIZATION

The linear initial boundary value differential problem, with constant coefficients, over the domain Ω with boundary Γ

$$\frac{\partial u}{\partial t} = L(u) \qquad \begin{array}{llll} u(\vec{x}, 0) = f(\vec{x}) & \text{for} & t = 0 & \text{and} & \vec{x} \in \Omega \\ u(\vec{x}, t) = g(\vec{x}, t) & \text{for} & t > 0 & \text{and} & \vec{x} \in \Gamma \end{array} \qquad (10.1.1)$$

can be transformed, after discretization of the space differential operator L, into the ordinary system of differential equations in time:

$$\frac{\mathrm{d}U}{\mathrm{d}t} = SU + Q \qquad (10.1.2)$$

370

where U is the vector of the mesh point values u_i and Q contains non-homogeneous terms and boundary values. Several examples of the structure of the S matrix have already been given in Table 8.1, but with exclusion of the boundary points and for one-dimensional problems. Note that the following developments are valid for an arbitrary number of space variables, when U is defined accordingly.

The stability analysis of the space discretization is based on the eigenvalue structure of the matrix S, since the exact solution of the system of equations (10.1.2) is directly determined by the eigenvalues and eigenvectors of S. Let Ω_j, $j = 1, \dots, N$ be the N eigenvalues of the $(N \times N)$ matrix S solution of the eigenvalue equation

$$\det | S - \Omega I | = 0 \tag{10.1.3}$$

and $V^{(j)}$, the associated eigenvectors, a solution of

$$S \cdot V^{(j)} = \Omega_j V^{(j)} \qquad \text{(no summation on } j) \tag{10.1.4}$$

N is the total number of mesh points and there are as many eigenvalues and associated eigenvectors as mesh points.

The matrix S is assumed to be of rank N and hence to have N linearly independent eigenvectors. Each eigenvector $V^{(j)}$ consists of a set of mesh point values $v_i^{(j)}$, that is,

$$V^{(j)} = \begin{vmatrix} \vdots \\ v_{i-1}^{(j)} \\ v_i^{(j)} \\ v_{i+1}^{(j)} \\ \vdots \end{vmatrix} \tag{10.1.5}$$

The $(N \times N)$ matrix T formed by the N columns $V^{(j)}$ diagonalizes the matrix S, since all equations (10.1.4) for the N eigenvalues can be grouped as

$$S \cdot T = T\Omega \tag{10.1.6}$$

where Ω is the diagonal matrix of the eigenvalues:

$$\Omega = \begin{vmatrix} \Omega_1 & & & \\ & \Omega_2 & & \\ & & \ddots & \\ & & & \Omega_N \end{vmatrix} \tag{10.1.7}$$

Hence

$$\Omega = T^{-1}ST \tag{10.1.8}$$

Since the eigenvectors $V^{(j)}$ form a complete set of basis vectors in the considered space of the mesh–point functions we can always write the exact solution $\bar{U}$ of equation (10.1.2) as a linear combination of the $V^{(j)}$ eigenvectors:

$$\bar{U} = \sum_{j=1}^{N} \bar{U}_j(t) \cdot V^{(j)} \qquad (10.1.9)$$

and similarly for Q, assumed independent of time:

$$Q = \sum_{j=1}^{N} Q_j \cdot V^{(j)} \qquad (10.1.10)$$

The time-dependent coefficients $\bar{U}_j(t)$ are obtained from the differential system (10.1.2) by inserting equation (10.1.9):

$$\frac{d}{dt} \bar{U}_j = \Omega_j \cdot \bar{U}_j + Q_j \qquad \text{(no summation on } j) \qquad (10.1.11)$$

and the general solution $\bar{U}$ of system (10.1.2) is the sum of the homogeneous solution (first term) and the particular solution (second term):

$$\bar{U}(t) = \sum_{j=1}^{N} \left(c_{0j} \, e^{\Omega_j t} - \frac{Q_j}{\Omega_j} \right) \cdot V^{(j)} \qquad (10.1.12)$$

where c_{0j} is related to the coefficients of the expansion of the initial solution U^0:

$$U^0 = \begin{vmatrix} \cdot \\ f_{i-1} \\ f_i \\ f_{i+1} \\ \cdot \end{vmatrix} \qquad (10.1.13)$$

in series of the basis vectors $V^{(j)}$:

$$U^0 = \sum_{j=1}^{N} U_{0j} \cdot V^{(j)} \qquad (10.1.14)$$

by

$$c_{0j} = U_{0j} + Q_j/\Omega_j \qquad (10.1.15)$$

and solution (10.1.12) can be written as

$$\bar{U}(t) = \sum_{j=1}^{N} \left[U_{0j} \, e^{\Omega_j t} + \frac{Q_j}{\Omega_j} (e^{\Omega_j t} - 1) \right] \cdot V^j \qquad (10.1.16)$$

The system of ordinary differential equations (10.1.2) will be *well posed*, or *stable*, if, according to definition (7.2.35), the solutions $\bar{U}(t)$ remain bounded. This requires that the real part of the eigenvalues be negative or zero:

$$Re(\Omega_j) \leqslant 0 \qquad \text{for all } j \qquad (10.1.17)$$

In addition, if an eigenvalue Ω_j is zero it has to be a simple eigenvalue.

This confirms that the solution of the initial boundary value problem

(10.1.2) is completely determined by the eigenvalues and eigenvectors of the space discretization matrix S.

All the properties of the space discretization are actually contained in the eigenvalue spectrum of the associated matrix. The analysis of a matrix, or an operator, through the eigenvalues and eigenvectors represents a most thorough investigation of their properties. The complete 'identity' of a matrix is 'X-rayed' through the eigenvalue analysis and its spectrum can be viewed as a unique form of identification. Therefore, if the space discretization, including the boundaries, lead to eigenvalues with non-positive real parts the scheme will be stable at the level of the semi-discretized formulation. A stable time integration scheme will always be possible. When stability condition (10.1.17) is not satisfied, no time-integration method will lead to a stable algorithm.

For time-independent or periodic non-homogeneous terms, equation (10.1.16) describes the general solution to the semi-discretized problem (10.1.2) as a superposition of eigenmodes of the matrix operator S. Each mode j contributes a time behaviour of the form $\exp(\Omega_j t)$ to the time-dependent part of the solution, called the *transient solution*, obtained from the homogeneous part of equation (10.1.2). The transient $\bar{U}_T$ is a solution of

$$\frac{d\bar{U}_T}{dt} = S\bar{U}_T \qquad \bar{U}_T = \sum_j U_{0j}\, e^{\Omega_j t} \cdot V^{(j)} \qquad (10.1.18)$$

The second part of the general solution is the modal decomposition of the *particular* solution of equation (10.1.2), and is called the *steady state* solution, since it is the only one remaining for large values of time when all the eigenvalues have real negative components. In this case the transient will damp out in time, and asymptotically we would have from equations (10.1.16) and (10.1.17)

$$\lim_{t\to\infty} \bar{U}(t) = -\sum_j \frac{Q_j}{\Omega_j} \cdot V^{(j)} = \bar{U}_S \qquad (10.1.19)$$

which is a solution of the stationary problem

$$S\bar{U}_S + Q = 0 \qquad (10.1.20)$$

For stationary problems, solved by a time-dependent formulation, we are interested in obtaining the steady-state solution in the shortest possible time, that is, with the lowest possible number of time steps. This will be the case if the eigenvalues Ω_j have large negative real parts, since $\exp(-\,|\,Re\,\Omega_j\,|\,t)$ will rapidly decrease in time. On the other hand, if $(Re\,\Omega_j)$ is negative but close to zero the corresponding mode will decrease very slowly and a large number of time steps will be necessary to reach the stationary conditions.

Unfortunately, in practical problems the spectrum of Ω_j is very wide, including very large and very small magnitudes simultaneously. In this case when the ratio $|\,\Omega_{max}\,|\,/\,|\,\Omega_{min}\,|$, called the *condition number* of the matrix, is very much larger than one, the convergence to the steady state is dominated by the eigenvalues close to the minimum Ω_{min}, and this could lead to very slow convergence. One method to accelerate convergence, consists of a *preconditioning* technique, whereby equation (10.1.20) is multiplied by a matrix M such that the preconditioned matrix $\bar{S} = MS$ has a spectrum of eigenvalues with a more favourable condition number and larger negative values of Ω_{min}. This is an important technique for accelerating the convergence of numerical algorithms to steady-state solutions, and several examples will be discussed in later chapters.

Single-mode analysis

Since the exact solution (10.1.16) is expressed as a contribution from all the modes of the initial solution, which have propagated or (and) diffused with the eigenvalue Ω_j, and a contribution from the source terms Q_j all the properties of the time-integration schemes, and most essentially their stability properties, can be analysed separately for each mode with the scalar equation (10.1.11), writing w for any of the U_j

$$\frac{\mathrm{d}w}{\mathrm{d}t} = \Omega w + q \qquad (10.1.21)$$

The space operator S is replaced by an eigenvalue Ω and the 'modal' equation (10.1.21) will serve as the basic equation for the analysis of the stability of a time-integration method as a function of the eigenvalues Ω of the space-discretization operators.

This analysis provides a general technique for the determination of the time-integration methods which lead to a stable algorithm for a given space discretization. For instance, it has been seen that a central discretization of the space derivative of the linear convection equation leads to an unstable scheme if the time derivative is discretized by an explicit, forward difference; the same space discretization is stable under a central difference in time as applied in the leapfrog scheme. The general rules behind these differences will be defined in the following.

10.1.1 Amplification factors and stability criteria

The stability of the semi-discretized equations (10.1.2) is determined by the time behaviour of the transient, and consequently it is sufficient for this purpose to investigate the time behaviour of the homogeneous part of equation (10.1.2). The exact transient solution of the homogeneous semi-discretized equation at time level $t = n\Delta t$ can be written as

$$\tilde{U}_T(n\Delta t) = \sum_J \bar{U}_{Tj}(n\Delta t) \cdot V^{(j)} = \sum_j U_{0j}\, e^{\Omega_j n \Delta t} \cdot V^{(j)} \qquad (10.1.22)$$

Hence the amplification factor of the exact solution for an arbitrary mode Ω is defined as $\bar{G}(\Omega)$:

$$U_T(n\Delta t) = \bar{G}(\Omega) \cdot U_T((n-1)\Delta t) = \bar{G}^n(\Omega) \cdot U_T(0) \qquad (10.1.23)$$

where $\bar{G}^n$ is $\bar{G}$ to the power n. By comparison with equation (10.1.22) we have

$$\bar{G}(\Omega) = e^{\Omega \Delta t} \qquad (10.1.24)$$

The stability condition (10.1.17) ensures that the transient will not be amplified and that the modulus $|\bar{G}|$ is lower than one.

If a two-level time integration scheme is applied to equation (10.1.2) the matrix form (8.2.6) is obtained:

$$U^{n+1} = CU^n + \bar{Q} \qquad (10.1.25)$$

The error e^n satisfies the homogeneous equation

$$e^n = Ce^n = C^n e^0 \qquad (10.1.26)$$

and, according to the stability condition (7.2.35), the matrix C should be uniformly bounded for all n, Δt and Δx, in particular for $n \to \infty$, $\Delta t \to 0$ with $n \, \Delta t$ fixed. That is, we should have, in a selected norm, for finite T

$$\| C^n \| < K \qquad \text{for } 0 < n \, \Delta t < T \qquad (10.1.27)$$

with K independent of n, Δt, Δx.

The norm of C^n is often very difficult to analyse, and instead a necessary but not always sufficient condition can be obtained from a local mode analysis on the eigenvalues of C. Significant examples of non-sufficient stability conditions, obtained as a result of this eigenvalue analysis, will be discussed in the following.

If the numerical solution of equation (10.1.25) is designated by U_j^n, the full solution U^n will be obtained at time level n by

$$U^n = \sum_j U_j^n \cdot V^{(j)} \qquad (10.1.28)$$

as an approximation to the exact solution $\bar{U}(n \, \Delta t)$ at $t = n \, \Delta t$. For an individual mode Ω the *amplification factor* $z(\Omega)$ of the *numerical* scheme (10.1.25) is defined by the behaviour of the homogeneous or transient part as

$$U_{Tj}^n = z(\Omega_j) U_{Tj}^{n-1} = z^n(\Omega_j) U_0 \qquad (10.1.29)$$

where z^n is z to the power n. This amplification factor of the scheme, associated with the mode Ω, is an approximation to the exact amplification factor $\bar{G}$, depending on the selected time integration.

Introducing equations (10.1.28) and (10.1.29) into the homogeneous component of equation (10.1.25) and considering each mode separately leads to

$$CV^{(j)} = z(\Omega_j) V^{(j)} \qquad (10.1.30)$$

This generalizes equation (8.2.17), which shows the Von Neumann amplification factor $G(\phi)$ to be the Fourier symbol of C:

$$Ce^{I\phi} = G(\phi)e^{I\phi} \qquad (10.1.31)$$

The single Fourier harmonic of the Von Neumann analysis is replaced in the matrix method by a more general modal representation as the eigenvectors of the space operator. As will be shown in the following chapter, when the boundary conditions are periodic, both representations are identical.

General time-integration method

Before we define the stability condition on the amplification factor z we consider first a more general family of time-integration methods which contains the two-level method (10.1.25) as a particular case. A large number of methods for the numerical resolution of ordinary differential equations are available, and general presentations can be found in Gear (1971), Lambert (1973) and Dahlquist and Björk (1974). A summary of some of the most useful methods will also be given in Chapter 11.

A general *multi-step* method of order K applied to the modal equation (10.1.21) can be written at time level n as

$$\sum_{k=0}^{K} \alpha_k w^{n+k} = \Delta t \sum_{k=0}^{K} \beta_k (\Omega w + q)^{n+k} \tag{10.1.32}$$

and the method is *explicit* if $\beta_k = 0$ for all $k \neq 0$. Otherwise the method is *implicit*. Note that the α_k and β_k have to satisfy consistency conditions, namely

$$\sum_{k=0}^{K} \alpha_k = 0, \qquad \sum_{k=0}^{K} k\alpha_k = \sum_{k=0}^{K} \beta_k \tag{10.1.33}$$

Introducing the time shift operator $\bar{E}$, ($\bar{E}^k$ is $\bar{E}$ to the power k):

$$\bar{E}w^n = w^{n+1}$$
$$\bar{E}^k w^n = w^{n+k} \tag{10.1.34}$$

equation (10.1.32) can be written as a polynomial in $\bar{E}$:

$$P_1(\bar{E})w^n = P_2(\bar{E})q^n \tag{10.1.35}$$

where

$$P_1(\bar{E}) = \sum_{k=0}^{K} (\alpha_k - \Delta t \Omega \beta_k)\bar{E}^k \tag{10.1.36}$$

$$P_2(\bar{E}) = \sum_{k=0}^{K} \beta_k \bar{E}^k q^n \tag{10.1.37}$$

Equation (10.1.35) generalizes formulation (10.1.25), which corresponds to $P_1 = \bar{E} - C$.

Stability condition on the time- and space-discretized equation

Since the stability of the scheme depends only on the (transient) solution of the homogeneous equation it is sufficient to investigate solutions of form (10.1.29) to

$$P_1(\bar{E})w^n = 0 \qquad (10.1.38)$$

Equation (10.1.38) will be satisfied by a solution of this form if z is a solution of the *characteristic polynomial*:

$$P_1(z) = 0 \qquad (10.1.39)$$

If P_1 is of order K, for a K-step method there will be K roots z_k of the polynomial of order K:

$$P_1(z) = \sum_{k=0}^{K} (\alpha_k - \Delta t \Omega \beta_k)z^k = 0 \qquad (10.1.40)$$

and the general solution of the homogeneous equation (10.1.38), defining the transient behaviour of the numerical solution w^n in time, for increasing n will be given by

$$w^n = \sum_{k=1}^{K} w_k^0 z_k^n \qquad (10.1.41)$$

The numerical solution of this scalar equation will be stable if w^n does not increase without bound in time for n tending to infinity at fixed Δt. From the above solution we can define the necessary stability condition for the time-discretization scheme as the requirement that all the roots z_k should be of modulus lower than, or equal to, one. That is,

$$|z_k| \leqslant 1 \text{ for all roots } k = 1,...,K \qquad (10.1.42)$$

When some roots lie on the unit circle, $z_k = 1$, they have to be simple. Otherwise the solution would increase with time as t^m, where $(m + 1)$ is the multiplicity of the root. This condition establishes a relation between the selected time discretization (defined by the α_k and β_k coefficients of equation (10.1.32)) and the space discretization as characterized by the eigenvalues Ω, since

$$z_k = z_k(\Omega) \qquad (10.1.43)$$

The condition (10.1.17) on the space discretization has to be always satisfied for the system of ordinary differential equations (10.1.2) to be stable (or well posed). For all the space discretizations, which satisfy this condition, the associated numerical discretization in time will be stable if condition (10.1.42) is satisfied. However, when applied to the matrix equation (10.1.25) this condition does not always guarantee convergence and stability, according to definitions (10.1.27), at fixed and finite times ($n\Delta t$). This will be discussed in more detail in Section 10.4.

The space discretization generates a representative set of eigenvalues Ω_j which cover a certain region of the complex Ω plane, to be situated on the left side of the plane including the imaginary axis for stability. If we consider the trace of every root $z_k(\Omega)$, as Ω covers the whole spectrum Ω_j, in a complex

378

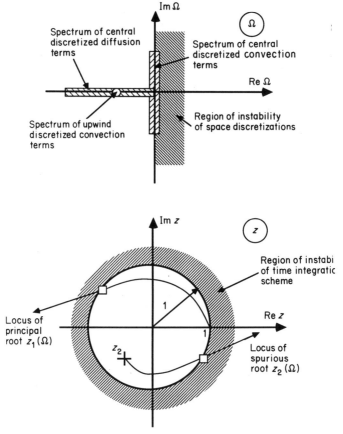

Figure 10.1.1 Representation of stability regions in the complex Ω (space-discretization) and z (time-discretization) planes

z-plane this trace will be represented by some line which has to remain inside a circle of radius one. If some roots come outside the stability circle when Ω covers the range of the spectrum Ω_j the scheme is unstable (Figure 10.1.1).

In Figure 10.1.1 the vertical imaginary axis of the Ω-plane contains the spectrum of the central discretized convection equation, while the negative real axis contains the spectrum of the central discretized diffusion equation. Note also that the spectrum of the first-order upwind scheme for the convection equation is concentrated in a single point on the real negative axis.

Multiple roots with multi-step methods

A method with two or more steps will generate more than one root; in particular a two-step method, which involves three time levels, will generate two roots, a three-step method, involving four time levels, will generate three roots, and so on, and this for the same value of Ω. When there is only one root the numerical time dependence is z^n and is to be considered as the approximation introduced by the time-discretization scheme to the exact time dependence. Hence, $z(\Omega)$ is the numerical approximation to $e^{\Omega \Delta t}$, since

$$z(\Omega) \approx e^{\Omega \Delta t} = 1 + \Omega \, \Delta t + \frac{(\Omega \, \Delta t)^2}{2} + \frac{(\Omega \, \Delta t)^2}{3} + \cdots \qquad (10.1.44)$$

Therefore, the first term in the expression of $z(\Omega)$ which differs from the Taylor development (10.1.44) defines the order of the method. For instance, for the explicit Euler method, $z(\Omega) = 1 + \Omega \, \Delta t$, which is clearly of first-order accuracy. The implicit Euler method is defined by

$$z = \frac{1}{1 - \Omega \, \Delta t} \approx 1 + \Omega \, \Delta t + (\Omega \, \Delta t)^2 - \cdots \qquad (10.1.45)$$

and since the quadratic term is not equal to the quadratic term in equation (10.1.44) the method is only first-order accurate.

When more than one root is present the consistency of the scheme requires that one of the roots should represent an approximation to the physical behaviour (10.1.44). This root, called the *principal* root, is to be recognized by the fact that it tends to one when $\Omega \, \Delta t$ goes to zero. Denoting by z_1 the principal ('physical') root we have

$$\lim_{\Omega \Delta t \to 0} z_1(\Omega) = 1 \qquad (10.1.46)$$

The other roots, called the *spurious roots*, represent a 'non-physical' time behaviour of the numerical solution introduced by the scheme and can destroy completely the stability of the scheme. For instance, we will see in the following section that the leapfrog scheme has two roots since it is a three-level scheme, and its first root z_1 is accurate up to second order but its spurious root z_2 starts at -1, and is the one that generates the instability when $(\Omega \, \Delta t) < 0$, that is, when dissipation is present in the scheme, either physically for diffusion equations or numerically for certain discretizations of the convection equations. Note that the spurious root is not always responsible for instability (see, for instance, Problem 10.8).

From equations (10.1.29) and (10.1.44) we can conclude that the time behaviour of the *numerical* solution w^n of the modal equation (10.1.21) is described by

$$w^n = w^0 z^n(\Omega) - \frac{q}{\Omega} (1 - z^n(\Omega)) \qquad (10.1.47)$$

Returning to the full solution U^n of equation (10.1.2) the above equation translates to, recalling that w represents an arbitrary mode U_j

$$U^n = \sum_{j=1}^{N} U_j^n \cdot V^{(j)}$$

$$= \sum_{j=1}^{N} \left[U_{0j} z^n(\Omega_j) - \frac{Q_j}{\Omega_j} (1 - z^n(\Omega_j)) \right] \cdot V^{(j)} \qquad (10.1.48)$$

This is to be compared with the exact solution (10.1.16).

The error in amplitude is defined by the ratio of the modulus of $|z|$ and the amplitude of the analytical solution (10.1.24) with $\Omega = \xi + I\eta$:

$$\varepsilon_D = \frac{|z|}{e^{\xi \Delta t}} \qquad (10.1.49)$$

and the phase error will be defined by the phase Φ of z, writing

$$z = |z| e^{I\Phi} \qquad (10.1.50)$$

and

$$\varepsilon_\phi = \Phi / (\eta \Delta t) \qquad (10.1.51)$$

10.2   THE SPECTRA OF SPACE-DISCRETIZED OPERATORS

The matrix S represents the space discretization including the boundary conditions, and in order to define the precise relation between space-discretization and time-integration methods it is essential to determine the spectrum of the eigenvalues Ω. This is investigated separately for the diffusion and the convection equation for different sets of boundary conditions.

10.2.1   The spectrum for the diffusion equation $u_t = \alpha u_{xx}$

Referring to Table 8.1, the central difference operator S is

$$S = \frac{\alpha}{\Delta x^2} (E - 2 + E^{-1}) \qquad (10.2.1)$$

and its matrix representation is given in the third column, without the inclusion of the boundary points. The complete matrix has to include these points and this is dependent on the type of boundary condition and their numerical implementation.

(1) If we assume Dirichlet conditions, that is,

$$u_0 = f(0) = a \qquad u_N = f(N) = b \text{ over the interval } 0 \leqslant x \leqslant L$$

the equation for the first mesh point, $i = 1$, will be

$$\frac{du_1}{dt} = \frac{\alpha}{\Delta x^2} (a - 2u_1 + u_2) \qquad (10.2.2)$$

and the equation for the last point $i = N - 1$

$$\frac{du_{N-1}}{dt} = \frac{\alpha}{\Delta x^2} (b - 2u_{N-1} + u_{N-2}) \tag{10.2.3}$$

This completes the determination of the matrix S representing the space discretization.

The system of ordinary differential equations can be written as

$$\frac{dU}{dt} = \frac{\alpha}{\Delta x^2} \begin{vmatrix} -2 & 1 & & & \\ 1 & -2 & 1 & & \\ & 1 & -2 & 1 & \\ & & \cdot & \cdot & \\ & & & \cdot & \\ & & & 1 & -2 \end{vmatrix} \begin{vmatrix} u_1 \\ \cdot \\ u_i \\ \cdot \\ \cdot \\ u_{N-1} \end{vmatrix} + \begin{vmatrix} \dfrac{\alpha a}{\Delta x^2} \\ \cdot \\ \cdot \\ \cdot \\ \dfrac{\alpha b}{\Delta x^2} \end{vmatrix} \equiv S \cdot U + Q \tag{10.2.4}$$

The introduction of the boundary conditions has also led to the definition of the non-homogeneous term Q containing the boundary contributions.

(2) When Von Neumann boundary conditions are imposed at an end-point, for instance,

$$\frac{\partial u}{\partial x} = a \text{ at } x = 0 \text{ and } u_N = b \tag{10.2.5}$$

the first equation, for mesh point $x_0 = 0$, could be written from a one-sided difference of the derivative condition as

$$\frac{1}{\Delta x} (u_1 - u_0) = a \tag{10.2.6}$$

leading to the equation for mesh point x_1, replacing u_0 in the second derivative:

$$\frac{du_1}{dt} = \frac{\alpha}{\Delta x^2} (-u_1 + u_2) - \frac{a\alpha}{\Delta x} \tag{10.2.7}$$

We obtain the matrix equation

$$\frac{dU}{dt} = \frac{\alpha}{\Delta x^2} \begin{vmatrix} -1 & 1 & & & \\ 1 & -2 & 1 & & \\ & 1 & -2 & 1 & \\ & & 1 & -2 & 1 \\ & & & 1 & -2 \end{vmatrix} \begin{vmatrix} u_1 \\ \cdot \\ \cdot \\ \cdot \\ u_{N-1} \end{vmatrix} + \begin{vmatrix} -\dfrac{a\alpha}{\Delta x} \\ \cdot \\ \cdot \\ \dfrac{\alpha b}{\Delta x^2} \end{vmatrix} \tag{10.2.8}$$

This matrix is different from the previous one in equation (10.2.4)) and will have *different eigenvalues and eigenvectors*. If other implementations of the boundary conditions are selected, other matrices will be obtained and some could lead to eigenvalues which are unstable, that is, which have positive real parts. This would be an indication that the problem is not well posed numerically or analytically.

(3) *Periodic boundary conditions* imply that

$$u_0 = u_N \text{ and } u_{-1} = u_{N-1} \qquad (10.2.9)$$

The first equation is written here for u_0 as

$$\frac{du_0}{dt} = \frac{\alpha}{\Delta x^2}(u_{N-1} - 2u_0 + u_1) \qquad (10.2.10)$$

and the last equation, written for u_{N-1}, is

$$\frac{du_{N-1}}{dt} = \frac{\alpha}{\Delta x^2}(u_0 - 2u_{N-1} + u_{N-2}) \qquad (10.2.11)$$

We have the same matrix structure as is obtained from Dirichlet conditions but with the addition of a coefficient 1 in the upper right and lower left corners:

$$\frac{dU}{dt} = \frac{\alpha}{\Delta x^2}
\begin{vmatrix}
-2 & 1 & \cdot & \cdot & & 1 \\
1 & -2 & 1 & & & \cdot \\
\cdot & 1 & -2 & 1 & & \cdot \\
& & & \ddots & & \\
& & & 1 & -2 & 1 \\
1 & & & & 1 & -2
\end{vmatrix}
\begin{vmatrix}
u_0 \\
\cdot \\
\cdot \\
u_i \\
\vdots \\
u_{N-1}
\end{vmatrix}
\qquad (10.2.12)$$

Banded matrix notation

The matrices obtained for the diffusion equation have a very simple form; they have a *banded* structure with three non-zero diagonals (and two non-zero corner elements for the periodic matrices). As can be seen from Table 8.1, this is also the case for the matrices associated with the convection equations, and results from the three-point difference schemes applied to the space derivatives. If higher-order formulas would have been applied (for instance, a fourth-order difference approximation for u_{xx}) we would obtain five non-zero diagonals but still maintain the same banded structure. In two and three dimensions, with finite differences or finite element space discretizations, we would also obtain a similar structure, constant coefficients along diagonals if the problem is linear with constant coefficients and the mesh Cartesian and uniform (see also Problems 10.2 and 10.3).

Following Lomax (1976) we introduce a compact notation for a banded matrix:

$$B(...b_{-2}, b_{-1}, b_0, b_1, b_2, ...) \equiv \begin{vmatrix} b_0 & b_1 & b_2 & \cdot & \cdot & & & & \\ b_{-1} & b_0 & b_1 & b_2 & & & & & \\ b_{-2} & b_{-1} & b_0 & b_1 & b_2 & & & & \\ & b_{-2} & b_{-1} & b_0 & b_1 & b_2 & & & \\ & & \cdot & \cdot & \cdot & \cdot & \cdot & & \\ & & & \cdot & \cdot & \cdot & \cdot & \cdot & \\ & & & & \cdot & \cdot & \cdot & \cdot & \cdot \\ & & & & & b_{-2} & b_{-1} & b_0 & b_1 & b_2 \\ & & & & & & b_{-2} & b_{-1} & b_0 & b_1 \end{vmatrix}$$

$$(10.2.13)$$

where $b_{\pm k}$ represent the elements on the kth diagonal above or below the main diagonal $k = 0$. With this notation the tridiagonal matrices of equation (10.2.4) can be written as $B(1, -2, 1)$, and equation (10.2.4) as

$$\frac{dU}{dt} = \frac{\alpha}{\Delta x^2} B(1, -2, 1)U + Q \qquad (10.2.14)$$

Hence this provides a short-hand notation for the matrix representation of the space-discretization operator S, which is associated here with a finite difference formula of second-order accuracy for the second derivative operator:

$$S = \frac{\alpha}{\Delta x^2}(E - 2 + E^{-1}) => \frac{\alpha}{\Delta x^2} B(1, -2, 1) \qquad (10.2.15)$$

The periodic matrices with 1 in the two corners will be indicated by $B_p(b_{-2}, b_{-1}, b_0, b_1, b_2, ...)$; the matrix in equation (10.2.12) is written as $B_p(1, -2, 1)$ and the equation as

$$\frac{dU}{dt} = \frac{\alpha}{\Delta x^2} B_p(1, -2, 1)U + Q \qquad (10.2.16)$$

The matrix in equation (10.2.8), arising from the imposition of Von Neumann boundary conditions, is tridiagonal but the coefficients are not equal on the diagonals. If all the elements of a given diagonal are grouped into a vector $\vec{b}$ we can write the matrix of equation (10.2.8) as

$$B(1, \vec{b}, 1) \text{ with } \vec{b} = (-1, -2, -2, ..., -2)^T \qquad (10.2.17)$$

Equation (10.2.12) reads

$$\frac{dU}{dt} = \frac{\alpha}{\Delta x^2} B(1, \vec{b}, 1)U + Q \qquad (10.2.18)$$

Eigenvalues and eigenvectors of tridiagonal matrices

It can be shown (Lomax, 1976) that the eigenvalues of a general tridiagonal matrix $B(a, b, c)$ of order N, solutions of

$$\det | B(a, b - \Omega, c) | = 0 \qquad (10.2.19)$$

are given by

$$\Omega_j = b + 2\sqrt{(ac)}\cos\left(\frac{j\pi}{N+1}\right) \qquad j = 1, ..., N \qquad (10.2.20)$$

and the corresponding eigenvectors $V^{(j)}$ with the components $v_i^{(j)}$ are defined up to an arbitrary normalization factor K_j by

$$v_i^{(j)} = K_j \left(\frac{a}{c}\right)^{(i-1)/2} \sin i\left(\frac{j\pi}{N+1}\right) \qquad i = 1, ..., N \qquad (10.2.21)$$

Observe that the eigenvectors are not dependent on the main diagonal term b.

For periodic matrices we have, for the eigenvalues Ω_j, solutions of

$$\det | B_p(a, b - \Omega, c) | = 0 \qquad (10.2.22)$$

$$\Omega_j = b + (a + c)\cos\frac{2\pi j}{N} - I(a - c)\sin\frac{2\pi j}{N} \qquad j = 1, ..., N \qquad (10.2.23)$$

or

$$\Omega_j = b + a\, e^{-I2\pi j/N} + c\, e^{I2\pi j/N} \qquad (10.2.24)$$

with the eigenvector $V^{(j)}$ with components $v_i^{(j)}$:

$$v_i^{(j)} = K_j\, e^{I(2\pi j/N)i} \qquad i = 1, ..., N \qquad (10.2.25)$$

Note the very remarkable fact that these eigenvectors are independent of a, b, c but depend only on the structure of the matrix. Actually, the periodic matrices (10.2.12) are a particular case of a *circulant* matrix, where two consecutive lines differ by a permutation of their elements such that the last element of line i is the first element of line $(i + 1)$, the others being shifted by one column position. The structure of such a matrix is

$$C \equiv \begin{vmatrix} c_1 & c_2 & c_3 & \cdot & \cdot & \cdot & c_{N-1} & c_N \\ c_N & c_1 & c_2 & \cdot & \cdot & \cdot & c_{N-2} & c_{N-1} \\ c_{N-1} & c_N & c_1 & \cdot & \cdot & \cdot & c_{N-3} & c_{N-2} \\ & & & & & & & \\ c_2 & c_3 & c_4 & \cdot & \cdot & \cdot & c_N & c_1 \end{vmatrix} \qquad (10.2.26)$$

and its eigenvalues, solutions of $\det | C - \Omega I | = 0$, are given by (Varga, 1962)

$$\Omega_j = \sum_{i=0}^{N-1} c_{i+1}\, e^{Ii2\pi(j-1)/N} \qquad j = 1, ..., N \qquad (10.2.27)$$

with the eigenvectors (10.2.25) independent of the c_{i+1} coefficients.

We recognize these eigenvectors as the Fourier components $e^{Ii\phi}$ with ϕ defined as in equation (8.1.12). The factor 2 appearing in equation (10.2.25) comes from the fact that the periodicity is taken here from $x = 0$ to $x = L$, while equation (8.1.12) results from a periodicity from $x = -L$ to $x = L$ after a reflection of the function $u(x)$ onto the negative axis. This demonstrates that the Fourier analysis and the matrix analysis are identical for periodic matrices, which was the assumption at the basis of the Von Neumann method.

Eigenvalues for the diffusion equation

If we consider equation (10.2.4) for the Dirichlet boundary conditions the eigenvalues are, with $a = c = 1$, $b = -2$,

$$\Omega j = \frac{-4\alpha}{\Delta x^2} \sin^2\left(\frac{\pi j}{2(N+1)}\right) \qquad j = 1, ..., N \qquad (10.2.28)$$

For the periodic boundary conditions we obtain from equation (10.2.24)

$$\Omega_j = \frac{-4\alpha}{\Delta x^2} \sin^2\left(\frac{\pi j}{N}\right) \qquad j = 1, ..., N \qquad (10.2.29)$$

For the matrix (10.2.8), obtained from Von Neumann boundary conditions at one end and Dirichlet conditions at the other boundary, we cannot apply the general formula (10.2.20) since all the diagonal elements are not equal. However, for this particular case a closed form for the eigenvalues and eigenvectors can still be obtained (Desideri and Lomax, 1981) as

$$\Omega_j = -\frac{4\alpha}{\Delta x^2} \sin^2\frac{(2j-1)\pi}{(2N+1)2} \qquad j = 1, ..., N \qquad (10.2.30)$$

and the eigenvectors $V^{(j)}$ have the components, with a normalization factor K_j,

$$v_i^{(j)} = K_j \sin\left(\frac{i(2j-1)\pi}{2N+1}\right) \qquad i = 1, ..., N \qquad (10.2.31)$$

As can be seen, all three types of boundary conditions lead to a stable system of ordinary differential equations for the second-order central difference approximation of u_{xx}, since the stability condition (10.1.17) is satisfied for $\alpha > 0$. Observe that all the *eigenvalues are real and negative* for this problem when α is considered as positive. A negative diffusion coefficient will lead to an unstable, exponentially increasing solution in time. The eigenfunction index j can be considered as a 'frequency parameter' in space, since for small j, say $j = 1$, the eigenfunction is smooth over the interval $i = 1, ..., N$, while for the highest value of j, $j = N$, the eigenfunction $V^{(N)}$ undergoes roughly $N/2$ oscillations, representing a high-frequency behaviour (see also Problem 10.6).

By now, it should be clear to the reader, by comparing equations (10.2.28)–(10.2.30), that the influence of the boundary conditions is an integral and undissociable part of the matrix method, since the eigenvalue and eigenvector structure is strongly dependent on the nature and the implementation of these conditions. Clearly, all types of conditions lead to stable schemes for the diffusion equation, but, as we will see next, this is no longer the case for the convective hyperbolic equations.

10.2.2 The spectrum for the convection equation $u_t + au_x = 0$

For convection equations the problem of the boundary conditions is much more severe and its impact on the schemes is often dominating. The reason for this strong influence is to be found in the physical nature of the convection or propagation phenomena, since they describe 'directional' phenomena in space. For instance, in one dimension, the propagation or convection of the quantity u in the positive x-direction, when $a > 0$, implies that the value of u at the downstream end-point of the computational region $0 \leqslant x \leqslant L$ is determined by the upstream behaviour of u and cannot be imposed arbitrarily, as was the case with the diffusion equation. However, from a numerical point of view, depending on the space discretization, we might need information on u_N in order to close the algebraic system of equations. In this case, the condition to be imposed on $u_N = u(x = L)$ cannot be taken from physical sources and has to be defined numerically. This is called a *numerical boundary* condition. The choice of this numerical boundary condition is critical for the whole scheme. Intuitively we suspect that a good choice should be compatible with the physics of the problem, but the stability analysis is the ultimate criterion.

In general, this is a difficult task since, as we will see next, even for very simple cases we cannot find analytically the eigenvalues of the matrix S. Actually, the analysis of initial boundary value problems for hyperbolic equations is very complex from a mathematical point of view, since the boundary conditions to be imposed are dependent on the structure of the equations. For $a > 0$, we have obviously to impose, on physical grounds, boundary values at the left end of the domain; while when a is negative, the propagation occurs in the negative x-direction and the physical boundary condition has to be defined at the right end of the domain. For a hyperbolic system of equations, with the simultaneous presence of positive and negative propagation speeds (as is the case for the Euler equations in subsonic flows) the problem becomes more complex. More detailed information in the framework of the Euler equations can be found in Volume 2. The reader interested in more mathematically oriented work can refer to Kreiss (1968, 1970), Gustafsson *et al.* (1972) and Gustafsson and Kreiss (1979), and to an interesting review by Yee (1981), where additional references can be found.

Upwind space discretization

Let us consider a first-order upwind scheme:

$$\frac{du_i}{dt} = -\frac{a}{\Delta x}(u_i - u_{i-1})$$ (10.2.32)

for the interior points (also called the *interior scheme*). A boundary value

$$u(0, t) = g(t) \qquad x = 0$$ (10.2.33)

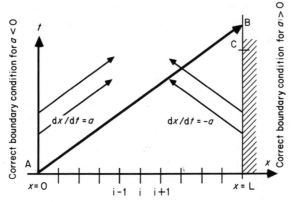

Figure 10.2.1 Boundary conditions for hyperbolic problems

will be imposed at the left end boundary, when $a > 0$. For the first mesh point $i = 1$ we would write (Figure 10.2.1)

$$\frac{du_1}{dt} = -\frac{a}{\Delta x}(u_1 - u_0) = -\frac{a}{\Delta x}u_1 + \frac{a}{\Delta x}g(t) \qquad (10.2.34)$$

At the downstream end the equation for $i = N$ is

$$\frac{du_N}{dt} = -\frac{a}{\Delta x}(u_N - u_{N-1}) \qquad (10.2.35)$$

and no additional, numerical condition is necessary. The semi-discretized system (10.1.2) becomes in this case,

$$\frac{dU}{dt} = \frac{-a}{\Delta x}\begin{vmatrix} 1 & 0 & & & & & \\ -1 & 1 & 0 & & & & \\ & -1 & 1 & & & & \\ & & & \cdot & & & \\ & & & & \cdot & & \\ & & & & -1 & 1 & 0 \\ & & & & & -1 & 1 \end{vmatrix}\begin{vmatrix} u_1 \\ \cdot \\ u_{i-1} \\ u_i \\ u_{i+1} \\ \vdots \\ u_N \end{vmatrix} + \begin{vmatrix} \dfrac{a}{\Delta x}g(t) \\ 0 \\ \cdot \\ \cdot \\ \cdot \\ \\ 0 \end{vmatrix}$$

$$(10.2.36)$$

The eigenvalues of this matrix are all identical and equal to

$$\Omega_j = \frac{-a}{\Delta x} \qquad (10.2.37)$$

Since these eigenvalues are negative, when $a > 0$, system (10.2.36) is stable and a suitable time-integration scheme is easily found, as seen by several examples in previous chapters. However, the matrix is unstable for $a < 0$.

In order to illustrate the dominating influence of boundary conditions let us look at what happens if we impose the physical condition at the downstream end. In this case, there is no information on u_0, to start equation (10.2.34), and the last equation for u_N is overruled by the boundary condition $u_N = g(t)$. However, the last equation

$$\frac{du_{N-1}}{dt} = \frac{-a}{\Delta x}(u_{N-1} - u_{N-2}) \qquad (10.2.38)$$

is separated from the physical information at u_N. Hence there is no way of ensuring that a numerical boundary condition at u_0 will lead to a downstream value satisfying the boundary condition $u_N = g(t)$, and the problem is not well posed. This is illustrated in Figure 10.2.1, where the characteristic $dx/dt = a$ is shown, along which u is constant. If A represents the value of u_0, the corresponding value of u_N is B and imposing $u_N = C$, for instance, is not compatible with $u_0 = A$. In order to stabilize the scheme for $a < 0$ we have to impose a boundary condition at $i = N$, $u_N = g(t)$ and apply a forward space discretization (see Problem 10.7).

Central space discretization

Applying a central space difference to the convection equation, and considering periodic boundary conditions, leads to the periodic version of the matrix structure obtained in Table 8.1 for the wave equation, written here as

$$\frac{dU}{dt} = \frac{-a}{2\Delta x} B_p(-1, 0, +1)U \qquad (10.2.39)$$

The eigenvalues are obtained from equation (10.2.23) and are purely imaginary:

$$\Omega_j = -\frac{a}{\Delta x} I \sin\left(\frac{2\pi j}{N}\right) \qquad j = 1, \ldots, N \qquad (10.2.40)$$

indicating that system (10.2.39) is stable for positive as well as negative values of a.

What is shown here is that certain time-integration methods for system (10.2.39) can lead to a stable numerical scheme. Equation (8.1.18) shows that the explicit Euler method is unstable for this space discretization, but others, such as the Euler implicit or the leapfrog method, will be stable for the same space discretization.

If we consider the boundary conditions (10.2.33) for $a > 0$, then the first equation would be written for $u = u_1$ as

$$\frac{du_1}{dt} = -\frac{a}{2\Delta x}(u_2 - u_0) = -\frac{a}{2\Delta x}u_2 + \frac{a}{2\Delta x}g(t) \qquad (10.2.41)$$

At the downstream end a numerical boundary condition has to be imposed in this case, since the central differenced equation for u_N would contain a contribution from a non-existent point u_{N+1}. A reasonable approach is to apply a backward difference in space at $u = U_N$ instead of a central difference. This leads to

$$\frac{du_N}{dt} = -\frac{a}{\Delta x}(u_N - u_{N-1}) \tag{10.2.42}$$

and the matrix equation (10.1.2) becomes

$$\frac{dU}{dt} = -\frac{a}{2\Delta x}
\begin{vmatrix}
0 & 1 & & & & \\
-1 & 0 & 1 & & & \\
 & -1 & 0 & 1 & & \\
 & & & 0 & & \\
 & & & & \cdot & \\
 & & -1 & 0 & 1 & \\
 & & & -1 & 0 & 1 \\
 & & & & -2 & 2
\end{vmatrix}
\begin{vmatrix}
u_1 \\ \vdots \\ u_{i-1} \\ u_i \\ u_{i+1} \\ \vdots \\ u_N
\end{vmatrix}
+
\begin{vmatrix}
\frac{a}{2\Delta x} g(t) \\ 0 \\ \\ \\ \\ \\ 0
\end{vmatrix}
\tag{10.2.43}$$

This matrix has no analytically defined eigenvalues, and they have to be computed numerically. Gary (1978) has shown that this matrix is stable. The distribution of the complex eigenvalues is shown for $N = 15$ on Figure 10.2.2 and the real part of the eigenvectors for the lowest and highest eigenvalues, in modulus, are illustrated in Figure 10.2.3.

As is seen on this example, all the eigenvalues have non-positive real parts and the system is stable. This has also been shown theoretically by Dahlquist (1978). If the boundary condition would have been taken at the downstream end, $i = N$ for $a > 0$ or if the boundary equation is applied for $a < 0$ (which is equivalent to implementing a non-physically justified boundary condition at $x = 0$) the real parts of the eigenvalues become positive and system (10.2.43) is unstable, as expected from the above-mentioned considerations.

It is interesting to observe that the upwind scheme has a (unique) eigenvalue (10.2.37), which is real and negative. This will generate a contribution $e^{-\Omega t}$ in solution (10.1.12), multiplied by a polynomial in t, due to the multiplicity of the eigenvalue. The important aspect to be noted is the damping of the solution of the space-discretized equations, introduced by the negative exponential there, where the analytical solution of the equation $u_t + au_x = 0$ is a pure wave with no damping. Hence the appearance of negative real parts in the eigenvalues of the space discretization operator of convective hyperbolic equations, indicates that this operator has generated a numerical dissipation. This is to be put in relation with the numerical viscosity appearing in the equivalent differential equation as well as to the presence of even-order

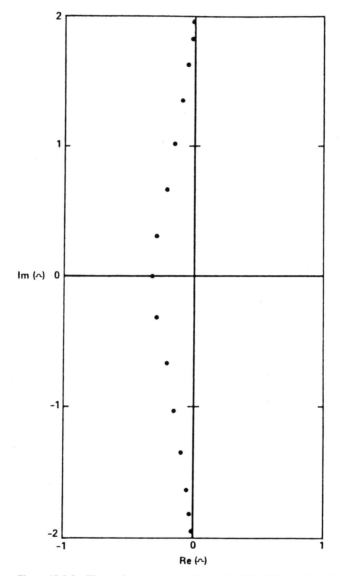

Figure 10.2.2 Eigenvalue spectrum of matrix (10.2.43) for $N = 15$
(Reproduced by permission of AIAA from Lomax *et al.*, 1981.)

derivatives in the development of the truncation error. For instance, the leapfrog scheme applied to the central differenced convective equation does not generate dissipation, as will be seen from Figure 10.3.2 below, and neither has even-order derivatives in the truncation error.

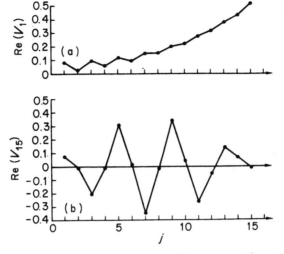

Figure 10.2.3 Real part of the eigenvectors of matrix
(10.2.43) for $N = 1$ and $N = 15$. (Reproduced by permission of
AIAA from Lomax *et al.*, 1981.)

10.3 THE STABILITY OF TIME-INTEGRATION SCHEMES

We can now investigate the stability of various time integration methods when
applied to the space discretizations introduced in the previous sections.

10.3.1 Euler explicit scheme

This scheme corresponds to a forward difference of the time derivative,
leading to the modal equation

$$w^{n+1} = w^n + \Omega \, \Delta t \, w^n + q^n \tag{10.3.1}$$

Hence

$$
\begin{aligned}
C &= 1 + \Omega \, \Delta t \\
P_1(\bar{E}) &= \bar{E} - (1 + \Omega \, \Delta t) = \bar{E} - C \\
P_2(\bar{E}) &= 1
\end{aligned}
\tag{10.3.2}
$$

The characteristic polynomial is of degree one, as for all explicit, two-time-
level schemes:

$$P_1(z) = z - (1 + \Omega \, \Delta t) = 0 \tag{10.3.3}$$

and a single root exists:

$$z = 1 + \Omega \, \Delta t \tag{10.3.4}$$

The scheme is stable for all space discretizations associated with an eigenvalue spectrum such that

$$|1 + \Omega \, \Delta t| \leqslant 1 \tag{10.3.5}$$

or

$$[1 + Re(\Omega \, \Delta t)]^2 + (Im \, \Omega \, \Delta t)^2 \leqslant 1 \tag{10.3.6}$$

Diffusion equation with scheme (10.2.4)

This is the scheme

$$u_i^{n+1} - u_i^n = \frac{\alpha \, \Delta t}{\Delta x^2} (u_{i+1}^n - 2u_i^n + u_{i-1}^n) \tag{10.3.7}$$

All Ω_j are real and negative, as seen from equations (10.2.28)–(10.2.30), and the stability condition is

$$-2 \leqslant - |Re(\Omega \, \Delta t)| \leqslant 0 \tag{10.3.8}$$

or

$$0 \leqslant \frac{\alpha \, \Delta t}{\Delta x^2} \leqslant \frac{1}{2} \tag{10.3.9}$$

as derived by the Von Neumann method in Chapter 8.

Convection equation with central scheme (10.2.39)

This is the scheme

$$u_i^{n+1} - u_i^n = -\frac{\sigma}{2} (u_{i+1}^n - u_{i-1}^n) \tag{10.3.10}$$

with periodic boundary conditions and

$$\sigma = a \, \Delta t / \Delta x \tag{10.3.11}$$

All eigenvalues Ω_j are purely imaginary, and the stability condition

$$1 + [Im(\Omega \, \Delta t)]^2 \leqslant 1 \tag{10.3.12}$$

is never satisfied. The scheme is therefore unstable.

Convection equation with upwind scheme (10.2.36)

This is the scheme

$$u_i^{n+1} - u_i^n = -\sigma(u_i^n - u_{i-1}^n) \tag{10.3.13}$$

The eigenvalues $\Omega_j = -a/\Delta x$ and the necessary stability condition is

$$0 \leqslant \frac{a \, \Delta t}{\Delta x} \leqslant 2 \tag{10.3.14}$$

This is in contradiction to the more severe CFL condition (8.1.21) obtained from a Von Neumann analysis. As shown in Figure 8.1.2, the scheme is indeed unstable for $a\,\Delta t/\Delta x > 1$. The reasons for this contradiction will be discussed in Section 10.4.

10.3.2 Leapfrog method

This is an explicit three-level, two-step method:

$$w^{n+1} - w^{n-1} = 2\,\Delta t\Omega w^n + q^n \tag{10.3.15}$$

Hence

$$P_1(\bar{E}) = \bar{E} - 2\Delta t\Omega - \bar{E}^{-1}$$
$$P_2(\bar{E}) = 1 \tag{10.3.16}$$

The characteristic polynomial is of second degree for this two-step method:

$$P_1(z) = z - 2\Omega\,\Delta t - 1/z = 0 \tag{10.3.17}$$

or

$$z^2 - 2\Omega\,\Delta t z - 1 = 0 \tag{10.3.18}$$

There are two roots:

$$z = +(\Omega\,\Delta t) \pm \sqrt{[(\Omega\,\Delta t)^2 + 1]} \tag{10.3.19}$$

which behave as follows, when $\Delta t \to 0$:

$$z_1 = \Omega\,\Delta t + \sqrt{[(\Omega\,\Delta t)^2 + 1]} = 1 + (\Omega\,\Delta t) + \frac{(\Omega\,\Delta t)^2}{2} + \cdots \tag{10.3.20}$$

$$z_2 = \Omega\,\Delta t - \sqrt{[(\Omega\,\Delta t)^2 + 1]} = -1 + \Omega\,\Delta t - \frac{(\Omega\,\Delta t)^2}{2} - \cdots \tag{10.3.21}$$

The first root is the physical one, while the second starts at -1 and is the spurious root.

Diffusion equation with scheme (10.2.4)

Since all Ω_j are real and negative the modulus of the second root z_2 becomes greater than one and the scheme is unconditionally *unstable*.

Convection equation with central scheme (10.2.39)

This is the scheme with periodic boundary conditions:

$$u_i^{n+1} - u_i^{n-1} = -\sigma(u_{i+1}^n - u_{i-1}^n) \tag{10.3.22}$$

All eigenvalues are purely imaginary and both roots are of modulus one when

$|\Omega \, \Delta t| \leqslant 1$:

$$|z_1| = |z_2| = 1 \text{ for } |\Omega \, \Delta t| \leqslant 1 \tag{10.3.23}$$

and the scheme is stable. With the eigenvalues (10.2.40) this is the CFL condition

$$\left| \frac{a \, \Delta t}{\Delta x} \right| \leqslant 1 \tag{10.3.24}$$

For $|\Omega \, \Delta t| > 1$, we have

$$|z_1| = |\Omega \, \Delta t \pm \sqrt{[\Omega^2 \, \Delta t^2 - 1]}| \tag{10.3.25}$$

and the scheme is unstable.

Convection equation with upwind scheme (10.2.36)

This is the scheme

$$u_i^{n+1} - u_i^{n-1} = -2\sigma(u_i^n - u_{i-1}^n) \tag{10.3.26}$$

Since all the eigenvalues are negative and real, the scheme is unstable.

10.3.3 Euler implicit (backward) scheme

This corresponds to the backward difference in time:

$$w^{n+1} = w^n + \Omega \, \Delta t w^{n+1} + q^{n+1} \tag{10.3.27}$$

leading to

$$C = \frac{1}{1 - \Omega \, \Delta t}$$
$$P_1(\bar{E}) = (1 - \Omega \, \Delta t)\bar{E} - 1 = C^{-1}\bar{E} - 1 \tag{10.3.28}$$
$$P_2(\bar{E}) = \bar{E}$$

The characteristic polynomial is of the form

$$P_1(z) = (1 - \Omega \, \Delta t)z - 1 = 0 \tag{10.3.29}$$

leading to the amplification factor

$$z = \frac{1}{1 - \Omega \, \Delta t} \tag{10.3.30}$$

This scheme will therefore be stable for all well-posed space discretizations for which $Re(\Omega \, \Delta t) \leqslant 0$.

10.3.4 Stability region in the complex Ω plane

Each time-discretization scheme is represented by a unique relation between z and Ω, determined by the solutions $z = z(\Omega)$ of $P_1(z) = 0$. When the stability

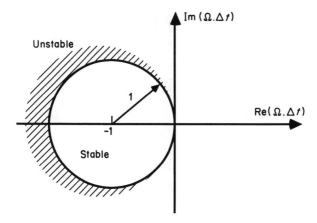

Figure 10.3.1 Stability region for the Euler explicit scheme in the complex ($\Omega \Delta t$) plane

limit for z (equation (10.1.42)) is introduced into this equation, under the form $z = e^{i\theta}$, $0 \leqslant \theta \leqslant 2\pi$, representing the stability circle of Figure 10.1.1, a corresponding stability region is defined in the ($\Omega \, \Delta t$)-plane through the mapping function $z = z(\Omega)$.

For the explicit Euler scheme we have $z = 1 + \Omega \, \Delta t$, and in the ($\Omega \, \Delta t$)-plane the stability region is a circle of radius 1 centred around $\Omega \, \Delta t = -1$ (Figure 10.3.1).

Schemes such as the central differenced convection equation, with their eigenvalue spectrum purely imaginary, are outside the stability circle and are unstable.

For the leapfrog scheme, we have

$$2(\Omega \, \Delta t) = z - \frac{1}{z} \qquad (10.3.31)$$

which becomes for $z = e^{i\theta}$

$$(\Omega \, \Delta t) = I \sin \theta \qquad (10.3.32)$$

The stability region of the leapfrog scheme is therefore a strip of amplitude ± 1 along the imaginary axis (Figure 10.3.2). It is seen immediately that the diffusion equation with its real negative eigenvalues or the upwind convection equation are unstable. Since negative values of ($\Omega \, \Delta t$) correspond to the presence of a dissipative mechanism (or numerical viscosity), for hyperbolic equations, it is seen that the leapfrog scheme is totally unadapted to the presence of dissipative terms, see however problem 10.18.

For the implicit Euler scheme we have

$$(\Omega \, \Delta t) = (z - 1)/z \qquad (10.3.33)$$

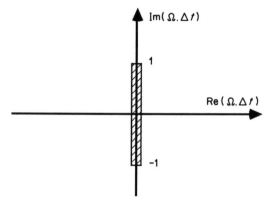

Figure 10.3.2 Stability region for the leapfrog scheme

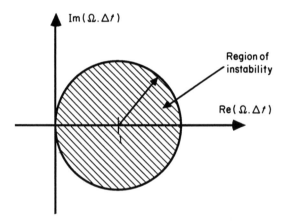

Figure 10.3.3 Stability region for implicit Euler scheme

For $z = e^{i\theta}$ the limit of the stability region is defined by

$$(\Omega \ \Delta t) = 1 - e^{i\theta} \qquad (10.3.34)$$

and represents a circle centred on $\Omega \ \Delta t = 1$ of radius 1 (Figure 10.3.3).

Since for $|z| < 1$, $|1 - \Omega \ \Delta t| > 1$ the stability region is outside the circle and is seen to cover even regions in the Ω plane where the space-discretized equations are unstable. Hence all the schemes seen so far will be stable with the implicit Euler time integration.

10.3.5 A realistic example (Eriksson and Rizzi, 1985)

The matrix method of analysis of stability and transient behaviour has been applied by Eriksson and Rizzi (1985) to a realistic computation of the full system of Euler equations for the flow around an airfoil at supercritical speeds, generating a shock wave on the suction surface of the airfoil. The purpose of the investigation was to analyse the influence of artificial dissipation and the least-damped transients which are responsible for the slowing down of the time-asymptotic convergence to steady state. As can be seen from equation (10.1.18), the eigenvalues with the lowest real negative part will remain at large times.

The system of Euler equations is discretized by central differences of the nature of equation (10.2.43), and a second-order accurate time-integration scheme is applied, with a single root z equal to

$$z = 1 + \Omega \, \Delta t + \frac{(\Omega \, \Delta t)^2}{2} + \frac{(\Omega \, \Delta t)^3}{4} \qquad (10.3.35)$$

All eigenvalues and eigenvectors were computed numerically.

As discussed previously, when the scheme does not generate any internal dissipation, as in the case of central differencing, artificial dissipation has to be introduced in order to counteract the non-linear instabilities, and the authors investigated the effect on the eigenvalues and eigenvectors of different forms of artificial viscosity.

The flow situation being considered is defined by an incident Mach number of $M = 0.8$ at $0°$ incidence on a NACA 0012 airfoil. The calculations

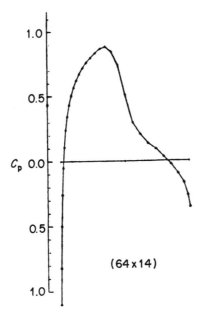

Figure 10.3.4 Pressure distribution on the NACA 0012 airfoil calculated on a 64×14 mesh at upstream Mach number of 0.8 and $0°$ incidence. (Reproduced by permission of Academic Press from Eriksson and Rizzi, 1985.)

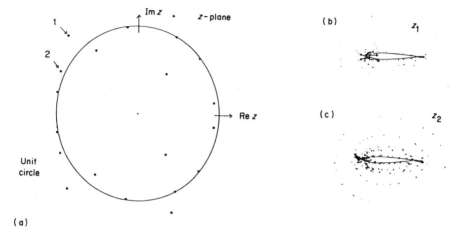

Figure 10.3.5 (a) Eigenvalue spectrum on a mesh of 32×7 points in the complex z-plane for a central discretization and no dissipation. (b) and (c) Imaginary part of eigenmodes of the velocity field associated with the unstable eigenvalues 1 and 2 on a mesh of 32×7 points. (Reproduced by permission of Academic Press from Eriksson and Rizzi, 1985)

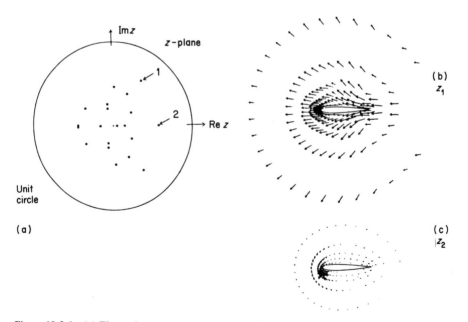

Figure 10.3.6 (a) Eigenvalue spectrum on a mesh of 32×7 points in the complex z-plane for a central discretization with dissipation. (b) and (c) Imaginary part of eigenmodes of the velocity field associated with the eigenvalues 1 and 2 on a mesh of 32×7 points with dissipation added. (Reproduced by permission of Academic Press from Eriksson and Rizzi, 1985.)

performed on a coarse mesh of (64×14) mesh points gave the pressure distribution shown in Figure 10.3.4. Two figures show here the effect of the dissipation on the otherwise unstable scheme. Figure 10.3.5 shows the eigenvalues in the z-plane when no dissipation is present, clearly indicating that some eigenvalues are outside the stability circle $|z| = 1$, on a mesh (32×7). The imaginary part of the velocity field modes, associated with the eigenvalues indicated 1 and 2, are shown in Figures 10.3.5(b) and 10.3.5(c). Adding dissipation stabilizes the scheme, and the modifications of the

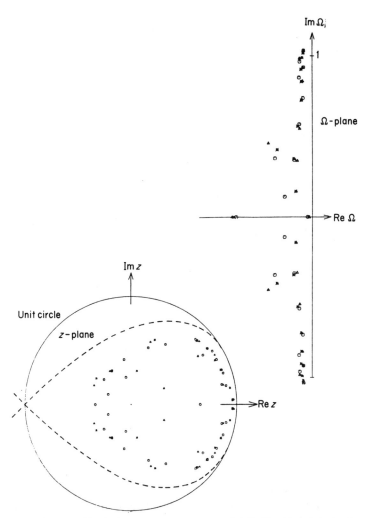

Figure 10.3.7 Eigenvalue spectrum on a mesh of 64×14 points in the complex Ω- and z-planes. For a central discretization with the addition of dissipation. (Reproduced by permission of Academic Press from Eriksson and Rizzi, 1985.)

spectrum and of the most unstable modes can be seen from Figure 10.3.6, which shows the new position and associated eigenvectors of the two least-damped eigenvalues 1 and 2. Comparing the structure of the eigenmodes it is seen that the unstable modes have a strong oscillatory structure, while the stabilized modes appear to behave very smoothly.

Similar behaviours can be investigated for finer meshes, and Figure 10.3.7 displays the spectra of eigenvalues in the Ω- and z-planes for a mesh 64×14 after addition of artificial viscosity terms. The reader will find more details and additional mode representations in the original reference.

10.4 EVALUATION OF STABILITY CRITERIA

The difference in implications of the two stability conditions (7.2.25) and (7.2.35) may appear, at first, rather academic to a non-mathematically oriented reader. They have, however, very deep and practically important consequences, particularly with regard to the matrix method, and these implications will be discussed in this section for two-level schemes. An example has already been shown with the necessary stability condition (10.3.14) for the upwind discretization on the convection equation, which is not sufficient for practical computations. According to the definition (7.2.25) of stability, the error e^n which is amplified by the matrix $(C)^n$, for a two-level scheme, should remain bounded at fixed Δt, for n going to infinity. This implies that

$$\| e^n \| \leqslant \| C^n \| \cdot \| e^0 \| < K \text{ at fixed } \Delta t, \, n \to \infty \qquad (10.4.1)$$

From the eigenmode decomposition (10.1.48) and the fact that z is to be considered as an eigenvalue of C, following equation (10.1.30), it is seen that the error behaves as

$$e^n = \sum_{j=1}^{N} e_{0j} z^n (\Omega_j) V^{(j)} \qquad (10.4.2)$$

and will tend to zero when n tends to infinity if $|z| < 1$. Hence since all eigenvalues have to be in absolute value lower than one, the stability condition (7.2.25) requires the spectral radius of C to be lower than one:

$$\rho(C) \leqslant 1 \qquad (10.4.3)$$

and eigenvalues $z = 1$ have to be simple. This condition therefore ensures that $\| e^n \| \to 0$ when $n \to \infty$.

Practical computations are, however, performed at finite values of n, and the above condition does not ensure that the norm $\| C^n \|$ does not become very large at finite values of n before decaying for n tending to infinity. In order to guarantee that this would not happen following equation (7.2.35) we have to require $\| C^n \|$ to remain uniformly bounded for all values of n, Δt (and Δx), such that

$$\| C^n \| < K \text{ for } 0 < n \, \Delta t \leqslant T \qquad (10.4.4)$$

In particular, this condition of bounded errors when $\Delta t \to 0$, $n \to \infty$ at fixed values of $(n \, \Delta t)$ is required by the equivalence theorem of Lax and the convergence condition (7.2.38). That is, the consistent difference scheme will converge to the exact solution when Δt, Δx tend to zero, *if and only if the difference scheme is stable for this refinement*.

Let us investigate the example of the upwind scheme applied to the convection equation. The associated matrix C is obtained from equations (10.2.36) and (10.3.2) as

$$C_u = \begin{vmatrix} 1-\sigma & 0 & & & \\ \sigma & 1-\sigma & 0 & & \\ 0 & \sigma & 1-\sigma & & \\ & & & & 0 \\ & & 0 & \sigma & (1-\sigma) \end{vmatrix} \qquad (10.4.5)$$

Its eigenvalues are $(1 - \sigma)$ and the condition (10.4.3) on the spectral norm leads to the necessary stability condition $0 < \sigma < 2$. On the other hand, condition (10.4.4) on the boundedness of the norm of the matrix C for all n, Δt and Δx is, in this particular case (Richtmyer and Morton, 1967),

$$\| C_u \|_{L_2} = \sigma + |1 - \sigma| + 0\left(\frac{1}{N}\right) < 1 \qquad (10.4.6)$$

if the matrix is of size $N \times N$.

The norm of C_u will remain uniformly bounded by one, *independently of Δt and Δx* for $\sigma \leqslant 1$. This condition is also obtained from the Von Neumann analysis, as shown in Chapter 8. A detailed investigation of the behaviour of the norm of matrix (10.4.5) in the range $1 < \sigma \leqslant 2$ at finite values of n has recently been performed by Hindmarsh *et al.* (1984). By evaluating analytically the maximum norm of C_u^n, $\| C_u^n \|_\infty$, it is found that, although $\| C_u^n \|_\infty$ tends to zero as $n \to \infty$ for $\sigma < 2$, it reaches very high values at some intermediate values of n. Typically, it is found that

$$\max_{n \geqslant 1} \| C_u^n \|_\infty \leqslant \left(\frac{\sigma}{2-\sigma}\right)^{N-1} \qquad (10.4.7)$$

which becomes very large at high values of N and σ close to two.

Figure 10.4.1 shows the behaviour of $\| C_u^n \|_\infty$ for $\sigma = 1.2$ as a function of n for $N = 40$ and $N = 100$. Observe the scale of the vertical axis and the Von Neumann amplification factor $G^n = (1.4)^n$, which follows the rise of $\| C_u^n \|_\infty$ quite closely. The values of the bounds for the maxima of $\| C_u^n \|_\infty$ as a function of N and σ are shown in Figure 10.4.2, and formula (10.4.7) is a good estimate. Hence for any value of σ in the range $1 < \sigma < 2$ errors will be considerably amplified before being damped when $n \to \infty$.

Referring to Section 10.3, the other examples do lead to the correct stability conditions, in agreement with the Von Neumann method. This is due to the fact that the considered matrices are symmetric or normal, so that in the L_2

402

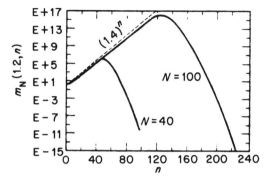

Figure 10.4.1 Growth rate of norm of $\| C_u^n \|_\infty$ as a function of n. (From Hindmarsh *et al.*, 1984)

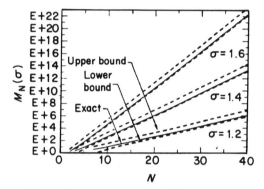

Figure 10.4.2 Maximum of norm $\| C_u^n \|_\infty$ as a function of N at different values of σ. (From Hindmarsh *et al.*, 1984)

norm

$$\| C^n \|_{L_2} = \rho^n(C) \tag{10.4.8}$$

and the condition on the spectral radius (equation (10.4.3)) is a necessary and sufficient one for stability according to equation (10.4.4). Otherwise we should apply definition (10.4.4) directly and attempt to estimate the norm $\| C^n \|$.

Although the Von Neumann stability condition (equation (8.2.24)) refers to the spectral radius of the amplification matrix G (which is the discrete Fourier symbol of $C(e^{I\phi})$, equation (8.2.17), or can be viewed as the eigenvalues of C for the Fourier modes), the Von Neumann analysis will mostly lead to the correct stability criteria even for non-periodic boundary conditions. The reason for this does not seem to be very clear. Most probably it is connected to the particular and unique properties of Fourier transforms and to the fact that conditions (8.2.24) apply to each Fourier mode separately, each mode being amplified independently from the others. As a consequence, a large range of situations can be defined for which the Von Neumann condition (8.2.24) is necessary and sufficient for stability according to definition (10.4.4) (see Richtmyer and Morton, 1967, for a review of some of these conditions).

Furthermore, it can be shown that in many circumstances the presence of non-periodic boundary conditions does not affect significantly the outcome of a Von Neumann analysis performed on the interior scheme (see, for instance, Godunov and Ryabenkii, 1963). Therefore if doubts arise from an analysis of the spectral radius of the discretization matrix C a Von Neumann analysis should be performed, and if a discrepancy occurs between the two results the Von Neumann condition is most likely to be more accurate.

10.4.1 The stability analysis of the convection–diffusion equation

Another spectacular example of the impact of the stability definition on its practical requirements are related to the controversial statements found in the literature for the space-centred, explicit Euler discretization of the time-dependent convection–diffusion equation. This apparently simple equation,

$$\frac{\partial u}{\partial t} + a\,\frac{\partial u}{\partial x} = \alpha\,\frac{\partial^2 u}{\partial x^2} \tag{10.4.9}$$

can be discretized with a second-order accurate scheme in space and first order in time:

$$u_i^{n+1} - u_i^n = -\frac{\sigma}{2}\,(u_{i+1}^n - u_{i-1}^n) + \beta(u_{i+1}^n - 2u_i^n + u_{i-1}^n) \tag{10.4.10a}$$

or

$$u_i^{n+1} = \left(\beta - \frac{\sigma}{2}\right) u_{i+1}^n + (1 - 2\beta)u_i^n + \left(\beta + \frac{\sigma}{2}\right) u_i^n \tag{10.4.10b}$$

with $\sigma = \alpha\,\Delta t/\Delta x$ and $\beta = \alpha\,\Delta t/\Delta x^2$.

The history of the stability conditions for scheme (10.4.10) is illustrative, on the one hand, of the difficulties of the Von Neumann analysis as soon as a scheme becomes more complex, even in one dimension, and, on the other, of the implications of the definitions of stability.

The mesh Reynolds number controversy

The first controversy is connected with the Von Neumann analysis and the erroneous statements sometimes found in the literature. Applying the standard technique of Chapter 8 we obtain the amplification factor

$$G - 1 = -I\sigma \sin\phi + 2\beta(\cos\phi - 1) \tag{10.4.11}$$

Historically, a first Von Neumann stability condition was incorrectly derived by Fromm (1964), and quoted in Roache (1972), as well as in more recent textbooks. From a polar plot of G, we see that $G(\phi)$ is on an ellipse centred at

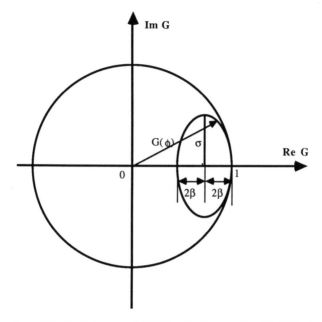

Figure 10.4.3 Polar plot of $G(\phi)$ for the discretization (10.4.10) of the convection–diffusion equation

$(1 - 2\beta)$ with semi-axes 2β and σ (Figure 10.4.3). By inspection of this plot we are tempted to define the stability condition

$$\sigma \leqslant 2\beta \leqslant 1 \qquad (10.4.12)$$

as deduced by Fromm (1964) in a two-dimensional case. This necessary condition is too restrictive, as will be shown next, and implies a limitation on the time step

$$\frac{\alpha \, \Delta t}{\Delta x^2} \leqslant \frac{1}{2} \qquad (10.4.13)$$

identical to that of the pure diffusion equation as well as a limitation on the space mesh size Δx, *independently of* Δt, namely

$$R = \sigma/\beta = \frac{a \, \Delta x}{\alpha} \leqslant 2 \qquad (10.4.14)$$

which is a limitation on the mesh Reynolds (or Peclet) number. This incorrect concept of a mesh size limitation for stability has generated considerable confusion (see, for instance, Thompson *et al.*, 1985, for an additional clarification).

The correct results were obtained initially by Hirt (1968), applying a different approach, but remained largely unnoticed, and an increase of interest in this subject has generated a variety of publications for one- and multi-dimensional stability analyses of the discretized convection–diffusion equation (Rigal, 1979; Leonard, 1980; Chan, 1984; Hindmarsh et al., 1984).

The intersections of the ellipse and the circle of radius one can be determined from their equation in the co-ordinates $\xi, \eta, \xi = ReG, \eta = ImG$:

$$\text{Ellipse: } \frac{[\xi - (1 - 2\beta)]^2}{4\beta^2} + \frac{\eta^2}{\sigma^2} = 1$$

$$\text{Circle: } \xi^2 + \eta^2 = 1$$

(10.4.15)

The stability condition $|G| \leq 1$ will be satisfied if the ellipse is wholly within the unit circle, implying that there is no intersection between the two curves next to the point $(1, 0)$. Eliminating η between the two equations (10.4.15) we obtain for the co-ordinates of the second intersection point:

$$\xi = \frac{(1 - 4\beta)\sigma^2 + 4\beta^2}{\sigma^2 - 4\beta^2} \quad \text{and} \quad \eta^2 = 1 - \xi^2$$

(10.4.16)

This value may not be smaller than one for stability, otherwise an intersection point with $\eta^2 = 1 - \xi^2 < 1$ would exist. Hence we should have

$$\sigma^2 \leq 2\beta$$

(10.4.17)

and the necessary and sufficient Von Neumann stability condition for $|G| \leq 1$ is given by

$$\sigma^2 \leq 2\beta \leq 1$$

(10.4.18)

These conditions can also be derived directly from equations (8.6.24).

The second condition $2\beta < 1$ also expresses that the coefficient of the u_i^n term in the right-hand side of equation (10.4.10b) has to be positive. A negative value indicates indeed that u_i^{n+1} decreases when u_i^n increases, which is contrary to physical diffusion effects.

Compared with condition (10.4.12), a second restriction on the time step appears as $\Delta t < 2\alpha/a^2$, which is independent of the mesh size Δx. Conditions (10.4.18) lead to

$$\Delta t \leq \text{Min}\left(\frac{2\alpha}{a^2}, \frac{\Delta x^2}{2\alpha}\right)$$

(10.4.19)

Observe that the scheme becomes unconditionally unstable when $\alpha \to 0$, as seen in Section 7.1. Hence the addition of a viscosity (or diffusion) term αu_{xx} has stabilized the otherwise unstable centrally discretized convection equation. The CFL condition $\sigma \leq 1$ is implied by equation (10.4.18) but is only necessary and certainly not sufficient. In terms of the cell Reynolds number R, the correct stability condition can be written, with $R = \sigma/\beta$, as

$$\sigma \leq \frac{R}{2} \leq \frac{1}{\sigma} \quad \text{or} \quad \sigma \leq \frac{2}{R} \leq \frac{1}{\sigma}$$

(10.4.20)

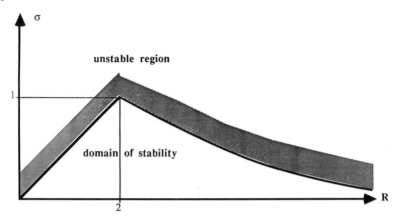

Figure 10.4.4   Stability region for the space-centred, explicit scheme (10.4.10)

and instead of the cell Reynolds number limitation (10.4.14) we obtain the condition

$$R \leqslant \frac{2}{\sigma} \tag{10.4.21}$$

showing that stable calculations can be performed with high values of the mesh Reynolds number. Figure 10.4.4 is a representation of the stability condition (10.4.20) in a (σ, R) diagram. The region below the curve is the region of stability. However, the value $R = 2$ plays a certain role, not with respect to stability but, as will be seen below, with regard to accuracy. If we consider scheme (10.4.10) as an iterative formulation for a steady-state problem it will be shown that oscillations in the numerical solution will appear when $R > 2$.

The stability criteria from the matrix method

In an investigation of the stability of the explicit central differenced convection–diffusion equation (10.4.10), Siemieniuch and Gladwell (1978) applied the matrix method with the criterion on the spectral radius $\rho(C) \leqslant 1$ for stability. The C matrix associated with this scheme is given, for Dirichlet boundary conditions, by

$$C = \begin{vmatrix} 1 - 2\beta & \beta - \sigma/2 & & & \\ \beta + \sigma/2 & 1 - 2\beta & \beta - \sigma/2 & & \\ & \beta + \sigma/2 & 1 - 2\beta & & \\ & & & \cdot & \cdot \\ & & & \beta + \sigma/2 & 1 - 2\beta \end{vmatrix} \tag{10.4.22}$$

Following equation (10.2.20) the eigenvalues $\lambda_j(C)$ are

$$\lambda_j(C) = 1 - 2\beta + 2\beta\sqrt{\left(1 - \frac{R^2}{4}\right)} \cos\left[j\pi/(N+1) \right] \tag{10.4.23}$$

and all eigenvalues are lower than one in modulus if $-1 \leqslant \lambda_j(C) \leqslant 1$.

Since the maximum and minimum values of $\lambda_j(C)$ are obtained when the cosine is $+1$ or -1, we have the two conditions, when λ_j is real,

$$-1 \leqslant 1 - 2\beta\left[1 + \sqrt{\left(1 - \frac{R^2}{4}\right)}\right]$$

and

$$1 - 2\beta\left[1 - \sqrt{\left(1 - \frac{R^2}{4}\right)}\right] \leqslant 1$$

That is,

$$0 < \beta \leqslant \frac{1}{1 + \sqrt{[1 - R^2/4]}} \qquad \text{for } R \leqslant 2 \qquad (10.4.24)$$

If $R > 2$, the eigenvalues (10.4.23) are complex and the condition $|\lambda| < 1$ is satisfied for

$$(1 - 2\beta)^2 + 4\beta^2\left(\frac{R^2}{4} - 1\right) < 1$$

or

$$0 < \beta \leqslant \frac{4}{R^2} \qquad R > 2 \qquad (10.4.25)$$

For $R = 2$, we have a multiple eigenvalue $\lambda_j(C) = 1 - 2\beta$, leading to the condition $\beta < 1$.

These conditions were derived by Siemieniuch and Gladwell (1978) and are clearly distinct from the Von Neumann conditions (10.4.20). Siemieniuch and Gladwell did not explain the discrepancies they observed between their derived stability limits and the inaccuracy or instability of their computed results in the range $1 < \sigma < 2$.

Figure 10.4.5 shows the stability region in the diagram (σ, R), according to the spectral norm criterion. Lines (c) and (d) are the conditions (10.4.24) and

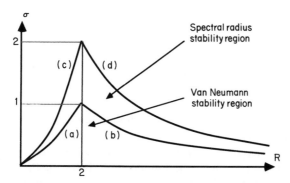

Figure 10.4.5 Stability region according to the spectral radius and Von Neumann conditions for equation (10.4.10) and Dirichlet boundary conditions

(10.4.25) with $\beta = \sigma/R$, while (a) and (b) represent the Von Neumann conditions (10.4.20).

Computations performed by Griffiths *et al.* (1980) as well as by Hindmarsh *et al.* (1984) show, without any doubt, that error modes are strongly amplified for finite values of n when the values of σ and R are chosen between the curves (a), (b) and (c), (d). This amplification is strong enough to dominate completely the solution u^n, although, as mentioned previously, the correct steady state could be obtained for sufficiently large values of n. A detailed investigation with other boundary conditions can be found in these two references, confirming the inadequacy of the spectral radius condition in this case and validating the Von Neumann conditions. The theoretical explanation behind these differences, which has been referred to at the beginning of this section, was strongly emphasized by Morton (1980).

10.5 NORMAL MODE REPRESENTATION

Another mode representation, the normal mode, offers a third alternative for the numerical representation of solutions to discretized equations. It constitutes perhaps the most powerful method for local analysis of the influence of boundary conditions. In addition, it allows us to derive exact solutions of stationary difference schemes and permits the analysis of the error propagation through the space mesh.

This representation is actually the most general one, since it looks for a solution of the following form at node point i and time step n:

$$u_i^n = \hat{u} z^n \varkappa^i \qquad (10.5.1)$$

where n is the time index when applied to u, but z^n and $\varkappa^i$ are, respectively, z to the power n and $\varkappa$ to the power i. Compared with the representation at the basis of the matrix method, we represent the contribution to the spatial behaviour of the numerical solution due to all the eigenvectors $v_i^{(j)}$ of the space operator S by $\hat{u}\varkappa^i$. When introduced into a numerical scheme formed by the interior scheme and the boundary conditions we obtain a characteristic equation which couples the time-amplification factor z with the space-amplification factor $\varkappa$. For a boundary scheme, involving points $N+1$, N, $N-1$ at a boundary, a quadratic equation is obtained for $\varkappa$, and if $\varkappa_1$ and $\varkappa_2$ are the two solutions, the general solution will be

$$u_i^n = \hat{u} z^n (A \varkappa_1^i + B \varkappa_2^i) \qquad (10.5.2)$$

where $z = z(\varkappa_1, \varkappa_2)$. Therefore this method allows an investigation of the effects of boundary conditions on the stability as developed by Godunov and Ryabenkii (1963), Kreiss (1968), Osher (1969) and Gustafsson *et al.* (1972).

An essential aspect of the normal mode analysis for the investigation of the influence of boundary conditions on the stability of a scheme is connected with a theorem of Gustafsson *et al.* (1972), stating that the initial boundary value problem for the linear hyperbolic equation is stable if it is stable separately for the following problems:

(1) Cauchy problem:
$$u_t + au_x = 0 \quad -\infty \leqslant x \leqslant \infty, \ t > 0, \ u(x,0) = f(x)$$

(2) Right quarter problem:
$$u_t + au_x = 0 \quad 0 \leqslant x \leqslant \infty \qquad u(0,t) = g(t), \ u(x,0) = f(x)$$

(2) Left quarter problem:
$$u_t + au_x = 0 \quad -\infty < x \leqslant 0 \qquad u(x,0) = f(x)$$

The Cauchy problem represents the interior scheme and is best analysed by a Von Neumann method. This theorem therefore states that the interior scheme has to be stable and that its stability could be destroyed by the boundary conditions, but the inverse is not possible.

When considered in the half-space $x \geqslant 0$, a mode $\varkappa^i$ with $|\varkappa| > 1$ will lead to an unbounded solution in space, that is, $\varkappa^i$ will increase without bound when i goes to infinity. Therefore $|\varkappa|$ should be lower than one, and the Godunov–Ryabenkii (necessary) stability condition states that all the modes with $|\varkappa| \leqslant 1$, generated by the boundary conditions, should correspond to $|z| < 1$. If, in addition, for $|z| = 1$ and $|\varkappa| = 1$ no solutions are found for which the amplification factor z tends to its position on the unit circle by the exterior, when $|\varkappa| \rightarrow 1$ from the interior of the unit circle the condition is also sufficient for stability, Kreiss (1968).

More details can be found in the references mentioned and also in Richtmyer and Morton (1967). A particular note is to be made of the work of Trefethen (1983, 1984), establishing a close relation between the boundary stability as defined by Gustafsson et al. (1972) and the sign of the group velocity of the scheme at the boundaries. The method is quite powerful, but often leads to very complex if not intractable calculations.

10.5.1 Exact solutions of a space difference scheme

The normal mode representation in space $\varkappa^i$ has been applied by Godunov and Ryabenkii (1964) to generate exact solutions of finite difference equations of stationary problems in a way that parallels the time mode z^n introduced with equation (10.1.29). Considering a stationary equation, such as the convection –diffusion equation, which we write here without source terms

$$a \frac{\partial u}{\partial x} = \alpha \frac{\partial^2 u}{\partial x^2} \qquad 0 \leqslant x \leqslant L$$

$$\begin{aligned} u(0) &= u_0 & x &= 0 \\ u(L) &= u_L & x &= L \end{aligned} \qquad (10.5.3)$$

A space-centred, second-order difference discretization will lead to the scheme

$$a \frac{u_{i+1} - u_{i-1}}{2\Delta x} = \alpha \frac{u_{i+1} - 2u_i + u_{i-1}}{\Delta x^2} \qquad i = 1, \dots, N-1 \qquad (10.5.4)$$

or

$$(2 - R)u_{i+1} - 4u_i + (2 + R)u_{i-1} = 0 \qquad (10.5.5)$$

where the mesh Reynolds or Peclet number R is introduced, defined by

$$R = \frac{a \, \Delta x}{\alpha} \qquad (10.5.6)$$

A solution $u_i = \varkappa^i$ to the difference equation (10.5.5) will exist for $\varkappa$ solution of the resolvent equation

$$(2 - R)\varkappa^2 - 4\varkappa + (2 + R) = 0 \qquad (10.5.7)$$

The two solutions, $\varkappa_1$, $\varkappa_2$ are

$$\varkappa_1 = 1 \qquad \varkappa_2 = \frac{2 + R}{2 - R} \qquad (10.5.8)$$

and the general solution

$$u_i = A\varkappa_1^i + B\varkappa_2^i \qquad (10.5.9)$$

becomes

$$u_i = A + B\left(\frac{2 + R}{2 - R}\right)^i \qquad (10.5.10)$$

The constants A and B are determined by the boundary conditions. For i ranging from 0 to N, we obtain

$$A = \frac{u_0 \varkappa_2^N - u_L}{\varkappa_2^N - 1} \qquad B = \frac{u_L - u_0}{\varkappa_2^N - 1} \qquad (10.5.11)$$

$$u_i = u_0 + (u_L - u_0) \frac{\varkappa_2^i - 1}{\varkappa_2^N - 1} \qquad (10.5.12)$$

This exact solution of the discretized equation should be a second-order approximation to the analytical solution of equation (10.5.3) in much the same way that the solution z of equation (10.1.40) is an approximation of the exact solution $\exp(\Omega \, \Delta t)$. Indeed, the exact solution of equation (10.5.3) is

$$\tilde{u} = u_0 + (u_L - u_0) \frac{e^{Re(x/L)} - 1}{e^{Re} - 1} \qquad (10.5.13)$$

where the Reynolds (or Peclet) number is introduced:

$$Re = \frac{aL}{\alpha} \qquad (10.5.14)$$

At point $x_i = i\Delta x$ the exact solution has the same structure as equation (10.5.12) and it is seen that $\varkappa_2$ is an approximation to

$$e^{Re\,\Delta x/L} = e^R$$

$$\equiv \frac{e^{R/2}}{e^{-R/2}} \approx \frac{1 + R/2}{1 - R/2}\left[1 - \left(\frac{a\,\Delta x}{\nu}\right)^3 \frac{1}{12} + \cdots\right] \qquad (10.5.15)$$

Hence we have

$$\varkappa_2^i = e^{Re\,x_i/L}\left[1 + x_i(a/\nu)^3 \frac{\Delta x^2}{12} + O(\Delta x^4)\right] \qquad (10.5.16)$$

An essential aspect of a space mode analysis is the possibility of investigating the numerical behaviour of the solution u_i. Solution (10.5.12) is a second-order approximation to the exact solution $\tilde{u}$ for values of the mesh Reynolds number R, which are sufficiently low (typically of the order of one), that is,

$$R/2 = 0(1) \qquad (10.5.17)$$

for expansions (10.5.15) to be valid. However, a deeper understanding of the behaviour of u_i is obtained from the value of $\varkappa_2$. For positive values of a, $\varkappa_2$ will become negative for $R > 2$ and $\varkappa_2^i$ will alternate in sign from one mesh point to the next. Hence the numerical scheme will generate an oscillatory behaviour of the computed solution (Figure 10.5.1).

This is best seen by investigating the spatial propagation of an error, for instance, a round-off error, $\varepsilon_i = u_i - \bar{u}_i$, where $\bar{u}_i$ is the exact solution of the discretized equation. Since ε_i also satisfies equation (10.5.4) the error $\varepsilon_i = \varepsilon \varkappa^i$ will behave as equation (10.5.12), with u_0 and u_L replaced by the errors at the

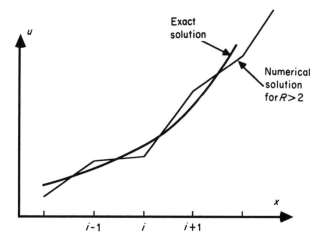

Figure 10.5.1 Typical behaviour of a numerical solution to the stationary, convection–diffusion equation for mesh Reynolds numbers larger than two

boundaries. The same approach can be applied to investigate the effects of other schemes and different implementations of boundary conditions than that applied in equation (10.5.11).

It is interesting to observe that the application of a Jacobi iterative method, as described in Chapter 12, to the solution of equation (10.5.4) leads to the following scheme, taking n as an iteration index:

$$\frac{2\alpha}{\Delta x^2}(u_i^{n+1} - u_i^n) = -\frac{a}{2\Delta x}(u_{i+1}^n - u_{i-1}^n) + \frac{\alpha}{\Delta x^2}(u_{i+1}^n - 2u_i^n + u_{i-1}^n)$$

$$(10.5.18)$$

Compared with the time-dependent scheme (10.4.10), with Δt chosen as $\Delta t = \Delta x^2/2\alpha$, that is, $2\beta = 1$, the time-dependent scheme becomes identical to the Jacobi iterative method (10.5.18) applied on the algebraic system obtained by the space-discretized stationary terms $(-au_x + \alpha u_{xx})$. Hence the considerations of the previous section apply with regard to the danger of an oscillatory solution when $R > 2$, so much that, as seen in Figure 10.4.4, when $2\beta = 1$ the stability condition reduces to $R \leqslant 2$.

10.5.2 Spatial propagation of errors in time-dependent schemes

The same analysis can be performed on time-dependent schemes, particularly implicit ones, in order to analyse the spatial propagation of errors and provide information on the influence of the solution algorithm on the stability of the scheme. Let us consider as an example the implicit upwind schemes for the convection equation:

$$u_i^{n+1} - u_i^n = -\sigma(u_i^{n+1} - u_{i-1}^{n+1}) \tag{10.5.19}$$

which requires the solution of the bidiagonal system

$$(1 + \sigma)u_i^{n+1} - \sigma u_{i-1}^{n+1} = u_i^n \tag{10.5.20}$$

This can be solved by an algorithm marching in the positive x-direction from $i = 1$ to $i = N$ or in the negative direction from $i = N$ to $i = 1$, according to the boundary condition. However, both marching directions cannot be simultaneously stable. Indeed, if we look at the way the error propagates in this space-marching algorithm:

$$\varepsilon_i^{n+1} = u_i^{n+1} - \bar{u}_i^{n+1} \tag{10.5.21}$$

where $\bar{u}_i^{n+1}$ is the exact solution of equation (10.5.20), ε_i^{n+1} satisfies the homogeneous part of equation (10.5.20):

$$(1 + \sigma)\varepsilon_i^{n+1} - \sigma\varepsilon_{i-1}^{n+1} = 0 \tag{10.5.22}$$

A solution of the form $\varepsilon_i^{n+1} = \hat{\varepsilon}\varkappa^i$ gives $\varkappa = \sigma/(1 + \sigma)$ and:

$$\varepsilon_i^{n+1} = \hat{\varepsilon}\left(\frac{\sigma}{1 + \sigma}\right)^i \tag{10.5.23}$$

If the boundary condition is imposed at the upstream end $i = 0$, then the error behaves as

$$\varepsilon_i^{n+1} = \varepsilon_0 \left(\frac{\sigma}{1 + \sigma} \right)^i \tag{10.5.24}$$

Since the interior scheme is unconditionally stable for $\sigma > 0$ and unstable for $\sigma < 0$ (as can be seen from a Von Neumann analysis), $\varkappa$ is lower than one and the initial error ε_0 will be damped while progressing in a postive marching sweep from $i = 1$ to $i = N$. However, if we impose a boundary condition at the downstream end, and solve the system by an algorithm marching in the negative direction, we would have, with ε_N the error at the downstream boundary,

$$\varepsilon_i^{n+1} = \varepsilon_N \left(\frac{\sigma}{1 + \sigma} \right)^{(i - N)} \equiv \varepsilon_N \left(\frac{1 + \sigma}{\sigma} \right)^{(N - i)} \tag{10.5.25}$$

The initial error ε_N is seen to be amplified by this algorithm, which has therefore to be rejected. This, of course, is in agreement with the physical properties of the convection equation with $a > 0$, as discussed previously.

Summary

The matrix method for stability analysis takes into account the effects of boundary conditions and is based on the eigenvalue spectrum of the space-discretization operators. In addition, through the numerical estimation of the eigenvalues we can evaluate effects such as non-uniform meshes or non-constant coefficients. When the boundary conditions are periodic, the matrix method becomes identical to the Von Neumann method, since the eigenvectors of all periodic matrices with constant coefficients are equal to the Fourier harmonics.

This approach also allows us to analyse *separately* the properties of the space discretization and of the time integration. For a given discretization in space we can select appropriate time integrations with stability regions containing the spatial eigenvalue spectrum. Hence a space discretization cannot be said to be unstable by itself when $Re\Omega \leqslant 0$. It is only when it is coupled to a time integration that we can decide upon stability. For instance, the example of the central discretization of the convection term shows that it is unstable with the explicit Euler method but stable with a leapfrog scheme or a Runge–Kutta method.

Note that the stability conditions on the spectral radius are only necessary but not always sufficient for stability, as seen in a few significant examples. It is also of interest to compare the representations of the numerical solution applied in the various stability investigations, as summarized in Table 10.1.

The Von Neumann method relies on a Fourier decomposition of the numerical solution in space, while the matrix method takes the eigenvectors of the space-discretization operator, including the boundary conditions, as a

Table 10.1. Comparison of various representations of a numerical solution

		Von Neuman method	Matrix method	Normal mode method	
Equation	$\dfrac{\partial u}{\partial t} = Lu$	$\dfrac{dU}{dt} = SU + Q$		$U^{n+1} = CU^n + \bar{Q}$	u^n is $u(t = n\,\Delta t)$,
Solution	$\bar{u} = u_0 e^{-I\omega t} e^{Ikx}$	Fourier modes	Eigenmodes of S	Normal modes	z^n, x^i: n and i are exponents
		$u_i^n = v^n e^{Ii\phi}, \; \phi = \dfrac{j\pi}{N}$	$SV^{(j)} = \Omega_j V^{(j)}$		
Amplification factor G	Exact amplification $\bar{G} = e^{-I\omega t}$ with $(-I\bar\omega)$ eigenvalue of differential operator L $L\bar{u} = -I\bar\omega\,\bar{u}$	$v^n = G^n v^0$ $G = e^{-I\omega \Delta t}$ $G = C(e^{I\omega})$ G: symbol of space operator C for harmonic space modes	$u_i^n = z^n v_i^{(j)}$ $\tilde{G} = e^{\Omega \Delta t}$ amplification of space discretized equations $z = z(\Omega)$ Ω: eigenvalue of space operator S z: symbol of operator C for modes $v_i^{(j)}$	$u_i^n = z^n x^i$ $z = z(x)$ z: symbol of C for arbitrary modes x^i	
	Examples: Diffusion equation $u_t = \alpha u_{xx}$ $-I\bar\omega = -\alpha k^2$ $\bar\omega$: pure imaginary	ω: complex dependent on scheme	Ω: real negative for central difference operators		
	Convection equation $u_t + a u_x = 0$ $\bar\omega = ka$ $\bar\omega$: real	ω: complex dependent on scheme	Ω: pure imaginary for central difference operators		

basis for the representation of the spatial behaviour of the solution. The associated time behaviour is represented by a local power law in an amplification factor z, with the time index as an exponent.

The third approach is the normal mode representation, which generalizes further the spatial eigenmode decomposition of the matrix method by replacing it with a local representation as a power law in a variable x, with the mesh point index i as an exponent. This is a powerful representation, due to its simplicity and generality, when linked to a similar representation in time, particularly for boundary conditions. It also allows investigations on the error propagation through a spatial mesh and leads to guidelines on the parameters of a scheme and (or) on resolution algorithms in order to avoid error growth.

References

Chan, T. F. (1984). 'Stability analysis of finite difference schemes for the advection–diffusion equation.' *SIAM Journal of Numerical Analysis*, **21**, 272–83.

Dahlquist, G. (1978). 'Positive functions and some application to stability questions for numerical methods.' In C. De Boor and G. Golub (eds), *Recent Advances in Numerical Analysis*, New York: Academic Press.

Dahlquist, G., and Björk, A. (1974). *Numerical Methods*, Englewood Cliffs, NJ: Prentice-Hall.

Desideri, J.-A., and Lomax, H. (1981). 'A preconditioning procedure for the finite difference computation of steady flows.' *AIAA Paper 81-1006*, AIAA 5th Computational Fluid Dynamics Conference.

Eriksson, L. E., and Rizzi, A. (1985). 'Computer-aided analysis of the convergence to steady state of discrete approximations to the Euler Equations.' *Journal of Computational Physics*, **57**, 90–128.

Fromm, J. (1964). 'The time dependent flow of an incompressible viscous fluid.' In *Methods in Computational Physics*, Vol. 3, New York: Academic Press, pp. 345–82.

Gary, J. (1978). 'On boundary conditions for hyperbolic difference schemes.' *Journal of Computational Physics*, **26**, 339–51.

Gear, G. W. (1971). *Numerical Initial Value Problems in Ordinary Differential Equations*, Englewood Cliffs, NJ: Prentice-Hall.

Godunov, S. K., and Ryabenkii, V. S. (1963). 'Spectral stability criteria of boundary value problems for non-self-adjoint difference equations.' *Russ. Math. Survey*, **18**, 1–12.

Godunov, S. K., and Ryabenkii, V. S. (1964). *The Theory of Difference Schemes*, Amsterdam: North-Holland.

Griffiths, D. F., Christie, I., and Mitchell, A. R. (1980). 'Analysis of error growth for explicit difference schemes in conduction–convection problems.' *Int. Journal for Numerical Methods in Engineering*, **15**, 1075–81.

Gustafsson, B., Kreiss, H. O., and Sundström, A. (1972). 'Stability theory of difference approximations for mixed initial boundary value problems.' *Mathematics of Computation*, **26**, 649–86.

Gustafsson, B., and Kreiss, H. O. (1979). 'Boundary conditions for time dependent problems with an artificial boundary.' *Journal of Computational Physics*, **30**, 333–51.

Hindmarsh, A. C., Gresho, P. M., and Griffiths, D. F. (1984). 'The stability of explicit Euler time integration for certain finite difference approximations of the multidimensional advection diffusion equation.' *Int. Journal for Numerical Methods in Fluids*, **4**, 853–97.

Hirt, C. W. (1968). 'Heuristic stability theory for finite difference equations.' *Journal of Computational Physics*, **2**, 339–55.

Kreiss, H. O. (1968). 'Stability theory for difference approximations of mixed initial boundary value problems, I.' *Mathematics of Computation*, **22**, 703–14.

Kreiss, H. O. (1970). 'Initial boundary value problem for hyperbolic systems.' *Comm. Pure and Applied Math.*, **23**, 273–98.

Lambert, J. D. (1973). *Computational Methods in Ordinary Differential Equations*, New York: John Wiley.

Leonard, B. P. (1980). 'Note on the Von Neumann stability of the explicit FTCS convection diffusion equation.' *Applied Mathematical Modeling*, **4**, 401–2.

Lomax, H. (1976). 'Recent progress in numerical techniques for flow simulation.' *AIAA Journal*, **14**, 512–18.

Lomax, H., Pulliam, T., and Jespersen, D. C. (1981). 'Eigensystems analysis techniques for finite difference equations: 1. Multilevel techniques.' *AIAA Paper 81-1027*, AIAA 5th Computational Fluid Dynamics Conference.

Morton, K. W. (1980). 'Stability of finite difference approximations to a diffusion–convection equation.' *Int. Journal for Numerical Methods in Engineering*, **15**, 677–83.

Roache, P. J. (1972). *Computational Fluid Dynamics*, Albuquerque, New Mexico: Hermosa.

Osher, S. (1969). 'Systems of difference equations with general homogeneous boundary conditions'. *Trans. Amer. Math. Soc.*, **137**, 177–201.

Richtmyer, R. D., and Morton, K. W. (1967). *Difference Methods for Initial Value Problems*, New York: John Wiley.

Rigal, A. (1978). 'Stability analysis of explicit finite difference schemes for the Navier–Stokes equations.' *Int. Journal for Numerical Methods in Engineering*, **14**, 617–20.

Siemieniuch, J., and Gladwell, I. (1978). 'Analysis of explicit difference methods for the diffusion–convection equation.' *Int. Journal for Numerical Methods in Engineering*, **12**, 899–916.

Thompson, H. D., Webb, B. W., and Hoffmann, J. D. (1985). 'The cell Reynolds number myth.' *Int. Journal for Numerical Methods in Fluids*, **5**, 305–10.

Trefethen, L. N. (1983). 'Group velocity interpretation of the stability theory of Gustafsson, Kreiss and Sundstrom.' *Journal of Computational Physics*, **49**, 199–217.

Trefethen, L. N. (1984). 'Instability of difference models for hyperbolic initial boundary value problems.' *Comm. Pure and Applied Mathematics*, **37**, 329–67.

Varga, R. S. (1962), *Matrix Iterative Analysis*, Englewood Cliffs, NJ: Prentice-Hall.

Yee, H. C. (1981), 'Numerical approximation of boundary conditions with applications to inviscid gas dynamics.' *NASA Report TM-81265*.

PROBLEMS

Problem 10.1

Derive the discretization matrix S for the diffusion equation $u_t = \alpha u_{xx}$, with conditions (10.2.5), by applying a central difference at $x = 0$, between $i = -1$ and $i = 1$. Use this equation to eliminate u_{-1} in the equation written for u_0. Obtain the matrix equation for the vector $U^T = (u_0, u_1, \ldots, u_{N-1})$.

Problem 10.2

Derive the matrix form of the discretization operator for the two-dimensional diffusion equation $u_t = \alpha(u_{xx} + u_{yy})$ on a Cartesian mesh $\Delta x = \Delta y = 1$, on the square $0 \leqslant x \leqslant 5$, $0 \leqslant y \leqslant 5$, for constant Dirichlet conditions $u = a$, on the boundaries and the five-point Laplace operator for the space discretization. Classify the vector U by selecting the unknowns u_{ij}, first by columns (or by rows) and also by diagonals.
Hint: In the first case: $U^T = (u_{11}, u_{12}, ..., u_{1N}; u_{21}, u_{22}, ..., u_{2N}; u_{31}, ...)$. In the second case: $U^T = (u_{11}; u_{21}, u_{12}; u_{31}, u_{22}, u_{13}; u_{41}, u_{32}, u_{23}, u_{14}, ...)$

Problem 10.3

Solve Problem 10.2 with a finite element space discretization and with bilinear elements.

Problem 10.4

Write the equations obtained in problems 10.2 and 10.3 under the form of banded matrices notation $B(-b_{-2}, b_{-1}, b_0, b_1, b_2, ...)$.

Problem 10.5

Derive the matrix structure for the diffusion equation $u_t = \alpha u_{xx}$, applying a fourth-order formula for the second derivative in space. Consider Dirichlet boundary conditions.
Hint: Obtain

$$\frac{dU}{dt} = \frac{\alpha}{12\Delta x^2} B(-1, 16, -30, 16, -1)U$$

Problem 10.6

Compute the eigenvalues and eigenvectors of the central, second-order derivative approximation $Su_i = 1/\Delta x^2(u_{i-1} - 2u_i + u_{i+1})$, whose matrix structure is $SU = 1/\Delta x^2 B(1, -2, 1)U$, for a mesh with 11 points and Dirichlet boundary conditions. Derive the diagonalization matrix T whose columns are the eigenvectors. Control your results by a direct computation of the right-hand side of equation (10.1.8) and plot the eigenfunctions as a function of the mesh point number i.
Hint: The matrix has $(N-2)$ lines and columns, starting at $i = 1$ to $i = 10$; the equations for $i = 0$ and $i = 11$ are removed from the system since the values u_0 and u_{11} are known.

Problem 10.7

Define the matrix structure for the convection equation $u_t + au_x = 0$ for $a < 0$, applying a forward space differencing of first order. Show that it is stable when a downstream boundary condition is applied.

Problem 10.8

Analyse the stability of the two-step (three-level) method

$$U^{n+1} + 4U^n - 5U^{n-1} = 2\Delta t\left[\left(\frac{dU_n}{dt}\right) + \left(\frac{dU_{n-1}}{dt}\right)\right]$$

for the modal equation

$$\frac{dw}{dt} = \Omega w + q$$

Calculate the characteristic polynomial and its roots and develop in a Taylor series for $\Delta t \to 0$. Show that the scheme is always unstable.

Problem 10.9

Repeat Problem 10.8 for the two-step (three-level) Adams–Bashworth scheme

$$U^{n+1} - U^n = -\frac{1}{2}\Delta t\left[\left(\frac{dU_{n-1}}{dt}\right) - 3\left(\frac{dU_n}{dt}\right)\right]$$

Apply to the diffusion equation with discretization (10.2.1) and to the convection equation with schemes (10.2.32) and (10.2.39). Calculate numerically the trace of the two roots for real negative values of Ω and for purely imaginary ones.
Hint: Obtain

$$z_\pm = \tfrac{1}{2}(1 + \tfrac{3}{2}\Omega\,\Delta t \pm \sqrt{[1 + \Omega\,\Delta t - \tfrac{9}{4}\Omega^2\,\Delta t^2]}$$

Show that the scheme is weakly unstable for the convection scheme (10.2.39) and conditionally stable for the other two schemes.

Problem 10.10

Consider the space operator of the Lax–Friedrichs scheme for the convection equation and the associated system:

$$\frac{du_i}{dt} = \frac{a}{2\Delta x}(u_{i+1} - u_{i-1}) + \frac{1}{2}(u_{i+1} - 2u_i + u_{i-1})$$

$$= \left[-\frac{a}{2\Delta x}(E - E^{-1}) + \frac{1}{2}(E - 2 + E^{-1})\right]u_i$$

Determine the space discretization matrix with periodic boundary condition and calculate its eigenvalues. Represent them in the complex Ω plane. Show that with an Euler explicit method the scheme is stable under the CFL condition. Observe, by comparing with equation (10.2.40), the influence of the added terms in the space operator, providing a negative real part to the eigenvalues.

Problem 10.11

Repeat Problem 10.10, applying Dirichlet conditions at both ends. It is assumed that the downstream conditions are known.

Problem 10.12

Analyse the error propagation for the Cranck–Nicholson scheme applied to the convection equation with second-order space-centred differences. Show that there is always one boundary error which is amplified, indicating that the Cranck–Nicholson scheme is not suitable for convection equations. Define the stability properties of the interior scheme by a Von Neumann analysis.
Hint: The scheme is written as

$$u_i^{n+1} - u_i^n = -\frac{\sigma}{4}\left[(u_{i+1}^{n+1} - u_{i-1}^{n+1}) + (u_{i+1}^n - u_{i-1}^n)\right]$$

The Von Neumann amplification factor is

$$G = \left(1 - \frac{\sigma}{2} I \sin \phi\right) \Big/ \left(1 + \frac{\sigma}{2} I \sin \phi\right)$$

and $|G| = 1$, showing neutral stability.

The error ε_i^{n+1} satisfies $\sigma\varepsilon_{i+1}^{n+1} + 4\varepsilon_i^{n+1} - \sigma\varepsilon_{i+1}^{n+1} = 0$.

The solution is $\varepsilon_{i+1} = ax_1^i + bx_2^i$ with x_1, x_2 solutions of $\sigma x^2 + 4x - \sigma = 0$ and $x_1 x_2 = -1$

Hence one always has $|x_1| > 1$, $|x_2| < 1$.

Problem 10.13

Determine the exact solutions of the scheme obtained by discretizing the convection term in the convection–diffusion equation $au_x = \alpha u_{xx}$, with a first-order backward formula (for $a > 0$). Show that the solution will always remain oscillation-free for $a > 0$. Analyse the exact numerical solutions for the boundary conditions $u(0) = u_0$, $u(L) = u_L$ for $N\Delta x = L$ and compare with equation (10.5.12). Calculate the error $\varepsilon_i = u_i - \bar{u}_i$ and show that the scheme is only first-order accurate.

Hint: The scheme is

$$a(u_i - u_{i-1}) = \frac{\alpha}{\Delta x} (u_{i+1} - 2u_i + u_{i-1})$$

Obtain $x_1 = 1$, $x_2 = (1 + R)$.

Problem 10.14

Apply the normal-mode analysis to the stationary convection–diffusion equation (10.5.4) with the following numerical implementation of the boundary conditions, resulting from a finite volume discretization where the boundaries $x = 0$ and $x = L$ are at the centre of the mesh cell:

$$2u_0 = u_{-1} + u_1 \quad \text{and} \quad 2u_L = (u_N + u_{N-1})$$

Obtain the exact numerical solution and compare with equation (10.5.12). Show that the error at node $i = 1$ is given by

$$\varepsilon_1 = \frac{(u_0 - u_L)}{8} \frac{[(u/\alpha)\Delta x]^2}{1 - e^{Re}} + 0(\Delta x^3)$$

Hint: Write the equations for nodes $i = 1$ and $i = N - 1$ by introduction of the boundary conditions in scheme (10.5.4). Obtain

$$(u_2 - 3u_1 + 2u_0) - \frac{R}{2} (u_2 + u_1 - 2u_0) = 0$$

$$(2u_L - 3u_{N-1} + u_{N-2}) - \frac{R}{2} (2u_L - u_{N-1} - u_{N-2}) = 0$$

Determine A and B in the general solution (10.5.9) from the above equations and obtain

$$u_i = u_0 + (u_L - u_0) \frac{x_2^i - 1}{x_2^N - 1} + \frac{R}{2} \frac{u_0 - u_L}{x_2^N - 1} x_2^i$$

where x_2 is given by equation (10.5.8). Observe the additional term in the solution u_i proportional to Δx in comparison with equation (10.5.12).

Problem 10.15

Obtain the equivalent differential equation for the convection-diffusion equation discretized by scheme (10.4.10). Show that the scheme is globally first-order accurate. Analyse the influence of the mesh Reynolds number on the accuracy of the representation of the diffusion term. Obtain the stability condition $\sigma < 2/R$.

Hint: The equivalent differential equation can be written, as a function of σ and R, as

$$u_t + au_x - \alpha u_{xx} = -\tfrac{1}{2}\alpha\sigma R u_{xx} + \alpha R\ \Delta x(\beta - \sigma^2/3 - 1/6)u_{xxx}$$

$$+ \frac{\alpha\ \Delta x^2}{12}(1 - 3\beta - 2\sigma R + 10\sigma^2 - 3\sigma^3 R)u_{xxxx} + \cdots$$

Observe that for $\sigma = 2/R$ the diffusion error is equal to the physical diffusion α of the problem and completely destroys its influence.

Problem 10.16

Proof the relations (10.5.12) and (10.5.13) and derive equation (10.5.16).

Problem 10.17

Analyse the behavior of the exact numerical solution (10.5.12) to the convection–diffusion scheme (10.5.4) for high values of the mesh Reynolds number R. Defining $\varepsilon = 2/R$, obtain by a Taylor expansion in powers of ε the following asymptotic approximation for large values of R,

$$\frac{u_i - u_0}{u_L - u_0} \approx \frac{1 - (-)^i(1 + 2i\varepsilon)}{1 - (-)^N(1 + 2N\varepsilon)}$$

Show that this numerical solution has an oscillatory character which depends on the parity of N. In particular, show that for N odd the right-hand side of the above solution oscillates between $[-i\varepsilon/(1 + N\varepsilon)]$ for i even and $[(1 + i\varepsilon)/(1 + N\varepsilon)]$ for i odd.

Plot this solution and observe that the even points are close to $u_i = u_0$ and satisfy the left boundary condition (at $i = 0$). Similarly, note that the odd points are close to $u_i = u_L$ and satisfy the right boundary condition (at $i = N$).

For N even the right-hand side of the above solution oscillates between i/N for i even and $-(1 + i\varepsilon)/N\varepsilon$ for i odd. Plot this solution and observe that the even points follow a linear variation and satisfy both boundary conditions, while the odd points are on another straight line which satisfies neither of the boundary conditions.

Problem 10.18

Consider the leapfrog scheme applied to the convection equation with the addition of a dissipation term of the form

$$\alpha\frac{\partial^4 u^{n-1}}{\partial x^4} = \frac{\alpha}{\Delta x^4}(u_{i+2} - 4u_{i+1} + 6u_i - 4u_{i-1} + u_{i-2})^{n-1}$$

which damps high frequency oscillations, leading to the scheme

$$u_i^{n+1} - u_i^{n-1} = -\sigma(u_{i+1}^n - u_{i-1}^n) - \frac{2\alpha\Delta t}{\Delta x^4}(u_{i+2} - 4u_{i+1} + 6u_i - 4u_{i-1} + u_{i-2})^{n-1}$$

Determine the amplification factor z for Fourier modes and show that the scheme is stable for $\sigma^2 < 1 - \gamma/2$ with $\gamma = 16\alpha\Delta t/\Delta x^4$ and that γ has to be limited by $0 < \gamma < 1/2$.

Show that the amplification factor is

$$z = -I\sigma \sin\phi \pm (1 - \sigma^2 \sin^2\phi - 2\gamma \sin^4\phi/2)^{1/2}.$$

PART IV: THE RESOLUTION OF DISCRETIZED EQUATIONS

In the last two chapters of this book we will investigate some of the most currently applied techniques for the resolution of either semi-discretized equations (that is, systems of ordinary differential equations in time) or algebraic systems of equations obtained generally after a space discretization of a stationary problem.

As mentioned in the introduction to Part II dealing with the computational approach, we have to make an essential decision as to the time dependence of the formulation. If the physical problem is time dependent there is obviously no choice; the mathematical initial value problem has to be discretized in time and the numerical solution must be time accurate. On the other hand, for physical stationary problems we can decide either to discretize the equations in space and deal with a time-independent numerical scheme or maintain the time dependency and discretize the equations in space and time, but aim only at the time-asymptotic, steady, numerical solution. For time-dependent formulations we deal with a system of ordinary differential equations in time, and a large body of techniques have been developed to define and analyse numerical schemes for time-dependent or, more precisely, initial value problems. The most widely applied of these techniques will be discussed in Chapter 11. With a stationary formulation in space we will generally deal with a boundary value problem (for elliptic equations), and an algebraic system of equations will be obtained. This algebraic system has to be solved in the lowest possible number of operations. Various techniques to be applied for achieving this will be discussed in Chapter 12.

It is important to observe that the methods developed for time-dependent formulations can also be applied to some stationary problems, particularly if the set of equations considered is parabolic or hyperbolic in one of the spatial directions. For instance, the parabolized Navier–Stokes approximation (discussed in Section 2.4) is parabolic in x; the stationary form of the potential equation (2.9.25) is hyperbolic in the x-direction for supersonic flows. Hence the x variable can be treated in these cases as a time-like co-ordinate and the methods for time-dependent problems can be applied.

Conversely, all iterative methods presented in Chapter 12 can be cast in a pseudo-time-dependent formulation by interpreting the iteration index as a time index. This forms a bridge between stationary and non-stationary formulations and enlarges the family of methods to be applied to obtain solutions to physical stationary problems. We can indeed select the pseudo-time operators in order to accelerate the convergence of the scheme, and an illustration is provided by the preconditioning methods. This bridge also has as an important consequence the fact that an iterative method can be analysed by the techniques used for the stability analysis of time-dependent schemes, for instance by a Von Neumann method.

From a spectral analysis of the errors of an iterative method appropriate operations can be defined in order to damp slectively certain frequency domains of the error spectrum. This philosophy is the basis of the most powerful of the iteration techniques currently available, namely the multi-grid method, to be presented in Chapter 12.

Chapter 11

Integration Methods for Systems of Ordinary Differential Equations

Many methods are available for the solution of the system of ordinary differential equations obtained from the semi-discretization technique, whereby the space operators are discretized separately from the time differentials. This method (also called the *method of lines*) leads to a system of time–dependent ordinary differential equations. As discussed in Chapter 10, the system obtained in this way is stable when the eigenvalues Ω of the space-discretization matrix S have non-positive real parts, that is, are located in the left half (imaginary axis included) of the complex Ω-plane.

The stability condition of the time-integration scheme has to be compatible with the particular form of the spectrum Ω, that is the stability region of the time-discretization method must include the whole spectrum of eigenvalues Ω. For instance, the explicit Euler scheme has a stability region which does not contain the imaginary axis of the Ω-plane, and will not be stable for a Ω-spectrum on the imaginary axis like that generated by the central discretization of the convection equation. However, it will be stable, conditionally, for the diffusion equation if *all* the real eigenvalues are contained within the range $-2 \leqslant (\Omega \Delta t) \leqslant 0$. Similarly, the leapfrog scheme is unstable in this latter case since its stability region is limited to a segment of the imaginary axis. Hence, as already noted, the leapfrog scheme is totally unadapted for dissipative problems, and the Euler explicit scheme is not suitable for non-dissipative space discretizations.

The group of methods suitable and adapted to all possible systems with a stable Ω spectrum (that is, whose region of stability coincides with the left half Ω-plane, including the imaginary axis) has been called *A-stable* (Dahlquist, 1963). In fact, it can easily be shown that A-stability can only be achieved with implicit methods, since A-stability is equivalent to unconditional stability as defined in Section 7.1. An essential distinction appears here between explicit and implicit methods, since explicit time-integration methods will nearly always lead to conditional stability, that is, with limitations on the maximum allowable time step Δt.

Clearly, a problem which leads to a spectrum in the $(\Omega \Delta t)$ plane with large imaginary values and widely distributed real (negative) parts will be a good candidate for an A-stable (implicit) method, where no Δt-restrictions will have

424

to be imposed. This is an important advantage, particularly for stationary problems, solved by a time-dependent formulation. In this case, the interest lies in the time-asymptotic steady-state solution, and we look for an approach allowing the largest possible time steps in order to reach the steady state in the lowest possible number of operations. Since, in this case, the time accuracy of the transient is of no importance, implicit methods are very well adapted. Explicit methods can also be applied to these situations, but with limitations in the maximum allowable time step. On the other hand, for time-dependent problems, time-step limitations have to be introduced for reasons of accuracy, and the additonal computational cost of implicit methods per time step is not always justified.

The debate between implicit and explicit methods for stationary problems is still open, and probably will remain so. Since, in practical problems, we deal with non-linear equations, time-step restrictions often appear in implicit formulations because of the generation of non-linear instabilities. Therefore we have to balance the higher allowable time step against the greater number of operations necessary to resolve the implicit algebraic system of equations. Linear multi-step methods for fluid mechanical problems have been introduced and analysed systematically by Beam and Warming (1976, 1980), and developed to operational computer codes for the Euler and Navier–Stokes equations (see, for instance, Pulliam, 1984, for a description of the structure and properties of these codes).

Methods based on a predictor–corrector sequence also have a wide spread application, the most popular being the explicit McCormack scheme. We will introduce them as an explicit approximation to the treatment of the non-linearities of implicit multi-step methods. Other linearization techniques do maintain the implicit character and have been applied by Briley and McDonald (1975) and Beam and Warming (1976).

For multi-dimensional problems the implicit matrices are too large for direct inversions and factorization techniques, reducing the problem to a succession of one-dimensional implicit oprators which can be defined. These methods are also known as alternating direction implicit (ADI) methods. The factorization process is by no means trivial, since it can destabilize the numerical scheme, as will be shown for the three-dimensional convection equation with central differenced space derivatives.

Finally, the classical Runge–Kutta methods will be summarized, since they appear to be well adapted to convection problems with central discretizations. We refer the reader to the literature for more details and general presentations, in particular, Gear (1971), Lambert (1973), Dahlquist and Björck (1974) and Beam and Warming (1982).

11.1 LINEAR MULTI-STEP METHODS

For a general system of ordinary differential equations such as equation (10.1.2) the right-hand side may, and often will, be non-linear. The general multi-step time integration will therefore be written for a system of the form

$$\frac{dU}{dt} = H(U, t) \tag{11.1.1}$$

where H is a non-linear function or matrix operator, acting on the vector U, as defined previously. The vector $H(U, t)$ represents the space-discretized operators $(SU + Q)$, equation (10.1.2), or can be considered as the differential space operator prior to discretization (see Section 11.2).

The general linear K-step integration method, already introduced in section 10.1, is defined by

$$\sum_{k=0}^{K} \alpha_k U^{n+k} = \Delta t \sum_{k=0}^{K} \beta_k H^{n+k} \tag{11.1.2}$$

where

$$H^{n+k} = H(U^{n+k}, t^{n+k}) \tag{11.1.3}$$

The consistency conditions are obtained from the requirement that a constant U should be a solution of the homogeneous system, and that the first terms of a Taylor expansion should be consistent with equation (11.1.1) when Δt tends to zero. Hence

$$\sum_{k=0}^{K} \alpha_k = 0 \tag{11.1.4}$$

$$\sum_{k=0}^{K} k\alpha_k = \sum_{k=0}^{K} \beta_k = 1 \tag{11.1.5}$$

the common value of the sums in equation (11.1.5) being set arbitrarily equal to one.

The stability is analysed by the following steps, following the developments of Chapter 10:

(1) Introduce the homogeneous part of the linear modal equation (10.1.21), written here as

$$\frac{dU}{dt} = \Omega U \tag{11.1.6}$$

(2) Form the characteristic polynomial (equation (10.1.40)):

$$P_1(z) = \sum_k \alpha_k z^k - (\Omega \, \Delta t) \sum_k \beta_k z^k \tag{11.1.7}$$

(3) The roots of $P_1(z) = 0$ should lie within the circle of radius 1 and the roots on the unit circle have to be simple.

It is customary to introduce two generating polynomials in the time shift operator $\bar{E}$, $\rho(\bar{E})$ and $\sigma(\bar{E})$, which define completely the multi-step integration method, by

$$\rho(\bar{E}) = \sum_{k=0}^{K} \alpha_k \bar{E}^k \qquad \sigma(\bar{E}) = \sum_{k=0}^{K} \beta_k \bar{E}^k \tag{11.1.8}$$

426

so that equation (11.1.3) can be written as

$$\rho(\bar{E})U^n = \Delta t \sigma(\bar{E})H^n \qquad (11.1.9)$$

The characteristic polynomial $P_1(z)$ becomes

$$P_1(z) = \rho(z) - \Omega \, \Delta t \sigma(z) = 0 \qquad (11.1.10)$$

In practical applications the vector U contains a large number of variables, namely $N = m \cdot M$; M is the total number of mesh points where the m-independent variables have to be determined. Generally, these variables have to be stored for each time level, and increasing the number of time levels could rapidly put severe restrictions on the allowable space variables and mesh points. It is therefore very exceptional to consider applications with more than three time levels (two-step methods) to fluid mechanical problems, the overwhelming majority of schemes being limited to one-step methods with two time levels.

Following Beam and Warming (1980) we introduce the most general consistent two-step method under the form

$$(1 + \xi)U^{n+2} - (1 + 2\xi)U^{n+1} + \xi U^n = \Delta t[\theta H^{n+2} + (1 - \theta - \phi)H^{n+1} - \phi H^n] \qquad (11.1.11)$$

The generating polynomials are

$$\rho(\bar{E}) = (1 + \xi)\bar{E}^2 - (1 + 2\xi)\bar{E} + \xi \qquad (11.1.12)$$

$$\sigma(\bar{E}) = \theta\bar{E}^2 + (1 - \theta + \phi)\bar{E} - \phi \qquad (11.1.13)$$

For second-order accuracy in time we should have the relation

$$\phi = \xi - \theta + 1/2 \qquad (11.1.14)$$

and if, in addition,

$$\xi = 2\theta - 5/6 \qquad (11.1.15)$$

the method is third-order accurate. Finally a unique fourth-order accurate method is obtained for

$$\theta = -\phi = -\xi/3 = 1/6 \qquad (11.1.16)$$

This is easily shown from a Taylor series expansion (see Problem 11.1).

For $\xi = \phi = 0$ we obtain a two-level, one-step scheme which is better known as the *generalized trapezoidal* method, replacing $(n + 1)$ by n:

$$U^{n+1} - U^n = \Delta t[\theta H^{n+1} + (1 - \theta)H^n] \qquad (11.1.17)$$

with

$$\rho(\bar{E}) = \bar{E} - 1 \qquad (11.1.18)$$

$$\sigma(\bar{E}) = \theta\bar{E} + 1 - \theta \qquad (11.1.19)$$

For $\theta = 1/2$ the method is second order in time and is known as the *trapezium formula* or the *Cranck–Nicholson* method when applied to the diffusion

Table 11.1. Partial list of one- and two-step methods (from Beam and Warming, 1982)

θ	ξ	ϕ	Method	Order	
0	0	0	Euler explicit	1	A-stable
1	0	0	Backward Euler	1	A-stable
1/2	0	0	One-step trapezoidal	2	A-stable
1	1/2	0	Backward differenciation	2	A-stable
3/4	0	−1/4	Adams type	2	A-stable
1/3	−1/2	−1/3	Lees type	2	A-stable
1/2	−1/2	−1/2	Two-step trapezoidal	2	A-stable
5/9	−1/6	−2/9	A-contractive	2	
0	−1/2	0	Leapfrog	2	
0	0	1/2	Adams–Bashforth	2	
0	−5/6	−1/3	Most accurate explicit (unstable)	3	
1/3	−1/6	0	Third-order implicit	3	
5/12	0	1/12	Adams–Moulton	3	
1/6	−1/2	−1/6	Milne	4	

equation

$$U^{n+1} - U^n = \frac{\Delta t}{2}(H^{n+1} + H^n) \tag{11.1.20}$$

A classification of several well-known methods are listed in Table 11.1. Note that the schemes with $\theta = 0$ are explicit.

A certain number of properties have been proven by Dahlquist (1963) with regard to the order of accuracy of A-stable methods:

(1) The two-step scheme (11.1.11) is A-stable if and only if

$$\begin{aligned} \theta &\geqslant \phi + 1/2 \\ \xi &\geqslant -1/2 \\ \xi &\leqslant \theta + \phi - 1/2 \end{aligned} \tag{11.1.21}$$

This is shown in Figure 11.1.1, for the family of second-order schemes, satisfying equation (11.1.14).
(2) An A-stable linear multi-step method cannot have an order of accuracy higher than two.
(3) The trapezoidal scheme (equation (11.1.20)) has the smallest truncation error of all second-order A-stable methods.

A particular family of schemes, extensively applied, is that with $\phi = 0$. The schemes are often written in incremental form for the unknowns $\Delta U^n = U^{n+1} - U^n$, as follows with a shift in the time index n:

$$(1 + \xi)U^{n+1} - (1 + 2\xi)U^n = \Delta t\,[\theta H^{n+1} + (1 - \theta)H^n] - \xi U^{n-1} \tag{11.1.22}$$

or

$$(1 + \xi)\Delta U^n - \xi\,\Delta U^{n-1} = \Delta t[\theta H^{n+1} + (1 - \theta)H^n] \tag{11.1.23}$$

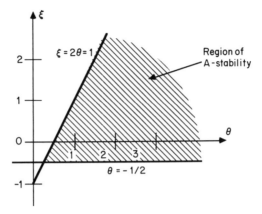

Figure 11.1.1 Domain of A-stability in the plane (ξ, θ) for second-order two-step methods

They are second-order accurate in time for $\xi = \theta - 1/2$. When applied to the Euler and Navier-Stokes equations with central space differencing these schemes have become known as the *Beam and Warming* schemes, when the linearization of Section 11.3 is introduced. The characteristic polynomial can be written as

$$P_1(z) = (1 + \xi)z^2 - (1 + 2\xi)z + \xi - (\Omega \, \Delta t)z(\theta z + 1 - \theta) = 0 \quad (11.1.24)$$

For the two-level schemes (11.1.17) with $\xi = 0$, one root is the trivial root $z = 0$ and the other is

$$z = \frac{1 + (1 - \theta)\Omega \, \Delta t}{1 - \theta \Omega \, \Delta t} \quad (11.1.25)$$

For fixed values of θ the stability domains in the Ω-plane are obtained from the condition $|z| \leqslant 1$. The stability curves displayed in Figure 11.1.2 are obtained

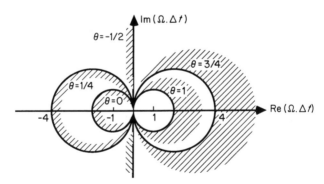

Figure 11.1.2 Stability regions for the two-step schemes (11.1.22) with $\xi = 0$

following the procedure described in Chapter 10. For $\theta = 0$ we recover the explicit Euler scheme and for $\theta = 1$ the implicit Euler scheme. The stability regions are inside the curves at the left ($\theta < 1/2$) and outside the curves at the right ($\theta > 1/2$). For $\theta = 1/2$ the line coincides with the imaginary axis and the stability region is the left half plane. This confirms that scheme (11.1.22) is A-stable for $\theta \geqslant 1/2$.

We can also deduce from Figure 11.1.2 that the generalized trapezoidal scheme (equation (11.1.17)) is conditionally stable for $0 \leqslant \theta < 1/2$ when the eigenvalue spectrum Ω is on the real negative axis. The stability condition $|z| \leqslant 1$ becomes with equation (11.1.25)

$$-\frac{2}{1-2\theta} \leqslant (\Omega \, \Delta t) \leqslant 0 \qquad \text{for } \theta < 1/2 \qquad (11.1.26)$$

Hence for the diffusion equation, with central space differencing, the scheme is implicit for $\theta \neq 0$, but only conditionally stable. It is therefore of very little use unless $\theta \geqslant 1/2$, where it becomes unconditionally stable. On the other hand, for the convection equation and central space differencing the eigenvalues are on the imaginary axis; the scheme will therefore be unstable for $\theta < 1/2$ and unconditionally stable for $\theta \geqslant 1/2$.

Example 11.1.1   Cranck–Nicholson scheme

Consider the diffusion equation $u_t = \alpha u_{xx}$ with the central difference scheme of Table 8.1. Equation (11.1.20) becomes for a single component u_i of U:

$$u_i^{n+1} - u_i^n = \frac{\alpha \, \Delta t}{2\Delta x^2}\left(u_{i+1}^{n+1} - 2u_i^{n+1} + u_{i-1}^{n+1}\right) + \frac{\alpha \, \Delta t}{2\Delta x^2}\left(u_{i+1}^n - 2u_i^n + u_{i-1}^n\right)$$
$$\text{(E11.1.1)}$$

The eigenvalues of the space operators have been determined in Chapter 10 for different boundary conditions, and have been shown to be real and negative.

The characteristic root is obtained from equation (11.1.24) as

$$P_1(z) = z - 1 - \tfrac{1}{2}(\Omega \, \Delta t)(z + 1) = 0 \qquad (E11.1.2)$$

$$z = \frac{1 + (\Omega \, \Delta t)/2}{1 - (\Omega \, \Delta t)/2} \qquad (E11.1.3)$$

or

$$\Omega \, \Delta t = 2\,\frac{z-1}{z+1} \qquad (E11.1.4)$$

The analysis of the amplification z for very large time steps, $\Delta t \to \infty$, shows that z tends to -1. Hence the solution u_i^n will tend to have an asymptotic behaviour of the form $u_i^n \approx (-1)^n \cdot u_i^0$, which represents an oscillatory decay. This can generate oscillations of the numerical solution. The stability condition, $|z| \leqslant 1$, is satisfied for all $\Omega \, \Delta t$ in the left half plane, including the imaginary axis. The method is therefore A-stable and scheme (E11.1.1) is unconditionally stable. Note that for imaginary eigenvalues Ω, $|z| = 1$ and the scheme is neutrally stable.

Example 11.1.2   Beam and Warming schemes for the convection equation

For the central second-order space discretization of $u_t + au_x = 0$, scheme (11.1.22) becomes

$$(1 + \xi)\, \Delta u_i^n - \xi\, \Delta u_i^{n-1} = -\frac{\sigma}{2}\, [\theta(u_{i+1}^{n+1} - u_{i-1}^{n+1}) + (1 - \theta)(u_{i+1}^n - u_{i-1}^n)]$$

$$(E11.1.5)$$

$$= -\frac{\sigma}{2}\, [\theta(\Delta u_{i+1}^n - \Delta u_{i-1}^n) + u_{i+1}^n - u_{i-1}^n]$$

The tridiagonal system for Δu_i^n becomes

$$(1 + \xi)\Delta u_i^n - \theta\frac{\sigma}{2}\, \Delta u_{i-1}^n + \theta\frac{\sigma}{2}\, \Delta u_{i+1}^n = -\frac{\sigma}{2}\, (u_{i+1}^n - u_{i-1}^n) + \xi\, \Delta u_i^{n-1}$$

$$(E11.1.6)$$

From a Von Neumann analysis the amplification matrix is readily obtained as

$$[1 + \xi + \theta\frac{\sigma}{2}(e^{I\phi} - e^{-I\phi})](G - 1) = -\frac{\sigma}{2}\, (e^{I\phi} - e^{-I\phi}) + \xi\left(1 - \frac{1}{G}\right) \quad (E11.1.7)$$

The three-level scheme leads to a quadratic form for G, with two solutions, corresponding to the two roots of the characteristic polynomial.

Since the method is A-stable the trapezoidal scheme, applied to the centrally discretized convection equation, will also be stable, since its eigenvalues are on the imaginary axis (see also Problems 11.3 and 11.4). For the two-level schemes $\xi = 0$ we obtain

$$G = 1 - \frac{I\sigma \sin \phi}{1 + I\sigma\theta \sin \phi} \quad (E11.1.8)$$

leading to an amplitude

$$|G|^2 = \frac{1 + \sigma^2(\theta - 1)^2\sin^2\phi}{1 + \sigma^2\theta^2 \sin^2\phi} \quad (E11.1.9)$$

and a phase error

$$\varepsilon_\phi = \frac{1}{\sigma\phi}\, \tan^{-1}\left(\frac{\sigma \sin \phi}{1 + \sigma^2\theta(\theta - 1)\sin^2\phi}\right) \quad (E11.1.10)$$

Observe that all the schemes have $G = 1$ at $\phi = \pi$, indicating a lack of dissipation of the high-frequency errors.

These curves are plotted in Figure 11.1.3 for two different schemes at different values of the Courant number. Note that they all have a large lagging phase error. The trapezium formula, $\theta = 1/2$, leads to $|G| = 1$ for all ϕ and is therefore not well suited for purely convective problems, that is, to problems without any dissipation.

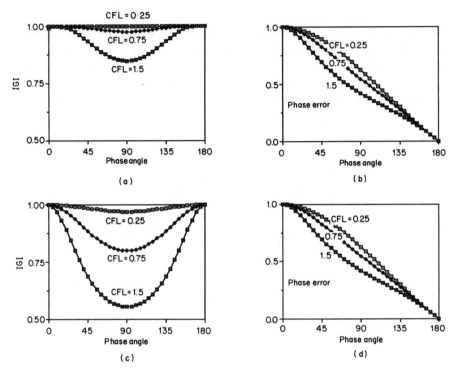

Figure 11.1.3 Polar plots of amplification factors and phase errors for different implicit Beam and Warming algorithms (E11.1.6). (a) Amplitude error for $\theta = 1, \xi = 1/2$ backward differentation scheme; (b) phase error for $\theta = 1, \xi = 1/2$ backward differentiation scheme; (c) amplitude error for $\theta = 1, \xi = 0$ backward Euler (implicit) scheme; (d) phase error for $\theta = 1, \xi = 0$ backward Euler (implicit) scheme

11.2 PREDICTOR–CORRECTOR SCHEMES

When applied to linear three-point schemes as in the previous examples, implicit two-step methods lead to a tridiagonal system which is easy to solve (see Section 12.1). However, an additional difficulty arises when $H(U)$ is non-linear in U and it is necessary to define either a linearization or an iterative procedure.

Considering scheme (11.1.22). $H^{n+1} = H(U^{n+1})$ has to be evaluated numerically in some way, since U^{n+1} is unknown. We can consider an iterative approach, whereby a first guess $\bar{U}^{n+1}$ of U^{n+1} is obtained (for instance, by applying an explicit scheme), called a *predictor-step*. Then equation (11.1.22) can be written as

$$(1 + \xi)\bar{\bar{U}}^{n+1} - (1 + 2\xi)U^n = \Delta t\,[\theta \overline{H^{n+1}} + (1 - \theta)H^n] - \xi U^{n-1} \quad (11.2.1)$$

and solved for $U^{n+1} = \bar{\bar{U}}^{n+1}$. This step is called a *corrector step*. defining what is called a *predictor–corrector* sequence, with $\overline{H^{n+1}} = H(\bar{U}^{n+1})$.

This approach can be pursued at the same time level by repeating equation

(11.2.1) for s steps, until some form of convergence between two consecutive estimations of $\bar{U}^{n+1}$ is achieved. This implies an evaluation of H at each 'local' iteration step and this procedure is generally not recommended, since the evaluation of H is often the most costly operation in the numerical simulation. Repeating the corrector sequence implies that the solution of equation (11.2.1) is designated by $\hat{U}^{n+1}$ and that the second corrector step is

$$(1 + \xi)\hat{U}^{n+1} - (1 + 2\xi)U^n = \Delta t\,[\theta\overline{\overline{H^{n+1}}} + (1 - \theta)H^n] - \xi U^{n-1} \quad (11.2.2)$$

where $\hat{U}^{n+1}$ is the new value for U^{n+1} and $\overline{\overline{H^{n+1}}} = H(\bar{\bar{U}}^{n+1})$.

A detailed discussion of various options and of the influence of the number of 'local' iterations can be found in Lambert (1974). One of the essential conclusions, from a comparison of the effect of different numbers of corrector steps coupled to the same predictor, is that the unique sequence of a predictor step followed by a single corrector step appears as being optimal in view of the reduction of the number of evaluations of the right-hand side.

In order to analyse the order of accuracy and the stability of a predictor corrector sequence we can combine the predictor $\bar{U}^n$ and the known value U^n in a single vector and write the sequence as a system of two equations. The characteristic polynomial is obtained by setting the determinant of the matrix to zero for the modal equation $H = \Omega U$. Let us illustrate this for the two-step scheme (11.1.22). By definition, we require the predictor to be explicit, hence we set $\theta = \xi = 0$ and obtain the first-order Euler method as predictor:

$$\overline{U^{n+1}} = U^n + \Delta t H^n \quad (11.2.3)$$

The corrector is equation (11.2.1), written in operator form as

$$[(1 + \xi)\bar{E} - (1 + 2\xi)]U^n = \Delta t\,[\theta\bar{H}^{n+1} + (1 - \theta)H^n] - \xi U^{n-1} \quad (11.2.4)$$

and can be treated as an explicit equation.

Inserting the modal equation $H = \Omega U$ we obtain the system

$$\begin{vmatrix} \bar{E} & -(1 + \Omega\Delta t) \\ -\theta\Omega\Delta t\bar{E} & \xi\bar{E}^{-1} + (1 + \xi)\bar{E} - (1 + 2\xi) - \Omega\Delta t(1 - \theta) \end{vmatrix} \begin{vmatrix} \overline{U^n} \\ U^n \end{vmatrix} = \begin{vmatrix} 0 \\ 0 \end{vmatrix} \quad (11.2.5)$$

leading to the polynomial equation obtained from the determinant of the system set to zero:

$$P_1(z) = (1 + \xi)z^2 - (1 + 2\xi)z + \xi - (\Omega\,\Delta t)(1 + \Omega\Delta t\theta)z = 0 \quad (11.2.6)$$

It is interesting to compare this equation with equation (11.1.25) derived for the implicit system, particularly when $\xi = 0$. The above equation becomes, removing a trivial root $z = 0$,

$$z = 1 + (\Omega\,\Delta t) + (\Omega\,\Delta t)^2\theta \quad (11.2.7)$$

Solution (11.2.7) is typical of explicit schemes, and the predictor–corrector sequence has become explicit for all values of θ. The sequence is only first order accurate in time, except or $\theta = 1/2$, where it is second order, since the quadratic term is equal to the corresponding term in the Taylor development of $\exp(\Omega \, \Delta t)$, as seen from equation (10.1.44). From the condition $|z| \leqslant 1$ we obtain the stability limit for real eigenvalues Ω:

$$0 \geqslant \Omega \, \Delta t \geqslant -\frac{1}{\theta} \qquad (11.2.8)$$

The scheme is conditionally stable, with condition (11.2.8). If Ω is purely imaginary (for instance, for the convection equation with central space differencing) the predictor–corrector scheme is conditionally stable, since we obtain for $\Omega = I\omega$ the condition

$$(\omega \, \Delta t)^2 \leqslant \frac{2\theta - 1}{\theta^2} \qquad (11.2.9)$$

The choice $\theta = 1/2$ corresponds to a second-order accurate sequence and is known as *Henn's* method (Dahlquist and Bjorck, 1974). It forms the basis of the McCormack scheme and can be written as

$$\begin{aligned}
\overline{U^{n+1}} &= U^n + \Delta t H^n \\
U^{n+1} &= U^n + \tfrac{1}{2} \Delta t (\overline{H^{n+1}} + H^n)
\end{aligned} \qquad (11.2.10)$$

This scheme is unstable for the convection equation, when central differences are applied, since equation (11.2.9) cannot be satisfied for $\theta = 1/2$. On the other hand, for $\theta = 1$ we obtain the following scheme:

$$\begin{aligned}
\overline{U^{n+1}} &= U^n + \Delta t H^n \\
U^{n+1} &= U^n + \Delta t \overline{H^{n+1}}
\end{aligned} \qquad (11.2.11)$$

which is stable for the convection equation with central differences under the condition $|\omega \, \Delta t| < 1$. Variants of this scheme have been applied by Brailowskaya (1965) to the Navier–Stokes equations by keeping the viscous diffusion terms in the corrector step at level n. In particular, for the linear convection equation, equation (11.2.11) becomes, with $H_i = -a(u_{i+1} - u_{i-1})/2 \, \Delta x$,

$$\begin{aligned}
\overline{u_i^{n+1}} &= u_i^n - \frac{\sigma}{2} (u_{i+1}^n - u_{i-1}^n) \\
u_i^{n+1} &= u_i^n - \frac{\sigma}{2} (\overline{u_{i+1}^{n+1}} - \overline{u_{i-1}^{n+1}})
\end{aligned} \qquad (11.2.12)$$

Combining the two steps leads to a scheme of second-order accuracy in space and first order in time, which involves points $i - 2$ and $i + 2$ and is stable under the CFL condition (see Problem 11.9).

When an upwind differencing is applied to the convection terms, then, according to equation (10.2.37), the eigenvalues are real and negative, and Henn's method, equation (11.2.10), is stable with the condition defined by equation (11.2.8) when the *same space discretization is applied on the predictor and the corrector.* This would lead to a necessary condition:

$$0 < \sigma = \frac{a \, \Delta t}{\Delta x} \leqslant 2 \tag{11.2.13}$$

which is twice the standard CFL limit of one.

Henn's method with the upwind differencing takes the following form for the linear convection equation:

$$\overline{u_i^{n+1}} = u_i^n - \sigma(u_i^n - u_{i-1}^n)$$

$$u_i^{n+1} = u_i^n - \frac{\sigma}{2} (\overline{u_i^{n+1}} - \overline{u_{i-1}^{n+1}} + u_i^n - u_{i-1}^n) \tag{11.2.14}$$

or

$$u_i^{n+1} = u_i^n - \sigma(u_i^n - u_{i-1}^n) + \frac{\sigma^2}{2} (u_i^n - 2u_{i-1}^n + u_{i-2}^n) \tag{11.2.15}$$

Compared with the only second-order accurate scheme, equation (9.3.13), this scheme differs by the coefficient in front of the last term. Therefore this scheme is only first-order accurate. A Von Neumann analysis shows, in addition, that the stability condition (11.2.13) is not sufficient, since the Von Neumann analysis leads to the CFL condition $0 < \sigma \leqslant 1$ (see Problem 11.10). Note that a similar situation has already been discussed in Section 10.4 in relation to the stability conditions of the explicit upwind scheme (equation (10.3.13)).

If scheme (11.2.11) is applied with upwind differences we obtain, instead of equation (11.2.15), the algorithm

$$u_i^{n+1} = u_i^n - \sigma(u_i^n - u_{i-1}^n) + \sigma^2(u_i^n - 2u_{i-1}^n + u_{i-2}^n) \tag{11.2.16}$$

In this case a necessary stability condition derived from equation (11.2.8) is $0 < \sigma \leqslant 1$, where the Von Neumann method gives the restriction $0 < \sigma \leqslant 1/2$ (see Problem 11.11).

McCormack's scheme

In order to obtain a second-order accurate scheme with Henn's predictor–corrector sequence and a first-order space differencing for H^n we could attempt to compensate the truncation errors in the combined sequence by applying a different space operator in the corrector step.

Considering scheme (11.2.10), and a space truncation error $a_{p+1} \, \Delta x^p$ in the predictor to the lowest order p, we could obtain a higher global accuracy for the corrector step if the corrector would generate an equal, but opposite in sign, truncation error. That is, we would have

$$\overline{U^{n+1}} = U^n + \Delta t H^n + \Delta t \Delta x^p a_{p+1} + \Delta t \Delta x^{p+1} a_{p+2} \tag{11.2.17a}$$

and

$$U^{n+1} = U^n + \tfrac{1}{2}\Delta t[\overline{H_1^{n+1}} + H^n + \Delta x^p a_{p+1} + a_{p+2}\,\Delta x^{p+1}$$
$$+ \overline{a_{p+1}}\,\Delta x^p + \overline{a_{p+2}}\,\Delta x^{p+1}] \tag{11.2.17b}$$

where $\overline{H_1}$ is different space operator from H. If $\overline{a_{p+1}} = -a_{p+1}$ the overall accuracy is increased by one unit, and becomes of order $(p+1)$. For the convection equation this would be realized if the corrector steps would contain a *forward* space difference, when the predictor contains a backward difference, or vice-versa.

This leads to the second-order accurate (in space and in time) *McCormack* scheme, which is widely applied for resolution of Euler and Navier–Stokes equations (McCormack, 1969, 1971):

$$\overline{u_i^{n+1}} = u_i^n - \sigma(u_i^n - u_{i-1}^n)$$
$$u_i^{n+1} = u_i^n - \frac{\sigma}{2}(\overline{u_{i+1}^{n+1}} - \overline{u_i^{n+1}} + u_i^n - u_{i-1}^n) \tag{11.2.18}$$

Since the unique second-order scheme for the linear wave equation is the Lax–Wendroff scheme, they should be identical (see Problem 11.12). However, this is no longer the case for non-linear problems, and we refer the reader to Volume 2 for applications of this scheme to the system of Euler and Navier–Stokes equations. Many other variants can be derived if we allow for the freedom of different space operators in the two steps of the predictor–corrector sequence.

11.3 LINEARIZATION METHODS FOR NON-LINEAR IMPLICIT SCHEMES

Another approach to the treatment of the non-linearity, which has the advantage of maintaining the fully implicit character of the schemes (when $\theta \neq 0$), is obtained from a local linearization of the non-linear terms. A Newton method is defined, based on

$$H^{n+1} = H(U^{n+1}) = H^n + \Delta H + 0(\Delta t^2)$$
$$= H^n + \left(\frac{\partial H}{\partial U}\right)\cdot(U^{n+1} - U^n) + 0(\Delta t^2) \tag{11.3.1}$$

where $(\partial H/\partial U)$ is the Jacobian of the operator H with respect to the dependent variable U. Writing

$$J(U) \equiv \frac{\partial H}{\partial U} \tag{11.3.2}$$

we have, up to order (Δt^2),

$$H^{n+1} = H^n + J(U^n)\cdot \Delta U^n \tag{11.3.3}$$

Scheme (11.1.23) becomes

$$[(1 + \xi) - \theta \, \Delta t J(U^n)] \, \Delta U^n = \Delta t H^n + \xi \, \Delta U^{n-1} \qquad (11.3.4)$$

This formulation is known as the *Beam and Warming* scheme (1976, 1978) and is often called the Δ-form (delta-form), although the first application of local linearization to flow problems was developed by Briley and McDonald (1975).

Example 11.3.1   Burger's equation $\partial u / \partial t + \partial (u^2/2)/\partial x = 0$

Applying a central difference to the space operator leads to

$$H_i = -\frac{1}{2\Delta x}\left(\frac{u_{i+1}^2}{2} - \frac{u_{i-1}^2}{2}\right) = \frac{-1}{2\Delta x}(E - E^{-1})\frac{u_i^2}{2} \qquad (\text{E11.3.1})$$

The Jacobian matrix J is obtained from

$$H_i^{n+1} - H_i^n = -\frac{1}{2\Delta x}(E - E^{-1})\left[\left(\frac{u_i^2}{2}\right)^{n+1} - \left(\frac{u_i^2}{2}\right)^n\right] \qquad (\text{E11.3.2})$$

as

$$J(u_i) = -\frac{1}{2\Delta x}(E - E^{-1})u_i^n \cdot \Delta u_i^n \qquad (\text{E11.3.3})$$

leading to the scheme, with $\Delta u_i^n = u_i^{n+1} - u_i^n$, from equation (11.3.4)

$$(1 + \xi)\Delta u_i^n + \frac{\theta \Delta t}{2\Delta x} u_{i+1}^n \, \Delta u_{i+1}^n - \frac{\theta \Delta t}{2\Delta x} u_{i-1}^n \Delta_{i-1}^n$$

$$= -\frac{\Delta t}{2\Delta x}\left[\left(\frac{u_{i+1}^n}{2}\right)^2 - \left(\frac{u_{i-1}^n}{2}\right)^2\right] + \xi \, \Delta u_i^{n-1} \qquad (\text{E11.3.4})$$

Application to the differential form

The whole development of this section did not specify the particular form of the right-hand side quantity H of equation (11.1.1). We have assumed here that it was already in space-discretized form in order to be able to establish the stability conditions. However, for the purpose of the scheme definition (11.3.4) we could consider that H is expressed as a function of the *differential* operators, and that a time integration is performed first, prior to the space discretization. Hence for the general equation in one dimension:

$$\frac{\partial U}{\partial t} + \frac{\partial f}{\partial x} = 0 \qquad (11.3.5)$$

where f is the flux function. We can write, with $H = -(\partial f / \partial x) \equiv -f_x$

$$f_x^{n+1} - f_x^n = \frac{\partial}{\partial x}(f^{n+1} - f^n) = \frac{\partial}{\partial x}\left[\frac{\partial f}{\partial U}\left(U^{n+1} - U^n\right)\right] + 0(\Delta t^2)$$

or

$$f_x^{n+1} - f_x^n = \frac{\partial}{\partial x} [A(U)\Delta U] + 0(\Delta t^2) \tag{11.3.6}$$

where the Jacobian A of the flux function has been introduced:

$$A(U) = \frac{\partial f}{\partial U} \tag{11.3.7}$$

Scheme (11.3.4) becomes, to order Δt^2,

$$\left[(1 + \xi) + \theta \Delta t \cdot \frac{\partial}{\partial x} A(U) \right] \Delta U^n = - \Delta t \frac{\partial f^n}{\partial x} + \xi \, \Delta U^{n-1} \tag{11.3.8}$$

written prior to the space discretization. This is the standard Δ-form of the Beam and Warming schemes. For Burger's equation, $f = u^2/2$ and $A(u) = u$; with a central difference for the space derivative we obtain exactly equation (E11.3.4).

All the above time-integration schemes can be written as a function of the *differential operators*, that is, *prior* to the space discretization by replacing f by $(\partial f/\partial x)$. For instance, McCormack's scheme, equation (11.2.18), becomes, for a general flux f,

$$\overline{u_i^{n+1}} = u_i^n - \frac{\Delta t}{\Delta x} (f_i^n - f_{i-1}^n)$$

$$\tag{11.3.9}$$

$$u_i^{n+1} = u_i^n - \frac{\Delta t}{2\Delta x} (\overline{f_{i+1}^{n+1}} - \overline{f_i^{n+1}} + f_i^n - f_{i-1}^n)$$

11.4 IMPLICIT SCHEMES FOR MULTI-DIMENSIONAL PROBLEMS: ALTERNATING DIRECTION IMPLICIT (ADI) METHODS

When the implicit schemes of form (11.1.11) are applied to multi-dimensional problems the resulting implicit matrix system is no longer tridiagonal, as it occurs for three-point discretizations on one-dimensional equations. For instance, the two-dimensional parabolic diffusion equation, with constant diffusion coefficient α,

$$u_t = \alpha(u_{xx} + u_{yy}) \tag{11.4.1}$$

discretized with a five-point finite difference Laplace scheme, leads to the system ($\Delta x = \Delta y$):

$$\frac{du_{ij}}{dt} = \frac{\alpha}{\Delta x^2} (u_{i+1,j} + u_{i-1,j} + u_{i,j+1} + u_{i,j-1} - 4u_{ij}) \tag{11.4.2}$$

With a backward Euler scheme, for instance, we obtain a pentadiagonal matrix system:

$$u_{ij}^{n+1} - u_{ij}^n = \frac{\alpha \Delta t}{\Delta x^2} (u_{i+1,j}^{n+1} + u_{i-1,j}^{n+1} + u_{i,j-1}^{n+1} + u_{i,j+1}^{n+1} - 4u_{ij}^{n+1}) \tag{11.4.3}$$

On an arbitrary mesh, with higher-order discretizations or with finite elements, additional mesh points appear in the Laplace discretization at point i, j, and the implicit matrix will have a more complicated structure than the pentadiagonal one.

The principle of the ADI method is to separate the operators into one-dimensional components and split the scheme into two (or three, for three-dimensional problems) steps, each one involving only the implicit operations originating from a single co-ordinate. This method has been introduced by Peaceman and Rachford (1955) and Douglas and Rachford (1956) and generalized by Douglas and Gunn (1964). Many developments and extensions have been brought to this approach by Russian authors and given different names, i.e. fractional step method by Yanenko (1971) or splitting method by Marchuk (1975). An excellent description of ADI methods can be found in Mitchell (1969) and Mitchell and Griffiths (1980).

If the matrix operator S on the right-hand side of

$$\frac{dU}{dt} = SU + Q \tag{11.4.4}$$

is separated into submatrices acting on the U components in a single direction, that is $S = S_x + S_y + S_z$, each operator in the right-hand side acting on the variable indicated as subscript, then equation (11.4.4) becomes

$$\frac{dU}{dt} = (S_x + S_y + S_z)U + Q \tag{11.4.5}$$

In equation (11.4.2) S_x and S_y represent the second-order derivatives in the x- and y-direction, respectively, with the shift operators E_x and E_y:

$$S_x U = \frac{\alpha}{\Delta x^2} (E_x - 2 + E_x^{-1})U \tag{11.4.6}$$

$$S_y U = \frac{\alpha}{\Delta y^2} (E_y - 2 + E_y^{-1})U \tag{11.4.7}$$

or explicitly:

$$S_x u_{ij} = \frac{\alpha}{\Delta x^2} (u_{i+1,j} - 2u_{ij} + u_{i-1,j}) \tag{11.4.8}$$

$$S_y u_{ij} = \frac{\alpha}{\Delta y^2} (u_{i,j+1} - 2u_{ij} + u_{i,j-1}) \tag{11.4.9}$$

With the implicit scheme (11.1.22) and, for instance, $\xi = 0, \theta = 1$, defining the first order in time, backward Euler method, we obtain

$$U^{n+1} - U^n = \Delta t(S_x + S_y + S_z)U^{n+1} + Q \, \Delta t \tag{11.4.10}$$

The implicit operators appear from

$$[1 - \Delta t(S_x + S_y + S_z)] U^{n+1} = Q \, \Delta t + U^n \tag{11.4.11}$$

The basic idea behind the ADI method consists of a factorization of the right-hand side operator in a product of one-dimensional operators. Equation (11.4.11) is replaced by

$$(1 - \tau \Delta t S_x)(1 - \tau \Delta t S_y)(1 - \tau \Delta S_z)U^{n+1} = \tau \Delta t Q + U^n \quad (11.4.12)$$

where τ is a free parameter. Developing equation (11.4.12) leads to

$$
\begin{aligned}
U^{n+1} - U^n = \tau \Delta t (S_x + S_y + S_z)U^{n+1} + \tau \Delta t Q \\
+ \tau^2 \Delta t^2 (S_x S_y + S_y S_z + S_z S_x)U^n + \tau^3 \Delta t^3 S_x S_y S_z U^n \quad (11.4.13)
\end{aligned}
$$

to be compared with equation (11.4.10).

The factorization (11.4.12) has introduced two additional terms which represent errors with respect to the original scheme to be solved. however, these are higher-order errors, proportional to Δt^2 and Δt^3, and since the backward Euler scheme is first order in time these error terms are of the same order as the trunctation error and do not affect the overall accuracy of the scheme.

The parameter τ appears as a relaxation parameter and has to be taken equal to one if this scheme is to be used for time-dependent simulations. However, the ADI technique is mostly of application for stationary problems, whereby we attempt to reach the convergence limit as quickly as possible. In this connection the relaxation parameter τ can be chosen to accelerate this convergence process, since $\tau \Delta t$ represents a scaling of the time step (see Chapter 12 for more details).

The factorized scheme is then solved in three steps:

$$
\begin{aligned}
(1 - \tau \Delta t S_x)\overline{\overline{U^{n+1}}} &= [1 + \tau \Delta t (S_y + S_z)] U^n + \tau \Delta t Q \\
(1 - \tau \Delta t S_y)\overline{U^{n+1}} &= \overline{\overline{U^{n+1}}} - \tau \Delta t S_y U^n \quad (11.4.14) \\
(1 - \tau \Delta t S_z)U^{n+1} &= \overline{U^{n+1}} - \tau \Delta t S_z U^n
\end{aligned}
$$

Introducing the variations $\Delta U^n = U^{n+1} - U^n$, $\overline{\Delta U^n} = \overline{U^{n+1}} - U^n$ and $\overline{\overline{\Delta U^n}} = \overline{\overline{U^{n+1}}} - U^n$ the ADI scheme can be rewritten as

$$
\begin{aligned}
(1 - \tau \Delta t S_x)\overline{\overline{\Delta U^n}} &= \tau \Delta t (SU^n + Q) \\
(1 - \tau \Delta t S_y)\overline{\Delta U^n} &= \overline{\overline{\Delta U^n}} \quad (11.4.15) \\
(1 - \tau \Delta t S_z)\Delta U^n &= \overline{\Delta U^n}
\end{aligned}
$$

This is sometimes called the Δ-*formulation*.

By recombining the factors the ADI approximation can be redefined by the formulation

$$(1 - \tau \Delta t S_x)(1 - \tau \Delta t S_y)(1 - \tau \Delta t S_z)\Delta U^n = \tau \Delta t (SU^n + Q) \quad (11.4.16)$$

For $\tau = 1$ we obtain the *Douglas–Rachford* scheme.

The *splitting* or *fractional step* method leads to another ADI formulation, based on a factorization of the Cranck–Nicholson scheme (equation (11.1.20)), and is therefore second order in time. Equation (11.4.10) is

440

replaced by

$$U^{n+1} - U^n = \Delta t(S_x + S_y + S_z)\frac{U^{n+1} + U^n}{2} + \Delta t Q \qquad (11.4.17)$$

or

$$\left[1 - \frac{\Delta t}{2}(S_x + S_y + S_z)\right]U^{n+1} = \left[1 + \frac{\Delta t}{2}(S_x + S_y + S_z)\right]U^n + \Delta t Q$$
$$(11.4.18)$$

This equation is factorized as follows:

$$\left(1 - \frac{\Delta t}{2}S_x\right)\left(1 - \frac{\Delta t}{2}S_y\right)\left(1 - \frac{\Delta t}{2}S_z\right)U^{n+1}$$

$$= \left(1 + \frac{\Delta t}{2}S_x\right)\left(1 + \frac{\Delta t}{2}S_y\right)\left(1 + \frac{\Delta t}{2}S_z\right)U^n + \Delta t Q \qquad (11.4.19)$$

and represents an approximation to equation (11.4.17) of second-order accuracy, since equation (11.4.19) is equal to

$$\left[1 - \frac{\Delta t}{2}(S_x + S_y + S_z)\right]U^{n+1} = \left[1 + \frac{\Delta t}{2}(S_x + S_y + S_z)\right]U^n + \Delta t Q$$

$$+ \frac{\Delta t^2}{4}(S_xS_y + S_yS_z + S_xS_z)(U^{n+1} - U^n) + \frac{\Delta t^3}{8}S_xS_yS_z(U^{n+1} + U^n) \quad (11.4.20)$$

Since $(U^{n+1} - U^n)$ is of order Δt, the error terms are $0(\Delta t^3)$. Equation (11.4.19) is then solved as a succession of one-dimensional Cranck-Nicholson schemes:

$$\left(1 - \frac{\Delta t}{2}S_x\right)\overline{U^{n+1}} = \left(1 + \frac{\Delta t}{2}S_x\right)U^n + \Delta t Q$$

$$\left(1 - \frac{\Delta t}{2}S_y\right)\overline{\overline{U^{n+1}}} = \left(1 + \frac{\Delta t}{2}S_y\right)\overline{U^{n+1}} \qquad (11.4.21)$$

$$\left(1 - \frac{\Delta t}{2}S_z\right)U^{n+1} = \left(1 + \frac{\Delta t}{2}S_z\right)\overline{\overline{U^{n+1}}}$$

When the S_i commute, we recover equation (11.4.19) by elimination of $\overline{\overline{U^{n+1}}}$ and $\overline{U^{n+1}}$. If the S_i operators do not commute, approximation (11.4.21) is still valid, but is reduced to first order in time.

11.4.1 Two-dimensional diffusion equation

Considering equation (11.4.1), with the central space discretizations, we can write the factorized ADI scheme, equation (11.4.14), as

$$(1 - \tau \Delta t S_x)\overline{U^{n+1}} = (1 + \tau \Delta t S_y)U^n$$
$$(1 - \tau \Delta t S_y)U^{n+1} = \overline{U^{n+1}} - \tau \Delta t S_y U^n \qquad (11.4.22)$$

Written out explicitly, we obtain the following tridiagonal systems, with $\Delta x = \Delta y$ and $\beta = \alpha\,\Delta t/\Delta x^2$:

$$\overline{u_{ij}^{n+1}} - \tau\beta\left(\overline{u_{i+1,j}^{n+1}} - 2\overline{u_{ij}^{n+1}} + \overline{u_{i-1,j}^{n+1}}\right) = u_{ij}^n + \tau\beta(u_{i,j+1}^n - 2u_{ij}^n + u_{i,j-1}^n)$$
$$u_{ij}^{n+1} - \tau\beta(u_{i,j+1}^{n+1} - 2u_{ij}^{n+1} + u_{i,j-1}^{n+1}) = \overline{u_{ij}^{n+1}} - \tau\beta(u_{i,j+1}^n - 2u_{ij}^n + u_{i,j-1}^n)$$

$$(11.4.23)$$

The first equation is solved as a tridiagonal system along all the horizontal j-lines, sweeping the mesh from $j = 1$ to $j = j_{max}$. After this first step, the intermediate solution $\overline{u_{ij}^{n+1}}$ is obtained at all mesh points. The second equation represents a succession of tridiagonal systems along the i-columns and the solution u_{ij}^{n+1} is obtained after having swept through the mesh along the vertical lines, from $i = 1$ to $i = i_{max}$.

A Von Neumann stability analysis can be performed for this system by defining an intermediate amplification factor, $\bar{G}$, by

$$\overline{U^{n+1}} = \bar{G}\,U^n \tag{11.4.24}$$

For the first step, with ϕ_x and ϕ_y being the Fourier variables in the x- and y-directions, respectively, we have

$$\bar{G} = \frac{1 - 4\tau\beta\,\sin^2\phi_y/2}{1 + 4\tau\beta\,\sin^2\phi_x/2} \tag{11.4.25}$$

$$G = \frac{\bar{G} + 4\tau\beta\,\sin^2\phi_y/2}{1 + 4\tau\beta\,\sin^2\phi_y/2} \tag{11.4.26}$$

and by combining these two equations:

$$G = \frac{1 + 16\tau^2\beta^2\,\sin^2\phi_x/2 \cdot \sin^2\phi_y/2}{(1 + 4\tau\beta\,\sin^2\phi_x/2)(1 + 4\tau\beta\,\sin^2\phi_y/2)} \tag{11.4.27}$$

The scheme is clearly unconditionally stable since $|G| \leqslant 1$. A similar calculation for three dimensions confirms this property.

Another version of the ADI technique is the *Peaceman–Rachford* method, based on the following formulation, with $\tau = 1/2$ in two dimensions:

$$\left(1 - \frac{\Delta t}{2}S_x\right)\overline{U^{n+1}} = \left(1 + \frac{\Delta t}{2}\,S_y\right)U^n$$
$$\left(1 - \frac{\Delta t}{2}\,S_y\right)U^{n+1} = \left(1 + \frac{\Delta t}{2}\,S_x\right)\overline{U^{n+1}}$$

$$(11.4.28)$$

We leave it to the reader to show that this scheme is also unconditionally stable for the diffusion equation.

In addition, this form of the ADI method is second-order accurate in time, as can be seen by eliminating the intermediate solution $\overline{U^{n+1}}$. We obtain

$$U^{n+1} - U^n = \Delta t(S_x + S_y)\frac{U^{n+1} + U^n}{2} + \frac{\Delta t^2}{4}\,S_x S_y(U^{n+1} - U^n) \tag{11.4.29}$$

442

and since $(U^{n+1} - U^n)$ is of order Δt, the last term is $0(\Delta t^3)$. The intermediate values $\overline{\Delta U}$ and $\overline{\overline{\Delta U}}$ have not necessarily a physical meaning, and boundary conditions have to be defined for these variables in accordance with the physical boundary conditions of U^n.

Boundary conditions for the intermediate steps can be obtained from the structure of system (11.4.15). In particular, for Dirichlet conditions we would write for the boundary values $(\ )_B$:

$$\overline{\overline{\Delta U}})_B = (1 - \tau \Delta t S_z)(U^{n+1} - U^n)_B$$
$$\overline{\Delta U})_B = (1 - \tau \Delta t S_y)(1 - \tau \Delta t S_z)(U^{n+1} - U^n)_B \qquad (11.4.30)$$

More details can be found in Mitchell and Griffiths (1980).

11.4.2 ADI method for the convection equation

Considering the three-dimensional convection equation

$$u_t + au_x + bu_y + cu_z = 0 \qquad (11.4.31)$$

the operators S_x, S_y and S_z can be written, for a second-order central difference of the space derivatives, as

$$\Delta t S_x u_{ijk} = -\frac{a\,\Delta t}{2\Delta x}(u_{i+1,jk} - u_{i-1,jk}) \equiv -\frac{\sigma_x}{2}(u_{i+1,jk} - u_{i-1,jk})$$

$$\Delta t S_y u_{ijk} = -\frac{b\,\Delta t}{2\Delta y}(u_{i,j+1,k} - u_{i,j-1,k}) \equiv -\frac{\sigma_y}{2}(u_{i,j+1,k} - u_{i,j-1,k}) \qquad (11.4.32)$$

$$\Delta t S_z u_{ijk} = -\frac{c\,\Delta t}{2\Delta z}(u_{ij,k+1} - u_{ij,k-1}) \equiv -\frac{\sigma_z}{2}(u_{ij,k+1} - u_{ij,k-1})$$

Considering first the *two-dimensional* case, the ADI scheme (equation (11.4.14)) can be written

$$\bar{u}_{ij}^{n+1} + \tau\frac{\sigma_x}{2}(\overline{u_{i+1,j}^{n+1}} - \overline{u_{i-1,j}^{n+1}}) = u_{ij}^n - \tau\frac{\sigma_y}{2}(u_{i,j+1}^n - u_{i,j-1}^n)$$

$$u_{ij}^{n+1} + \tau\frac{\sigma_y}{2}(u_{i,j+1}^{n+1} - u_{i,j-1}^{n+1}) = \overline{u_{i,j}^{n+1}} + \tau\frac{\sigma_y}{2}(u_{i,j+1}^n - u_{i,j-1}^n)$$

$$(11.4.33)$$

The amplification factors are given by

$$\bar{G} = \frac{1 - I\tau\sigma_y \sin\phi_y}{1 + I\tau\sigma_x \sin\phi_x} \qquad (11.4.34)$$

$$G = \frac{\bar{G} + I\tau\sigma_y \sin\phi_y}{1 + I\tau\sigma_y \sin\phi_y} = \frac{1 - \tau^2\sigma_x\sigma_y \sin\phi_x \sin\phi_y}{(1 + I\tau\sigma_x \sin\phi_x)(1 + I\tau\sigma_y \sin\phi_y)} \qquad (11.4.35)$$

Since G is of the form

$$G = \frac{1 - \tau^2 \sigma_x \sigma_y \sin \phi_x \sin \phi_y}{1 - \tau^2 \sigma_x \sigma_y \sin \phi_x \sin \phi_y + I\tau(\sigma_x \sin \phi_x + \sigma_y \sin \phi_y)} \qquad (11.4.36)$$

we have $|G|^2 \leqslant 1$, and the scheme is unconditionally stable.

However, in three dimensions this is no longer the case. The ADI scheme applied to the three-dimensional convection equation, centrally discretized (without dissipation contributions) is *unconditionally unstable*, as shown by Abarbanel *et al.* (1982).

In three dimensions the above procedure leads to the following amplification matrix, writing s_x for $\sigma_x \sin \phi_x$ and similarly for the y and z components:

$$G = \frac{1 - \tau^2(s_x s_y + s_z s_x + s_y s_z) - I\tau^3 s_x s_y s_z}{(1 + I\tau s_x)(1 + I\tau s_y)(1 + I\tau s_z)} \qquad (11.4.37)$$

It can be shown that there are always values of s_x, s_y, s_z such that $|G| > 1$. The following proof is due to Saul Abarbanel and Eli Turkel (private communication).

The amplification matrix G is written as

$$G = \frac{\alpha_1 + I\beta_1}{\alpha_1 + I\beta_2} \qquad (11.4.38)$$

and since the real parts are equal, stability is obtained if

$$|\beta_1| < |\beta_2| \qquad (11.4.39)$$

With

$$\beta_2 = \beta_1 + \tau(s_x + s_y + s_z) \equiv \beta_1 + \gamma$$

the condition $|\beta_1| \leqslant |\beta_2|$ implies

$$(\beta_1 + \gamma)^2 \geqslant \beta_1^2$$

or

$$\gamma(2\beta_1 + \gamma) \geqslant 0$$

If we select values of ϕ_x, ϕ_y, ϕ_z, such that $\gamma > 0$, assuming $\tau > 0$, the stability condition (11.4.39) becomes

$$2\tau^2 s_x s_y s_z \leqslant (s_x + s_y + s_z) = \gamma/\tau \qquad (11.4.40)$$

However, we can always find values for these variables which do *not* satisfy this inequality. For instance, taking $\gamma = \varepsilon$ through

$$s_x = -1/4, \qquad s_y = -1/4, \qquad s_z = \tfrac{1}{2} + \varepsilon \qquad (11.4.41)$$

the above condition becomes

$$\varepsilon \geqslant \frac{\tau^2}{16}(1 + 2\varepsilon)$$

444

or

$$\tau^2 \leqslant \frac{16\varepsilon}{1 + 2\varepsilon} \tag{11.4.42}$$

We can always select a value of ε sufficiently small such that this condition is never satisfied *for any fixed finite value of* τ, since the right-hand side goes to zero with ε. For instance, for $\tau = 1, \varepsilon < 1/14$ leads to an unstable scheme.

This instability is associated with low frequencies, since equation (11.4.41) implies that $s_x + s_y + s_z = \varepsilon/\tau$ is a small quantity, and can only be satisfied by small values of the wavenumbers ϕ_x, ϕ_y, ϕ_z. In addition, it can be considered as a weak instability since, from equation (11.4.38), for ε small:

$$|G|^2 = \frac{\alpha_1^2 + \beta_1^2}{\alpha_1^2 + (\beta_1 + \varepsilon)^2} \approx 1 - \frac{2\varepsilon\beta_1}{\alpha_1^2 + \beta_1^2} \tag{11.4.43}$$

where β_1 is defined, in assumption (11.4.41), by

$$\beta_1 = -\frac{\tau^3}{16} \left(\frac{1}{2} + \varepsilon \right) < 0$$

and is a small negative quantity. Hence $|G|$ is higher than one by an amount proportional to ε and therefore the amplification of errors should remain limited. This is confirmed by recent computations performed by Compton and Whitesides (1983) with the full system of Euler equations. However, the addition of adequate damping terms can remove the instability, as shown by Abarbanel *et al.* (1982).

Actually, the three-dimensional fractional step method, equation (11.4.21), is stable in three dimensions for the convection equation, although neutrally stable since it is a product of one-dimensional Crank–Nicholson schemes.

An essential difference between the fractional step method and the ADI method in its form (11.4.15) is connected with their behaviour for stationary problems. In this case convergence towards steady-state $\Delta U^n = 0$ is sought, implying $(SU^n + Q) = 0$, according to equation (11.4.16), when this limit is reached. Hence the steady-state limit resulting from the computation will be independent of the time step Δt, since S is a pure space discretization. Equation (11.4.20), on the other hand, shows that it in the limit $\Delta U^n \to 0$, we solve, with $(U^{n+1} + U^n)/2 \approx U^n$,

$$\left[1 - \frac{\Delta t^2}{4} (S_x S_y + S_y S_z + S_x S_z) \right] \Delta U^n = \Delta t (SU^n + Q) + \frac{\Delta t^3}{4} S_x S_y S_z U^n \tag{11.4.44}$$

which produces a stationary solution with a vanishing right-hand side. That is, the obtained stationary solution satisfies

$$SU^n + Q = -\frac{\Delta t^2}{4} S_x S_y S_z U^n \tag{11.4.45}$$

and is function of the time step. In addition, for large time steps, which is what we aim at with implicit methods, the right-hand side might become unacceptably high.

11.5 THE RUNGE–KUTTA SCHEMES

An important family of time-integration techniques whch are of a high order of accuracy, explicit but non-linear, and limited to two time levels is provided by Runge–Kutta methods. Compared with the linear multi-step method the Runge–Kutta schemes achieve high orders of accuracy by sacrificing the linearity of the method but maintaining the advantages of the one-step method, while the former are basically of a linear nature but achieve great accuracy by involving multiple time steps. A detailed description of the Runge–Kutta method can be found in Gear (1971), Lambert (1974) and Van der Houwen (1977). These methods have recently been applied to the solution of Euler equations by Jameson et al. (1981) and further developed to highly efficient operational codes (Jameson and Baker, 1983, 1984).

The basic idea of the Runge–Kutta methods is to evaluate the right-hand side of the differential system (11.1.1) at several values of U in the interval between $n \, \Delta t$ and $(n + 1)\Delta t$ and to combine them in order to obtain a high-order approximation of U^{n+1}. The general form of a K-stage Runge–Kutta method is as follows:

$$U^{(1)} = U^n$$
$$U^{(2)} = U^n + \Delta t \alpha_2 H^{(1)}$$
$$U^{(3)} = U^n + \Delta t \alpha_3 H^{(2)} \qquad (11.5.1)$$
$$\vdots$$
$$U^{(K)} = U^n + \Delta t \alpha_K H^{(K-1)}$$
$$U^{n+1} = U^n + \Delta t \sum_{k=1}^{K} \beta_k H^{(k)}$$

where, for consistency,

$$\sum_{k=1}^{K} \beta_k = 1 \qquad (11.5.2)$$

The notation $H^{(k)}$ implies

$$H^{(k)} = H(U^{(k)}) \qquad (11.5.3)$$

written here for the case where H is explicitly independent of time, which is generally the case in fluid mechanical problems. Note that equation (11.5.1) is not the most general form of the Runge–Kutta schemes, since $H^{(k)}$ is only a function of $U^{(k)}$. In more general schemes, $H^{(k)}$ is a function of a linear combination of all the $U^{(m)}$ with $m \leqslant k$. The most popular version is the fourth-order Runge–Kutta method, defined by the coefficients

$$\alpha_2 = \tfrac{1}{2} \qquad \alpha_3 = \tfrac{1}{2} \qquad \alpha_4 = 1$$
$$\beta_1 = \tfrac{1}{6} \qquad \beta_2 = \beta_3 = \tfrac{1}{3} \qquad \beta_4 = \tfrac{1}{6}$$

(11.5.4)

leading to

$$U^{(1)} = U^n$$
$$U^{(2)} = U^n + \tfrac{1}{2} \, \Delta t H^{(1)}$$
$$U^{(3)} = U^n + \tfrac{1}{2} \, \Delta t H^{(2)}$$
$$U^{(4)} = U^n + \Delta t H^{(3)}$$

(11.5.5)

$$U^{n+1} = U^n + \frac{\Delta t}{6} \, (H^n + 2H^{(2)} + 2H^{(3)} + H^{(4)})$$

(11.5.6)

where $H^{(1)}$ has been written as H^n.

A well-known two-stage Runge–Kutta method (Henn's method) is defined by the predictor–corrector scheme of equation (11.2.10). With the restriction to order two there exists an infinite number of two-stage Runge–Kutta methods with order two but none with an order higher than two. They all can be considered as predictor–corrector schemes.

Another popular, second-order scheme in this family is defined by

$$\overline{U^{n+1}} = U^n + \frac{\Delta t}{2} \, H^n$$
$$U^{n+1} = U^n + \Delta t \, \overline{H^{n+1}}$$

(11.5.7)

Note that for each number of stages K there is an infinite number of possible Runge–Kutta schemes, with maximum order of accuracy. An even larger member of free parameters can be selected when the requirements on the order of accuracy are relaxed.

Stability analysis for the fourth-order Runge–Kutta method

In the linear case of the model problem $H = \Omega U$, equation (11.5.5) becomes

$$U^{(2)} = (1 + \tfrac{1}{2} \, \Omega \, \Delta t) U^n$$
$$U^{(3)} = [1 + \tfrac{1}{2} \, \Omega \, \Delta t + \tfrac{1}{4} (\Omega \, \Delta t)^2] U^n$$
$$U^{(4)} = [1 + \Omega \, \Delta t + \tfrac{1}{2} (\Omega \, \Delta t)^2 + \tfrac{1}{4} (\Omega \, \Delta t)^3] U^n$$

(11.5.8)

and

$$U^{n+1} = \left[1 + \Omega \, \Delta t + \frac{(\Omega \, \Delta t)^2}{2} + \frac{(\Omega \, \Delta t)^3}{6} + \frac{(\Omega \, \Delta t)^4}{24}\right] U^n \equiv z U^n$$

(11.5.9)

where z is the root of the characteristic polynomial.

The stability region in the $\Omega \, \Delta t$ plane is shown in Figure 11.5.1, together with the stability regions for the second-order method (11.2.10) and the third-order Runge–Kutta method, defined by $\alpha_2 = \tfrac{1}{3}, \alpha_3 = \tfrac{2}{3}, \beta_1 = \tfrac{1}{4}, \beta_2 = 0$,

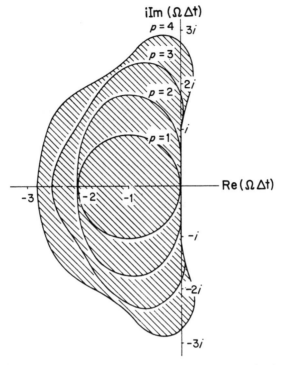

Figure 11.5.1 Stability regions for Runge–Kutta methods of
orders 2, 3 and 4 in the $\Omega\Delta t$ plane

$\beta_3 = \frac{3}{4}$. Actually, all methods with K stages and order K have the same domain
of stability.

It is seen from equation (11.5.9) that the scheme is fourth-order accurate,
since z is the Taylor expansion of the exact amplification $\exp(\Omega \Delta t)$ up to
fourth order. Observe that the fourth-order method contains the imaginary
axis up to a maximum value of $2\sqrt{2}$. Therefore the fourth-order Runge–Kutta
integration scheme will be conditionally stable for central differenced convection equations, in contrast to the second-order method (11.2.10), which does
not include any portion of the imaginary axis. The limits on the real axis are
$-2, -2.51, -2.78$ for the second, third and fourth order, respectively
(Lambert, 1974). The first-order Runge–Kutta method is the Euler explicit
scheme.

For stationary problems it is important to be able to allow the greatest
possible time steps, and therefore the extension of the stability region is more
important than their order. By relaxing the order we can obtain Runge–Kutta
methods of a given stage number, with higher stability regions; for instance, a
third-order Runge–Kutta method with second-order accuracy can be defined
which cuts the real axis at $(-4.52, 0)$, Lambert (1974).

448

Example 11.5.1   Diffusion equation $u_t = \alpha_{xx}$ with second-order central space differences

If the system

$$\frac{du_i}{dt} = \frac{\alpha}{\Delta x^2} (u_{i+1} - 2u_i + u_{i-1})$$ (E11.5.1)

is solved by a Runge–Kutta method whose stability domain cuts the real axis at $(-B, 0)$ then from equation (10.2.28), for Dirichlet conditions, the method will be stable for

$$0 \leqslant \frac{\alpha \, \Delta t}{\Delta x^2} \leqslant \frac{B}{4}$$ (E11.5.2)

For instance, for the fourth-order method the limit is $2.78/4$, which is only marginally larger than the single-step, first order Euler explicit method.

Example 11.5.2   Convection equation $u_t + au_x = 0$ with second order central space difference

The system

$$\frac{du_i}{dt} = -\frac{a}{2\Delta x} (u_{i+1} - u_{i-1})$$ (E11.5.3)

solved by a fourth-order Runge–Kutta method has a stability range on the imaginary axis limited to $\pm I 2\sqrt{2}$. Hence, with periodic boundary conditions we have, from equation (10.2.40),

$$\Omega \, \Delta t = -I\sigma \sin \phi$$ (E11.5.4)

and the stability condition becomes

$$|\sigma| < 2\sqrt{2}$$ (E11.5.5)

This scheme therefore allows a Courant number close to three times the usual CFL limit of one.

However, for $\phi = \pi$, that is, for high frequencies, $\Omega = 0$ and the amplification factor z becomes equal to one. Therefore it is to be expected that this scheme will become unstable when applied to non-linear hyperbolic problems with central differences if some dissipation is not added to the scheme (Problems 11.18 and 11.19).

These two examples show that there is little to be gained in applying high-order Runge–Kutta methods to pure diffusion problems with regard to maximum allowable time steps when compared with the one-step explicit Euler method. However, this is not the case for convection equations where high-stage Runge–Kutta methods have an increasing segment of the imaginary axis of the Ω-plane in their stability region. The length of this segment

$(-I\beta, +I\beta)$ is increasing with the stage number K of the method. When the order p is left as a parameter (not being necessarily equal to its maximum value) the limit value $\beta = K - 1$ can be obtained for K odd and a certain combination of coefficients, Van der Houwen (1977). For an even number of stages (for instance, $K = 4$) Vichnevetsky (1983) has shown that the value $(K - 1)$ is also an upper limit, and a recent proof by Sonneveld and Van Leer (1984) indicates that this limit can indeed be reached for first-order accurate methods $p = 1$ and $K \geqslant 2$. Hence certain variants of the standard Runge–Kutta method could increase the stability limit of equation (E11.5.5) to $|\sigma| \leqslant 3$.

Other applications of the large number of degrees of freedom available in the selection of the α_k and β_k coefficients of equation (11.5.1) can be directed towards the selective damping of high-frequency error components, suitable for integration into multi-grid iterative methods, Jameson and Baker (1983).

References

Abarbanel, S., Dwoyer, D. D., and Gottlieb, D. (1982). 'Stable implicit finite difference methods for three dimensional hyperbolic systems.' *ICASE Report 82-39*, NASA Langley RC, Hampton, Virginia, USA: ICASE.

Beam, R. M., and Warming, R. F. (1976). 'An implicit finite-difference algorithm for hyperbolic systems in conservation law form.' *Journal Computational Physics*, **22**, 87–109.

Beam, R. M., and Warming R. F. (1978). 'An implicit factored scheme for the compressible Navier–Stokes equations.' *AIAA Journal*, **16**, 393–402.

Beam, R. M., and Warming, R. F. (1980). 'Alternating direction implicit methods for parabolic equations with a mixed derivative.' *SIAM Journal of Sci. Stat, Comp.*, **1**, 131–59.

Beam, R. M., and Warming, R. F. (1982). 'Implicit numerical methods for the compressible Navier–Stokes and Euler equations.' *Von Karman Institute Lecture Series 1982-04*, Rhode Saint Genese, Belgium.

Brailovskaya, I. (1965). 'A difference scheme for numerical solutions of the two-dimensional nonstationary Navier–Stokes equations for a compressible gas.' *Sov. Phys. Dokl.*, **10**, 107–10.

Briley, W. R., and McDonald, H. (1975). 'Solution of the three-dimensional compressible Navier–Stokes equations by an implicit technique.' *Proc. Fourth International Conference on Numerical Methods in Fluid Dynamics, Lecture Notes in Physics*, Vol, 35, Berlin: Springer Verlag.

Briley, W. R., and McDonald, H. (1980). 'On the structure and use of linearized block implicit schemes.' *Journal Computational Physics*, **6**, 428–53.

Compton II, W. B., and Whiteside, J. L. (1983). 'Three-dimensional Euler solutions for long duct nacelles.' *AIAA Paper 83-0089*, AIAA 21st Aerospace Sciences Meeting.

Dahlquist, G. (1963). 'A special stability problem for linear multistep methods.' *BIT*, **3**, 27–43.

Dahlquist, G., and Björk, A. (1974). *Numerical Methods*, Englewood Cliffs, NJ: Prentice-Hall.

Douglas, J., and Gunn, J. E. (1964). 'A general formulation of alternating direction methods: parabolic and hyperbolic problems.' *Numerische Mathematik*, **6**, 428–53.

Douglas J., and Rachford, H. H. (1956). 'On the numerical solution of heat conduction problems in two and three space variables.' *Transactions American Math. Society*, **82**, 421–39.

Jameson, A., Schmidt, W., and Turkel, E. (1981). 'Numerical simulation of the Euler equations by finite volume methods using Runge–Kutta time stepping schemes.' *AIAA Paper 81-1259*, AIAA 5th Computational Fluid Dynamics Conference.

Jameson, A., and Baker, T. J. (1983). 'Solutions of the Euler equations for complex configurations.' *AIAA Paper 83-1929*, AIAA 6th Computational Fluid Dynamics Conference.

Jameson, A., and Baker, T. J. (1984). 'Multigrid solution of the Euler equations for aircraft configurations.' *AIAA Paper 84-0093*, AIAA 22nd Aerospace Sciences Meeting.

McCormack, R. W. (1969). 'The effect of viscosity in hypervelocity impact cratering.' *AIAA Paper 69-354*.

McCormack, R. W. (1971). 'Numerical solution of the interaction of a shock wave with a laminar boundary layer.' *Proc. Second International Conference on Numerical Methods in Fluid Dynamics*, pp. 151–63, *Lecture Notes in Physics*, Vol. 8, Berlin: Springer Verlag.

Mitchell, A. R. (1969). *Computational Methods in Partial Differential Equations*, New York: John Wiley.

Peaceman, D. W., and Rachford, H. H. (1955). 'The numerical solution of parabolic and elliptic differential equations.' *SIAM Journal*, **3**, 28–41.

Pulliam, T. H. (1984). 'Euler and thin layer Navier–Stokes codes: ARC2D, ARC3D.' *Proc. Computational Fluid Dynamics User's Workshop*, Tullahoma, Tennessee: The University of Tennessee Space Institute.

Sonneveld, P., and Van Leer, B. (1985). 'A minimax problem along the imaginary axis.' *Nieuw Archief voor Wiskunde*, **3**, pp 19–22.

Van der Houwen, P. J. (1977). *Construction of Integration Formulas for Initial Value Problems*, Amsterdam: North-Holland.

Vichnevetsky, R. (1983). 'New stability theorems concerning one-step numerical methods for ordinary differential equations.' *Mathematics and Computers in Simulation*, **25**, 199–211.

Yanenko, N. N. (1979). *The Method of Fractional Steps*, New York: Springer Verlag.

PROBLEMS

Problem 11.1

Show that the trapezoidal scheme (11.1.17) is conditionally stable for $0 \leqslant \theta < 1/2$ when the eigenvalue spectrum Ω is on the real negative axis as for the diffusion equation with central differences. Proof equation (11.1.26) as well as the unconditional stability for $\theta > 1/2$.

Define the stability conditions for convection equations when Ω is purely imaginary.

Problem 11.2

Apply the techniques of Chapter 10, through Taylor series development, to the general two-step time integration (11.1.11) and proof relations (11.1.14) and (11.1.15).

Problem 11.3

Apply the Cranck–Nicholson scheme (equation (E11.1.1)) and show its unconditional stability by a Von Neumann analysis for $\alpha > 0$. Compare with formula (E11.1.3) and notice the relation with the amplification factor for the periodic eigenvalues (10.2.28).

Derive also the equivalent differential equation. Draw a plot of the dissipation error

$$\varepsilon_D = G/\tilde{G} = G/e^{-\alpha k^2 \Delta t} = G/e^{-\beta \phi^2}$$

and compare with the dissipation errors (8.3.24) and (8.3.25) generated by the explicit scheme.

Hint: Obtain the amplification factor, $\beta = \alpha \Delta t/\Delta x^2$:

$$G = \frac{1 - 2\beta \sin^2\phi/2}{1 + 2\beta \sin^2\phi/2}$$

and the equivalent equation

$$u_t - \alpha u_{xx} = \frac{\alpha \Delta x^2}{12}\left(\frac{\partial^4 u}{\partial x^4}\right) + \frac{\alpha \Delta x^4}{12}\left(\beta^2 + \frac{1}{30}\right)\left(\frac{\partial^6 u}{\partial x^6}\right) + \cdots$$

Observe the absence of phase errors.

Problem 11.4

Repeat problem 11.3 for the convection equation $u_t + a u_x = 0$ with a central difference discretization and periodic boundary conditions. Calculate the amplitude and phase errors as a function of ϕ and draw a polar plot of both quantities.

Hint: The scheme is written as $\sigma = a \Delta t/\Delta x$:

$$u_i^{n+1} - u_i^n = \frac{\sigma}{4}(u_{i+1}^{n+1} + u_{i-1}^{n+1} + u_{i+1}^n - u_{i-1}^n)$$

and the amplification matrix is

$$G = \frac{1 - I\sigma/2 \sin \phi}{1 + I\sigma/2 \sin \phi}$$

The equivalent differential equation is

$$u_t + a u_x = -\frac{a \Delta x^2}{6}\left(\frac{\sigma^2}{2} + 1\right)u_{xxx} - \frac{a \Delta x^4}{8}\left(\frac{1}{15} + \frac{\sigma^2}{2} + \frac{\sigma^4}{10}\right)\left(\frac{\partial^5 u}{\partial x^5}\right) + \cdots$$

Observe that the scheme is not dissipative and therefore oscillations might appear in non-linear problems.

Problem 11.5

Apply the Euler implicit scheme to the second-order space discretized diffusion equation

$$\frac{dU}{dt} = \frac{\alpha}{\Delta x^2}(E - 2 + E^{-1})U$$

where

$$Eu_i = u_{i+1}$$

Determine the root of the characteristic polynomial and compare with the amplification factor from a Von Neumann analysis. Draw a plot of the dissipation error and compare with the results of Problem 11.5. Determine the truncation error.

Hint: Write the scheme as

$$u_i^{n+1} - u_i^n = \beta(u_{i+1}^{n+1} - 2u_i^{n+1} + u_{i-1}^{n+1})$$

The amplification factor is

$$G = \frac{1}{1 + 4\beta \sin^2\phi/2}$$

The equivalent differential equation is

$$u_t - \alpha u_{xx} = \frac{1}{2}\alpha(\Delta x)^2\left(\beta + \frac{1}{6}\right)\left(\frac{\partial^4 u}{\partial x^4}\right) + \frac{1}{3}\alpha\,\Delta x^4\left(\beta^2 + \frac{1}{4}\beta + \frac{1}{120}\right)\frac{\partial^6 u}{\partial x^6} + \cdots$$

Problem 11.6

Apply the Euler implicit scheme to the convection equation with upwind discretization. Repeat Problem 11.5 and compare.

Hint: Write the scheme as

$$u_i^{n+1} - u_i^n = -\sigma(u_i^{n+1} - u_{i-1}^{n+1})$$

The amplification matrix is

$$G = \frac{1}{1 + 2\sigma I\,e^{-I\phi/2}\sin\phi/2} = \frac{1}{1 + I\sigma\sin\phi + 2\sigma\sin^2\phi/2}$$

and the scheme is stable for $\sigma > 0$.

   Note that the scheme has a bidiagonal structure and observe that, in order to start the resolution, in a sweep from $i = 1$ to N we have to know the solution from the left. Hence the boundary condition to be imposed is at the left boundary. Refer to the discussions of Section 10.2.

Problem 11.7

Solve Problem 9.8 with the upwind implicit Euler scheme of Problem 11.6 for $\sigma = 0.5, 1, 2$, after $10, 20, 50$ time steps. Generate a plot of the numerical solution and compare with the exact solution.

Problem 11.8

Solve the moving shock of Problem 9.9 with the upwind implicit Euler scheme of Problem 11.6 for $\sigma = 0.5, 1, 2$, after $10, 20, 50$ time steps. Generate a plot of the numerical solution and compare with the exact solution.

Problem 11.9

Consider the predictor–corrector scheme (11.2.12) and show that it is equivalent to the following one-step scheme with five-point support $i, i \pm 1, i \pm 2$:

$$u_i^{n+1} = u_i^n - \frac{\sigma}{2}(u_{i+1}^n - u_{i-1}^n) + \frac{\sigma^2}{4}(u_{i+1}^n - 2u_i^n + u_{i-1}^n)$$

Shows that condition (11.2.9) leads to the CFL condition $|\sigma| \leqslant 1$ and confirm this result from the matrix method by a Von Neumann analysis.

Problem 11.10

Consider equation (11.2.15) obtained from the first-order predictor–corrector method (equation (11.2.14)) and perform a Von Neumann analysis. Show, by investigating the behaviour of the amplification matrix for the phase angle $\phi = \pi$, that the stability condition has to be restricted to the CFL condition $0 < \sigma \leqslant 1$.

   Calculate the amplification factor and plot the dissipation and dispersion error as a function of ϕ. Determine also the equivalent differential equation and compare with the properties of the second-order upwind scheme (9.3.13).

Problem 11.11

Show that equation (11.2.16) is obtained from the first-order predictor–corrector method (11.2.11) when a first-order upwind difference is applied to the linear convection equation. Obtain the necessary condition from the stability condition (11.2.8). Show, by investigating the behaviour of the Von Neumann amplification matrix for the phase angle $\phi = \pi$, that this condition is not sufficient and obtain the Von Neumann necessary and sufficient condition $0 < \sigma \leqslant 1/2$.

Calculate the amplification factor and plot the dissipation and dispersion errors as a function of ϕ. Determine also the equivalent differential equation and compare with the properties of the second-order upwind scheme (9.3.13).

Problem 11.12

Show that the McCormack scheme (11.2.18) is identical to the Lax–Wendroff scheme. Write also the scheme by taking a forward space difference in the predictor and a backward difference in the corrector. Analyse the stability by the method of the characteristic polynomial and obtain the CFL condition for stability.

Problem 11.13

Solve Burger's equation with McCormack's scheme, where

$$H^n = \left(\frac{u_i^n}{2} \right)^2 \quad \text{and} \quad \overline{H^{n+1}} = \left(\overline{\frac{u_{i+1}^{n+1}}{2}} \right)^2 - \left(\overline{\frac{u_i^{n+1}}{2}} \right)^2$$

for a stationary discontinuity $(+1, -1)$ and a moving discontinuity $(1, -0.5)$. Consider the same condition as applied to Figures 8.5.1, 8.5.2 and 9.4.1.

Problem 11.14

Repeat Problem 11.13 with the Euler implicit scheme and with the trapezoidal scheme. Observe the appearance of strong oscillations and explain their origin by the analysis of the Von Neumann amplification function and by the truncation error structure.

Problem 11.15

Solve Problem 9.8 with McCormack's scheme (11.2.18). Compare with the previous results.

Problem 11.16

Solve Problem 9.9 with McCormack's scheme (11.2.18). Compare with the previous results.

Problem 11.17

Show the unconditional stability of the Peaceman–Rachford method, equation (11.4.28), for the two-dimensional diffusion equation (11.4.1) with central differences.

Problem 11.18

Define the stability conditions for the convection equation $u_t + au_x = 0$, discretized with central differences and an added second-order artificial viscosity $\Delta x \varepsilon u_{xx}$, when solved with a fourth-order Runge–Kutta method. Determine in the $(\Omega \, \Delta t)$ plane the curves of

454

constant ($\Omega \Delta t$) for different Courant numbers σ and for constant values of ε. What is the maximum allowable ε?

Hint: Consider the equation

$$u_t + au_x = \varepsilon \Delta x u_{xx}$$

and discretize centrally. Obtain the system

$$\frac{du_i}{dt} = \frac{-a}{2\Delta x}(u_{i+1} - u_{i-1}) + \frac{\varepsilon}{\Delta x}(u_{i+1} - 2u_i + u_{i-1})$$

Determine the eigenvalues of the right-hand side tridiagonal matrix for periodic conditions. Show that we obtain, with $\phi = 2\pi j/N$, referring to equation (10.2.40)

$$\Omega \Delta t = -I\sigma \sin \phi - \frac{4\varepsilon \Delta t}{\Delta x} \sin^2 \phi/2$$

The artificial viscosity parameter, ε, is restricted by the condition

$$0 \leqslant \frac{4\varepsilon \Delta t}{\Delta x} \leqslant 2.78$$

Problem 11.19

Repeat Problem 11.18 for a fourth-order dissipation term added to the central discretized convection equation of the form $-\varepsilon \Delta x^3 (\partial^4 u/\partial x^4)$.

Hint: Apply a central-second-order difference formula for the fourth derivative of

$$u_t + au_x = \varepsilon \Delta x^3 \cdot u_{xxxx}$$

For instance,

$$\frac{du_i}{dt} = -\frac{a}{2\Delta x}(u_{i+1} - u_{i-1}) - \frac{\varepsilon}{\Delta x}(u_{i+2} - 4u_{i+1} + 6u_i - 4u_{i-1} + u_{i-2})$$

Obtain the eigenvalues of the space operator on the left-hand side for a Fourier harmonic $e^{Ii\phi}$ and show that we have

$$\Omega \Delta t = -I\sigma \sin \phi - \frac{4\varepsilon \Delta t}{\Delta x}(1 - \cos \phi)^2$$

Determine the limit value of the dissipation parameter ε as

$$0 \leqslant 16\left(\frac{\varepsilon \Delta t}{\Delta x}\right) \leqslant 2.78$$

Problem 11.20

Obtain the characteristic roots ξ for the Beam and Warming scheme of Example 11.1.2 with the eigenvalues (10.2.40). Compare with equation (E11.1.8)

Problem 11.21

Consider the two-step method formed by an explicit predictor followed by an implicit corrector, applied to the one-dimensional diffusion equation with periodic boundary conditions. In general terms the scheme is written as

$$\bar{U} - U^n = \Delta t H(U^n)$$
$$L\Delta U = \overline{\Delta U} \equiv \bar{U} - U^n$$

Take $L = 1 - \varepsilon \Delta x^2 \partial^2/\partial x^2$ as implicit operator, where ε is a free parameter, and perform a central-second-order discretization.

Determine the amplification factor of the combined scheme and show that unconditional stability is obtained for all $\varepsilon > 0$.

Show that the implicit step increases the time step limit of the explicit corrector step, without modifying its spatial order of accuracy.

Hint: The amplification factor is

$$z = 1 - \frac{4\beta \sin^2 \phi/2}{1 + 4\varepsilon \sin^2 \phi/2}$$

Chapter 12

Iterative Methods for the Resolution of Algebraic Systems

Algebraic systems of equations are obtained either as a result of the application of implicit time-integration schemes to time-dependent formulations (as shown in the previous chapter) or from space discretizations of steady-state formulations.

Two large families of methods are available for the resolution of a linear algebraic system: the *direct* and the *iterative* methods. Direct methods are based on a finite number of arithmetic operations leading to the exact solution of a linear algebraic system. Iterative methods, on the other hand, are based on a succession of approximate solutions, leading to the exact solution after an infinite number of steps. In practice, however, the number of arithmetic operations of a direct method can be very high, and often larger than the total number of operations in an iterative sequence limited to a finite level of convergence, as a consequence of the finite arithmetic (number of digits) of the computer. So much that most of the flow problems lead to sparse matrices, a situation which would tend to favour the iterative method, particularly when the size of the matrices is very large.

In non-linear problems we have to set up an iterative scheme to solve for the non-linearity, even when the algebraic system, at each iteration step, is solved by a direct method. In these cases it is often more economical, for large problems, to insert the non-linear iterations into an overall iterative method for the algebraic system. For problems with a smaller number of mesh points (say, of the order of 1000) direct methods would generally not penalize the global cost of the iterative process.

The subject of resolutions of algebraic systems has been treated extensively in the literature, and is still an area of much research with the aim of improving the algorithms and reducing the total number of operations. Some excellent expositions of direct and iterative methods can be found in Varga (1962), Wachspress (1966), Young (1971), Dahlquist and Björk (1974), Marchuk (1975), Hageman and Young (1981), Meis and Morcowitz (1981) and Golub and Meurant (1983). A general presentation of matrix properties can be found, for instance, in Strang (1976) and Berman and Plemmons (1979).

Since fluid mechanical problems mostly require fine meshes in order to obtain sufficient resolution, direct methods are seldom applied, with the exception of tridiagonal systems. For these often-occurring systems a very efficient direct solver, known as the Thomas algorithm, is applied, and, because of its importance, the method and FORTRAN subroutines are presented in the Appendix for the scalar case with Dirichlet or Neumann boundary conditions as well as for the case of periodic boundary conditions, leading to a matrix of the form of equation (10.2.12). Otherwise, we will restrict this presentation to iterative methods.

The basis of iterative methods is to perform a small number of operations on the matrix elements of the algebraic system in an iterative way, with the aim of approaching the exact solution, within a preset level of accuracy, in a finite and, hopefully, small number of iterations. A large number of methods are available with different rates of convergence and complexity. The classical *relaxation* methods of Jacobi and Gauss–Seidel are the simplest to implement, including their variants known as *block-relaxation*, and are widely used. Another family of iterative methods are known as the *alternating direction implicit* (ADI) methods which, when optimized, have excellent convergence properties. They are, however, difficult to optimize for general configurations. High rates of convergence can be obtained with *preconditioning* techniques, particularly when coupled to *conjugate gradient* methods. These methods are more complex, but appreciable savings in computer time can often be achieved. The most efficient method, but also the most delicate to program, is the *multigrid method*, which appears to be the most efficient of all the iterative techniques. It leads to the same asymptotic convergence rate, or operation count, as the best fast Poisson solvers, without being limited by the same restricted conditions of application.

12.1 BASIC ITERATIVE METHODS

The classical example on which the iterative techniques are applied and evaluated is Poisson's equation on a uniform mesh. We will follow this tradition here, which allows us to clearly demonstrate the different approaches, their properties, structure and limitations.

12.1.1 Poisson's equation on a Cartesian, two-dimensional mesh

Let us consider the Poisson equation

$$\Delta u = f \qquad 0 \leqslant x \leqslant L, \qquad 0 \leqslant y \leqslant L \qquad (12.1.1)$$

with

$$u = g \qquad \text{on the boundary of the rectangle} \qquad (12.1.2)$$

and discretized with a second-order, five-point scheme as follows, with $\Delta x = \Delta y = L/M$, $i, j = 0, \ldots, M$:

$$(u_{i+1,j} - 2u_{ij} + u_{i-1,j}) + (u_{i,j+1} - 2u_{ij} + u_{i,j-1}) = f_{ij}\Delta x^2 \qquad (12.1.3)$$

If the vector U is set up with the u_{ij} classified *line by line*, that is,

$$U^T = (u_{11}, u_{21}, \ldots, u_{M1}; \; u_{12}, u_{22}, u_{32}, \ldots, u_{M2}; \; \ldots; \; u_{1j}, u_{2j}, \ldots, u_{ij}, \ldots, u_{Mj}; \; \ldots$$

$$(12.1.4)$$

The matrix system obtained from equation (12.1.3) is, with S being a $(M^2 \times M^2)$ matrix,

$$SU = F \, \Delta x^2 + G \equiv -Q \qquad (12.1.5)$$

$$
\begin{vmatrix}
& \text{M spaces} & & & & & \\
-4 & 1 & \ldots & 1 & & & \\
\vdots & \vdots & & \vdots & & & \\
& 1 & -4 & 1 & \ldots & 1 & \\
\vdots & & & & & & \\
1 \ldots & 1 & & -4 & 1 & \ldots & 1 \\
& & & 1 & \ldots & 1 & -4 & 1 & \ldots & 1 \\
& & & \text{M spaces} & & & \text{M spaces} \\
& & & & & 1 & \ldots & 1 & -4
\end{vmatrix}
\begin{vmatrix}
u_{11} \\
\cdot \\
\\
u_{i1} \\
\vdots \\
u_{ij} \\
\vdots \\
u_{i,j+1} \\
\vdots \\
u_{M-1,M-1} \\
\\
\cdot
\end{vmatrix}
$$

$$
= \Delta x^2
\begin{vmatrix}
f_{11} \\
\cdot \\
f_{i1} \\
\vdots \\
f_{ij} \\
\vdots \\
f_{i,j+1} \\
\vdots \\
f_{M-1,M-1}
\end{vmatrix}
-
\begin{vmatrix}
g_{10} + g_{01} \\
\cdot \\
g_{i0} \\
\vdots \\
0 \\
\vdots \\
0 \\
\vdots \\
g_{M,M-1} + g_{M-1,M}
\end{vmatrix}
$$

$$(12.1.6)$$

The first equation for $i = j = 1$ becomes, with the Dirichlet conditions (12.1.2),

$$u_{21} - 4u_{11} + u_{12} = \Delta x^2 f_{11} - g_{10} - g_{01} \equiv -q_{11} \qquad (12.1.7)$$

Along $j = 1$ we have

$$u_{i+1,1} - 4u_{i1} + u_{i-1,1} + u_{i2} = \Delta x^2 f_{i1} - g_{i0} \equiv -q_{i1} \qquad (12.1.8)$$

Similar equations can be written for the other boundaries (see Problem 12.1). The vectors F, G are defined by equation (12.1.5). The sum of the right-hand side vectors will be represented by the vector Q. This matrix is conveniently written in block form by introducing the $(M \times M)$ tridiagonal

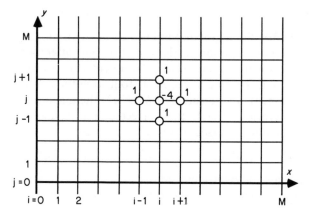

Figure 12.1.1 Cartesian mesh for Laplace operator

matrix $B(1, -4, 1)$ and the unit $(M \times M)$ matrix I:

$$S = \begin{vmatrix} B & +I & & \\ I & +B & +I & \\ & I & B & I \\ & & \cdot & \cdot \\ & & I & B \end{vmatrix} \qquad (12.1.9)$$

where formally S now has M lines and columns, each element being itself an $(M \times M)$ matrix. This corresponds to a grouping of all the u_{ij} *of the same line* into a subvector v_j, with the superscript T indicating the transpose:

$$\begin{aligned} v_j^{\mathrm{T}} &= (u_{1j}, \ldots, u_{ij}, \ldots, u_{Mj}) \\ U^{\mathrm{T}} &= (v_1^{\mathrm{T}}, \ldots, v_j^{\mathrm{T}}, \ldots, v_M^{\mathrm{T}}) \end{aligned} \qquad (12.1.10)$$

It is seen that S has the following properties:

(1) S is irreducible diagonal dominant (see the first line); and
(2) $(-S)$ is symmetric and hence non-singular and positive definite.

In order to relate the notation in system (12.1.6) to the classical notations of linear algebra we consider the vector U, with elements u_I, $I = 1, \ldots, N = M^2$, where $N = M^2$ is the total number of mesh points. Hence to each component u_{ij} we associate the component u_I, where I is, for instance, defined by $I = i + (j - 1)M$, as would be the case in a finite element discretization. Equation (12.1.5) will be written as

$$\sum_{J=1}^{N} s_{IJ} u_J = -Q_I \qquad I = 1, \ldots, N \qquad (12.1.11)$$

The coefficients s_{IJ} represent the space discretization of the L-operator; in a finite element discretization, s_{IJ} are the stiffness matrix elements.

12.1.2 Point Jacobi Method—Point Gauss–Seidel Method

In order to solve for the u_{ij} (or u_I) unknowns in equation (12.1.3) we could define an initial approximation to the vector U and attempt to correct this approximation by solving equation (12.1.3) sequentially, sweeping through the mesh, point by point, starting at $i = j = 1$, following the mesh line by line or column per column. If we indicate by u_{ij}^n (or u_I^n) the assumed approximation (n will be an iteration index), the corrected approximation is u_{ij}^{n+1} (or u_I^{n+1}), and can be obtained (see Figure 12.1.2(a)) as

$$u_{ij}^{n+1} = \tfrac{1}{4}(u_{i+1,}^n + u_{i-1,}^n + u_{i,j-1}^n + u_{i,j+1}^n) + \tfrac{1}{4}q_{ij}^n \qquad (12.1.12)$$

where q_{ij} represents the right-hand side.

In a general formulation for system (12.1.11) the point Jacobi method is defined by the algorithm

$$u_I^{n+1} = \frac{1}{s_{II}}\left(-q_I^n - \sum_{\substack{J=1 \\ J \neq I}}^{N} s_{IJ}u_J^n\right) \qquad (12.1.13)$$

The general formulation is best represented in matrix form if we decompose S into a sum of three matrices containing the main diagonal D, the upper triangular part F and the lower triangular part E. That is, we write

$$S = D + E + F \qquad (12.1.14)$$

with

$$
D = \begin{vmatrix} s_{11} & & 0 \\ & \cdot\; s_{II}\; \cdot & \\ 0 & & s_{NN} \end{vmatrix}
\qquad
E = \begin{vmatrix} 0 & 0 & & & 0 & 0 \\ s_{21} & 0 & & & 0 & \\ s_{31} & s_{32} & 0 & & & \\ s_{I1} & \cdot & s_{IJ} & 0^{\cdot} & & \\ \cdot & & \cdot & \cdot & & \\ s_{N1} & \cdot & \cdot & s_{N,N-1} & 0 \end{vmatrix}
$$

$$
F = \begin{vmatrix} 0 & s_{12} & \cdot & s_{1J} & \cdot & s_{1N} \\ & 0 & & & & \vdots \\ & & 0 & \cdot & \cdot\; \cdot & s_{IN} \\ & & & 0 & & \vdots \\ & & & & & s_{N-1,N} \\ 0 & & & & & 0 \end{vmatrix} \qquad (12.1.15)
$$

For the Laplace operator the splitting (12.1.14) is obvious from the matrix (12.1.9).

Equation (12.1.13) can be written as

$$DU^{n+1} = -Q^n - (E + F)U^n \qquad (12.1.16)$$

defining the iterative point Jacobi method for system (12.1.11), written as

$$(D + E + F)U = -Q^n \qquad (12.1.17)$$

The iterative Jacobi scheme can also be written in Δ-form by introducing the *residual* R^n at iteration n:

$$R^n \equiv R(U^n) = SU^n + Q^n = (D + E + F)U^n + Q^n \qquad (12.1.18)$$

Equation (12.1.16) can be rewritten, after subtraction of DU^n on both sides,

$$D\Delta U^n = -R^n \qquad (12.1.19)$$

where

$$\Delta U^n = U^{n+1} - U^n \qquad (12.1.20)$$

The residual is an important quantity, since it is a measure of the error at a given iteration number n. Obviously, the iterative method will have to generate a sequence of decreasing values of ΔU^n and R^n when the number of iterations increases, since the converged solution corresponds to a vanishing residual. For equation (12.1.12) the residual form (12.1.19) is obtained by subtracting u_{ij}^n from both sides, leading to

$$4(u_{ij}^{n+1} - u_{ij}^n) = (u_{i+1,j}^n + u_{i-1,j}^n + u_{i,j+1}^n + u_{i,j-1}^n - 4u_{ij}^n) + q_{ij}^n$$

$$(12.1.21)$$

Gauss–Seidel method

In Figure 12.1.2(a) we observe that the points $(i, j-1)$ and $(i-1, j)$ have already been updated at iteration $(n+1)$ when u_{ij} is calculated. We are therefore tempted to use these new values in the estimation of u_{ij}^{n+1} as soon as they have been calculated. We can expect thereby to obtain a higher convergence rate, since the influence of a perturbation on u^n is transmitted more rapidly. With the Jacobi method a perturbation of $u_{i-1,j}^{n+1}$ will be felt on u_{ij} only after the whole mesh is swept, since it will occur for the first time at the next iteration through the equation for u_{ij}^{n+2}. With the Gauss–Seidel method this influence already appears at the current iteration, since u_{ij}^{n+1} is immediately affected by $u_{i-1,j}^{n+1}$ (see Figure 12.1.2(b)).

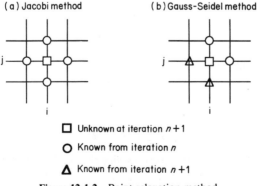

(a) Jacobi method   (b) Gauss-Seidel method

☐ Unknown at iteration $n+1$

○ Known from iteration n

△ Known from iteration $n+1$

Figure 12.1.2 Point relaxation method

As a byproduct it can be observed that as soon as a new value u_{ij}^{n+1} is calculated, the 'old' value u_{ij}^n is no longer needed. Hence the new value can be stored in the same location and can overwrite, in the coding, the local value u_{ij}^n. Therefore only one vector U of length N has to be stored, while two vectors U^{n+1} and U^n have to be saved in the Jacobi method.

The Gauss–Seidel method is, from every point of view, more advantageous than the Jacobi method, and is defined by the iterative scheme

$$u_{ij}^{n+1} = \tfrac{1}{4}(u_{i+1,j}^n + u_{i-1,j}^{n+1} + u_{i,j-1}^{n+1} + u_{i,j+1}^n) + \tfrac{1}{4}q_{ij}^n \qquad (12.1.22)$$

For the general system (12.1.11) the Gauss–Seidel method takes all the variables associated with the lower diagonal of the matrix S, at the new level $(n+1)$. Hence instead of equation (12.1.13), we have

$$u_I^{n+1} = \frac{1}{s_{II}}\left(-q_I^n - \sum_{J=1}^{I-1} s_{IJ}u_J^{n+1} - \sum_{j=I+1}^{N} s_{IJ}u_J^n\right) \qquad (12.1.23)$$

In operator form this scheme can be written as

$$(D+E)U^{n+1} = -Q^n - FU^n \qquad (12.1.24)$$

and in residual form:

$$(D+E)\Delta U^n = -R^n \qquad (12.1.25)$$

By comparing this equation with equation (12.1.19) it can be seen that the Gauss–Seidel method corresponds to another choice for the matrix which 'drives' ΔU^n to zero, that is, to convergence or, equivalently, to another splitting of the matrix S.

12.1.3 Convergence analysis of iterative schemes

An iterative method is said to be convergent if the error tends to zero as the number of iterations goes to infinity. If $\bar{U}$ is the exact solution of system (12.1.5) the error at iteration n is

$$e^n = U^n - \bar{U} \qquad (12.1.26)$$

Let us consider an arbitrary splitting of the matrix S:

$$S = P + A \qquad (12.1.27)$$

and an iterative scheme

$$PU^{n+1} = -Q^n - AU^n \qquad (12.1.28)$$

or equivalently:

$$P\Delta U^n = -R^n \qquad (12.1.29)$$

The matric P will be called a *convergence* or *conditioning matrix (operator)*, and is selected such that system (12.1.28) is easily solvable. It is important to observe here that the iterative scheme (12.1.29) replaces the full iterative scheme, corresponding to a direct method:

$$SU^{n+1} = -Q^n \quad \text{or} \quad S\,\Delta U^n = -R^n \qquad (12.1.30)$$

Subtracting from equation (12.1.28) the relation $(P+A)\bar{U} = -Q$ we obtain, assuming that Q is independent of U,

$$Pe^{n+1} = -Ae^n$$

or

$$e^{n+1} = (-P^{-1}A)e^n = (-P^{-1}A)^n e^1 = (1 - P^{-1}S)^n e^1 \qquad (12.1.31)$$

where $(P^{-1}A)^n$ is the matrix $P^{-1}A$ to the power n. If $\| e^{n+1} \|$ is to go to zero for increasing n, the matrix $P^{-1}A$ should be a convergent matrix and its spectral radius has to be lower than one.

The iterative scheme will be convergent if and only if the matrix G, called the *iteration* or *amplification* matrix (operator),

$$G = 1 - P^{-1}S \qquad (12.1.32)$$

is a convergent matrix, that is, satisfying the condition on the spectral radius:

$$\rho(G) \leqslant 1 \qquad (12.1.33)$$

or

$$\lambda_J(G) \leqslant 1 \quad \text{for all } J \qquad (12.1.34)$$

All the eigenvalues $\lambda_J(G)$ of $G = (1 - P^{-1}S)$ have to be lower than one in modulus for the iterative scheme to converge. Hence it is seen that we can replace equation (12.1.30) and the matrix S, acting on the corrections ΔU^n, by another operator P, which is expected to be easier to invert, provided the above conditions are satisfied. Except for this condition, the choice of P is arbitrary. If $P = S^{-1}$ we obtain the exact solution, but this corresponds to a direct method, since we have to invert the matrix S of the system. (See Section 12.4 for a generalization of these considerations to non-linear problems.)

The residual R^n is connected to the error vector e^n by the following relation obtained by subtracting R^n (equation (12.1.18)) from $\bar{R} = S\bar{U} + Q = 0$, assuming Q to be independent of U:

$$R^n = Se^n \qquad (12.1.35)$$

This shows the quantitative relation between the error and the residual and proves that the residual will go to zero when the error tends to zero.

In practical computations, the error is not accessible since the exact solution is not known. Therefore we generally use the norm of the residual $\| R^n \|$ as a measure of the evolution towards convergence, and practical convergence criteria will be set by requiring that the residual drops by a predetermined number of orders of magnitude. Observe that even when the residual is reduced to machine accuracy (machine zero) it does *not* mean that the solution U^n is within machine accuracy of the exact solution. When SU results from a space discretization, achieving machine zero on the residual will produce a solution U^n, which differs from the exact solution $\tilde{U}$ of the differential problem by the amount of the truncation error. For example, a second-order accurate space discretization, with $\Delta x = 10^{-2}$, will produce an error of the order of 10^{-4} on the solution, which cannot be reduced further even if the residual equals 10^{-14}.

Equation (12.1.28) can also be written as

$$
\begin{aligned}
U^{n+1} &= (1 - P^{-1}S)U^n - P^{-1}Q \\
&= GU^n - P^{-1}Q
\end{aligned}
\tag{12.1.36}
$$

Comparing equation (12.1.31), written here as

$$
e^{n+1} = G\, e^n = (G)^{n+1}e^0
\tag{12.1.37}
$$

with equation (10.1.26) (and equation (12.1.36) with (10.1.25)), the similitude between the formulation of an iterative scheme of a stationary problem and a time-dependent formulation of the same problem clearly appears. The matrix G, being the amplification matrix of the iterative method, can be analysed along the lines developed in Chapters 8 and 10. We will return to this very important link below.

Convergence conditions for the Jacobi and Gauss–Seidel methods

For the Jacobi methods $G = D^{-1}(E + F) = 1 - D^{-1}S$ and for the Gauss–Seidel method $G = -(D + E)^{-1}F = 1 - (D + E)^{-1}S$. From the properties of matrices we can show that the Jacobi and Gauss–Seidel methods will converge if S is irreducible diagonal dominant. When S is symmetric it is sufficient for convergence that S, or $(-S)$, is positive definite for both methods. In addition, for the Jacobi method, $(2D - S)$, or $(S - 2D)$ in our case, also has to be positive definite.

Estimation of the convergence rate

An important problem of iterative methods is the estimation of the rate of reduction of the error with increasing number of iterations. The *average* rate of error reduction over n iterations can be defined by

$$
\overline{\Delta e_n} = \left(\frac{\| e^n \|}{\| e^0 \|} \right)^{1/n}
\tag{12.1.38}
$$

From equation (12.1.37) we have

$$
\overline{\Delta e_n} \leqslant \|(G)^n\|^{1/n}
\tag{12.1.39}
$$

Asymptotically it can be shown (Varga, 1962) that this quantity tends to the spectral radius of G for $n \to \infty$:

$$
s \equiv \lim_{n \to \infty} \|(G)^n\|^{1/n} = \rho(G)
\tag{12.1.40}
$$

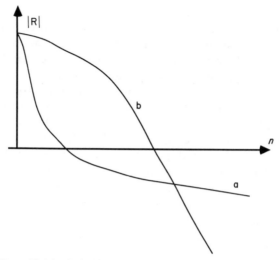

Figure 12.1.3 Typical residual history of a relaxation method

The logarithm of s is the *average convergence rate*, measuring the number of iterations needed to decrease the error by a given factor.

If we consider s to be a valid measure of $\overline{\Delta e_n}$ the norm of the error will be reduced by an order of magnitude (factor 10), in a number of iterations n, such that

$$\left(\tfrac{1}{10}\right)^{1/n} \leqslant s = \rho(G) \tag{12.1.41}$$

or

$$n \geqslant -1/\log \rho(G) \tag{12.1.42}$$

This relation is valid asymptotically for large n, and is generally not to be relied on at small values of n. In the early stages of the iterative process it is not uncommon, particularly with non-linear problems, to observe even an increase in the residual before it starts to decrease at larger n. Also we can often detect different rates of residual reduction, usually an initial reduction rate much higher than s, slowing down gradually to the asymptotic value.

The explanation for this behaviour is to be found in the frequency distribution of the error and in the way a given iterative method treats the different frequencies in the spectrum. The residual history of curve (a) of Figure 12.1.3 is typical for a method which damps rapidly the high-frequency components but damps poorly the low-frequency errors, while the opposite is true for curve (b). The reason for this behaviour is explained by an eigenvalue analysis.

12.1.4 Eigenvalue analysis of an iterative method

The frequency response, as well as the asymptotic convergence rate of an iterative method, are defined by the eigenvalue spectrum of the amplification matrix G and by the way the matrix P^{-1} treats the eigenvalues of the matrix S. Referring to Section 10.1, equation (10.1.48), the solution U^{n+1} can be written as a linear combination of the eigenmodes $V^{(J)}$ of the matrix S (which represents the space discretization of the differential operator L). Each term represents the contribution from the initial component U_{0J} damped by the factor (λ_J^n), where λ_J is the eigenvalue of the matrix G, associated with the eigenvalue Ω_J of the matrix S. *We assume here that S and G commute and have the same set of eigenfunctions*, although this will not always be true, even in simple cases. For the present example of Poisson's equation on a Cartesian mesh this propety is satisfied for the Jacobi iteration but *not* for the Gauss–Seidel method, since the matrices $(D + E)^{-1}$ and S do not commute. However, the conclusions of this section with regard to the convergence properties of iterative methods and their relation to the eigenvalues of the space operator will remain largely valid, independently of this hypothesis, as can be seen from the analysis of Section 12.1.5. Hence we assume

$$U^n = \sum_{J=1} \lambda_J^n \, U_{0J} \cdot V^{(J)} - \sum_{J=1} \frac{Q_J}{\Omega_J} \cdot V^{(J)}(1 - \lambda_J^n) \qquad (12.1.43)$$

Note that λ_J^n is λ_J to the power n, and that λ_J represents the variable $z(\Omega_J)$ of equation (10.1.48), since both are to be considered as eigenvalues of the amplification matrix.

The first term is the 'transient' behaviour of the homogeneous solution while the last is a transient contribution from the source term. Since, for convergence, all the λ_J are in modulus lower than one, after a sufficiently large number of iterations we are left with the 'converged' solution:

$$\lim_{n \to \infty} U^n = -\sum_J \frac{Q_J}{\Omega_J} \cdot V^{(J)} \qquad (12.1.44)$$

which is the eigenmode expansion of the solution $\bar{U} = -S^{-1}Q$. Indeed, by writing $\bar{U}$:

$$\bar{U} = \sum_J U_J \cdot V^{(J)} \qquad (12.1.45)$$

we obtain

$$S\bar{U} = \sum_J U_J \Omega_J \cdot V^{(J)} = -\sum_J Q_J \cdot V^{(J)} \qquad (12.1.46)$$

Since the $V^{(J)}$ form a basis of the U-space:

$$U_J = -Q_J/\Omega_J \qquad (12.1.47)$$

which shows that equation (12.1.44) is indeed the solution. Equation (12.1.43) shows that the transient will, for large n, be dominated by the largest eigenvalue λ_J; this explains relation (12.1.42).

It is seen from equation (12.1.31) that the error e^n behaves like the homogeneous part of (12.1.43). Therefore, if e_J^0 is the initial error distribution, after n iterations the error is reduced to

$$e^n = \sum_J \lambda_J^n e_J^0 \cdot V^{(J)} \qquad (12.1.48)$$

The residual R^n can also be expanded in a series of eigenfunctions $V^{(J)}$, since from equation (12.1.35) we can write

$$R^n = \sum_J \lambda_J^n e_J^0 S \cdot V^{(J)}$$

$$\qquad (12.1.49)$$

$$= \sum_J \lambda_J^n \Omega_J e_J^0 \cdot V^{(J)}$$

If we assume the eigenvectors $V^{(J)}$ to be orthonormal, then the L_2-norm of the residual will be

$$\| R^n \|_{L2} = \left[\sum_J (\lambda_J^n \Omega_J e_J^0)^2 \right]^{1/2} \qquad (12.1.50)$$

When the iteration number n increases, the low values of λ_J will be damped more rapidly, and at large values of n we will be left with the highest λ_J, particularly those which are the closest to the upper limit of one. Hence if the iterative scheme is such that the highest eigenvalues $\lambda_J = \lambda(\Omega_J)$ are close to one, the asymptotic convergence rate will be poor. Since generally the high frequencies of the operator S produce the lowest values of λ_J, we will tend to have a behaviour such as shown on curve (a) of Figure 12.1.3, where the initial rapid decrease of the residual is due to the damping of the high frequencies (low λ), and where we are left, at large n, with the low-frequency end (high λ) of the error spectrum.

On the other hand, if such an iterative method is combined with an algorithm which damps effectively the low frequencies of S, that is, the large λ_J region, then we would obtain a behaviour of the type shown on curve (b) of Figure 12.1.3. This is the principle at the basis of the multi-grid method described in Section 12.6.

The relation $\lambda_J = \lambda(\Omega_J)$ can be computed explicitly only in simple cases. For instance, for the Jacobi method we have, representing by G_J the amplification matrix,

$$G_J = 1 - D^{-1}S \qquad (12.1.51)$$

and since D is a diagonal matrix,

$$\lambda_J = 1 - \frac{1}{d_J} \Omega_J(S) \qquad (12.1.52)$$

Note that we define S as the matrix representation of the discretized differential operator, which has to satisfy the condition (10.1.17) of stability of the time-dependent counterpart, and therefore have eigenvalues with negative real parts. Hence S will be *negative* definite when all the eigenvalues are real.

12.1.5 Fourier analysis of an iterative method

The Fourier analysis allows a simple estimation of eigenvalues, when periodic conditions are assumed, such that the Fourier modes are eigenvectors of the S matrix and of the amplification matrix G. This is the case for the Jacobi method but not for the Gauss–Seidel iterations.

With a representation such as equation (8.4.6) and the discretization (12.1.3) with Dirichlet boundary conditions, the eigenfunctions of S reduce to $\sin(i\phi_x) \cdot \sin(j\phi_y)$ and the eigenvalues are equal to (dropping the index J)

$$\Omega = -4(\sin^2\phi_x/2 + \sin^2\phi_y/2) \qquad (12.1.53)$$

$$\phi_x = l\pi/M, \qquad \phi_y = m\pi/M, \qquad J = l + (m-1)M, \qquad l, m = 1, ..., M-1$$

The eigenvalues of G_J are therefore, with $d_j = -4$,

$$\lambda(G_J) = 1 - (\sin^2\phi_x/2 + \sin^2\phi_y/2) = \tfrac{1}{2}(\cos\phi_x + \cos\phi_y) \qquad (12.1.54)$$

Observe that this can be viewed as resulting from a Von Neumann stability analysis on the pseudo-time scheme (12.1.12), where n is interpreted as a time index.

At the low-frequency end, $\phi_x \approx 0$, $\phi_y \approx 0$, the damping rate is very poor, since $\lambda \approx 1$. The intermediate frequency range is strongly damped, that is, for the frequencies corresponding to high values of $(\sin^2\phi_x/2 + \sin^2\phi_y/2)$. The spectral radius $\rho(G_J)$ is defined by the highest eigenvalue, which corresponds to the lowest frequencies of the Ω spectrum. These are obtained for $\phi_x = \phi_y = \pi/M$:

$$\rho(G_J) = \cos\pi/M \qquad (12.1.55)$$

Hence the convergence rate becomes, for small mesh sizes $\Delta x = \Delta y$, that is, for increasing number of mesh points,

$$\rho(G_J) \approx 1 - \frac{\pi^2}{2M^2} = 1 - \frac{\pi^2}{2N} = 1 - 0(\Delta x^2) \qquad (12.1.56)$$

and, from equation (12.1.42), the number of iterations needed to reduce the error by one order of magnitude is, asymptotically, $(2N/0.43\pi^2)$. Since each iteration (12.1.12) requires $5N$ operations, the Jacobi method will require roughly $2kN^2$ operations for a reduction of the error by k orders of magnitude.

Gauss–Seidel method

Applying a straightforward Von Neumann analysis to the homogeneous part of equation (12.1.22) leads to

$$(4 - e^{-I\phi_x} - e^{-I\phi_y})\lambda_{GS} = e^{I\phi_x} + e^{I\phi_y} \qquad (12.1.57)$$

However, this does *not* give the eigenvalues of the matrix G_{GS}, since this matrix does not commute with S. It can be shown, Young, (1971), that the corresponding eigenvalues of the amplification matrix of the Gauss–Seidel method are equal to the square of the eigenvalues of the Jacobi method:

$$\lambda(G_{GS}) = \tfrac{1}{4}(\cos \phi_x + \cos \phi_y)^2 = \lambda^2(G_J) \qquad (12.1.58)$$

This eigenvalue does not satisfy equation (12.1.57). Hence the spectral radius is

$$\rho(G_{GS}) = \rho^2(G_J) = \cos^2 \pi/M \qquad (12.1.59)$$

and for large M,

$$\rho(G_{GS}) \approx 1 - \frac{\pi^2}{M^2} = 1 - \frac{\pi^2}{M^2} \qquad (12.1.60)$$

and the method converges *twice* as fast as the Jacobi method, since the number of iterations for one order of magnitude reduction in error is $N/(0.43\pi^2)$.

For a more general case, it can be shown, Varga, (1962), that the Gauss–Seidel method converges if S is an irreducible diagonal dominant matrix, or if $(-S)$ is symmetric, positive definite. In the latter case we have $\lambda(G_{GS}) = \lambda^2(G_J)$. Interesting considerations with regard to the application of Fourier analysis to iterative methods can be found in LeVeque and Trefethen (1986).

12.2 OVERRELAXATION METHODS

We could be tempted to increase the convergence rate of an iterative method by 'propagating' the corrections $\Delta U^n = U^{n+1} - U^n$ faster through the mesh. This idea is the basis of the overrelaxation method, introduced independently by Frankel and Young in 1950 (see Young, 1971; Mageman and Young, 1981).

The overrelaxation is formulated as follows. If $\overline{U^{n+1}}$ is the value obtained from the basic iterative scheme the value introduced at the next level U^{n+1} is defined by

$$U^{n+1} = \omega\overline{U^{n+1}} + (1 - \omega)U^n \qquad (12.2.1)$$

where ω is the overrelaxation coefficient. Alternatively, a new correction is defined:

$$\Delta U^n = U^{n+1} - U^n \text{ with } \Delta U^n = \omega\overline{\Delta U^n} \qquad (12.2.2)$$

When appropriately optimized for maximum convergence rate a considerable gain can be achieved, as will be seen next.

12.2.1 Jacobi overrelaxation

The Jacobi overrelaxation method becomes, for the Laplace operator,

$$u_{ij}^{n+1} = \frac{\omega}{4}(u_{i+1,j}^n + u_{i-1,j}^n + u_{i,j-1}^n + u_{i,j+1}^n + q_{ij}^n) + (1 - \omega)u_{ij}^n \qquad (12.2.3)$$

In operator form, equation (12.2.3) becomes

$$DU^{n+1} = -\omega Q^n - \omega(E+F)U^n + (1-\omega)DU^n = -\omega Q^n - \omega SU^n + DU^n \tag{12.2.4}$$

or

$$D\,\Delta U^n = -\omega R^n \tag{12.2.5}$$

instead of equation (12.1.19). The new amplification matrix of the iterative scheme $G_J(\omega)$ is, from equation (12.2.4),

$$G_J(\omega) = (1-\omega)I + \omega G_J = 1 - \omega D^{-1}S \tag{12.2.6}$$

The convergence will be ensured if the spectral radius of $G_J(\omega)$ is lower than one, that is, if

$$\rho(G_J(\omega)) \leqslant |(1-\omega)| + \omega\rho(G_J) < 1$$

This will be satisfied if

$$0 < \omega < \frac{2}{1+\rho(G_J)} \tag{12.2.7}$$

The eigenvalues are related by

$$\lambda(G_J(\omega)) = (1-\omega) + \omega\lambda(G_J) \tag{12.2.8}$$

and this allows us to select optimum values of ω for maximal damping in a given frequency range. As a function of ω, the eigenvalue $\lambda(G_J(\omega))$ will vanish for the particular choice $\omega = 1/(1 - \lambda(G_J))$ and the corresponding frequency component in developments, (12.1.43) or (12.1.48) will not contribute. Hence if we attempt to damp selectively the high frequencies, corresponding to the values of λ towards the end-point of -1, we could select ω values lower than one. On the other hand, the low frequencies generally span the λ-region close to the other end-point of the permissible λ-range, namely $\lambda \leqslant 1$, and we would select values of ω towards the higher end of its range if the goal would be to damp the low frequencies more selectively.

We can also define an optimum value of ω, ω_{opt}, which produces an optimal damping over the whole range. From Figure 12.2.1, where $|\lambda(G_j(\omega)|$ is plotted against ω at constant values of $\lambda(G_J)$, the intersection of the curves associated, respectively, with the minimum value of $\lambda(G_J)$, λ_{min}, and the maximum eigenvalue λ_{max}, appears to improve best the damping over the whole range. Hence ω_{opt} is defined by

$$-1 + \omega_{opt}(1 - \lambda_{min}) = 1 - \omega_{opt}(1 - \lambda_{max})$$

or

$$\omega_{opt} = \frac{2}{2 - (\lambda_{min} + \lambda_{max})} \tag{12.2.9}$$

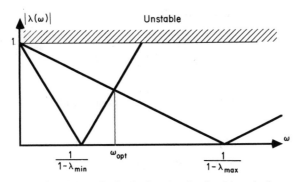

Figure 12.2.1 Optimal relaxation for Jacobi method

For the Laplace operator, from equation (12.1.54)

$$\lambda_{max} = -\lambda_{min} = \cos \pi/M \qquad (12.2.10)$$

and $\omega_{opt} = 1$. Hence the Jacobi method is optimal by this definition.

12.2.2 Gauss–Seidel overrelaxation—successive overrelaxation (SOR)

The benefits of the overrelaxation concept assume their full value when applied to the Gauss–Seidel method, and in this case the methods are usually called successive overrelaxation methods, indicated by the symbol SOR. Applying definition (12.2.1) to the Gauss–Seidel method (12.1.22) for the Laplace operator leads to the following iterative scheme:

$$\overline{u_{ij}^{n+1}} = \tfrac{1}{4}(u_{i+1,j}^n + u_{i-1,j}^{n+1} + u_{i,j-1}^{n+1} + u_{i,j+1}^n) + \tfrac{1}{4} q_{ij}^n \qquad (12.2.11)$$
$$u_{ij}^{n+1} = \omega\overline{u_{ij}^{n+1}} + (1-\omega)u_{ij}^n$$

In operator form this corresponds to the general residual formulation

$$D\,\overline{\Delta U^n} = -R^n - E\,\Delta U^n \qquad (12.2.12)$$

and the overrelaxation method is defined by the iteration operator $(D+\omega E)$, since equation (12.2.12) reads

$$(D+\omega E)\,\Delta U^n = -\omega R^n \qquad (12.2.13)$$

The amplification matrix $G_{SOR}(\omega)$ is now

$$G_{SOR}(\omega) = 1 - \omega(D+\omega E)^{-1}S$$
$$= (D+\omega E)^{-1}[(1-\omega)D - \omega F] \qquad (12.2.14)$$

Condition on the relaxation parameter ω

Since $G_{SOR}(\omega)$ is a product of triangular matrices its determinant is the

472

product of its diagonal elements. Hence

$$\det\, G_{SOR}(\omega) = \det(I + \omega D^{-1}E)^{-1} \cdot \det[(1-\omega)I - \omega D^{-1}F]$$
$$= 1 \cdot \det[(1-\omega))I - \omega D^{-1}F] \qquad (12.2.15)$$
$$= (1-\omega)^N$$

On the other hand, the determinant of a matrix is equal to the product of its eigenvalues, and hence

$$\rho(G_{SOR}(\omega))^N \geqslant |\det\, G_{SOR}(\omega)| = (1-\omega)^N \qquad (12.2.16)$$

The spectral radius will be lower than one if

$$|(1-\omega)| \leqslant \rho(G_{SOR}(\omega)) < 1 \qquad (12.2.17)$$

or

$$0 < \omega < 2 \qquad (12.2.18)$$

For irreducible diagonal dominant matrices it can be shown (Young; 1971) that the SOR method will converge if

$$0 < \omega < \frac{2}{1 + \rho(G_{GS})} \qquad (12.2.19)$$

More information can be obtained for symmetrical matrices. The eigenvalues of $G_{SOR}(\omega)$ satisfy the following relations, for symmetrical matrices of the form

$$S = \begin{vmatrix} D_1 & F \\ F^T & D_2 \end{vmatrix} \qquad (12.2.20)$$

where D_1 and D_2 are block diagonal, writing $\lambda(\omega)$ for $\lambda(G_{SOR}(\omega))$:

(1) The eigenvalues of the SOR amplification factor satisfy the relation

$$\lambda(\omega) = 1 - \omega + \omega\lambda^{1/2}(\omega) \cdot \lambda(G_J) \qquad (12.2.21)$$

(2) For $\omega = 1$ we recover relation (12.1.58):

$$\lambda(\omega = 1) \equiv \lambda(G_{GS}) = \lambda^2(G_J) \qquad (12.2.22)$$

(3) An optimal relation coefficient can be defined, rendering $\rho(G_{SOR}(\omega))$ minimum, as

$$\omega_{opt} = \frac{2}{1 - \sqrt{(1 - \rho^2(G_J))}} \qquad (12.2.23)$$

with

$$\rho(G_{SOR}(\omega_{opt})) = \omega_{opt} - 1 \qquad (12.2.24)$$

Note the role played by the eigenvalues of the Jacobi point iteration matrix G_J.
For Laplace's operator:

$$\rho(G_J) = \cos\, \pi/M \qquad (12.2.25)$$

$$\omega_{opt} = \frac{2}{1 + \sin(\pi/M)} \approx 2\left(1 - \frac{\pi}{M}\right) \qquad (12.2.26)$$

and the spectral radius at optimum relaxation coefficient ω_{opt} is

$$\rho(G_{SOR}(\omega_{opt})) = \frac{1 - \sin(\pi/M)}{1 + \sin(\pi/M)} \approx 1 - \frac{\pi}{M} + 0\left(\frac{1}{M^2}\right) \qquad (12.2.27)$$

The number of iterations for a one order of magnitude reduction in error norm is of the order of $2.3M/\pi \approx \sqrt{N}$. Compared with the corresponding values for Jacobi and Gauss–Seidel iterations, the optimal SOR will require $kN\sqrt{N}$ operations for a reduction of the error of k order of magnitude. This represents a considerable gain in convergence rate.

For non-optimal relaxation parameters, however, this rate can seriously deteriorate. Since, for more general problems, the eigenvalues are not easily found, we can attempt to find numerical estimates of the optimal relaxation parameter. For instance, for large n, the error being dominated by the largest eigenvalue, we have, from equation (12.1.37) and (12.1.48),

$$\rho(G) = \lim_{n \to \infty} \| e^{n+1} \| / \| e^n \| \qquad (12.2.28)$$

and when this ratio stabilizes we can deduct $\rho(G_J)$ from equation (12.2.21) by introducing this estimation of $\rho(G)$ for $\lambda(\omega)$ and find ω_{opt} from equation (12.2.23). Other strategies can be found in Wachspress (1966) and Hageman and Young (1981).

12.2.3 Symmetric successive overrelaxation (SSOR)

The SOR method, as described by equation (12.2.11), sweeps through the mesh from the lower left corner to the upper right corner. This gives a bias to the iteration scheme which might lead to some error accumulations, for instance if the physics of the system to be solved is not compatible with this sweep direction. This can be avoided by alternating SOR sweeps in both directions. We obtain a two-step iterative method (a form of predictor–corrector approach), whereby a predictor $\overline{U^{n+1/2}}$ is obtained from a SOR sweep (12.2.11) followed by a sweep in the reverse direction starting at the end-point of the first step. The SSOR can be described in residual form by

$$(D + \omega E)\, \overline{\Delta U^{n+1/2}} = -\omega R^n \qquad (12.2.29)$$

$$(D + \omega F)\, \overline{\Delta U^n} = -\omega R(\overline{U^{n+1/2}}) \qquad (12.2.30)$$

with

$$\begin{aligned}
\overline{U^{n+1/2}} &= U^n + \overline{\Delta U^{n+1/2}} \\
U^{n+1} &= \overline{U^{n+1/2}} + \overline{\Delta U^n} = U^n + \overline{\Delta U^{n+1/2}} + \overline{\Delta U^n}
\end{aligned} \qquad (12.2.31)$$

This method converges under the same conditions as the SOR method, and an optimal relaxation parameter can only be estimated approximately as

$$\omega_{opt} \approx \frac{2}{1 + \sqrt{[2(1 - \rho^2(G_J))]}} \qquad (12.2.32)$$

Hence the SSOR method converges twice as fast as SOR, but since it requires twice as much work there is not much to be gained in convergence rate. However, SOR might lead to better error distributions and is a good candidate for preconditioning matrices. Also, when applied in conjunction with non-stationary relaxation, significant improvements in convergence rate over SOR can be achieved, Young (1971).

12.2.4 Successive line overrelaxation methods (SLOR)

We could still accelerate the convergence of the SOR method by increasing the way the variations ΔU^n are transmitted throughout the field if we are willing to increase the workload per iteration by solving small systems of equations. Referring to Figure 12.2.2, for the Laplace operator, we could consider the three points on a column i, or on a line j, as simultaneous unknowns, and solve a tridiagonal system along each vertical line, sweeping through the mesh column by column from $i = 1$ to $i = M$ (Figure 12.2.2(a)) or along horizontal lines as in Figure 12.2.2(b), from $j = 1$ to $j = M$. This is called *line Gauss–Seidel* ($\omega = 1$) or *line successive overrelaxation* (generally abbreviated as SLOR) when $\omega \neq 1$.

For the Laplace operator the SLOR iterative scheme along vertical lines (VLOR) is defined by

$$\overline{u_{ij}^{n+1}} = \tfrac{1}{4}(u_{i+1,j}^n + u_{i-1,j}^{n+1} + \overline{u_{i,j-1}^{n+1}} + \overline{u_{i,j+1}^{n+1}}) + \tfrac{1}{4}q_{ij}^n$$
$$u_{ij}^{n+1} = \omega \overline{u_{ij}^{n+1}} + (1 - \omega)u_{ij}^n \tag{12.2.33}$$

or, in incremental form:

$$4\Delta U_{ij}^n - \Delta U_{i,j-1}^n - \Delta U_{i,j+1}^n - \omega\,\Delta U_{i-1,j}^n = \omega R_{ij}^n \tag{12.2.34}$$

which is solved by the tridiagonal Thomas algorithm in $5M$ operations for each line. Hence $5M^2 = 5N$ operations are required for all the tridiagonal

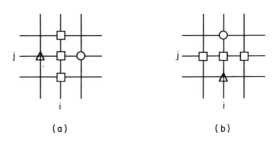

(a)   (b)

☐ Unknown at iteration $n + 1$
◯ Known from iteration n
△ Known from iteration $n + 1$

Figure 12.2.2 Line relaxation method

systems. This number is to be roughly doubled to take into account the computation of the right-hand side terms.

A Jacobi line relaxation method can also be defined and written as

$$\overline{u_{ij}^{n+1}} = \tfrac{1}{4}(u_{i+1,j}^{n} + u_{i-1,j}^{n} + \overline{u_{i,j-1}^{n+1}} + \overline{u_{i,j+1}^{n+1}}) + \tfrac{1}{4}q_{ij}^{n}$$
$$u_{ij}^{n+1} = \omega\overline{u_{ij}^{n+1}} + (1-\omega)u_{ij}^{n} \tag{12.2.35}$$

or, in incremental form:

$$4\Delta U_{ij}^{n} - \Delta U_{i,j-1}^{n} - \Delta U_{i,j+1}^{n} = \omega R_{ij}^{n} \tag{12.2.36}$$

and corresponds in Figure 12.2.2 to a replacement of the triangles by a circle symbol.

In order to put line relaxation methods in operator form we need a splitting of the S matrix, different from equation (12.1.15), since the grouping of points is no longer done with respect to the diagonal but along lines. With the definition of U in equation (12.1.4) as a classification line by line, the block form of the matrix S, given by equation (12.1.9), has precisely the appropriate form. The block diagonal elements $B(1, -4, 1)$ operate on the element u_{ij} along the constant j-lines, while the unit block matrices I operate on the corresponding points of the i-column. Hence we introduce a splitting for the matrix S:

$$S = D + V + H \tag{12.2.37}$$

where D is the diagonal matrix equal to $-4I$ in the present example. In block form:

$$H + D = \begin{vmatrix} B(1,-4,1) & & & \\ & B(1,-4,1) & & 0 \\ & & B(1,-4,1) & \\ & & \cdot & B(1,-4,1) \\ 0 & & & \end{vmatrix} \qquad V = \begin{vmatrix} 0 & I & & \\ I & 0 & I & \\ & I & 0 & I \\ & & & \cdot \\ & & I & 0 \end{vmatrix}$$

$$\tag{12.2.38}$$

or explicitly:

$$H = \begin{vmatrix} 0 & 1 & & & & & \\ 1 & 0 & 1 & & & & \\ & 1 & 0 & 1 & & & \\ & & 1 & 0 & & & \\ \hline & & & & 0 & 1 & \\ & & & & 1 & 0 & 1 \\ & & & & & 1 & 0 & 1 \\ & & & & & & 1 & 0 \\ \hline & & & & & & & \cdot \end{vmatrix} = \begin{vmatrix} B(1,0,1) & & & \\ & B(1,0,1) & & \\ & & \cdot & \\ & & & B(1,0,1) \\ & & & & \cdot \end{vmatrix}$$

$$\tag{12.2.39}$$

More generally, referring to equation (12.1.9) H is defined as the matrix formed by the block diagonal elements where the main diagonal D has been subtracted and V is defined by the off-diagonal block elements. Each $(M \times M)$ block represents the coefficients of the space-discretization scheme along a horizontal j-line for the diagonal blocks, and along the associated points on the vertical i-lines for the off-diagonal block elements of V.

With the above splitting of A the Jacobi line relaxation along the successive vertical lines is defined by the following generalization of equation (12.2.36):

$$(D + V)\Delta U^n = -\omega R^n \qquad (12.2.40)$$

The iteration matrix is

$$G_L(\omega) = 1 - \omega(D + V)^{-1}S \qquad (12.2.41)$$

For the Gauss–Seidel overrelaxation method, as illustrated in Figure 12.2.2, we have to introduce splitting between the upper and lower triangular parts of H and V. Defining by subscripts U and L these submatrices, we have with

$$H = H_U + H_L, \qquad V = V_U + V_L \qquad (12.2.42)$$

the operator form for the classical vertical line relaxation method (VLOR), generalizing equation (12.2.34):

$$[D + V + \omega H_L]\, \Delta U^n = -\omega R^n \qquad (12.2.43)$$

This amplification matrix of this VLOR scheme is

$$G_{VL} = 1 - \omega[D + V + \omega H_L]^{-1}S \qquad (12.2.44)$$

The optimization of ω can be shown to satisfy relations (12.2.23)–(12.2.25) for the point relaxation method. The horizontal line relaxation (HLOR) is obtained by interchanging H and V.

Many variants of the relaxation method can be derived by combining point or line relaxations on alternate positions or on alternate sweep directions. A few of them are listed here and their matrix structure and eigenvalues for periodic conditions can easily be derived. It can be shown that the line relaxations for the Laplace operator have a convergence rate twice as fast as the point relaxations. The overrelaxation method (SLOR) is $\sqrt{2}$ times faster than SOR at the same value of the *optimal* ω.

Red–black point relaxation

In this approach the Jacobi relaxation method is applied to two distinct series of mesh points, called red and black points, where the red points can be considered as having an even number $I = i + (j - 1)M$ and the black points an odd number (see Figure 12.2.3). If $G_J(\omega)$ is the iteration matrix of the point Jacobi relaxation method (equation (12.2.6)) we have the two-step algorithm, where b_{ij} designates the non-homogeneous terms:

$$
\begin{aligned}
\overline{u_{ij}^{n+1/2}} &= G_J(\omega)u_{ij}^n + b_{ij}^n \qquad (i+j) \text{ even} \qquad \text{(red points)} \\
\overline{u_{ij}^{n+1/2}} &= u_{ij}^n \qquad\qquad\quad\ (i+j) \text{ odd} \qquad \text{(black points)}
\end{aligned}
\qquad (12.2.45)
$$

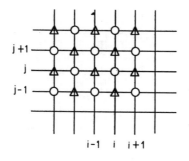

O Red point

▲ Black point

Figure 12.2.3 Red–Black ordering

followed by the second step:

$$u_{ij}^{n+1} = G_J(\omega) \overline{u_{ij}^{n+1/2}} + \overline{b_{ij}^{n+1/2}} \qquad (i+j) \text{ odd} \qquad \text{(black points)}$$
$$u_{ij}^{n+1} = \overline{u_{ij}^{n+1/2}} \qquad\qquad\qquad (i+j) \text{ even} \qquad \text{(red points)} \qquad (12.2.46)$$

The relaxation factors may be different in the two steps. We can also introduce the SOR iteration matrix $G_{SOR}(\omega)$ instead of $G_J(\omega)$.

Zebra line relaxation

Here Jacobi line relaxations are performed on alternating lines. If $G_L(\omega)$ is the Jacobi line relaxation operator (equation (12.2.41)) then the Zebra algorithm along vertical lines is

$$\overline{u_{ij}^{n+1/2}} = G_L(\omega)u_{ij}^n + b_{ij}^n \qquad i \text{ odd}$$
$$\overline{u_{ij}^{n+1/2}} = u_{ij}^n \qquad\qquad\quad i \text{ even} \qquad (12.2.47)$$

and

$$u_{ij}^{n+1} = G_L(\omega) \overline{u_{ij}^{n+1/2}} + \overline{b_{ij}^{n+1/2}} \qquad i \text{ even}$$
$$u_{ij}^{n+1} = \overline{u_{ij}^{n+11/2}} \qquad\qquad\qquad i \text{ odd} \qquad (12.2.48)$$

An alternative variant obviously exists for the horizontal lines, and various combinations can be defined such as interchanging of horizontal and vertical lines, alternating the sweeping directions or considering multiple zebras with different families of lines. Note that Gauss–Seidel zebra relaxation can be defined, whereby the iteration matrix of equation (12.2.44) is used instead, and that these schemes can be used by themselves as well as 'smoothers' in multigrid iterations.

An important aspect of the iterative techniques is their ability to vectorize on parallel computers. For instance, line Jacobi vectorizes much easier than line Gauss–Seidel overrelaxation (see Figure 12.1.5), and, although less efficient

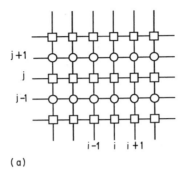

(a)

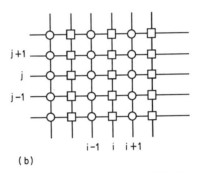

(b)

Figure 12.2.4   Zebra line relaxation along (a) horizontal lines
and (b) vertical lines

than the standard SLOR, it might be more attractive for certain applications
on parallel computers. In that respect, the advantage of the Zebra method
appears clearly.

12.3   PRECONDITIONING TECHNIQUES

All the iterative schemes of the previous sections can be put under the residual
form (12.1.29), with various forms for the convergence or conditioning
operator P. As shown in Section 12.1.3, this emphasizes a clear connection
between the behaviour of iterative schemes with the iteration number n and
that of time-dependent formulations where n is a time-step index. Hence if we
view n in this way, we can consider the general iterative scheme

$$P \frac{\Delta U^n}{\tau} = -\omega(SU^n + Q^n) \equiv -\omega R^n \qquad (12.3.1)$$

as the explicit Euler integration, with time step τ, of the differential system

$$P \frac{dU}{dt} = -\omega(SU + Q) \qquad (12.3.2)$$

where we are only interested in the asymptotic steady state. The operator P can be chosen in an arbitrary way, with the only restriction that the amplification matrix G

$$G = 1 - \omega\tau P^{-1}S \qquad (12.3.3)$$

should have all its eigenvalues lower than one, that is

$$\rho(G) \leqslant 1 \qquad (12.3.4)$$

In addition, we know from equation (10.1.17) that the operator $(\omega P^{-1}S)$ must have eigenvalues with positive parts for stability. The parameter τ appears as a pseudo-time step, which can eventually be absorbed by ω, left as an additional optimization parameter or simply set equal to one.

It is seen from equation (12.3.3) that for $P = \omega\tau S$, $G = 0$, implying that we have the exact solution in one iteration, but, of course, this amounts to the solution of the stationary problem. Therefore the closer $P/\omega\tau$ approximates S, the better the iterative scheme, but, on the other hand P has to be easily solvable. Let us consider a few well-known examples.

12.3.1 Richardson method

This is the simplest case, where $P/\tau = -1$. The choice (-1) is necessary because S has eigenvalues with real *negative* parts. For the Laplace equation all the eigenvalues of S are negative. The iterative scheme is

$$U^{n+1} = U^n + \omega(SU^n + Q^n) = (1 + \omega S)U^n + \omega Q^n$$
$$\equiv G_R U^n + \omega Q^n \qquad (12.3.5)$$

defining the iteration or amplification operator G_R as $G_R = 1 + \omega S$. The eigenvalues of G_R satisfy the relations (where the eigenvalues Ω_J of S are negative and real)

$$\lambda_J(G_R) = 1 + \omega\Omega_J \qquad (12.3.6)$$

The parameter ω can be chosen to damp selectivity certain frequency bands of the Ω-spectrum but has to be limited for stability by the following condition, for Ω_J real:

$$0 < \omega < \frac{2}{|\Omega_J|_{max}} = \frac{2}{\rho(S)} \qquad (12.3.7)$$

An optimal relaxation parameter can be selected, following Figure 12.3.1, as

$$\omega_{opt} = \frac{2}{|\Omega_J|_{min} + |\Omega_J|_{max}} \qquad (12.3.8)$$

The corresponding spectral radius of the Richardson iteration operator is

$$\rho(G_R) = 1 - \frac{2|\Omega_J|_{max}}{|\Omega_J|_{min} + |\Omega_J|_{max}} = \frac{\varkappa(S) - 1}{\varkappa(S) + 1} = 1 - \frac{2}{\varkappa(S) + 1} \qquad (12.3.9)$$

480

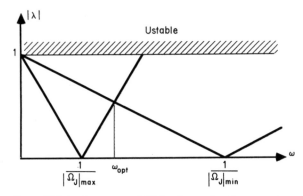

Figure 12.3.1 Spectrum of Richardson relaxation operator for real negative Ω

where the condition number of the S matrix has been introduced:

$$\varkappa(S) = \frac{|\Omega_J|_{\max}}{|\Omega_J|_{\min}} \tag{12.3.10}$$

For the Laplace operator the eigenvalue (12.1.53) leads to

$$|\Omega_J|_{\max} = 8 \qquad |\Omega_J|_{\min} = 8 \sin^2 \frac{\pi}{2M} \approx \frac{2\pi^2}{N}$$

and the condition number is

$$\varkappa(S) = \frac{1}{\sin^2(\pi/2M)} \approx \frac{4N}{\pi^2} \tag{12.3.11}$$

The approximate equality refers to the case of larger values of M. On a mesh of 50×50 points, $\varkappa(S) \approx 1000$.

The convergence rate, equation (12.1.40), becomes, for large M,

$$\rho(G_R) \approx 1 - \frac{\pi^2}{2N} \approx 1 - 2/\varkappa(S) \tag{12.3.12}$$

giving the same (poor) convergence rate as the point Jacobi method (see equation (12.1.56)).

We could improve the convergence properties by adopting a series of different values of ω, such as to cover the whole range from $1/|\Omega_J|_{\max}$ to $1/|\Omega_J|_{\min}$, giving complete damping of the correspondence frequency component Ω_J when $\omega = 1/|\Omega_J|$ (for real Ω). Such a process, whereby ω changes from one iteration to another, is called *non-stationary relaxation*. When ω is constant, we deal with a *stationary* relaxation.

The selection of ω in a sequence of p values covering this range can be optimized to yield the maximal convergence rate. This is realized when the intermediate values of ω are distributed between the two extreme values $1/|\Omega_J|_{\min}$ and $1/|\Omega_J|_{\max}$, as the zeros of the Chebyshev polynomials (Young,

1971; Marchuk, 1975):

$$\frac{1}{\omega_k} = \frac{|\Omega_J|_{\min} + |\Omega_J|_{\max}}{2} + \frac{|\Omega_J|_{\max} - |\Omega_J|_{\min}}{2} \cos\left(\frac{2k-1}{p}\frac{\pi}{2}\right) \qquad k = 1, 2, \ldots, p$$

(12.3.13)

Observe that for anti-symmetric S matrices ($S = -S^T$) the eigenvalues are purely imaginary; for instance, the centrally discretized first space derivative in the convection equations. In this case the eigenvalues of the Richardson iteration matrix become equal to $\lambda(G_R) = 1 + I\omega\Omega$, and the iterative method will diverge, since $|\lambda(G_R)| > 1$. This is in full agreement with the instability of the Euler explicit scheme for the central differenced convection equation.

12.3.2 Alternating direction implicit (ADI) method

An effective iterative method is obtained by using, as conditioning operator P, the ADI-space factorized operators described in Chapter 11. Referring to equation (11.4.16), the ADI iterative method is defined by

$$(1 - \tau S_x)(1 - \tau S_y)(1 - \tau S_z) \, \Delta U^n = \tau\omega(SU^n + Q^n) \qquad (12.3.14)$$

and is solved in the sequence

$$\begin{aligned}
(1 - \tau S_x) \, \overline{\Delta U} &= \tau\omega(SU^n + Q^n) \\
(1 - \tau S_y) \, \overline{\overline{\Delta U}} &= \overline{\Delta U} \\
(1 - \tau S_z) \, \Delta U^n &= \overline{\overline{\Delta U}}
\end{aligned} \qquad (12.3.15)$$

With the matrix notations and conventions of equation (12.2.37)–(12.2.39) we have, for a two-dimensional problem, $S_x = D/2 + H$, $S_y = D/2 + V$, and the ADI iteration can be written as

$$\begin{aligned}
\left[1 - \tau\left(\frac{D}{2} + H\right)\right] \overline{\Delta U} &= \tau\omega(SU^n + Q^n) \\
\left[1 - \tau\left(\frac{D}{2} + V\right)\right] \Delta U^n &= \overline{\Delta U}
\end{aligned} \qquad (12.3.16)$$

The parameters ω and τ have to be optimized in order to make the ADI iterations efficient. For optimized parameters the convergence rate can be of the order of $N \log N$ and is an order of magnitude faster than SOR. However, the ADI method is difficult to optimize for general problems. A discussion of this topic is to be found in Wachspress (1966), Mitchell (1969) and Mitchell and Griffiths (1980).

It can be shown that an optimal value of ω is $\omega_{\text{opt}} \approx 2$. A guideline to selections of τ is

$$\tau_{\text{opt}} = \frac{1}{\sqrt{(\Omega_{\min}\Omega_{\max})}}$$

or to distribute τ in a sequence of p values covering the range of frequencies in a selected direction. If the S matrix has the Fourier harmonics as eigenvectors (for periodic conditions) then the eigenvalues of the different submatrices can be obtained by the Von Neumann analysis. For the Laplace equation the eigenvalues $\lambda(G_{ADI})$ are given by equations (11.4.27) for $\omega = 1$, but with $\omega = 2$, we obtain, with $\tau_1 = \tau/\Delta x^2$,

$$\lambda(G_{ADI}) = \frac{(1 - 4\tau_1\sin^2\phi_x/2)(1 - 4\tau_1\sin^2\phi_y/2)}{(1 + 4\tau_1\sin^2\phi_x/2)(1 + 4\tau_1\sin^2\phi_y/2)} \tag{12.3.17}$$

The ADI iterative method is clearly unconditionally stable for positive values of τ. If the frequencies in a given direction, say y, are selected to be damped we can distribute τ between the minimum and maximum values associated, respectively, with the high-frequency $\phi_y \approx \pi$ and the lowest frequency $\phi_y = \pi/M$:

$$(\tau_1)_{\min} = \frac{1}{4} \qquad (\tau_1)_{\max} = \frac{1}{4\,\sin^2(\pi/2M)} \approx \frac{M^2}{\pi^2} \tag{12.3.18}$$

The sequence of τ values could be distributed by a Chebyshev optimization, for instance following equation (12.3.13). Another distribution law has been applied by Ballhaus et al. (1979) to the computation of solutions to the transonic potential flow equation

$$\tau = \tau_{\min}\left(\frac{\tau_{\max}}{\tau_{\min}}\right)^{k-1/p-1} \qquad k = 1, \ldots, p \tag{12.3.19}$$

where p is taken between 5 and 10. Even with an approximate optimization, the ADI method is very efficient and generally requires less computational time than point or line overrelaxations.

12.3.3 Other preconditioning techniques

All the matrices representing the relaxation methods can be considered as preconditioning matrices. An essential guideline towards the selection of the operator P is to be obtained from the generalization of equation (12.3.1) defining the Richardson iteration. If we write the general form of the iteration scheme as

$$U^{n+1} = U^n - \omega\tau P^{-1}(SU^n + Q^n) \tag{12.3.20}$$

we can consider this method as a Richardson iteration for the operator $B = \tau P^{-1}S$, which is positive definite by definition of P. Hence the eigenvalues of the iteration matrix $G = 1 - \tau\omega P^{-1}S$ satisfy the relation

$$\lambda(G) = 1 - \omega\lambda(\tau P^{-1}S) \tag{12.3.21}$$

An optimal relaxation parameter can be defined by

$$\omega_{opt} = \frac{2}{\lambda_{\min}(\tau P^{-1}S) + \lambda_{\max}(\tau P^{-1}S)} \tag{12.3.22}$$

and the spectral radius of G becomes

$$\rho(G) = \frac{\kappa(\tau P^{-1}S) - 1}{\kappa(\tau P^{-1}S) + 1} \qquad (12.3.23)$$

The Richardson iteration has a poor convergence rate because of the very high value of the condition number. In practical problems the mesh is often dense in certain regions and, more importantly, not uniform. A large variation in mesh spacing contributes to an increase in the condition number, and values of $10^5 - 10^6$ are not uncommon in large-size problems. On the other hand, a condition number close to one leads to an excellent convergence rate. Therefore various methods have been developed which aim at choosing P such that the eigenvalues of $\tau P^{-1}S$ are much more concentrated. We mention here a few of them, referring the reader to the literature cited for more details.

Incomplete Choleski factorization (Meijerink and Van Der Vorst, 1977, 1981; Kershaw, 1978)

During the LU factorization of the banded matrix S, elements appear in L and U at positions where the S matrix has zero-elements. If an approximate matrix P is defined formed by $\bar{L}$ and $\bar{U}$, $P = \bar{L}\bar{U}$, where $\bar{L}$ and $\bar{U}$ are derived from L and U by maintaining only the elements which correspond to the location of non-zero elements of S, we can expect P to be a good approximation to S and that $\rho(P^{-1}S)$ will be close to one. Hence $\bar{L}$ is defined by

$$\begin{aligned} \bar{l}_{IJ} &= l_{IJ} \quad && \text{if } s_{IJ} \neq 0 \\ &= 0 \quad && \text{if } s_{IJ} = 0 \end{aligned} \qquad (12.3.24)$$

and similarly for $\bar{U}$. Since P is very close to S, the matrix $P^{-1}S = (\bar{L}\bar{U})^{-1}LU$ will be close to the unit matrix and the condition number of $P^{-1}S$ will be strongly reduced compared with the condition number $\kappa(S)$ of the S-matrix. Hence $\lambda(G) = 1 - P^{-1}S$ will be small and high convergence rates can be expected. When used with a conjugate gradient method, very remarkable convergence accelerations on symmetric, positive definite matrices, have been obtained, Kershaw (1978).

The coupling with the conjugate gradient method is essential, since the P matrix obtained from the incomplete Cholesky decomposition, although close to unity, still has a few extreme eigenvalues which will be dominant at large iteration numbers. The conjugate gradient method, when applied to this preconditioned system, rapidly eliminates the extreme eigenvalues and corresponding eigenmodes. We are then left with a system where all the eigenvalues are close to one and the amplification matrix becomes very small.

Strongly implicit procedure (SIP) (Stone, 1968; Schneider and Zedan, 1981)
This is also an incomplete factorization technique, based on the matrix structure arising from the discretization of elliptic operators. The obtained matrix is block tridiagonal or block pentadiagonal for the most current

discretizations. The SIP method defines a conditioning matrix P as $P = L \cdot U$, where $U = L^T$ for symmetrical matrices such that L and U have the non-zero elements in the same location as the S matrix. A procedure is derived allowing the determination of L and L^T and the resulting system is solved by a forward and backward substitution as

$$L\overline{\Delta U} = -R^n$$
$$L^T \Delta U^n = \overline{\Delta U} \qquad (12.3.25)$$

The L and L^T matrices are obtained by a totally different procedure from the Choleski factorization.

Conjugate gradient method (Reid, 1971; Concus et al., 1976; Kershaw, 1978)

This method, originally developed by Hestenes and Stiefel, is actually a direct method, which did not produce the expected results with regard to convergence rates. However, as an acceleration technique, coupled to preconditioning matrices, it produces extremely efficient convergence accelerations. When coupled to the incomplete Choleski factorization the conjugate gradient method converges extremely rapidly (see Golub and Meurant, 1983, and Hageman and Young, 1981 for some comparisons on elliptic test problems). However, they require more storage than the relaxation methods, and this might limit their application for three-dimensional problems.

12.4 NON-LINEAR PROBLEMS

In a non-linear problem we can define a Newton linearization in order to enter an iterative approximation sequence. If the non-linear system is of the form

$$S(U) = -Q \qquad (12.4.1)$$

the Newton iteration for U^{n+1} is

$$S(U^{n+1}) = S(U^n + \Delta U) = S(U^n) + \left(\frac{\partial S}{\partial U}\right) \cdot \Delta U = -Q \qquad (12.4.2)$$

where higher-order terms in ΔU have been neglected. The Jacobian of S with respect to U (also called the tangential stiffness matrix, $K_T = \partial S/\partial U$, in the literature on finite elements) defines the iterative system

$$K_T \Delta U^n = -R^n \qquad (12.4.3)$$

In the linear case $K_T = S$ and equation (12.4.3) is identical to the basic system (12.1.30). In the non-linear case the Newton formulation (12.4.3) actually replaces the basic system (12.4.1). Indeed, close enough to the converged solution, the higher-order terms can be made as small as we wish, and equation (12.4.3) can therefore be considered as an 'exact' equation.

If equation (12.4.3) is solved by a direct method we would obtain the exact solution $\bar{U}$ as

$$\bar{U} = U^n - K_T^{-1} R^n \tag{12.4.4}$$

The error $e^n = U^n - \bar{U}$, with respect to this exact solution, satisfies the relation

$$K_T e^n = R^n \tag{12.4.5}$$

If system (12.4.3) is solved by an iterative method, represented by the conditioning operator P/τ,

$$\frac{P}{\tau} \Delta U = -R^n \tag{12.4.6}$$

the error e^n will be amplified by the operator G, such that

$$\begin{aligned} e^{n+1} = U^{n+1} - \bar{U} &= e^n + \Delta U \\ &= e^n - \tau P^{-1} R^n \\ &= (1 - \tau P^{-1} K_T) e^n \equiv G e^n \end{aligned} \tag{12.4.7}$$

Hence for a non-linear system of equations we can consider that the conditioning operator should be an approximation to the Jacobian matrix of system (12.4.1).

The iterative scheme (12.4.6) will converge if the spectral radius of

$$G = 1 - \tau P^{-1} K_T \tag{12.4.8}$$

is lower than or equal to one. Equation (12.4.8) therefore replaces equation (12.3.3) for a non-linear system. Actualy, since $K_T = S$ for linear equations, equation (12.4.8) is the most general form of the amplification matrix of an iterative scheme with operator P/τ, applied to linear or non-linear equations. Definition (12.4.8) therefore allows an analysis of convergence conditions for non-linear systems when the Jacobian operator K_T can be determined. An example of such an investigation will be given in Chapter 13 of Volume 2, dealing with the non-linear potential equation.

Constant and secant stiffness method

This denomination originates from finite element applications where the following conditioning matrices are often used, essentially for elliptic equations.

Constant stiffness

$$P = S(U^0) \tag{12.4.9}$$

The convergence matrix P is taken as the operator S at a previous iteration and kept fixed.

Secant stiffness

$$P = S(U^{n-1}) \tag{12.4.10}$$

The Jacobian is approximated by the previous value of the matrix $S(U^{n-1})$. The systems obtained can be solved by any of the methods mentioned, direct or iterative. In the former case we have a *semi-direct* method in the sense that the iterations treat only the non-linearity.

12.5 THE DIFFERENTIAL EQUATION REPRESENTATION OF A RELAXATION SCHEME

Since the matrix S originates from a space discretization of a differential operator the submatrices E, F or H, V, H_L, V_L, H_U, V_U can also be considered as representing differential operators acting on U or ΔU. Therefore the left-hand side $P \Delta U$ can be considered as the representation of some space operator acting on ΔU or on the pseudo-time derivative $dU/dt = \Delta U^n/\tau$. This idea has been developed by Garabedian (1956) and leads to an interesting interpretation of an iterative scheme as a modified time-dependent partial differential equation, which can be investigated analytically for stability, well-posedness and other properties.

Let us illustrate this on the Gauss–Seidel method (equations (12.1.22) and (12.1.25)). The conditioning operator $P = D + E$ represents a combination of x and y derivatives whose detailed form depends on the particular problem considered. For the Laplace operator, equation (12.1.22) is recast into the form (12.1.25):

$$
\begin{aligned}
&2\Delta u_{ij}^n + (\Delta_{i,j}^n - \Delta u_{i-1,j}^n) + (\Delta u_{i,j}^n - \Delta u_{i,j-1}^n) \\
&= (u_{i+1,j}^n + u_{i,j+1}^n + u_{i-1,j}^n + u_{i,j-1}^n - 4u_{ij}^n + q_{ij}^n)
\end{aligned}
\tag{12.5.1}
$$

The residual in the right-hand side is Δx^2 times the discretized Laplace operator and represents the differential equation $Lu = f$. On the left-hand side the first bracket can be viewed as a backward space discretization in the x-direction and the second as a similar operation in the y-direction. Therefore introducing the pseudo-time derivative $u_t = \Delta u^n/\tau$ we can write this iterative scheme as the following partial differential equation, with $\Delta x = \Delta y$:

$$
\left(\frac{2\tau}{\Delta x^2}\right) u_t + \left(\frac{\tau}{\Delta x}\right) u_{xt} + \left(\frac{\tau}{\Delta x}\right) u_{yt} = Lu - f
\tag{12.5.2}
$$

representing the Gauss–Seidel method, where it is understood that the space derivatives on the left-hand side are backward differenced.

This equation can be analysed by the methods of Chapter 3, via a transformation to a system of three first-order equations, by introducing the variables $v = u_x$, $w = u_y$, $z = u_t$ and adding the equations $v_y = w_x$ and $v_t = z_x$ (the subscripts indicate partial derivatives). The differential equation of the iterative scheme is hyperbolic, as can be seen from the real eigenvalues of the system (see Problem 12.6). However, a more straightforward analysis can be performed by introducing the transformation (Jameson, 1974)

$$
T = t + \alpha x + \beta y
\tag{12.5.3}
$$

where $\alpha = \tau/2\Delta x$ and $\beta = \tau/2\Delta y$. Equation (12.5.2) is rewritten here as follows, with $\gamma = 2\tau/\Delta x^2$:

$$\gamma u_t + 2\alpha u_{xt} + 2\beta u_{yt} = u_{xx} + u_{yy} - f \qquad (12.5.4)$$

Performing the transformation to the variables (T, x, y), via a transformation such as

$$u_x => u_x + \alpha u_T$$
$$u_{xx} => u_{xx} + 2\alpha u_{xT} + u_{TT} \qquad (12.5.5)$$

we obtain

$$\gamma u_T + (\alpha^2 + \beta^2)u_{TT} = u_{xx} + u_{yy} - f \qquad (12.5.6)$$

Since this equation is clearly hyperbolic in the T-direction (which is a characteristic direction) the coefficient γ of the first-order T derivative, which provides the damping, must be positive.

A similar analysis can be performed for the SOR relaxation method (equation (12.2.11)), leading to the differential equation

$$2(2 - \omega)\,\Delta u_{ij}^n + \omega(\Delta u_{i,j}^n - \Delta u_{i-1,j}^n) + \omega(\Delta u_{ij}^n - \Delta u_{i,j-1}^n) = \omega\,\Delta x^2(Lu - f) \qquad (12.5.7)$$

or

$$\frac{2\tau(2 - \omega)}{\Delta x^2}\, u_t + \left(\frac{\omega\tau}{\Delta x}\right) u_{xt} + \left(\frac{\omega\tau}{\Delta x}\right) u_{yt} = \omega(Lu - f) \qquad (12.5.8)$$

Compared to the above analysis we see that the positiveness of the coefficient of u_t requires the relaxation coefficient to satisfy $0 < \omega < 2$.

From a differential point of view the conditioning matrix P represents the differential operator

$$P_{SOR} = \frac{2\tau(2 - \omega)}{\Delta x^2} + \left(\frac{\omega\tau}{\Delta x}\right)\frac{\partial}{\partial x} + \left(\frac{\omega\tau}{\Delta y}\right)\frac{\partial}{\partial y} \qquad (12.5.9)$$

For the Jacobi line relaxation method along vertical y-lines, equation (12.2.36), we have

$$4(\Delta u_{i,j}^n - \Delta u_{i,j+1}^n - \Delta u_{i,j-1}^n) = \omega(Lu - f)\,\Delta x^2 \qquad (12.5.10)$$

or

$$\frac{2\tau}{\Delta x^2}\, u_t - \tau u_{yyt} = \omega(Lu - f) \qquad (12.5.11)$$

The line relaxation has lead to the appearance of a second derivative along the selected lines in the·conditioning operator P_{JLR}:

$$P_{JLR} = \frac{2\tau}{\Delta x^2} - \tau\frac{\partial^2}{\partial y^2} \qquad (12.5.12)$$

Since the presence of the second derivative in P makes it closer to the differential Laplace operator, which also contains a second derivative term, we can expect the line relaxation to be more effective in iteration counts than the point overrelaxation. This is indeed the case in practice.

Applied to the VLOR method (equation (12.2.33)) an additional u_{xt} term appears in the left-hand side (see Problem 12.7). Actually, Figures 12.1.2 and 12.2.2 also explain the background of the derivative terms in P by the location of the point taken at level $(n + 1)$.

12.6 THE MULTI-GRID METHOD

The multi-grid method is the most efficient and general iterative technique known today. Although originally developed for elliptic equations, such as the discretized Poisson equation (Fedorenko, 1962, 1964), the full potential of the multi-grid approach was put forward and systemized by Brandt (1972, 1977, 1982). It has recently been applied with great success to a variety of problems, ranging from potential equations to Euler and Navier–Stokes discretizations, and is still subject to considerable research and development. A collection of reports on the basics of multi-grid methods and some selected applications can be found in Hackbusch and Trottenberg (1982), Hackbusch (1985).

The multigrid method has its origin in the properties of conventional iterative techniques, discussed in the previous section, in particular, their asymptotic slow convergence because of the poor damping of the low frequency errors (long wavelengths). Indeed, as can be seen from equations (12.1.48) and (12.1.50), the asymptotic behaviour of the error (or of the residual) is dominated by the eigenvalues of the amplification matrix close to one (in absolute value). These are associated with the eigenvalues of the space-discretization operator S with the lowest absolute value. For the Laplace operator, for instance, the lowest $|\Omega_J|_{\min}$ is obtained from equation (12.1.53) for the low-frequency component $\phi_x = \phi_y = \pi/M$.

We can therefore consider, on a fairly general basis, that the error components situated in the low-frequency range of the spectrum of the space-discretization operator S are the slowest to be damped in the iterative process. On the other hand, the higher frequencies are the first to be reduced and, after a few iterations, a large part of the high-frequency error components will generally be damped (Figure 12.6.1). As a consequence, an iterative method with these properties will act as a *smoother* of the error in such a way that after one or more iteration sweeps through the mesh (also called *relaxation* sweeps) *the error behaviour is sufficiently smooth for it to be adequately represented on a coarser mesh.*

The basic idea of multi-grid methods is as follows:

(1) Apply one or more sweeps of an iterative method with good smoothing properties of the higher-frequency components.
(2) Transfer the problem to a coarser grid, where an approximation to the correction ΔU is obtained at a reduced computational cost, since there are fewer mesh points.

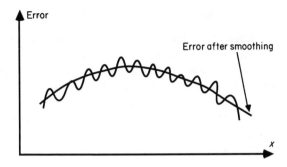

Figure 12.6.1 Smoothing of the error by relaxation sweeps

(3) Transfer the corrections obtained to the fine grid in order to generate a new approximation of the solution.

These different steps will be defined more precisely in the following, but it should be clear that error smoothing on the fine mesh is the essential property of multi-grid methods.

Another view of this essential requirement to the effectiveness of multi-grid methods can be obtained by noting that the high-frequency components are *not* visible on the coarse mesh. Therefore they have to be reduced in magnitude, compared to the low-frequency components, in such a way that the remaining, smoothed error can be represented by the frequency modes visible on the coarse mesh. In a one-dimensional space of length L, divided into M cells, with $\Delta x = L/M$, the visible frequencies range from $\phi = \pi/M$ to $\phi = \pi$, associated with the wavelengths from $2\Delta x$ to $2L$. If a coarse mesh is defined by removing every second point, the shortest wave that can be represented on this mesh is $4\Delta x$, corresponding to a highest frequency defined by $\phi = \pi/2$. Hence the whole range of wavenumber variables $\phi = 2\pi\,\Delta x/\Lambda$, where Λ is the wavelength, situated between $\pi/2$ and π, cannot be represented on the coarse mesh of mesh size $(2\Delta x)$. On the other hand, the components associated with the low-frequency range $\phi = [\pi/M, \pi/2[$ are well represented on the coarse mesh (Figure 12.6.2).

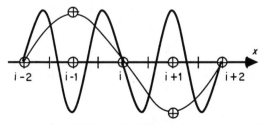

Figure 12.6.2 Visible and invisible modes on coarse mesh

12.6.1 Smoothing properties

The smoothing qualities of an iterative (relaxation) method can be estimated by evaluating the maximum eigenvalues of the amplification matrix in the high-frequency domain of the associated wavenumber variable $\phi[\pi/2, \pi]$. In two dimensions the high-frequency domain is defined by the region $\pi/2 \leqslant |\phi_x| \leqslant \pi$ and $\pi/2 \leqslant |\phi_y| \leqslant \pi$ (Figure 12.6.3) when the Fourier modes are considered, with a coarsening whereby every second point is removed along the co-ordinate lines. This is the so-called standard coarsening. Hence the smoothing factor μ is defined by

$$\mu = \operatorname*{Max}_{\frac{\pi}{2} \leqslant |\phi| \leqslant \pi} |\lambda(G)| \qquad (12.6.1)$$

For instance, the Jacobi method applied to the Laplace operator has the eigenvalues (12.1.54), which take on the value -1 for $\phi_x = \phi_y = \pi$. Therefore

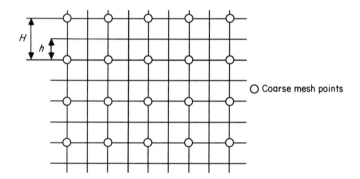

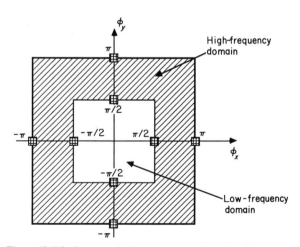

Figure 12.6.3 Low- and high-frequency domains for two-dimensional problems

the smoothing factor (12.6.1) is equal to one, indicating that some high-frequency error components are not damped. As a consequence, the Jacobi method is not a valid smoother. However, with relaxation methods we can select the relaxation coefficient, ω, in order to optimize the high-frequency damping properties. With equation (12.2.8) the smoothing factor for Jacobi overrelaxation becomes, introducing the extreme eigenvalues $\lambda(G_J) = -1$ and $+1/2$ in the high-frequency domain,

$$\mu(G_J(\omega)) = \text{Max}[|1 - \omega/2|, |1 - 2\omega)] \qquad (12.6.2)$$

This is illustrated in Figure 12.6.4, where the optimum relaxation coefficient $\omega = 4/5$ gives the lowest smoothing factor of 0.6, showing that after four iterations the high-frequency components will be damped by nearly one order of magnitude.

The Gauss–Seidel iterative method leads to a still better smoothing factor for the Laplace operator, since we obtain $\mu_{GS} = 0.5$ (Hackbusch and Trottenberg, 1982). Here, an order of magnitude reduction in high-frequency error components is achieved after three iterations. All the iterative methods can be analysed for their smoothing properties, and the reader will find more information in the above-mentioned literature.

In general, the smoothing factor may depend on the number of relaxation sweeps performed and also on the definition of the coarse mesh points with respect to the fine mesh. For the Laplace operator on a uniform mesh we can compare various line relaxation methods for Jacobi or SOR, as well as Red–Black, Zebra and other variants. The following values are obtained (Hackbusch and Trottenberg, 1982):

Line Jacobi relaxation	0.6
Line Gauss–Seidel	0.447
Red–Black point relaxation ($\omega = 1$)	0.25
Zebra line relaxation	0.25
Alternating Zebra	0.048

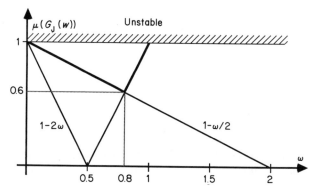

Figure 12.6.4 Smoothing factor for Jacobi overrelaxation

492

Red–Black relaxation has a smoothing factor which increases with the number of iterations. This is due to the coupling between high and low frequencies introduced by the red–black sequences, and consequently there is no gain in increasing the number of relaxation sweeps in this case. Note the excellent smoothing properties of alternating zebra line relaxation.

12.6.2   The coarse grid correction (CGC) method for linear problems

After a few relaxation sweeps the remaining error is expected to be sufficiently smooth to be approximated on a coarser grid, with $2\Delta x$, for instance, as basic step size. The procedures for generating this coarse grid correction are the building blocks of the multi-grid method.

The fine mesh, on which the solution is sought, will be represented by h, which will also be used as a sub- or superscript in order to indicate the mesh on which the operators or the solutions are defined. Similarly, the coarse mesh will be designated by H. Both h and H also designate representative mesh sizes, with $H > h$. For instance, $H = 2h$ when every second mesh point is removed in the standard coarsening, as shown in Figure 12.6.3.

We consider the linear problem on the fine mesh h:

$$S_h U_h = -Q_h \qquad (12.6.3)$$

and an iterative scheme under the residual form (12.1.29):

$$P \, \Delta U_h = -R_h \qquad (12.6.4)$$

Actually, the coarse grid correction method consists of selecting P proportional to the basic operator S *defined on the coarse mesh*. This process involves three steps:

(1)   The transfer from the fine to the coarse grid, characterized by a *restriction* operator I_h^H. This operator defines the way the mesh values on the coarse grid are derived from the surrounding fine mesh values. In particular, the residual on the coarse grid, R_H, is obtained by

$$R_H = I_h^H R_h \qquad (12.6.5)$$

(2)   The *solution* of the problem on the coarse mesh:

$$S_H U_H = -Q_H \qquad (12.6.6)$$

or

$$S_H \, \Delta U_H = -R_H \qquad (12.6.7)$$

(3)   The transfer of the *corrections* ΔU_H from the coarse mesh to the fine one. This defines an *interpolation* or *prolongation* operator I_H^h:

$$\Delta U_h = I_H^h \, \Delta U_H \qquad (12.6.8)$$

Restriction operator

Let us consider a one-dimensional mesh and a second-order derivative discretized by the central difference operator, $h = \Delta x$:

$$R_h = \frac{1}{h^2}(u_{i+1} - 2u_i + u_{i-1}) + q_i \qquad (12.6.9)$$

where i represents the fine mesh points (Figure 12.6.5). In order to relate i to the coarse mesh points, obtained by removing every second point, we will set $i = 2k$. The coarse mesh points are then designated by $k - 1, k, k + 1, ...$ Hence equation (12.6.9) becomes, for constant q,

$$R_h = \frac{1}{h^2}(u_{2k+1} - 2u_{2k} + u_{2k-1}) + q \qquad (12.6.10)$$

The full weighting restriction operator is defined by the distribution of Figure 12.6.5 such that each coarse mesh point receives a contribution from itself and the two neighbouring fine mesh points, weighted, respectively, by the factors $1/2, 1/4, 1/4$. For consistency, the sum of the weighting factors should always be equal to one.

The coarse mesh residual R_H is obtained by

$$R_H = I_h^H R_h \qquad (12.6.11)$$

leading to

$$R_H = \frac{1}{4h^2}(u_{k-1} - 2u_k + u_{k+1}) + q \qquad (12.6.12)$$

which is the discretization of the second derivative on the coarse mesh. This restriction operator is represented here by a matrix $M/2 \times M$, where M is the number of cells in the fine mesh:

$$I_h^H = \begin{array}{c} \\ k-1 \\ k \\ k+1 \\ \end{array} \left| \begin{array}{ccccccccc} & & & 2k-1 & 2k & 2k+1 & & & \\ \cdot & \cdot & \cdot & \cdot & & \cdot & & \cdot & \cdot \\ 1/4 & 1/2 & 1/4 & \cdot & & \cdot & & \cdot & \cdot \\ \cdot & \cdot & 1/4 & 1/2 & 1/4 & \cdot & & \cdot \\ \cdot & \cdot & \cdot & \cdot & 1/4 & 1/2 & 1/4 \\ & & & & \cdot & \cdot & \cdot \end{array} \right| \qquad (12.6.13)$$

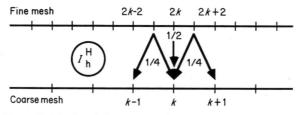

Figure 12.6.5 Restriction operator for one-dimensional problem

494

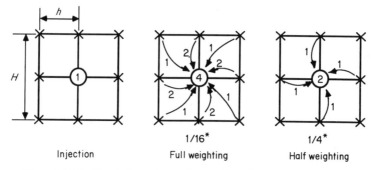

Figure 12.6.6 Restriction operator in two dimensions on a uniform mesh

Another choice is referred to as *simple injection*, whereby the coarse mesh functions are taken equal to their fine mesh value:

$$I_h^H = 1 \qquad \text{for all } 2k \text{ fine mesh points}$$

$$I_h^H = \begin{vmatrix} \cdot & \cdot & \cdot & \cdot & \cdot & \cdot & \cdot \\ 0 & 1 & 0 & \cdot & \cdot & \cdot & \cdot \\ & 0 & 1 & 0 & \cdot & \cdot & \cdot \\ & & 0 & 1 & 0 & \cdot \\ & \cdot & \cdot & \cdot & \cdot & \cdot \end{vmatrix} \qquad (12.6.14)$$

In two dimensions the simple injection is represented by the molecule in Figure 12.6.6(a), where the central coarse mesh point is shown surrounded by its fine mesh neighbours. The number inside the circle shows the local weight factor between the fine and coarse mesh functions in this point. Full weighting restriction is shown by the molecule in Figure 12.6.6(b), while a half-weighting restriction operator is indicated in Figure 12.6.6(c).

Prolongation operator

Prologation or interpolation operators are generally defined by tensor products of one-dimensional interpolation polynomials of odd order, such as the linear (or cubic) finite element interpolation functions. For the previous one-dimensional example of Figure 12.6.5 a linear interpolation would lead to the representation shown in Figure 12.6.7. This is represented by a matrix $M \times M/2$:

$$I_H^h = \begin{vmatrix} & & & & k-1 & k & k+1 & & \\ \cdot & \cdot & \cdot & \cdot & \cdot & \cdot & & \cdot \\ \cdot & \cdot & \cdot & 0 & 1/2 & 1/2 & \cdot & \cdot \\ \cdot & 0 & 0 & 1 & 0 & \cdot \\ \cdot & \cdot & \cdot & 1/2 & 1/2 & \cdot \end{vmatrix} \begin{matrix} \\ \\ 2k-1 \\ 2k \\ 2k+1 \end{matrix} \qquad (12.6.15)$$

Generalized to a uniform two-dimensional mesh, we obtain the molecule shown in Figure 12.6.8.

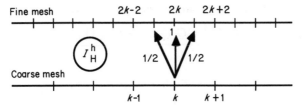

Figure 12.6.7 One-dimensional prolongation operator

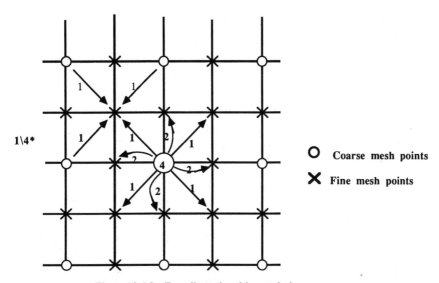

Figure 12.6.8 Two-dimensional interpolation operator

Coarse-grid correction operator

Combining the three steps mentioned above, the conditioning operator for the coarse-grid correction method P_{CGC}:

$$P_{\text{CGC}} \, \Delta U_h = -R_h \tag{12.6.16}$$

is obtained by combining equations (12.6.5), (12.6.7) and (12.6.8), leading to

$$\begin{aligned}
\Delta U_h &= I_H^h \, \Delta U_H \\
&= I_H^h \, S_H^{-1} R_H \\
&= -I_H^h \, S_H^{-1} I_h^H R_h
\end{aligned} \tag{12.6.17}$$

Hence

$$P_{\text{CGC}}^{-1} = I_H^h \, S_H^{-1} I_h^H \tag{12.6.18}$$

The amplification matrix associated with equation (12.6.16) is defined by

$$U_h^{n+1} = G_{CGC} U_h^n - P_{CGC}^{-1} Q \qquad (12.6.19)$$

with

$$\begin{aligned} G_{CGC} &= 1 - P_{CGC}^{-1} S_h \\ &= 1 - I_H^h S_H^{-1} I_h^H S_h \end{aligned} \qquad (12.6.20)$$

It is assumed here that the problem is solved exactly on the coarse grid H, that is, that S_H^{-1} can be evaluated exactly.

12.6.3 The two-grid iteration method for linear problems

The multi-grid method on the two grids h, H is now completely defined when the smoothing is combined with the coarse-grid correction. Hence the two-grid iteration method is obtained by the following sequence:

(1) Perform n_1 relaxation sweeps with smoother S_1 on the fine mesh solution U_h^n.
(2) Perform the coarse-grid correction, obtaining $U_h^{n+1} = U_h^n + \Delta U_h$.
(3) Perform n_2 relaxation sweeps with a smoother S_2 on the fine-mesh solution U_h^{n+1}.

Denoting by $G_{h,H}$ the amplification matrix of the two-grid iteration method and G_S the amplification matrix of the relaxation method selected as a smoothing operator, we obtain

$$G_{h,H} = G_{S2}^{n_2} \cdot G_{CGC} \cdot G_{S1}^{n_1} = G_{S2}^{n_2}[1 - I_H^h S_H^{-1} I_h^H S_h] G_{S1}^{n_1} \qquad (12.6.21)$$

Typical values for n_1, n_2 are one or two, and the smoothers S_1 and S_2 may be different. For instance, when an alternating line relaxation technique (one of the most efficient smoothers for elliptic equations) is applied $S1$ could be connected to the horizontal lines and $S2$ to the vertical ones. Another option is to modify the values of some relaxation coefficients between $S1$ and $S2$.

Convergence properties

The two-grid method will converge, if the spectral radius of $G_{h,H}$ is lower than one:

$$\rho(G_{h,H}) < 1 \qquad (12.6.22)$$

The convergence properties of the two-grid method can be analysed on simple problems, such as the Poisson equation and uniform meshes by a Fourier–Von Neumann method, whereby the eigenvalues of the different operators in equation (12.6.21) are determined. Examples of this approach can be found in Hackbusch and Trottenberg (1982), where the spectral radius $\rho(G_{h,H})$ is obtained for various combinations and choices of the operators in equation (12.6.21).

The two-grid method has remarkable convergence properties, which have been proved by Hackbusch (see Hackbusch and Trottenberg, 1982, Hackbusch, 1985), for a rather large class of problems. Namely, the asymptotic convergence rate is independent of the mesh size h, that is, independent of the number of mesh points. Hence the computational work in a two-grid method is determined essentially by the work required to perform the smoothing steps. Since the relaxation sweeps require a number of operations proportional to the number of fine mesh points N, the total work will vary linearly with N.

12.6.4 The multi-grid method for linear problems

In the two-grid method the solution step on the coarser grid, that is, obtaining ΔU_H from equation (12.6.7), is assumed to have been performed accurately. In the multigrid approach this solution is obtained by application of another two-grid iteration on a coarser grid, for instance a grid of size $2H = 4h$. This can be repeated for the solution of equation (12.6.7) on the grid $2H$, and so on. The multi-grid method is therefore defined on a succession of increasingly coarser grids, applying recursively the two-grid iteration, whereby the 'exact' solution of equation (12.6.7) has only to be obtained on the last, very coarse, grid.

Usually, four to five grids are used and various strategies can be chosen in the sequence of transfer and smoothing between successive grids. The V-cycle, shown in Figure 12.6.9(a), consists of a succession of smoothing (symbol 0) and transfer to the next coarser grid, with a unique exact solution (symbol □) on the coarsest grid, followed by a unique sequence of transfer and smoothing, back to the finest grid. In the W-cycles intermediate V-cycles are performed on the coarser grids, as shown in Figure 12.6.4(b), where γ indicates the number of internal V-cycles.

The amplification matrix of the multi-grid operator is obtained from

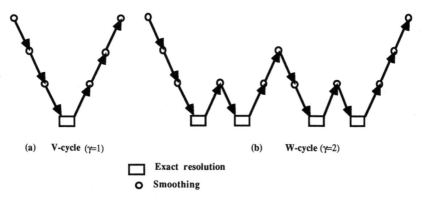

(a)   V-cycle ($\gamma=1$)   (b)   W-cycle ($\gamma=2$)

□   Exact resolution

O   Smoothing

Figure 12.6.9   Multigrid strategies on a four-grid method

498

equation (12.6.21) by replacing S_H^{-1} in $G_{h,H}$ by

$$S_H^{-1} => (1 - G_{H,2H})S_H^{-1} \qquad (12.6.23)$$

and performing the same replacement for the operator S_{2H}^{-1} appearing in $G_{H,2H}$ until the coarsest grid is reached.

It can be shown that the convergence properties of the two-grid method also apply to the multigrid approach, under fairly general conditions, namely the mesh-size independence of the convergence rate. Therefore the computational work of a multi-grid cycle is essentially dominated by the number of operations required for the two-grid computation from the finest to the next grid. Hence the total work will remain proportional to the number of mesh points N of the fine grid. Note that W-cycles are roughly 50% more expensive than V-cycles, but are generally more robust and should be preferred, particularly for sensitive problems.

Other iterative techniques require a number of operations for an error reduction of one order of magnitude, of the order of $N^{3/2}$ (optimal SOR) or $N^{5/4}$ for preconditioned conjugate gradient methods, against N for the

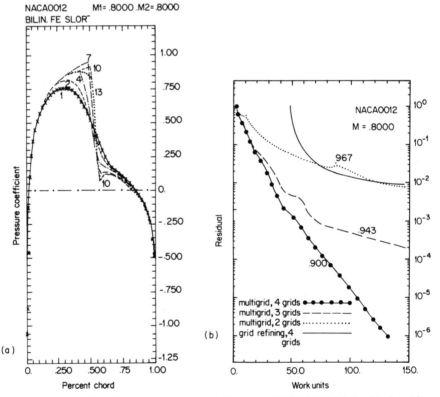

Figure 12.6.10 Multigrid solutions for transonic flow over a NACA 0012 airfoil at Mach number 0.8. (a) Convergence history of the solution; (b) convergence history of the residuals. The numbers indicate the average convergence rate

multi-grid approach. Hence multi-grid methods are optimal in convergence rates and operation count and are considerably more general than the so-called fast solvers, which are the best performing direct methods, when they apply.

An additional advantage of multi-grid methods, which appears in practice, is the extremely rapid way the solution approaches its final shape after only a very few multi-grid cycles. This is actually a consequence of the fact that the errors are damped over the whole frequency range in every cycle, when passing from the finest to the coarsest grid and back. With other methods the high frequencies are generally first damped, then progressively, as the number of iterations increases, an increasing number of frequencies in the medium and lower range of the spectrum are treated and removed. With the multi-grid method the whole spectrum is scanned during each cycle. An example of this behaviour is shown in Figure 12.6.10 (from Deconinck and Hirsch, 1982) for a potential flow computation over an airfoil in the transonic regime, with multi-grid cycles of four grids.

Figure 12.6.10(a) shows the evolution of the surface Mach number, containing a shock, from the first to the thirteenth multi-grid cycle, after which it cannot be distinguished on the drawing from the final converged solution. After one single cycle we are already very close to the overall shape of the converged solution. Figure 12.6.10(b) compares the convergence rates of the residual for a line relaxation SLOR iterative method and the multi-grid with two, three and four grids. The numbers indicated represent the average convergence rate of the residuals and are an approximation of the spectral radius of the multi-grid amplification matrix. One work unit on the horizontal scale corresponds to one SLOR iteration on the fine grid. The improvement with respect to SLOR is spectacular, and we can also notice the influence of the number of grids on the overall convergence rates. Note that these computations have been performed for a non-linear problem by the method presented in the following section.

12.6.5 The multi-grid method for non-linear problems

For non-linear problems the multi-grid method can be applied in association with a linearization process such as the Newton linearization (equation (12.4.3)). However, we can also adapt the multi-grid approach directly to the non-linear equations by operating simultaneously on the residuals R_h and on the solution U_h itself, instead of the corrections ΔU_h. This is known as the full approximation scheme (FAS), Brandt (1977).

Considering the non-linear problem on mesh h,

$$S_h(U_h) = -Q_h \qquad (12.6.24)$$

the two-grid non-linear FAS solves for $U_h^{\text{new}} = (U_h + \Delta U_h)$, solution of

$$S_h(U_h + \Delta U_h) - S_h(U_h) = -R_h, \qquad (12.6.25)$$

by a transfer to a coarser grid H.

After n_1 relaxation sweeps on U_h a transfer of residuals and of the solution U_h to the coarse grid is defined by two restriction operators I_h^H and $\hat{I}_h^H$:

$$R_H = I_h^H R_h \qquad (12.6.26)$$

$$U_H = \hat{I}_h^H U_h \qquad (12.6.27)$$

These restriction operators may be different, since they are operating on different spaces. Hence on the coarse grid H, equation (12.6.25) becomes

$$S_H(U_H + \Delta U_H) - S_H(U_H) = -R_H \qquad (12.6.28)$$

or

$$S_H(U_H^{new}) = -I_h^H Q_h + S_H(\hat{I}_h^H U_h) - I_h^H S_h(U_h) \qquad (12.6.29)$$

In the multi-grid method equation (12.6.29) is solved by subsequent transfer to a next coarser grid. On the final grid this non-linear equation has to be solved exactly.

When transferring from the coarse to the fine grid a prolongation operation is applied to the *corrections* ΔU_h, defined by

$$\Delta U_H = U_H^{new} - U_H \qquad (12.6.30)$$

leading to

$$\Delta U_h = I_H^h \, \Delta U_H \qquad (12.6.31)$$

and to the fine grid solution

$$U_h^{new} = U_h + I_H^h \, \Delta U_H \qquad (12.6.32)$$

The multi-grid cycle is closed after performing n_2 smoothing relaxation sweeps on the new solution U_h^{new}.

In the full approximation scheme no global linearization is required, except on the coarsest mesh and for the local linearizations of the relaxation sweeps.

References

Ballhaus, W. F., Jameson, A., and Albert T. J. (1978). Implicit approximate factorization schemes for steady transonic flow problems.' *AIAA Journal*, **16**, 573–79.

Berman, A., and Plemmons, R. J. (1979). *Non-Negative Matrices in the Mathematical Sciences*, New York: Academic Press.

Brandt, A. (1972). 'Multilevel adaptive technique for fast numerical solution to boundary value problems.' In *Proc. 3rd Int. Conf. on Numerical Methods in Fluid Dynamics, Lecture Notes in Physics*, Vol, 18, pp. 82–9, New York: Springer Verlag.

Brandt, A. (1977). 'Multilevel adaptive solutions to boundary value problems.' *Math. of Computation*, **31**, 333–90.

Brandt, A. (1982). 'Guide to multigrid development.' In *Multigrid Methods, Lecutre Notes in Mathematics*, Vol. 960, New York: Springer Verlag.

501

Concus, P., Golub, G. H., and O'Leary, D. P. (1976). 'A generalized conjugate gradient method for the numerical solution of elliptic partial differential equations.' In J. R. Bunch and D. J. Rose (eds), *Space Matrix Computations*, New York: Academic Press.

Dahlquist, G., and Björk, A. (1974). *Numerical Methods* Englewood Cliffs, NJ: Prentice-Hall.

Deconinck. H., and Hirsch, Ch. (1982). 'A multigrid method for the transonic full potential equation discretized with finite elements on an arbitrary body fitted mesh.' *Journal Computational Physics*, **48**, 344–65.

Fedorenko, R. P. (1962). 'A relaxation method for solving elliptic differential equations.' *USSR Computational Math. and Math. Physics*, **1**, 1092–6.

Fedorenko, R. P. (1964). 'The speed of convergence of an iterative process.' *USSR Computational Math. and Math. Physics*, **4**, 227–35.

Garabedian P. R. (1956). 'Estimation of the relaxation factor for small mesh sizes.' *Math. Tables Aids Computation*, **10**, 183–5.

Golub, G. H., and Meurant, G. A. (1983). *Résolution Numérique des Grands Systèmes Linéaires*, Paris: Eyrolles.

Hackbusch, W. (1985). *Multigrid Methods*, Berlin, Springer-Verlag.

Hackbusch, W., and Trottenberg, U. (eds) (1982). *Multigrid Methods, Lecture Notes in Mathematics*, Vol. 960, New York: Springer Verlag.

Hageman, P., and Young, D. M. (1981). *Applied Iterative Methods*, New York: Academic Press.

Jameson, A. (1974). 'Iterative solutions of transonic flows over airfoils and wings, including flows at Mach 1.' *Communications Pure and Applied Math.*, **27**, 283–309.

Kershaw, D. S. (1978). 'The incomplete Cholesky conjugate gradient method for the iterative solution of linear equations.' *Journal of Computational Physics*, **26**, 43–65.

LeVeque, R. A., and Trefethen, L. N. (1986). 'Fourier analysis of the SOR iteration.' *NASA-CR 178191, ICASE Report 86-63*

Marchuk, G. I. (1975). *Method of Numerical Mathematics*, Berlin: Springer Verlag.

Meijerink, J. A., and Van Der Vorst, H. A. (1977). 'An iterative solution method for linear systems of which the coefficient matrix is a M-matrix.' *Mathematics of Computation*, **31**, 148–62.

Meijerink, J. A., and Van Der Vorst, H. A. (1981). 'Guidelines for the usage of incomplete decompositions in solving sets of linear equations as they occur in practice.' *Journal of Computational Physics*, **44**, 134–55.

Meis, T., and Marcowitz, U. (1981). *Numerical Solution of Partial Differential Equations*, New York: Springer Verlag.

Mitchell, A. R. (1969). *Computational Methods in Partial Differential Equations*, New York: John Wiley.

Mitchell, A. R., and Griffiths, D. F. (1980). *The Finite Difference Method in Partial Differential Equations*, New York: John Wiley.

Reid, J. K. (1971). 'On the method of conjugate gradients for the solution of large sparse systems of linear equations.' In Reid, J. K. (ed.), *Large Sparse Sets of Linear Equations*, New York: Academic Press.

Schneider, G. E., and Zedan, M. (1981). 'A modified strongly implicit procedure for the numerical solution of field problems.' *Numerical Heat Transfer*, **4**, 1–19.

Stone, H. L. (1968). 'Iterative solution of implicit approximations of multidimensional partial differential equations.' *SIAM Journal of Numerical Analysis*, **5**, 530–58.

Strang, G. (1976). *Linear Algebra and its Applications*, New York: Academic Press.

Varga, R. S. (1962). *Matrix Iterative Analysis*, Englewood Cliffs, NJ: Prentice-Hall.

Wachspress, E. L. (1966). *Iterative Solution of Elliptic Systems*, Prentice-Hall.

Young, D. M. (1971). *Iterative Solution of Large Linear Systems*, Englewood Cliffs, NJ: New York: Academic Press.

PROBLEMS

Problem 12.1

Write the discretized equations for the nodes close to all the boundaries for the Poisson equations (12.1.1) and (12.1.2) based on Figure 12.1.1. Repeat the same exercise for Von Neumann conditions along the boundary $j = 0$ ($y = 0$), of the form

$$\frac{\partial u}{\partial y} = g \qquad y = 0$$

by applying a one-sided difference.

Problem 12.2

Consider the stationary diffusion equation $\alpha u_{xx} = q$ in the domain $0 \leqslant x \leqslant 1$ and the boundary conditions

$$x = 0 \qquad u(0) = 0$$
$$x = 1 \qquad u(1) = 0$$

with $q/\alpha = -4$. Apply a central second-order difference and solve the scheme with the tridiagonal algorithm for $\Delta x = 0.1$ and $\Delta x = 0.02$ and compare with the analytical solution.

Problem 12.3

Solve Problem 12.2 with the Jacobi method. Write the matrices D, E, F as tridiagonal matrices $B(a, b, c)$.

Problem 12.4

Solve Problem 12.2 with the Gauss–Seidel method and compare the convergence rate with the Jacobi method.

Problem 12.5

Obtain the eigenvalue $\lambda(G_{\text{ADI}})$ of equation (12.3.17) for the ADI preconditioning operator.

Problem 12.6

Analyse the properties of the differential equation (12.5.2) representing the Gauss-Seidel iterative method, following the method described in Chapter 3.

Problem 12.7

Derive the differential equation representing the VLOR method of equation (12.2.33). Show that

$$P_{\text{SLOR}} = \frac{\tau(2 - \omega)}{\Delta x^2} + \left(\frac{\omega \tau}{\Delta x}\right)\frac{\partial}{\partial x} - \tau \frac{\partial^2}{\partial y^2}$$

Problem 12.8

Consider the steady-state convection–diffusion equation

$$a \frac{\partial u}{\partial x} = \alpha \frac{\partial^2 u}{\partial x^2}$$

discretized with second-order central differences. Write explicitly the tridiagonal system obtained and apply a Gauss–Seidel method. Derive the equivalent differential equation and define the associated stability conditions.

Hint: Show that we obtain $Pu_t = \alpha u_{xx} - a u_x$ with

$$P = \frac{\alpha \tau}{\Delta x^2} \left(1 - \frac{a \, \Delta x}{2\alpha} \frac{\partial}{\partial x} \right)$$

and explain that $P > 0$ is necessary for stability, leading to $\alpha > 0$ and $a \, \Delta x / \alpha < 2$.

Problem 12.9

Repeat Problem 12.8 with a SOR method. Show that in this case

$$P = \frac{\alpha \tau \omega}{\Delta x^2} \left[\frac{2 - \omega}{\omega} - \frac{a \, \Delta x}{2\alpha} \frac{\partial}{\partial x} \right]$$

for SOR.

Problem 12.10

Solve the Poisson equation

$$\Delta u = -2\pi^2 \sin \pi x \sin \pi y$$

on a square $0 \leqslant x \leqslant 1$ with the homogeneous Dirichlet boundary conditions, $u = 0$, on the four sides. Define a five-point discretization on a rectangular mesh. Consider a 11×11 mesh and solve with a Jacobi iteration. Compare the convergence rate with the same computation on a 21×21 mesh. Compare the results with the exact solution $u = \sin \pi x \sin \pi y$.

Problem 12.11

Repeat Problem 12.10 with a Guass–Seidel iteration and compare the convergence rates and computational times.

Problem 12.12

Repeat Problem 12.10 with a SOR method and compare the convergence rates and computational times. Try different values of the overrelaxation parameter.

Problem 12.13

Repeat Problem 12.10 with a line SOR method and compare the convergence rates and computational times. Try different values of the overrelaxation parameters.

Problem 12.14

Repeat Problem 12.10 with a SSOR method and compare the convergence rates and computational times. Try different values of the overrelaxation parameter.

Problem 12.15

Repeat Problem 12.10 with a Zebra line relaxation method and compare the convergence rates and computational times. Try different values of the over-relaxation parameter.

Problem 12.16

Repeat problem 12.10) with an ADI method and compare the convergence rates and computational times. Try different values of the overrelaxation parameter.

Problem 12.17

Consider the alternative definition of a line overrelaxation method, where the intermediate values are obtained with the fully updated values in the right-hand side. Instead of equations (12.2.33) and (12.2.34), define

$$\overline{u_{ij}^{n+1}} = \tfrac{1}{4}\left(u_{i+1,j}^{n} + u_{i-1,j}^{n+1} + u_{i,j-1}^{n+1} + u_{i,j+1}^{n+1}\right) + \tfrac{1}{4}\, q_{ij}^{n}$$
$$u_{ij}^{n+1} = \omega \overline{u_{ij}^{n+1}} + (1-\omega)\, u_{ij}^{n}$$

or, in incremental form,

$$4\Delta U_{ij}^{n} - \omega\,\Delta U_{i,j-1}^{n} - \omega\,\Delta U_{i,j+1}^{n} - \omega\,\Delta U_{i-1,j}^{n} = \omega R_{ij}^{n}$$

Derive the generalized form of this scheme and compare with equation (12.2.44).

Derive the differential equation for this scheme and compare with Problem 12.7. Show that

$$P = \frac{\tau(4-3\omega)}{\Delta x^{2}} + \left(\frac{\omega\tau}{\Delta x}\right)\frac{\partial}{\partial x} - \omega\tau\,\frac{\partial^{2}}{\partial y^{2}}$$

Appendix

Thomas Algorithm for Tridiagonal Systems

A.1 SCALAR TRIDIAGONAL SYSTEMS

For tridiagonal systems the LU decomposition method leads to an efficient algorithm, known as *Thomas's* algorithm. For a system of the form

$$a_k x_{k-1} + b_k x_k + c_k x_{k+1} = f_k \qquad k = 1, \dots, N \qquad (A.1)$$

with

$$a_1 = c_N = 0 \qquad (A.2)$$

the following algorithm is obtained.

Forward step

$$\beta_1 = b_1 \qquad \beta_k = b_k - a_k \frac{c_{k-1}}{\beta_{k-1}} \qquad k = 2, \dots, N$$

$$\gamma_1 = \frac{f_1}{\beta_1} \qquad \gamma_k = \frac{(-a_k \gamma_{k-1} + f_k)}{\beta_k} \qquad k = 2, \dots, N \qquad (A.3)$$

Backward step

$$x_N = \gamma_N$$

$$x_k = \gamma_k - x_{k+1} \frac{c_k}{\beta_k} \qquad k = N-1, \dots, 1 \qquad (A.4)$$

This requires, in total, $5N$ operations.

It can be shown that the above algorithm will always converge if the tridiagonal system is diagonal dominant, that is, if

$$|b_k| \geqslant |a_k| + |c_k| \qquad k = 2, \dots, N-1$$

$$|b_1| > |c_1| \text{ and } |b_N| > |a_N| \qquad (A.5)$$

If a, b, c are matrices we have a *block-tridiagonal* system, and the same algorithm can be applied. Due to the importance of triadiagonal system, we present here a subroutine which can be used for an arbitrary scalar tridiagonal system.

Subroutine TRIDAG

```
      SUBROUTINE TRIDAG (AA,BB,CC,FF,N1,N)
C
C*******************************************************************
C     SOLUTION OF A TRIDIAGONAL SYSTEM OF N-N1+1 EQUATIONS OF THE FORM
C
C     AA(K)*X(K-1) + BB(K)*X(K) + CC(K)*X(K+1) = FF(K)      K=N1,...,N
C
C     K RANGING FROM N1 TO N
C     THE SOLUTION X(K) IS STORED IN FF(K)
C     AA(N1) AND CC(N) ARE NOT USED
C     AA,BB,CC,FF ARE VECTORS WITH DIMENSION N, TO BE SPECIFIED IN THE
C     CALLING PROGRAM
C*******************************************************************
C
      DIMENSION AA(1),BB(1),CC(1),FF(1)
      BB(N1)=1./BB(N1)
      AA(N1)=FF(N1)*BB(N1)
      N2=N1+1
      N1N=N1+N
      DO 10 K=N2,N
      K1=K-1
      CC(K1)=CC(K1)*BB(K1)
      BB(K) =BB(K)-AA(K)*CC(K1)
      BB(K) =1./BB(K)
      AA(K) =(FF(K)-AA(K)*AA(K1))*BB(K)
   10 CONTINUE
C
C     BACK SUBSTITUTION
C
      FF(N)=AA(N)
      DO 20 K1=N2,N
      K=N1N-K1
      FF(K)=AA(K)-CC(K)*FF(K+1)
   20 CONTINUE
      RETURN
      END
```

A.2 PERIODIC TRIDIAGONAL SYSTEMS

For periodic boundary conditions, and a tridiagonal matrix with one in the extreme corners as in equation (10.2.12), the above method does not apply. The following approach leads to an algorithm whereby two tridiagonal systems have to be solved.

If the periodic matrix $B_p(\vec{a}, \vec{b}, \vec{c})$ has $(N+1)$ lines and columns resulting from a periodicity between points 1 and $N+2$, the solution X is written as a linear combination $X = X^{(1)} + x_{N+1}X^{(2)}$, or

$$x_k = x_k^{(1)} + x_k^{(2)} \cdot x_{N+1} \tag{A.6}$$

where $x_k^{(1)}$ and $x_k^{(2)}$ are solutions of the tridiagonal systems obtained by removing the last line and last column of B_p, containing the periodic elements. If this matrix is called $B^{(N)}(\vec{a}, \vec{b}, \vec{c})$ we solve successively, where the right-hand side terms f_k are put in a vector F:

$$B^{(N)}(\vec{a}, \vec{b}, \vec{c})X^{(1)} = F \tag{A.7}$$

and

$$B^{(N)}(\vec{a}, \vec{b}, \vec{c})X^{(2)} = G \tag{A.8}$$

with

$$G^{\mathrm{T}} = (-a_1, ..., 0, -c_N) \tag{A.9}$$

The last unknown x_{N+1} is obtained from the last equation by back-substitution:

$$x_{N+1} = \frac{f_{N+1} - c_{N+1}x_1^{(1)} - a_{N+1}x_N^{(1)}}{b_{N+1} + a_{N+1}x_N^{(2)} + c_{N+1}x_1^{(2)}} \tag{A.10}$$

The periodicity condition determines x_{N+2} as

$$x_{n+2} = x_1 \tag{A.11}$$

The subroutine TRIPER, based on this algorithm is included here. Note that if the periodicity condition is

$$x_{N+2} = x_1 + C \tag{A.12}$$

then the periodicity constant C has to be added to the right-hand side of the last instruction, defining $FF(N + 2)$.

```
Subroutine TRIPER

      SUBROUTINE TRIPER(AA,BB,CC,FF,N1,N,GAM2)
C
C*****************************************************************
C     SOLUTION OF A TRIDIAGONAL SYSTEM OF EQUATIONS WITH PERIODICITY
C     IN THE POINTS K=N1 AND K=N+2
C
C     AA(K)*X(K-1) + BB(K)*X(K) + CC(K)*X(K+1) = FF(K)     K=N1,...,N+1
C
C     THE ELEMENT IN THE UPPER RIGHT CORNER IS STORED IN AA(N1)
C     THE ELEMENT IN THE LOWER LEFT  CORNER IS STORED IN CC(N+1)
C     AA,BB,CC,FF,GAM2 ARE VECTORS WITH DIMENSION N+2, TO BE SPECIFIED
C     IN THE CALLING PROGRAM
C     GAM2 IS AN AUXILIARY VECTOR NEEDED FOR STORAGE
C     THE SOLUTION IS STORED IN FF
C*****************************************************************
C
      DIMENSION AA(1),BB(1),CC(1),FF(1),GAM2(1)
      BB(N1)=1./BB(N1)
      GAM2(N1)=-AA(N1)*BB(N1)
      AA(N1)=FF(N1)*BB(N1)
```

```
      N2=N1+1
      N1N=N1+N
      DO 10 K=N2,N
      K1=K-1
      CC(K1)=CC(K1)*BB(K1)
      BB(K) =BB(K)-AA(K)*CC(K1)
      BB(K) =1./BB(K)
      GAM2(K)=-AA(K)*GAM2(K1)*BB(K)
      AA(K) =(FF(K)-AA(K)*AA(K1))*BB(K)
   10 CONTINUE
      GAM2(N)=GAM2(N)-CC(N)*BB(N)
C
C     BACK SUBSTITUTION
C
      FF(N)=AA(N)
      BB(N)=GAM2(N)
      DO 20 K1=N2,N
      K=N1N-K1
      K2=K+1
      FF(K)=AA(K)-CC(K)*FF(K2)
      BB(K)=GAM2(K)-CC(K)*BB(K2)
   20 CONTINUE
C
      K1=N+1
      ZAA=FF(K1)-CC(K1)*FF(N1)-AA(K1)*FF(N)
      ZAA=ZAA/(BB(K1)+AA(K1)*BB(N)+CC(K1)*BB(N1))
      FF(K1)=ZAA
      DO 30 K=N1,N
      FF(K)=FF(K)+BB(K)*ZAA
   30 CONTINUE
C
      FF(N+2)=FF(N1)
      RETURN
      END
```

Index

512